Biosemiotics

Volume 10

Combining research approaches from biology, semiotics, philosophy and linguistics, the field of biosemiotics studies semiotic processes as they occur in and among living systems. This has important implications and applications for issues ranging from natural selection to animal behaviour and human psychology, leaving biosemiotics at the cutting edge of the research on the fundamentals of life.

The Springer book series *Biosemiotics* draws together contributions from leading scholars in international biosemiotics, producing an unparalleled series that will appeal to all those interested in the origins and evolution of life, including molecular and evolutionary biologists, ecologists, anthropologists, psychologists, philosophers and historians of science, linguists, semioticians and researchers in artificial life, information theory and communication technology.

Jan Toman • Jan Havlíček
Tomáš Hermann • Radek Kundt • Jana Švorcová
Editors

Culture Evolving

Bridging Life Sciences and Humanities

Springer

Editors
Jan Toman
Department of Philosophy and History of Sciences, Faculty of Science
Charles University
Prague, Czech Republic

Jan Havlíček
Department of Zoology
Faculty of Science
Charles University
Prague, Czech Republic

Tomáš Hermann
Department of Philosophy and History of Sciences, Faculty of Science
Charles University
Prague, Czech Republic

Radek Kundt
LEVYNA Laboratory for the Experimental Research of Religion
Masaryk University
Brno, Czech Republic

Jana Švorcová
Department of Philosophy and History of Sciences, Faculty of Science
Charles University
Prague, Czech Republic

ISSN 1875-4651 ISSN 1875-466X (electronic)
Biosemiotics
ISBN 978-3-032-16123-9 ISBN 978-3-032-16124-6 (eBook)
https://doi.org/10.1007/978-3-032-16124-6

This Springer imprint is published by the registered company Springer Nature Switzerland AG
The registered company address is: Gewerbestrasse 11, 6330 Cham, Switzerland

Preface

Evolutionary thinking about culture has a long but controversial history. On the one hand, some thinkers have been fascinated by the idea that principles designed to explain biological processes can also be applied to cultural phenomena. As noted by Duda in his chapter (Chap. 12), Darwin (1871, p. 89) himself had observed that

> The formation of different languages and of distinct species, and the proofs that both have been developed through a gradual process, are curiously parallel.

On the other hand, there still resonates the postwar horror at the consequences of uncritically accepting the idea that culture follows the same laws as living beings do. As Hermann et al. neatly show in their chapter (Chap. 3), this sentiment was poignantly expressed in the memoirs of Max Brod (1960, p. 324) who, reflecting on his time in Prague, recalled how

> Weber confidently steered our study of Darwinism, Lamarck's 'direct adaptation', Mendel's investigations into the race theory [i.e. of heredity], August Weismann's 'germ plasm continuity' and his rejection of the 'inheritance of acquired characteristics' (which became a matter of dispute again with the racial teachings of the Third Reich and later, in the opposite direction, with the theories of the Soviet scholar Lysenko). We talked about the national community and its generative health – about problems that have so painfully and unscientifically affected all of our lives several decades later. (...) Looking back, I feel that neither the teacher nor the students were really aware of the urgency of the issues we were discussing. We were petting the hounds of hell who were already rattling their chains.

The main goal of this book is twofold. First, we explore how evolutionary principles can offer novel insights into traditional research questions concerning culture and its dynamics; the core of the book is then dedicated to this perspective, as illustrated by Darwin's quote above. But we are also aware of the potential limitations of such approaches, including the interpretative nature of cultural transmission. Our second goal is therefore to address the concerns raised by Brod's quote: while the first section considers historical misuses of the theory of evolution, the final section critically examines these issues from several theoretical standpoints, including biosemiotics.

The book is organised in four sections. The first, *The Human Condition and Evolutionary Thought in the History of Ideas*, investigates the concepts required for evolutionary theorising, including the notion of linear time and deep time, from a historical perspective. The authors argue that no evolutionary theory is possible without these fundamental assumptions. Moreover, the historical examples discussed in this section demonstrate the perilous consequences of an ill-considered straightforward application of evolutionary principles to human society. The second section, *Cultural Evolution and the Human Sciences: Archaeological and Anthropological Perspectives*, looks at how the theory of evolution has been used and misused in social anthropology and archaeology, while also demonstrating that various novel evolutionary approaches can offer unexpected insights into some traditional topics such as religious practices. The third section, *Culture from the Perspective of Evolutionary Behavioural Science*, provides an overview of key theoretical perspectives to the study of cultural evolution. It starts with an examination of traditions in non-human animals and moves on to discussions of various conceptual frameworks that have been applied to understanding cultural transmission and change, including the cognitive mechanisms involved. Finally, the fourth section, *Biosemiotic Reflections upon the Relationships Between Nature and Culture*, discusses the main issues introduced earlier and adds the biosemiotic perspective to highlight potential isomorphisms between biological and cultural evolution.

Each section begins with a brief section introduction, which outlines the content of the chapters included in that section and situates them within broader theoretical perspectives. Recognising that not all readers will be familiar with every concept discussed, the individual chapters also serve to introduce key ideas and disciplinary approaches. Although this may lend some chapters a somewhat encyclopaedic or descriptive tone, we believe such approach is necessary to minimise misconceptions, which are all too common in cross-disciplinary debates. Moreover, no chapter serves a merely introductory purpose: an overview of the key concepts and developments is always complemented by novel insights into the subject at hand. To ensure clarity, each chapter explicitly states its main aims. Alongside full-length chapters that deal with broader subjects, each section also includes a case study that offers an in-depth view of some specific aspect of cultural evolution. While individual chapters reflect different perspectives, they have been carefully edited and cross-referenced to create a complementary and cohesive whole.

The target audience of this volume includes scholars and postgraduate students across multiple fields concerned with cultural phenomena, ranging from evolutionary biology and behavioural sciences all the way to social anthropology, archaeology, biosemiotics, and the philosophy of science. In view of the broad disciplinary scope, all of the contributions aim at being broadly accessible: the authors clarify the key terms and concepts and minimise the use of jargon and overly specialised language.

The idea for this book emerged from the Charles University Research Centre programme entitled 'Nature and Culture: Historical, Cultural, and Biological Concepts of Human Nature', which operated between 2018 and 2023. This project brought together scholars from various departments of the Charles University, encompassing disciplines such as evolutionary biology, behavioural science, philosophy, the history of science, and biosemiotics. During numerous—sometimes intense—discussions at our regular meetings, workshops, and conferences, we were fascinated by the diverse and often unfamiliar ideas and perspectives held by our colleagues in neighbouring departments. We also came to fully appreciate the importance of intellectual openness and patience needed to grasp unfamiliar concepts and worldviews. To our great surprise, we found similarly minded who shared our need for epistemic exchange on the topic of cultural evolution, but we have also realised that there is still little academic literature on this subject—and that prompted us to begin this collaborative endeavour.

This brief historical background explains why many contributors to this volume come from a single institution. While outlining the book's concept, we invited colleagues from various Central European academic institutions. Although geographically concentrated, the resulting team of contributors represents a wide range of disciplines, each with its own 'culture' – and differences between these 'cultures' can sometimes be greater than any difference between the views of scholars from a single discipline who live in different regions. Moreover, several key contributors to this volume represent a distinctive school of thought that is sometimes overlooked in the global perspective, namely the Prague biosemiotics school (formerly known as biohermeneutics), whose influence spans across various areas, as demonstrated throughout the volume's chapters and especially in the entire last section of this volume. Because this book is part of Springer's *Biosemiotics* series, we have made a point of integrating semiotic perspectives wherever relevant. Nonetheless, the authors adopt diverse theoretical frameworks, some of which may be common within their field but they seldom appear in such proximity to one another. What unifies them here is the overarching theme of cultural evolution. We believe that our somewhat unorthodox approach will prove valuable to our fellow scholars in biosemiotics by broadening debate on the issues that concern us.

Many readers might suppose that the misuse of evolutionary thought to justify racism and social injustice—well known from the standard social science curricula—would serve as a lasting cautionary example, but the second decade of the twenty-first century has shown this assumption to be overly optimistic. Some radical social constructivists maintain that elimination of gender and ethnic distinctions is merely a question of sufficient effort, while the rapid rise of technologies such as artificial intelligence has prompted some techno-optimists with longtermist ethical views to propose radical solutions for dismantling the current social order.

Although these are extreme positions, even their moderate versions uncomfortably echo certain twentieth-century utopias examined in this book in depth. Coming from Central Europe, with its history of political and ideological upheavals, we feel compelled to question whether the hard lessons of history have truly been learned. Perhaps naively, we hope that this book might not only contribute to a more nuanced comprehension of cultural phenomena but also inspire vigilance against alluring ideologies that can have devastating consequences.

Department of Zoology, Faculty of Science
Charles University
Prague, Czech Republic
Jan Havlíček

Department of Philosophy and History of Sciences
Faculty of Science
Charles University
Prague, Czech Republic
Jan Toman

References

Brod, M. (1960). *Streitbares Leben. Autobiographie*. Kindler.
Darwin, C. (1871). *Sexual selection and the descent of man*. John Murray.

Acknowledgments

We would like to thank Anna Pilátová for translating a large portion of the texts and for assistance with editing the entire publication. We also want to thank colleagues at our home institutions for providing a stimulating environment that fosters our work and creativity. We extend our gratitude to Springer Publishing for facilitating the publication of this book, especially Alexei Sharov, Kalevi Kull, Claus Emmeche, and Donald F. Favareau for accepting it into their series. Finally, we also wish to thank the Charles University, within whose University Research Centers (UNCE) programme, specifically projects '204004 Nature and Culture – Interaction of Cultural and Biological Evolution in an Interdisciplinary Perspective' and '204056 Nature and Culture: Historical, Cultural, and Biological Concepts of Human Nature', most of the topics and ideas presented in the individual chapters were developed and discussed. Radek Kundt also acknowledges the support of the European Regional Development Fund project 'Beyond Security: Role of Conflict in Resilience-Building' (reg. no.: CZ.02.01.01/00/22_008/0004595). This book complements, expands upon, and thoroughly updates topics discussed in the book *O původu kultury: Biologické, antropologické a historické koncepce kulturní evoluce* [On the Origin of Culture: Biological, Anthropological, and Historical Concepts of Cultural Evolution], published by Academia in Prague in 2017.

Contents

Editors and Contributors

Editors

Jan Toman Department of Philosophy and History of Sciences, Faculty of Science, Charles University, Prague, Czech Republic

Jan Havlíček Department of Zoology, Faculty of Science, Charles University, Prague, Czech Republic

Tomáš Hermann Department of Philosophy and History of Sciences, Faculty of Science, Charles University, Prague, Czech Republic

Radek Kundt LEVYNA Laboratory for the Experimental Research of Religion, Masaryk University, Brno, Czech Republic

Jana Švorcová Department of Philosophy and History of Sciences, Faculty of Science, Charles University, Prague, Czech Republic

Contributors

Tomáš Alušík Institute for History of Medicine and Foreign Languages, First Faculty of Medicine, Charles University, Prague

Pavel Duda Department of Zoology, Faculty of Science, University of South Bohemia, České Budějovice, Czech Republic

Terezie Dvořáčková LEVYNA Laboratory for the Experimental Research of Religion, Masaryk University, Brno, Czech Republic

Jaroslav Flegr Department of Philosophy and History of Sciences, Faculty of Science, Charles University, Prague, Czech Republic

Eliška Fulínová Center for Theoretical Study, Charles University and Czech Academy of Sciences, Prague, Czech Republic

Petr Hampl Department of Philosophy and History of Sciences, Faculty of Science, Charles University, Prague, Czech Republic

Jan Havlíček Department of Zoology, Faculty of Science, Charles University, Prague, Czech Republic

Tomáš Hermann Department of Philosophy and History of Sciences, Faculty of Science, Charles University, Prague, Czech Republic

Pavlína Hillerová Department of Philosophy and History of Sciences, Faculty of Science, Charles University, Prague, Czech Republic

Vojtěch Hladký Department of Philosophy and History of Sciences, Faculty of Science, Charles University, Prague, Czech Republic

Jan Horský Department of Historical Studies, Faculty of Humanities, Charles University, Prague

Václav Hrnčíř Institute of Archaeology, Czech Academy of Sciences, Prague, Czech Republic

Department of Linguistic and Cultural Evolution, Max Planck Institute for Evolutionary Anthropology, Leipzig, Germany

Danijela Jerotijević Institute of Social Anthropology, Faculty of Social and Economic Sciences, Comenius University, Bratislava, Slovakia

Karel Kleisner Department of Philosophy and History of Sciences, Faculty of Science, Charles University, Prague, Czech Republic

Radek Kundt LEVYNA Laboratory for the Experimental Research of Religion, Masaryk University, Brno, Czech Republic

Eva Kundtová Klocová LEVYNA Laboratory for the Experimental Research of Religion, Masaryk University, Brno, Czech Republic

Petr Květina Institute of Archaeology, Czech Academy of Sciences, Prague, Czech Republic

Jitka Lindová Department of Psychology and Life Sciences, Faculty of Humanities, Charles University, Prague, Czech Republic

Anton Markoš Department of Philosophy and History of Sciences, Faculty of Science, Charles University, Prague, Czech Republic

Jan Musil Department of Philosophy and History of Sciences, Faculty of Science, Charles University, Prague, Czech Republic

Lenka Ovčáčková Department of Philosophy and History of Sciences, Faculty of Science, Charles University, Prague, Czech Republic

Lucie Strnadová Department of Philosophy and History of Sciences, Faculty of Science, Charles University, Prague, Czech Republic

Michal V. Šimůnek Institute of Contemporary History, Czech Academy of Sciences, Prague, Czech Republic

Jana Švorcová Department of Philosophy and History of Sciences, Faculty of Science, Charles University, Prague, Czech Republic

Jan Toman Department of Philosophy and History of Sciences, Faculty of Science, Charles University, Prague, Czech Republic

Petr Tureček Department of Philosophy and History of Sciences, Faculty of Science, Charles University, Prague, Czech Republic

Center for Theoretical Study, Charles University and Czech Academy of Sciences, Prague, Czech Republic

Part I
The Human Condition and Evolutionary Thought in the History of Ideas

Tomáš Hermann

Modern thinking of the past two centuries has been so fundamentally steeped in the idea of evolution that it makes sense to posit something like an 'evolutionary a priori'. In parallel, Modern thought has also come to view the human society – including all its cultural, scientific, technological, philosophical, religious, and political artefacts – as necessarily shaped by history, so we could, by analogy, speak of a 'historical a priori'. These seemingly necessary and 'natural' assumptions had become so deeply integrated into the structure of our perception and thought that we approach *all phenomena* in nature and culture *a priori* as related to their evolutionary and historical origins, conditions, and contexts. A closer look, however, reveals that this fundamental conceptual setting is evolutionarily and historically relative – and therefore neither necessary nor 'natural'. That brings to the fore certain basic questions: What does this world of evolutionarily and historically conditioned phenomena and facts *mean*? How did we *arrive at* this understanding of the world, under what conditions did it emerge, and what consequences does it have? How did the basic concepts determining the relationship between nature and culture, the world and human beings, change – and what new fundamental concepts have emerged in the process (Bowler, 2003; Larson, 2004; Schnädelbach, 1974; Wittkau, 1992)?

We cannot fully cover, much less answer, these broad but fundamental questions, but they form the shared background of all the contributions in this section. Through their focus on particular subjects, the four chapters in this section explore the assumptions, pathways of reception, and the impact of the concepts of evolution and historicity. While dealing with different subjects, they all address the epistemological question of the *relation between nature and people.* Their ambition is to trace

T. Hermann
Department of Philosophy and History of Sciences, Faculty of Science, Charles University, Prague, Czech Republic

how the human situation has been interpreted, by rational means, within the available knowledge of life, and how this has co-determined our image of the world.

While it is possible to grasp the concepts of evolution and human historicity using rational terms and scientific theories, their interpretation has not been straightforward. These two concepts carry internal tensions which gave rise to numerous disputes, and so even their formation has been nonlinear, questionable, and at variance with the Enlightenment notion of 'progress'. On top of that, the understanding of evolution and human historicity has been influenced by religious, theological, philosophical, and generally valuative assumptions much more than the advocates of their rationalised form would care to admit.

Generally speaking, any analysis of evolution and human historicity must consider the mutual interactions of four distinct topics:

1. The assumption (or concept) of *evolution* as the process of the development of life, of our planet, and of the universe;
2. Both the development and the concept of development of the *human mind* and *human society*;
3. Ideas regarding human origins, *anthropogenesis*, and an interpretation of traces of prehistoric forms of human presence on our planet;
4. A search for *analogies* between human culture and other forms of life on Earth, together with the reciprocal projections between them.

These four areas are distinct but cannot be strictly separated in a given historical context because they are interrelated: while focusing on one, the others are always to some extent implicitly present. This constant semantic interference forms the basis of the semiotic and hermeneutic approach to the history of ideas. By focusing on the mutual relationships between the assumptions of our thoughts and the world that is being conceptualised, we find ourselves in the dynamic field of *semiotics* and *hermeneutics*. Most generally, semiotics is the study of meaning, which centres around one general question: What is 'something'? We draw inspiration from Umberto Eco (1976, 1990, 2014), who understood semiotics as a comprehensive theory of culture. He studied various concepts, phenomena, and doctrines from the perspective of their specific deep history in the sense of the history of thought and knowledge.

When addressing the concept of evolution and the relationship between nature and culture, we also engage in *biosemiotics* in the sense of exploring certain basic assumptions (Cobley, 2016) (while specific theoretical concepts of biosemiotics are addressed in other sections of this book). Central to this exploration is hermeneutics, which had, since the nineteenth century, examined the structure of presuppositions and of our prior understanding, that is, historical consciousness (Grondin, 1991). The historical accounts of science presented in our chapters serve this purpose: their aim is to uncover the hidden yet necessary presuppositions of our thinking about nature and culture.

Although the chapters are ordered chronologically according to their subject, they do not offer a historical presentation *sensu stricto*. Rather, they present a kind of archaeology of ideas and concepts, each highlighting certain motifs which determine, or latently underpin, our concepts or experiences as their formative roots. The aim of contributions in this section is to highlight the historical and theoretical background of topics which are in their own right the subject of extensive research and much discussion. One of the leitmotifs of this section is the tension between two trends: on the one hand, with increasing differentiation of scientific knowledge, one can observe a clear tendency to a *de-anthropocentric* view of the human situation in the world. In other words, humans are perceived as an increasingly small part of nature and the world. On the other hand, modernity, progress, and scientific self-confidence went hand in hand with attempts to define a new concept of *human uniqueness* in a world full of evolutionary complexity and its new hierarchy of values. This tension is still urgent and relevant.

The evolutionary perspective is, moreover, closely related to *time and temporality*, and this has been reflected in the various ideas regarding the age and development of nature and culture. This is the second leitmotif of this section, which investigates thoughts on the evolution of culture within the ancient (archaic) concept of *cyclic* time, the discovery of the evolutionary *past*, and projections of the evolutionary *future*. The first three chapters in this section of the book have synthetising ambitions regarding these three time modes, while the fourth chapter is a separate case study.

The Chap. 1 analyses several proto-evolutionary conceptions of nature and society that survived as legacies of ancient Greek and Roman thought; ever since the Renaissance, these approaches have also influenced the Early Modern discourse. New interpretations of ancient thinkers, especially Lucretius, have had a truly formative impact. In general, these pre-evolutionary conceptions provided an alternative to the traditional theological and biblical perspective by viewing humans and their culture as part of the development of nature. The ancient ideas of progress, decline, and renewal are nevertheless part of a different paradigm, that of *cyclical time*, which assumed much longer periods of time than those envisaged in traditional biblical chronologies.

The Chap. 2 investigates the 'discovery' of deep time and the conceptual and other obstacles this idea had to overcome to become a universal *a priori* that enabled further changes in how we think about nature and society. It describes the new knowledge, experiences, and theoretical steps that enabled the conceptualisation of long historical development in four areas: Earth's geological past and fossils, the ancient past against the background of literary legacy and tradition, cultural prehistory based on material artefacts, and multilingualism along with the history of human speech. The underlying temporal scheme is one of *linear deep time*, which later, in Modern evolutionary thinking, fundamentally expanded the traditional biblical temporal scale.

The Chap. 3 investigates the *evolutionary future*: it presents Modern evolutionary 'new utopias' and their para-religious aspects as something that had disastrous real-life consequences, which made us reconsider further exploration of cultural evolution. This chapter focuses on *eugenics* as the context for interpreting the concept of the *gene* and its politicisation. It outlines the motivations of evolutionism, the scientific tools of eugenics, and the changes brought about by the advent of the *new genetics* through two concrete literary examples of the utopian extrapolation of these ideas by Francis Galton and Alexis Carrel. These striking examples of scientific overconfidence and ambitions of scientific power went hand in hand with a blending of discourses about our understanding of nature and culture. On the one hand, there was the view of humans as part of nature and thus subject to natural laws, and on the other hand, there was the idea of humans as part of society and civilisation, and as such subject to laws created (and possibly changed) by humans themselves. Eventually, these ideas were applied to humanity as an object of controlled manipulation and the repercussions were so disastrous that they came to form an internal corrective for all efforts aimed at applying experimental research to humans.

The Chap. 4 presents a case study that elaborates on the topics of the previous contributions from a different angle. The *evolutionist* conceptions of the origins and development of nature and culture proposed by Ernst Haeckel and Christian von Ehrenfels were contemporary with and in many ways closely linked to the eugenic movement, its socially reformist efforts, and para-religious ambitions exemplified by the foundation of new 'scientific religions'. Nevertheless, a comparison of two different contemporary interpretations of Darwinism in this chapter sheds more light on their scientific and spiritual motivations and underpinnings. In the context of various efforts to identify the source of human uniqueness, Haeckel's and Ehrenfels's proposals are presented as clearly distinct in their contrasting emphasis on *monist* and *dualist* foundations, respectively, thereby highlighting the potential for alternative, either rationalist or metaphysical, approaches to interpreting the evolutionist paradigm.

References

Bowler, P. J. (2003). Evolution. *The history of an idea*. University of California Press.

Cobley, P. (2016). *Cultural Implications of Biosemiotics. Biosemiotics* (Vol. 15). Springer.

Eco, U. (1976). *A theory of semiotics*. Indiana University Press.

Eco, U. (2014). *From the tree to the labyrinth: Historical studies on the sign and interpretation*. Harvard University Press.

Eco, U. (1990). *The limits of interpretation*. Indiana University Press.

Grondin, J. (1991). *Einführung in die philosophische Hermeneutik*. Wissenschaftliche Buchgesellschaft.

Larson, E. J. (2004). *Evolution. The remarkable history of a scientific theory*. Modern Library.

Schnädelbach, H. (1974). *Geschichtsphilosophie nach Hegel. Die Probleme des Historismus*. Karl Alber.

Wittkau, A. (1992). *Historismus. Zur Geschichte des Begriffs und des Problems*. Vandenhoeck & Ruprecht.

Chapter 1
Progress or Decline? The Evolution of Culture According to Some Ancient Thinkers

Vojtěch Hladký and Eliška Fulínová

Abstract Some highly influential ideas about the gradual development and transformation of living beings and human culture were first articulated in antiquity but re-discovered and reformulated centuries later, from the Renaissance and until the Modern Era. These views can be broadly divided in two contrasting approaches: according to the first, the cosmos, living organisms, and human civilisation are characterised by a progressive development, while according to the other view, the world once experienced a Golden Age, from which it has since declined all the way to the present, rather unfortunate, times. In this context, Empedocles and Lucretius are often cited as ancient precursors to modern evolutionary thought, as the authors of early conceptualisations of a gradually increasing complexity of living beings. The following text, however, examines specifically views on culture and its transformations and will therefore focus on Hesiod and Lucretius, who represent two contrasting visions of time and progress. Because their views in many respects differ, they capture this subject in unsettling ambiguity.

Keywords Classical antiquity · Hesiod · Empedocles · Plato · Lucretius

V. Hladký (✉)
Department of Philosophy and History of Sciences, Faculty of Science, Charles University, Prague, Czech Republic
e-mail: vojtech.hladky@natur.cuni.cz

E. Fulínová
Center for Theoretical Study, Charles University and Czech Academy of Sciences, Prague, Czech Republic
e-mail: fulinova@cts.cuni.cz

J. Toman et al. (eds.), *Culture Evolving*, Biosemiotics 10,
https://doi.org/10.1007/978-3-032-16124-6_1

1.1 Introduction[1]

According to the oldest texts that survive from ancient civilisations, the world was not created all at once. It emerged gradually, in several stages that can be understood as a kind of evolution from the initial chaos and turmoil to the current, more ordered state. This chapter explores proto-evolutionary thought in early Greek cosmogonical traditions with emphasis on early theories of cultural evolution.

The notion of a progressive constitution of the world-order is apparent even in the well-known account presented in the Book of Genesis, where the world is created over the course of 6 days. As the Book of Genesis continues, one can observe that the creator of the world, who initially stands radically outside it, also gradually changes, at least in relation to the world. The strict and punishing god becomes God of the Covenant and assumes the role of a guarantor of order and law. It is nowadays commonly accepted that the text of Book of Genesis originated in about the sixth century BCE. It is also generally accepted that the story it tells emerged in the context of various similar stories of creation of the world. The oldest of these stories that survives to our times is recorded in *Enūma Eliš*, a Babylonian myth dated to the beginning of the second millennium BCE, but one can also find parallels with Hesiod's *Theogony*, composed around 700 BCE, which describes how three subsequent generations of gods divided among themselves the rule over the world (Fig. 1.1).

This gradual search for a proper world order, or equilibrium among the actors who participate in its formation, enables us to follow—at first glance somewhat surprisingly—various parallels between mythical accounts of creation of the world and the living beings that inhabit it on the one hand and the modern theory of evolution on the other hand. Moreover, cosmogony also forms the background for anthropology because according to some ancient accounts, the development of human beings and their culture (which takes place either in one or in multiple stages) broadly mirror the development of the world (cf. Chap. 14) (Fig. 1.1).[2]

[1]Eliška Fulínová is the author of passages on Hesiod, Aeschylus, and Plato's dialogues *The Statesman* and *Protagoras*. Vojtěch Hladký authored the remaining sections of this contribution and served as its general editor. Along with the book *O původu kultury* (see acknowledgements), this English text—especially the passages written by Hladký—also draws on Hladký et al. (2012).

[2]See the interpretation of Genesis as a story of evolutionary creation in Schmid (2015). Editors of an excellent volume of Hardt's *Entretiens* dedicated to cosmology and cosmogony in antiquity point to similar 'evolutionary' structures in the Babylonian myth *Enūma Eliš* and in Hesiod; cf. Fuhrer & Erler (2015, esp. 4, 8) and 'evolutionary' terminology is used in describing the sequence of generations of gods and creation of our world also for instance by Clay (2003: 1, 4, 13, 31, 86). Herren (2017: 19–24), on the other hand, emphasises other aspects and draws attention to differences between Hesiod's evolutionary description, where the gods who establish the world order emerge by birth, and the book of Genesis, where everything in the world has one divine creator, so that—to use current terminology—the overall vision is creationist. In relation to the emergence of human beings, however, even Hesiod according to Herren abandons the narrative of gradual evolution, because he claims that people were created by the gods directly. Scully (2015, esp. 23–25), on the other hand (especially in *ibid.*, 22–29, 61, and 77), compares the course of creation of the world according to Hesiod's *Theogony* and according to Genesis.

Fig. 1.1 The Monnus Mosaic, featuring a detail of Hesiod (inscribed as ESIO-DVS), comes from a Roman domus in Augusta Treverorum (modern-day Trier). This artwork, dating to the late third century AD, is housed in the Rheinisches Landesmuseum Trier, Germany. Photograph by Carole Raddato, via Wikimedia Commons (https://commons.wikimedia.org/wiki/File:Monnus_Mosaic,_detail_of_Hesiod_(ESIODVS)_from_a_Roman_Domus_in_Augusta_Treverorum_(Trier),_end_of_the_3rd_century_AD,_Rheinisches_Landesmuseum_Trier,_Germany_(30038562895).jpg)

Hesiod's epic composition known as the *Theogony* describes the process of formation of the world on several interconnected levels (see Chap. 6). The leitmotif is a gradual increase in and differentiation of both body parts and spatial constituents of the world. In connection with it, one can observe a similar development of genetic and power dynamics, a gradual transformation of how new living beings appear in the world and changing forms of rule over the world. This complex process takes place along two independent genealogical lines, which remain strictly separate throughout.

The first is derived from the primordial, further unspecified Chaos, from which Nyx (Night) is born, while the other evolves from Gaia (Earth). In terms of establishment of the world order, the genealogical line of powers derived from the Earth is the more important one. Powers that emerge from the line of Chaos represent rather the limits or boundaries. What is crucial is that in the initial stages of cosmogonies, the female divinities (Nyx, Gaia) play a key role: they are the primordial mothers who give birth to their offspring on their own, from themselves. Male gods appear only in the next phase, when they acquire an important status.

In the central line of Gaia (Earth), a key role is played by Ouranos (Sky), who is her son and subsequently also her husband. The offspring that emerged from this union were oppressed by their father: they could not be fully born and had to remain in the confines of their mother's womb. Gaia, therefore, with the help of her son Kronos, tricked her husband and this plot resulted in his castration. Ouranos then withdrew to the heights above Earth's surface, stopped descending to his wife,

Earth, to copulate with her, and that is how the basic spatial layout of the universe was established.

By violently attacking his father, Kronos reversed the genealogical order of filial subservience and became the first true ruler of the world. But he, too, failed in his efforts to establish a lasting rule. Like his father, he oppressed the offspring conceived with his sister/wife Rhea, not allowing them to emerge into the world. Rhea therefore tricked her husband, who then instead of devouring his youngest child swallowed only a stone wrapped in nappies, and the youngest son, Zeus, was secretly brought up with the help of Kronos's mother Gaia. This 'rebellion of the mothers' was brought to a successful conclusion when Zeus overpowered his father, forced him to disgorge his siblings, and defeated the older gods in a mighty battle for dominance over the world. It was thus Zeus and the generation of Olympic gods, his relations, who finally established a stable form of rule over the world. Zeus did not oppress his offspring. Rather, he swallowed his first pregnant wife Metis (Cunning Wisdom) outright, because it had been prophesied that she would one day give birth to Zeus's challenger. By incorporating Metis, Zeus acquired unprecedented insight into what is good and what is bad, and later gave birth to his daughter Athena from his own head.

The cosmogonic dynamics thus switched from the initial self-fertilisation of mothers to the very opposite, to a father who appropriated the maternal ability to give birth. Nevertheless, as noted above, Zeus did not suppress his offspring (and more generally his potential competitors and challengers) by brutal power. Instead, he included them into his network of power: he put them in charge of particular areas of cosmic relations and divided among them particular co-ruling tasks (*timai*), while remaining in charge of the entire power structure. This established a dynamic equilibrium capable of flexibly reacting to new elements, integrating them, and thus managing the risks that could threaten the overall order of the world.

Regarding the emergence of the first human beings in a thus constituted cosmic whole, it is notable that narratives which could be viewed as early physical anthropology played in Greek mythology—unlike in other, for instance Near Eastern traditions—generally speaking a rather marginal and merely local role. They were mostly limited to locally specific myths of autochthony, i.e., myths aimed at legitimising the right of a certain group of people to rule over a particular area. All the more important in Greek thought was since is earliest beginning cultural anthropology, which aimed at answering questions about the origin of the specifically human cultural way of life (Heath, 2005: 28).

Especially important in this respect are two kinds of narratives which express the opposite poles of possible approaches to this subject. On the one side, there are the myths about Prometheus and his gifts to the mortals. Their message is apparently progressivist[3]: they tell us that in the beginning people lived like animals and even worse, because they were not as well physically equipped. In short, humans were

[3] For more on conceptions of progress in antiquity, see especially Guthrie (1957, ch. 5), Dodds (1973), and Burkert (1997).

barely surviving until the Titan Prometheus brought them fire, which opened the door to various cultural skills whose acquisition elevated people from the level of animals to full humanity. At the opposite end of the spectrum, there are the no less impressive stories about the 'lost paradise' of the Golden Age during the rule of Kronos. These stories are clearly primitivist: they tell of generations that were morally pure and lived in ancient times in perfect bliss, because the Earth itself fed them and satisfied their modest needs. In contrast to that, current people, who live under the rule of Zeus, must work hard to get by, and in this world of human culture their initially unspoilt nature was replaced by widespread arrogance and self-indulgence.[4]

Progress or decline? The development of human cultural sphere was in ancient Greek thought viewed both ways, which rather than excluding one another are complementary. It is not unusual to find that one thinker in different contexts developed both approaches: the Greeks were keenly aware of the inherently ambivalent nature of human culture, and that went hand in hand with an ambivalent evaluation of human prehistory (can we decide whether the predecessors of current people were brutish primitives or noble savages?). The oldest literary source for both approaches, and therefore also for the abovementioned ambivalence in terms of evaluation, are Hesiod's epic compositions *Theogony* and *Works and Days.*[5]

1.2 The Myth of the Five Ages of Humanity

Let us first take a closer look at Hesiod's version of the story of a Golden Age. It is part of the so-called myth of the five ages of man from *Works and Days* (110–201). Strictly speaking, the poet does not speak about a golden 'age' but rather about a (1) golden race of mortals, followed by (2) the silver race, by (3) humans who sprang from ash trees and wielded bronze weapons. After them, the earth was inhabited by (4) heroes, known especially from the Homeric tradition (warriors from the war for Troy but also heroes of the Theban cycle; cf. Chap. 6), and (5) by the iron race, which denotes people of the current kind.

Jean-Pierre Vernant (1960, 1966, 1985) had convincingly shown that this myth can be interpreted as describing the process of establishment of a specifically human way of life. Neither the world nor the cultural sphere in its complexity can be founded by one single act. Both evolve gradually, in a series of steps that follow one after another. By differentiating between five 'races' of mortals, Hesiod shows that although the chronological aspect naturally plays an important role, his scheme is not just about the progression of time. Perhaps even more importantly, it is about the

[4] For more on the notion of primitivism in antiquity and in the Middle Ages, see Lovejoy and Boas (1935), and Boas (1948).

[5] This text summarises and further elaborates material assembled in Luhanová [Fulínová] (2014). For more on Hesiod, cf. also West (1966, 1978), and Clay (2003).

deep transformation of the nature of mortal beings, which is then linked to their separation from and self-definition in contrast to the rest of the world.[6]

The oldest mortals of the Golden Age (*Works and Days*, 109–120) were created directly by the gods and lived, under Kronos's rule of the world, somewhat like gods. In a sense, they were thus closer to gods than to the mortals of the current type: they did not suffer from any evils, were not afflicted with ageing or disease, lived in constant, unchanging youth, and died easily, as if overcome by sleep. Moreover, they did not have to work because the earth provided them with plenty of food of its own accord: they thus did not kill animals and did not eat meat (Vidal-Naquet, 1981b). We also find no references to families or any other forms of social organisation, which has traditionally been interpreted as implying that these creatures were born directly from the earth. It is thus probable that they did not reproduce sexually and possibly were not even sexually differentiated. It also seems that after living their allotted time, this 'race' of people had simply died out. The 'golden' mortals did not need to strive for a better life in either material or moral sense: in the descriptions of their life, we find no mention of cultural skills and achievements.

These themes were later elaborated for instance by Plato: the myth told by the Stranger from Elea in the *Statesman* plays on the motifs of *Works and Days*[7] but also emphasises that the ancient mortals from the age of Kronos were in their way of life akin not only to gods but also to animals. Under the direct care of the god/shepherd, they lived naked in the meadows where they happily grazed on various 'fruits of the earth' which grew there on their own (271d–272a). They had no knowledge of agriculture or of living in houses (272a–b). Born directly from the earth and fed by it, they did not form families or sizeable communities, and they had no political establishment. In fact, various animals at that time did not eat each other and people and animals were capable of mutual communication (271e, 272b–c), which also implies that there was no difference between rational human speech (*logos*) and animal sounds.

In sum, during the age of Kronos, people exhibited no specifically human cultural traits, which is expressly contrasted with the abundance of material necessities of life (272a). It is thus not surprising that the question of whether this ancient way of life would have been happier than the current one does not have a clearcut answer (272b–d). These first mortals, who were not fully distinct from either the gods or the animals, may have known no suffering or moral decline but they also lacked any specifically human characteristics.

Let us now return to Hesiod. The 'races' that followed after the golden one gradually came to resemble the current kind of people by establishing increasingly clear distinctions between gods, humans, and animals. The second, silver, kind of mortals (*Works and Days*, 127–142), was still created by the gods but differed from the golden race both in stature and in mind. At this point, there already existed families

[6] Cf. also a comprehensive exposition of the various levels of the myth of five ages as presented in Most (1998).

[7] See also Vidal-Naquet (1981a), and Mattéi (1996).

as elementary units of social organisation and people did not take all nourishment from the earth: mothers nursed their children themselves. The importance of this novelty is emphasised by noting that the childhood of these mortals was extremely long. They also already knew moral decay. In fact, Hesiod claims that their mutual relations were full of arrogance and even worse, people of the silver race did not maintain correct relations with the gods, because they did not bring fitting sacrifices. Because of these failings, the highest of gods had decided to wipe them all out. At that point, though, it was not Kronos but—and this is explicitly stated—his son Zeus, who thus apparently at some point during the transition between the Golden and Silver Age overthrew his father and assumed the rule of the world (*Theogony*, 453ff).

In the next stage, Zeus himself created the third race, the gigantic humans that sprung out of ash-trees (*Works and Days*, 143–155). These mortals, however, continued in distancing themselves from the gods and even took it to extremes. The first sign of their depravity was their frighteningly huge stature, coarse body, and mighty arms, which distinguished them from both the golden and the silver races. Even their minds were more deprived and their hearts harder. They knew nothing but brazenness and violence. They constantly fought with bronze weapons and even built their houses of bronze. As the ultimate sign of their difference from gods but also current people, Hesiod notes that they did not eat grains, which implies that they did not practice agriculture. One can therefore assume that they lived on a purely meaty diet of hunters, which would have been in line with their war-like way of life.

The following race of heroes (156–173) represents in a sense a return to the gods: it was a divine race, more just and deserving than its predecessors. Its representatives are referred to as demigods, because they had one divine parent. Hesiod does not say much about their way of life, because this race is represented by warriors well known from the epic heroic tradition, especially the stories of the siege of Troy and war against Thebes (cf. Chap. 6). There is thus no need to stress that, at that time, people already knew agriculture and seafaring and lived in cities with developed social organisation. Moreover, they felt a moral obligation to behave properly to both people and gods, whereby it is especially their relationship with gods that forms the basis of positive evaluation of the heroes.

The current iron race (*Works and Days*, 174–201) followed with no major interruption after the age of heroes, who were in some important families still viewed as direct ancestors. When describing their life, Hesiod emphasises the labour and toil people must engage in to survive. This replaced the erstwhile effortless bliss but even in the hard lives of current mortals, one finds some slivers of good, albeit only few and far between, and one cannot achieve them without considerable effort, which is thus characteristic of the Iron Age.

Hesiod's description culminates with a series of warnings of what awaits humanity if it fails to live up to the demands placed upon it and succumbs to moral decay. He explains that if family ties and friendly relations between people become devalued, justice scorned, if violence and right of the stronger prevails, modesty and justice are replaced by envy and brazenness, and people are reduced to poverty and pain, Zeus will rather wipe them all out. Nevertheless, Hesiod clearly believes that if we try, it is within our power to act against this tendency to general decline that would inevitably lead to ill ends. This appeal to active efforts to overcome one's own imperfections, which would have been quite alien to the Golden Age, is a specific feature of people of the Iron Age. According to Hesiod, this is also what sets people apart from animals and this difference is also apparent in food taboos: while animals, whose mutual relations do not follow the precepts of justice (*dikē*), can feed on the bodies of their fellows, among people this is forbidden. Their mutual relations are governed by a just order ordained by Zeus (*Works and Days*, 276–279; Clay, 2003: 127).

1.3 Promethean Myths

Many of the motifs outlined above appear also in the stories linked to Prometheus, a Titan who became the 'cultural hero' of humanity for bringing them the gift of craft skills and other achievements. The story of Prometheus has been resonating since antiquity to this day although, as Carol Dougherty (2006) shows, it has over time undergone various transformations. For Hesiod, this mythological theme seems to have had a special importance: he treats it both in his *Works and Days* and in the *Theogony*, where the Promethean myths form a kind of complement and counterpoint to the myth of the five ages of man. In these myths, human prehistory and development towards the current state of humanity are not viewed as a decline from an initial state of bliss and moral perfection, but rather as progression from animality to fully developed cultural humanity. Even here, though, the poet's rendition is not free of ambivalence.

In the *Theogony* (521–616), the story of Prometheus is set into a cosmological framework: its function is to explain the impact of the fundamental transformation that took place during the transition from Kronos's rule to the rule of Zeus. Prometheus is here an ambivalent rather than a clearly positive figure (which is why some scholars place him in the anthropological category of tricksters).[8] As a representative of Titans, the old generation of gods, he is trying to outsmart the wise Zeus, head of the new generation of Olympians (534),[9] but it is a match he cannot win. His actions eventually lead not only to a cruel punishment for him but also to many evils for the mortals he was trying to help.

In the *Works and Days* (42–105), we find an emphasis on the importance of Promethean novelties in the lives of mortals: the new skills and abilities have

[8] Cf. Dougherty (2006: 33–35), but the same term is also used in Dodds (1973: 7), and Burkert (1997: 23).

[9] This is a clash between two types of temperament: the ancient Titanic cunning, guile, and trickiness, and the new wise foresight of Zeus, who by absorbing Metis, the goddess of cunning wisdom, acquired unprecedented insight into the nature of reality. For more on this, see Luhanová [Fulínová] (2014: 99–101, 104–110).

significantly contributed to the establishment of a fully human (social, political, moral, and cultural) way of life. But all these achievements are also in a sense a reaction to a radical *worsening* of the human condition because, as Hesiod points out at the beginning of his story, Zeus made people work for their living by the sweat of their brow and sent onto them grief and troubles as a revanche for being tricked by Prometheus (42–49). Initially, all people lived their lives free of evils, toil, and illness (90–92). This primordial state of bliss, which refers to the age of Kronos, had under Zeus's rule passed away but that had also opened the possibility of cultural growth and progress.

The first step in this process was taken in Mekone (*Theogony*, 535–560), where sacrifice was established as the basic element that divided the sphere of mortals from the sphere of gods but also enabled communication between the two. Prometheus at that time interfered in favour of the mortals and divided the sacrificed bull unevenly: on the one side he put meat, which he wrapped in a hide and covered with unappetising stomach, while on the other side he placed bare bones, which he covered with fat. With tempting and misleading words, he then invited Zeus to choose the portion his heart desires—and the highest of gods, although he saw through the ruse, opted for the masked pile of bones. Since that time, mortals burn on their altars only bones, while the meat of sacrificed animals is left to them as food. Despite the abovementioned prohibition on eating one another, people thus thanks to bringing this sacrifice do not depend solely on the vegetarian fare provided by earth. Sacrifice is thus pivotal to the distinction between humans and gods but also the distinction between the human sphere and the world of animals (Vernant, 1974).

The establishment of sacrifices is closely linked to the next step, Prometheus's gift of fire. The deceived Zeus had decided to punish people and not give them fire for their free use (*Theogony*, 562–564; *Works and Days*, 47–52), so that although people had acquired meat thanks to Prometheus's ruse, without fire it would be indigestible, and they would have to survive on raw plant foods. In reaction to this, Prometheus stole fire from the gods, hid it in a large hollow stem, and brought it to people (which should be understood as meaning that he revealed to them the secrets of starting and maintaining it).

In the Greek tradition, the mastery of fire was understood as not just enabling the preparation of food and sacrifice of meat to the gods, but also as a technological precondition of achieving further cultural knowledge and skills. This is clearly expressed for instance in *Prometheus Bound* (254), a tragedy traditionally ascribed to Aeschylus,[10] where Prometheus, punished for the theft of fire and shacked to the top of a mountain, has a long speech in which he explains all this. He claims that Zeus did not much care for the mortals and would have certainly soon killed them, which is why a Titan (228–236), who belonged to the previous generation of gods, was moved by compassion and took them under his wing. According to this description, people were at that point barely surviving; like phantoms from dreams,

[10] Some scholars doubt Aeschylus's authorship; cf. Dougherty (2006: 66–67).

bumbling unreasonable children or ants that inhabit holes in the ground, they lived in caves instead of houses. Prometheus gave them the gift of fire (252–254), which enabled them to improve their lives. But equally important was that he bestowed on them the gift of reason (444) and taught them various crafts, which in many cases rely on the mastery of fire. Prometheus thus taught the mortals how to build houses, revealed to them the heavenly phenomena that enable orientation in the calendar, gave them mathematical knowledge, alphabet and with it memory, taught them how to harness the strength of animals for work, thus giving them agriculture, and invented for their benefit seafaring, medicine, mining, and work with metals, as well as various forms of prophesy. Prometheus concludes, quite immodestly, that all cultural knowledge of the mortals comes from him (442–471, 476–506). In short, in this story Prometheus is no longer an ambivalent figure who determines the fate of humanity by placing it in-between the gods and the animals: he becomes a suffering rebel who rose against the tyrant Zeus, a compassionate benefactor of humanity who enabled people to extricate themselves from their initially most primitive existence. This shift doubtless reflects the optimism that prevailed in Athens of Aeschylus's times, when after the defeat of vastly stronger Persian aggressor there came the brilliant epoch of Pericles. In modern times, Prometheus is depicted along similar lines in various works, most notably in Percy Bysshe Shelley's 1820 drama *Prometheus Unbound*.[11]

After Aeschylus, the motif of gods as teachers of humanity was also elaborated by Plato. In the abovementioned myth presented in the *Statesman*, Plato describes how after the end of Kronos's rule and the time of shepherds, there came the age of Olympians, when people were initially left to their own devices. The peaceful cohabitation of living creatures had ended and, compared to other animals, people turned out to be biologically much less well equipped for survival: they were weak and could not properly fend for themselves, neither in terms of defence against predators, nor in terms of satisfying their basic needs (274b–c). They had no skills and did not know how to take care of themselves. During the previous cosmic phase, they did not learn such things because they did not need them (274c). At this point, the gods, who were no longer the shepherds but rather teachers of the mortals, entered the scene. Prometheus had revealed to people the secret of fire but Hephaistos and Athena taught them various crafts (274c–d). Crucially, at this point—to paraphrase the well-known adage—gods no longer gave people fish but taught them how to fish: the mortals were taught what they needed and henceforth had to fend for themselves, to assume responsibility for their survival and wellbeing (274a, d). That is how the form of life known from our times came into being (272b).

Rather similar is the message of the myth presented in Plato's *Protagoras* (320c–323a). While one can only speculate whether it actually presents the views of the famous intellectual after whom the dialogue is named, there are good reasons to believe that Plato uses him to express views which were in Athens of fifth century BCE discussed by many thinkers, from playwrights to sophists—headed by

[11] Cf. Rossi (1970: 174–186), and Dougherty (2006: 91–141).

Protagoras.[12] In this version of the story, Epimetheus, whom tradition depicts as the wayward brother of the far-seeing Prometheus (*Theogony*, 511–513), equipped various kinds of animals with different advantages that helped them survive (warm fur, firm hoofs, the ability to fly or swim under water, etc.) but nothing was left for people. When Prometheus saw a human being, miserable, naked, defenceless, and without a shelter, his first thought was to take this pitiful creature away from Hephaistos and Athena, and to share with him the secret of various cultural skills. First of all, though, he gave people fire, because—and at this point Plato follows Aeschylus's version—that is the basic precondition of any cultural progress and, without fire, no craft can be practiced properly (321d). Plato subsequently develops the Promethean myth by emphasising the establishment of a complex political organisation based on the sense of shame (*aidōs*) and justice (*dikē*). Nevertheless, these gifts, which enabled the existence of communities and peaceful human cohabitation, did not come from Prometheus but from Zeus himself. Elsewhere, Zeus is Prometheus's opponent, but he is also viewed as the guarantor of justice in the world. This divine instruction separated humans from animals and brought them closer to the gods, whom they now started to properly respect. Still, even here one can see that while without divine intervention people would have likely perished, once properly equipped and instructed they must fend for themselves and overcome their shortcomings with their own efforts.

In the Promethean myths, the mortals must always pay a price for the achievements they acquired with the help of Prometheus. This is the main motif of Hesiod's exposition. In the *Prometheus Bound*, the choir listening to the Titan's self-aggrandisement chastises the punished god, who fails to see the error of his ways. He is told that in his blindness he transgressed (259–260) because—and that is a key moment—humans should not be provided with goods above a proper measure (507). It may well be that people are too weak to be left to their own devices, with no divine help but, ultimately, they must overcome their essential weaknesses themselves, through their own hard work. This is the foundation of living a good life and shaping one's own fate (Nelson, 1998: 61). From this perspective, fire and other cultural achievements given to (or taken away from) people by the gods are, like progress itself, inherently ambivalent: they help people survive and live well but can also become the instrument of humanity's downfall if they make human lives too easy.

During the period that elapsed between the times of Aeschylus and Plato, various debates in Athens have touched upon and explored the origin of human culture but they are rather difficult for us to reconstruct and the conceptions formulated at that time, while apparently novel and groundbreaking, survive mostly just in fragments and in the reflections of later scholars. The thinker of key importance for our topic was Archelaus, known mostly as the teacher of Socrates. Evidence indicates that he

[12] Cf. Kerferd (1981: 125–126, 140–148), who points to numerous parallels with the thoughts of other contemporary authors although it must be noted that those other authors cannot be identified with the sophists, much less with Protagoras in particular. Cf. also Guthrie (1969: 63–68), Barney (2006: 86, note 16), Taylor (2020).

may have been the first thinker to define the difference between what is given by nature (*physei*) and what is given by custom or convention (*nomō*). In other words, he was probably the first to contrast nature versus nurture (DK 60 A 1, 16; 2).

In the second half of the fifth century BCE, the relation of these two areas in-between which spans the human life became a key topic of Greek thought, as apparent perhaps most clearly in discussions of the sophists (Guthrie, 1969, chap. 4; Taylor, 2020). Archelaus, too, presented his vision of the origin of the world and, in the next step, of living creatures who according to him originated in life-giving slime that gushed forth from warm earth (DK 60 A 1, 16–17; 4). His purely naturalistic description of the processes of formation of our cosmos and life in it continues, with no significant break, with an exposition on the origins of culture as if human culture was an extension of nature. 'And the humans separated from the other ones [scil. animals] and established leaders, laws (*nomoi*), arts (*technai*), cities and the rest' (DK 60 A 4, 5–6, trans. Laks & Most, 2016: 26.Arch. D2). As far as we can judge based on surviving sources, this is the first formulation of what came to be known as the social contract.[13] In any case, we are clearly at a point when, in the history of European thought, this entire issue moves to another level. In this type of explanations, culture is presented as based on its own specific laws and viewed as ultimately a human creation that reflects natural human needs and inclinations. It is no longer perceived as a gift from the gods, nor is it dependent on them: rather, it evolves within the framework of cosmos, which at an opportune moment in its development creates suitable conditions for it. Archelaus thus comes surprisingly close to current biosemiotic discussions (see, e.g., Sonesson & Dunér, 2016; Sonesson, 2016).

Such explanations of an autonomous emergence of the world and human civilisation, which were based on natural science (and assumed no divine intervention, at least not in a traditional sense), had naturally provoked also some dissent, with Plato being the most vocal of the critics. Not only his cosmology, presented mainly in the *Timaeus*, rests on the notion of a divine creator (*demiurge*) but even his ideal community, outlined in the *Republic* and the *Laws*, rests on theological footings. Building on the ideas of his teacher Socrates, Plato sets the origin of the world, the emergence of living creatures in it, and the laws that would guide an ideal community, in a decidedly teleological framework, thus becoming in a sense the first proponent of creationism (Sedley, 2007).

[13] For more on the abovementioned conceptual motifs in Archelaus and on his position in early Greek thought, see also Guthrie (1969: 58–59), Dodds (1973: 11), Kahn (1981: 102–103), Burkert (1997: 29), Campbell (2006: 22–23, 27, 47–48), Rechenauer (2013: 803, 806–808), and Betegh (2016).

1.4 Empedocles and the Cycle of Life

In the previous section, we focused mainly on how thinkers in antiquity explained the emergence of human culture, leaving aside (with the exception of Archelaus) the issue of explaining the formation of the world and living creatures. Nevertheless, various motifs that form the constitutive elements of the modern theory of evolution appeared already in the work of the earliest Greek thinkers. Already Anaximander claimed in the second half of the sixth century BCE that the first living creatures were born due to the effect of warmth in a moist environment and while they resembled fish, they eventually evolved into a variety of animals including people (DK 12 A 30; see also A 10, 11, 6, cf. Campbell, 2000: 152–154).[14] At around the same time, Xenophanes of Colophon extrapolated from the presence of various marine fossils to cyclic changes of various parts of the Earth (DK 21 A 33, 5–6). But most notable among thinkers who in antiquity developed some proto-evolutionary motifs to higher levels of complexity and detail were Empedocles and Lucretius.

Empedocles of Acragas (fifth century BCE) became a classic of natural philosophy chiefly because he was the first to formulate the theory of the four elements (fire, air, water, and earth), from which all other things are composed (Fig. 1.2).[15] Two independent forces, Love (*Philotēs*) and Strife (*Neikos*), then act upon these elements, whereby Love is the uniting and merging principle, while Strife is the principle of differentiation and fragmentation. These two opposing divine principles do not act upon the four elements with the same intensity over time: one is always dominant and their 'rule' alternates, so that the elements merge and divide again in a cosmic cycle. First, Strife is at the pinnacle of its power, the four elements are separate and exist in a pure form. Then Love enters and starts uniting them, which gives rise to various things, including living beings. At the point of culmination of Love, the four elements form a spherical body, the Sphairos, in which the elements are perfectly merged. But this state is just temporary. Eventually, Strife again starts penetrating among the elements, the Sphairos falls apart, and things that were created under the rule of Love gradually dissolve until the elements become once again separate and re-emerge in a pure form.

The Sphairos is not, however, an undifferentiated mixture, a 'perfect cocktail' of the four elements. It should rather be understood as a structured whole, a flawless spherical body with clearly differentiated individual 'limbs'. One could thus view it as a sort of gigantic organism reminiscent of Lovelock's Gaia or its predecessors in the earlier tradition of European philosophical and scientific thinking that conceived of the cosmos as a living body. Importantly, living beings, which had emerged in the course of the particular stages of the development of life, can survive within the Sphairos. Aside from that, it should be noted that Empedocles does not seem to

---

[14] Numerous similar motifs, which seem to go back to the Presocratics, can also be found in the work of Diodorus of Sicily, *Historical Library*, I,7.

[15] The following exposition summarises the conclusions of studies Hladký (2017) and Hladký (2025), where the author argues for this interpretation.

Fig. 1.2 The death of Empedocles, painted by Salvator Rosa (1615–1673), portrays the putative suicide of the philosopher as he leaps into the fiery depths of Mount Etna in Sicily. Wikimedia Commons; (https://commons.wikimedia.org/wiki/File:The_Death_of_Empedocles_by_Salvator_Rosa.jpg)

assume any divine plan or purpose followed by the gods regarding the overall order of the cosmos or particular living beings in it. Love and Strife act in a sense as blind forces, which on a cosmic level unite the elements and then separate them again.[16]

In Empedocles's proto-evolutionary scheme of the origin of living beings (zoogony), one can reconstruct five stages, which correspond to the five generations of living creatures, whereby we are currently at the fourth, penultimate stage (DK 31 A 72/1):

1. The first generation of living beings was born at the beginning of Empedocles's cosmic cycle, when Love started to act as a unifying force and to create, using all four elements, various living forms. First to emerge were tissues such as blood or flesh (DK 31 B 98), but then there also emerged various separate body parts or limbs of living creatures, for instance bones, separate eyes, or heads (DK 31 B 57–58, 86–87, 95–96, cf. A 72/2). These body parts were born in the earth, from which they literally grew like plants, which are likewise rooted in the soil. Empedocles says that these separate limbs then 'wandered' around the world, which implies some level of autonomous existence, but they moved around aimlessly and chaotically.

[16] In contrast to the exposition above (see further Campbell, 2000: 184–185), elements of teleological thinking in Empedocles are emphasised by Sedley (2007: 31–62).

2. The second generation of living beings emerged thanks to the growing influence of Love, which led to a gradual merging, growing-together, of the limbs formed during the previous stage of zoogony (DK 31 B 20, 59). The limbs thus functioned as building blocks, which at this stage united to form higher wholes—organisms. Just like in the previous generation, living beings still grew from the earth (DK 31 B 61.2–3; cf. B 62.4), which is also where their merging probably took place. It is important to note that this process was accidental, that is, not purposefully directed.[17] During more intensive action of Love, limbs that happened to be close to each other started growing together, which led to the occasional emergence of various imperfect beings (DK 31 B 60) that were not viable and eventually perished. Already Charles Darwin had noted that we find here an anticipation of natural selection (see Chap. 15).[18] Some beings that emerged in this way resembled 'phantoms from a dream' (DK 31 B 61),[19] which is how Empedocles elegantly deals with the tales of various 'composite' beings from Greek myths.
3. The third generation of living beings had according to Empedocles consisted of 'whole-natured' forms (*oulophyeis typoi*), which like in the previous stages grew from the earth (DK 31 B 62). It seems they were like the androgynes known from Plato's *Symposium* (189c–193d). These androgynous beings first appeared during the previous stage of zoogony, and they not only turned out to be viable but at this stage reached the pinnacle of their development. These whole-natured beings with their perfect, round shape corresponded to the equally spherical Sphairos and like the Sphairos, they were formed during the greatest ascendancy of Love. Empedocles describes this period as a lost paradise, a time when people and animals lived in perfect harmony but he deviates from Hesiod's description of the Golden Age in claiming that this period was not ruled by Kronos or Zeus (or even Ares, the god of war) but by Aphrodite, to whom people brought non-bloody sacrifices and meticulously refrained from shedding blood (DK 31 B 128, 130). Moreover, it is also clear that the abovementioned ambivalence connected with evaluation of the Golden Age had disappeared: compared to further developments, Empedocles perceives this period as clearly and unambiguously positive.
4. The time of perfect harmony, during which the Sphairos was in existence, nevertheless came to an end: eventually, Strife once again started to grow in power and with this turn of the cosmic cycle, the individual elements again began to separate. With the Sphairos, there also disintegrated the whole-natured forms, which Strife split in two sexes, thus creating the state of affairs known to us today (DK

[17] This is also how this part of zoogony is explained by Aristotle, *Physics*, II,8 198b16–32, where he also quotes part of Empedocles's fragment DK 31 B 61, and Aristotle's commentator Simplicius, *In Physica*, 371.33–372.11.

[18] Darwin (1866, xiii, with a footnote). Darwin added his observation to the fourth edition of *On the Origin of Species*. In fact, though, he ascribes a primitive formulation of the principle of natural selection to Aristotle, who quotes Empedocles' fragment DK 31 B 61 in his *Physics*, II,8 198b23–31.

[19] Empedocles, *Strasbourg Papyrus*, d.11–18 Martin and Primavesi = 341–348 Janko.

31 B 62–63, 124). After differentiation of the sexes, people were no longer born from the earth but from each other (*di' allēlōn*) (DK 31 A 72/1). Even so, Love is still present in the world and although its influence is waning, it is still responsible for the two sexes coming together (DK 31 B 64). In fact, when contributions from both parents unite in the mother's body, the embryo grows much like plants used to grow from the earth during the previous phase (A 70/1).
5. In the next round of the cosmic cycle, Strife will again increase in power and the development will head to a point where the four elements will once again fully separate. But living beings will disintegrate into separate limbs even before that takes place (DK 31 B 20.4–5).[20]

1.5 The Evolution of Humans and Culture According to Lucretius's *De rerum natura*

The Epicurean philosopher and poet Titus Lucretius Carus (died around 50 BCE) in many ways went in Empedocles's footsteps, albeit primarily in poetic form (Fig. 1.3).[21] His large didactic epos in six books known as *De rerum natura* (*On the Nature of Things*) is the first Latin poetic work to survive as a whole and it is undoubtedly the work of an artistic genius. The main subject of Lucretius's exposition is Epicurean philosophy, but he was also importantly inspired by Empedocles, and not only in poetry. Given his philosophical starting points, Lucretius represents a sort of alternative undercurrent in Western culture. He has often been perceived as a thinker who stands closer to the modern view of the world than to the great classics of the past.[22]

According to the basic philosophical tenets of Epicureanism, everything that comes into being is of a material origin because everything consists of miniscule and further indivisible particles, atoms, which are immutable and eternal. The other

[20] As noted above in the text, [Aëtius], V,19,6 Mansfeld and Runia = DK 31 A 72/1, in his summary of Empedocles's zoogony does not describe this stage. His exposition ends with the phase we are in now, that is, one that includes the emergence of sexually differentiated men and women. Nevertheless, the final phase can be reconstructed from Plato's description of the androgynes in *Symposium*: 188c–193d, especially 190c–d, which can be based on similarity of expressions viewed as a reflection of the third phase of the development of life as described by Empedocles. Further, in DK 31 B 20.4–5, Empedocles describes how—in contrast to the merging of limbs and formation of more complex organisms during the rule of Love—Strife separates the limbs, which then roam around separately. It also seems that this stage is described in the *Strasbourg Papyrus*, d.1–10 Martin and Primavesi = 331–340 Janko.

[21] The key commentary on Lucretius's understanding of the origin of life and culture, based among other things on the following text, is Campbell (2003), cf. also Schrijvers (1999). For more on the relation between Empedocles and Lucretius, see Garani (2007).

[22] Cf. Greenblatt (2011), whose work is, however, an example of reading Lucretius from a modern perspective rather than a historical study aiming to map the reception of Lucretius's work during the Renaissance in all of its complexity.

Fig. 1.3 Lucretius pointing to casus, the downward movement of the atoms. Drawn and engraved by Michael Burghers (1650–1727). Wikimedia Commons (https://commons.wikimedia.org/wiki/File:Lucretius_pointing_to_the_casus.jpg)

key principle of Epicurean philosophy is the void, which is identical with the universe and in all directions endless, thus reminiscent of the modern conception of space in Newtonian physics. According to the Epicureans, an infinite number of atoms move in this infinite void over an infinite length of time, occasionally uniting and forming individual worlds and everything in them. These worlds periodically come into being and perish again, and they can assume various forms, including that of our world. Chance clearly plays a key role in this conception of the universe, which also excludes the possibility of action of a divine creator or organisation of the world according to some higher purpose (teleology). Everything in the world has a naturalistic explanation, including the soul, which consists of atoms and is therefore immortal.

In the fifth book of his epic poem, Lucretius presents a detailed explanation of various aspects of our world that had traditionally been the source of wonder and reverence, of phenomena whose origins were thought to be divine. His aim is to explain the origin of the world and living beings in it, as well as the emergence of language, religion, and the nature of heavenly bodies whose regular movements have been viewed as evidence of divine action in this world, in a purely naturalistic fashion (V.55–90). Especially the explanation of the origin of living beings, including humans, in many ways anticipates later evolutionary perspective.

Lucretius strictly rejects the idea that living beings were made by some divine creator, claiming instead that they emerged at the beginning of our world from the earth, which is therefore the mother of all living. During this period, one encounters this metaphor repeatedly but for Lucretius, who was an atomist, it was rather an analogy: he does not really claim that the earth is a living organism (cf. Schrijvers, 1999: 199–201; Garani, 2007: 81–93; see also Guthrie, 1957, ch. 1–2) although he admits that the origin of living creatures is to this day apparent in the case of small creatures, such as insects, which are born by spontaneous generation (autogenesis). In general, though, during the formation of our world this phenomenon was much more widespread and even large living creatures were born in this way. First, plants and trees sprung up from the earth, then birds and other animals. While birds were born from eggs, which the earth had produced during this ancient fruitful phase, animals were born from embryonic sacks in the earth, literally from wombs (*uteri*) filled with 'milk', that is, a nutritious juice similar to Archelaus's life-giving slime. Lucretius adopts the assumption—repeatedly found in antiquity in both mythical and scientific thought—of a close analogy between embryogenesis and phylogenesis (the emergence of living creatures) and between plants and animals (Schrijvers, 1999: 10–12; Campbell, 2003: 78–84). From our perspective, his rather fantastic idea that early living creatures sprang from the earth is in effect an attempt to pursue the parallel between the plant and animal kingdoms, an implementation of a parallel between the role of the womb in sexually reproducing animals and the earth, from which plants grow. On a deeper level, though, all living creatures are, according to Lucretius, composed of the same physical components, namely atoms.

During this early period of spontaneous generation of living creatures from the earth, there also emerged various strange and fantastic beings, such as androgynes, or imperfect creatures whose organs were insufficiently developed or defective and they were therefore incapable of independent existence and further procreation (V.837–854). According to the analogy which Lucretius clearly perceives between the emergence of living creatures and embryonic development, such beings seem to correspond to embryos which for various reasons fail to develop into viable individuals (Schrijvers, 1999, ch. 2). Unlike Empedocles, he does not believe in the existence of beings composed of parts belonging to different species but he concurs with Empedocles in believing that various myths about fantastical monsters record real recollections of creatures that formed in the beginning of the world and died out because their physical form was ultimately not viable (Campbell, 2003: 107–111). The survival of other creatures, which turned out to be well-formed, was enabled by certain characteristics they happened to have, such as strength, intelligence, or speed, while other animals started to associate with people, who subsequently protected them as their domestic animals (V.855–870).

In this way, living creatures evolved into their current form. Like Empedocles, with his explanation of the origins of life, Lucretius clearly in various ways anticipates Darwinian concept of evolution, including for instance natural selection. There is, however, one crucial difference: according to Lucretius, the living beings that spontaneously emerged from the earth and subsequently proved fit to survive no longer developed. After this initial phase, the species were established in their definitive form. In contrast to Empedocles, the individual beings did not merge into more complex entities: they were born and died as whole organisms. This difference between Lucretius and Empedocles is probably due to their different views regarding the physical foundations of the world because, according to Lucretius, poor structure and resulting lack of viability are determined already at the level of atoms (cf. Campbell, 2003: 107–109).

Nevertheless, one finds in Lucretius's exposition one exception, one case where a species developed further: this exception is the evolution of humans and their culture. Lucretius explains the origins of life based on purely natural principles and it is thus unsurprising that his explanation of gradual emergence of human civilisation does not assume any external, much less divine intervention. According to him, the *genus* of the first people was originally much rougher (*durius*) because they were born from earth that was likewise rough. Under these initial conditions, human constitution was also more robust (V.925–930; Schrijvers, 1999: 83–85; Campbell, 2003: 185–187, 261). The first people led a 'wandering life' and depended on what they found in nature. They had no agriculture and built no shelters. They lived in the forests and caves, wore no clothing, had no fire, families, or laws and their lives were more immediate and free from fear of what the future may bring. True, they often became prey to wild animals but they also waged no wars and did not have to deal with the perils of the modern world (V.931–1010). One cannot say this was Golden Age of humanity, because at this stage people were not protected from the external influences and their lives were harsh. They lived in a much closer contact with nature and had to deal with a much more demanding climate than nowadays. One can thus see that Lucretius in his description of the lives of early humans brought together various primitivist and progressivist motifs in such a way that the result is even more ambivalent than what we find in the work of the other thinkers mentioned above (cf. Schrijvers, 1999, ch. 6; Campbell, 2003: 240–243, 2006: 53–57).

Over time, people gradually learnt how to build houses, and they started to form families, villages, and larger communities based on mutual help and collaboration. They learned to use fire and thanks to it, their nature over time became more refined and civilisation could emerge (V.1011–1027). Even so, according to Lucretius the progress of culture is not straightforward, it does not head towards an ever more optimal organisation of society. He was convinced that our civilisation culminated in the rural, village-based way of life, when people no longer lived in isolation and left behind their initial dependence on nature, but their lives and thoughts were not yet spoilt by some of the luxuries of civilisation. We shall see that with further concentration of people in cities, civilisation peaked and started to descend into a phase of decline (Campbell, 2003: 252–262, 2006: 57–60).

Rather remarkable is also Lucretius's explanation of the origin of language. He is convinced that, like society and justice, language emerged 'bottom-up', spontaneously, without the aid of any divine creator or lawgiver who defined the meaning of words right in the beginning. The Epicurean explanation sought a compromise within the abovementioned polarity of things determined by nature (*physei*) and those determined by convention (*nomō*), between nature and culture. Just like animals in accordance with their nature vocalise in different ways, so people, too, initially reacted to things in their surroundings by various sounds. But human nature differs depending on one's physical constitution (including racial type) and the area one inhabits, which is why already at this primitive level different ethnicities used different utterances. In the next phase, there emerged conventions, that is, an agreement among the speakers of a language regarding a unified version of their language, and individual sounds became words. Finally, during the third stage, intellectuals devised names for abstract entities (V.1028–1090; Campbell, 2003: 289–292; cf. Eco, 1995: 88). The origin of language and abstract thought is thus explained in purely naturalist terms as a progressive development of semiotic capacity of human beings, without any divine intervention.[23]

Further in his exposition, Lucretius focuses on other important aspects of civilisation, which he tries to explain in a similarly nonreligious, natural, and rationalist way. For instance, fire was not given to the humankind by a divine being, such as Prometheus or some mythical hero; rather, people acquired it after wind blew it to them or in a forest fire, where fire is created by the force of the wind rubbing branches against each other (Fig. 1.4).[24] The mastery of fire enabled people to cook, which is a classical theme of transition from a wild state to civilisation in both modern and ancient anthropology (V.1091–1104; Campbell, 2003: 322–323, 327–329). Fire and other achievements of civilisation subsequently led to the foundation of cities, which is when the first kings started to emerge and grasp power, and some people started to accumulate possessions and pursue their particular interests. Later, the kings were toppled by popular uprisings and power was grasped by the mob. To overcome the general lawlessness and omnipresent violence, people had to establish

Fig. 1.4 The forest fire, painted by Piero di Cosimo (1462–1522), drew inspiration from Book 5 of Lucretius's *De rerum natura*. Wikimedia Commons (https://commons.wikimedia.org/wiki/File:Piero_di_Cosimo_%E2%80%94_The_Forest_Fire.jpg)

[23] It is instructive to compare Lucretius' account of the origin of language with contemporary biosemiotic approaches, which share many common points (see, e.g., van Heusden, 2004, Zimmer, 2004, Carling, 2016, or Zlatev, 2016).

[24] See further Lucretius, *De rerum natura*, V.897–900, Campbell (2003: 325–326). Cf. also Vitruvius, *De architectura*, II,1,1.

laws and law-enforcing institutions, which is a motif that in the Modern Era took the form of considerations about a 'social contract' (V.1105–1160). In parallel, there also emerged religion, which is according to Lucretius based on an incorrect interpretation of certain notions that come to our minds: people placed the gods in the heaven because that is where they observed various astronomical and meteorological phenomena they sought to explain. That is also how the fear of gods was born, and that fear is according to the Epicureans the root of all evil (V.1161–1240). This is also why Lucretius believes that the transition from the initially simple way of life on the level of village communities to a more complex social organisation was a change for the worse: in his view, the fear of gods went hand in hand with the emergence of social instability, which disrupted the initially harmonious human communities (Campbell, 2006: 59–60).

In the subsequent exposition, we can nevertheless observe something like social progress. People thus for instance discovered metals after finding them accidentally melted by heat (V.1241–1296). Gradually, and also in part by accident, people have developed the art of war. Lucretius further explains how people discovered the production of textiles, agriculture, gardening, and acquired the knowledge of music, astronomy, geometry, seafaring, letters, and other skills (V.1297–1411). At the end of fifth book, Lucretius concludes that people over time achieved utter mastery in many of their skills (V.1412–1467) but the apparent progress in some areas only underscores the fact that, according to Lucretius, the development of civilisation is ambivalent and does not lead to an unambiguously better state of the world. This is perhaps best exemplified by the abovementioned progress in the art of war, which Lucretius mentioned above.

1.6 Conclusions

To sum up, the explanations of the emergence of life presented by Empedocles and Lucretius contain many elements familiar from the much later Darwin's theory of evolution. According to the ancient thinkers, not only our world had formed in a series of steps—which is a motif known from even the oldest myths—but living creatures, too, evolved through a number of phases (as described by Empedocles in his zoogony) and the same applies to human civilisation (as presented by Lucretius). Both Empedocles and Lucretius anticipated the notion of natural selection, noting that some lifeforms turned out to be nonviable and perished.[25] Generally speaking, they both believed the process through which particular species are established includes adaptation to current living conditions but also mutual competition, where chance inherently plays an important role and decides about the survival of living beings. This is closely linked to the fact that the various forms of life did not emerge

[25] In today's discussion of evolutionary theory, their concepts could be rather defined as negative selection (viability-based sorting); cf. Toman and Flegr (2017).

to fulfil some predetermined purpose: they appeared largely spontaneously, in reaction to the prevailing conditions and general processes that were taking place in the world. According to both Empedocles and Lucretius, the world as such is not organised teleologically. It has no wider purpose and, unsurprisingly, also no divine creator who had fashioned the world in accordance with a predetermined plan. On the contrary: living beings (and according to Lucretius also human culture) had emerged autonomously, based on processes that took place in the world in accordance with natural laws, processes that have their continuity and analogy in the gradual development of humans and their culture. People are thus not exempt from the natural order of things: they are in nature and fully connected with it.

Still, it would be wrong to believe that Empedocles and Lucretius in their descriptions of the origin of life presented a full-fledged theory of evolution. One should rather view their explanations as ‘proto-evolutionary’ because they lack some constitutive features of the modern theory of evolution (most notably the elements of positive selection, such as differential reproduction based on fitness, and the cumulative nature of adaptations). Although Empedocles believed that living beings assumed their current complexity in several subsequent steps, neither he nor Lucretius considered the possibility of species evolving from one another or the idea of such processes continuing even in our times (cf. Campbell, 2000, 2003: 6–8, 98–101, 107–109). Both Empedocles and Lucretius lacked a suitably large time frame to be able to conceive of a series of changes leading to a transformation of one species into another in an infinite series.

Unfortunately, we do not know how long Empedocles’s cosmic cycle was supposed to take. Based on surviving testimonies, one could speculate perhaps about tens of thousands of years (myriads).[26] The duration of Lucretius’s individual worlds is likewise extremely difficult to determine because it can significantly differ depending on the character of each world. What we do learn, though, is that our world is relatively young: this is apparent from the utter absence of reports by poets about any events that preceded the fighting at Thebes or Troy (cf. Chap. 6). Moreover, human skills still develop, and that applies not only to crafts and technical skills, but even to arts, including Lucretius’s poem as such (V.324–337).[27] Based on these

[26] In DK 31 B 115.6, Empedocles claims that a daimon that becomes tainted by shedding the blood of living creatures must reincarnate for 30,000 ‘seasons (*hōrai*)’, but it is not clear how large a part of a year one season represents; cf. Wright (1995b: 274). The Florentine scholia do give some numbers regarding the length of Empedocles’ cycle but they unfortunately do not allow for a more precise quantification. For instance, Sedley (2007: 67–70), a Rashed (2018, chap. 1), have tried to use this information to reconstruct a cosmic cycle and in their work, one can also find a recent reconstruction of these scholia. For more on the cosmic cycles in antiquity and their length, see Wright (1995a: 138–144); cf. also Burkert (1997: 34–38).

[27] Elsewhere, Lucretius says about our world that it has already passed the initial phase when life was spontaneously emerging from the earth. The earth eventually grew old, like women do, and exhausted its ability to give birth (II.1105–1174, V.821–829). Elsewhere, though, he claims that the world is still young (V.330, 780, 818, 943). Cf. Campbell (2003: 92–93). According to Lucretius, myths contain the memories of formation of the world and life in it, but their distorted stories should be rectified by a rational explanation, one that takes into account the laws of atomistic physics. Myths are told by humans, whose culture—we as could see—had gradually evolved and keeps evolving, although the span of our cultural memory is limited.

hints, we can estimate that our world is at most a few thousands of years old (tradition dated the Troy war to about one thousand years before the time when Lucretius was active). Other ancient authors estimated the duration of cosmic cycles in tens of thousands of years, and it is likely that Lucretius assumed a similar scale. As we shall be able to follow in detail in the following chapter, the modern notion of the evolution of life assumes the existence of 'deep time' (Gould, 1987; Rossi, 1984).

When it comes to the notion of progress in antiquity, that is, the idea of evolution of things and their improvement over time, scholars agree that, save for some brief periods, progress was considered mainly in relation to science and technology: the evolution of our knowledge and instruments into more elaborate and fuller form is, after all, easiest to follow. On the other hand, ancient philosophers, writers, intellectuals, and apparently even the general population at large, seem to have thought about progress in the world only rarely.[28]

It is clear from the above that in antiquity, there emerged some theories of evolution and development of the world, life, and culture. These theories had, among other things thanks to the conceptual influence of their authors, a notable impact for a very long time. We find here some clearly progressivist conceptions of the development of human culture (Hesiod, Aeschylus, the sophists) and some of these conceptions clearly rejected the idea of divine intervention in favour of basing their explanations on long-term human effort and work. Lucretius's rationalistic story of a gradual formation of human civilisation without any divine intervention in many ways resembles even some current conceptions of the evolution of culture (see Chap. 12 or Sonesson & Dunér, 2016, and Sonesson, 2016 for a more specific biosemiotic discussion). On the other hand, there also appeared highly influential conceptions that described a gradual decline of civilisation and return to the primitive forms of human life (once again Hesiod, Plato's *Statesman*, and others). In many of these cases, one can observe in connection with the development of culture a degree of ambiguity, so that for instance the Golden Age is described as a time of plenty and of innocence but also a time of close connection with nature and ensuing blissful primitivism. In antiquity, but also later, the idea of progress from some initial state of humanity was thus viewed with notable ambivalence—and in that, we can see a profound difference from more modern perceptions (cf. Chap. 3).

[28] Cf. Dodds (1973: 18–20, 24), Burkert (1997: 38–43).

References

Barney, R. (2006). The sophistic movement. In M. L. Gill & P. Pellegrin (Eds.), *A companion to ancient philosophy* (pp. 77–97). Wiley-Blackwell.

Betegh, G. (2016). Archelaus on cosmogony and the origins of social institutions. *Oxford Studies in Ancient Philosophy, 51*, 1–40.

Boas, G. (1948). *Primitivism and related ideas in the middle ages*. The John Hopkins Press.

Burkert W. (1997). Impact and limits of the idea of progress in antiquity. In A. Burgen, P. McLaughlin, & J. Mittelstrass (Eds.), *The idea of progress* (pp. 19–46). de Gruyter. Reprinted in *Kleine Schriften* (Vol. 8): *Philosophica* (pp. 240–259). Vandenhoeck & Ruprecht, 2008.

Campbell, G. (2000). Zoogony and evolution in Plato's *Timaeus*: The Presocratics, Lucretius and Darwin. In M. R. Wright (Ed.), *Reason and necessity: Essays on Plato's Timaeus* (pp. 145–180). Duckworth/Classical Press of Wales.

Campbell, G. L. (2003). *Lucretius on creation and evolution: A commentary on De rerum natura: Book five, lines 772–1104*. Oxford University Press.

Campbell, G. L. (2006). *Strange creatures: Anthropology in antiquity*. Duckworth.

Carling, G. (2016). Language: The role of culture and environment in proto-vocabularies. In D. Dunér & G. Sonesson (Eds.), *Human lifeworlds: The cognitive semiotics of cultural evolution* (pp. 83–96). Peter Lang.

Clay, J. S. (2003). *Hesiod's cosmos*. Cambridge University Press.

Darwin, C. (1866). *On the origin of species by means of natural selection, or the preservation of favoured races in the struggle for life* (4th ed.). John Murray.

Dodds, E. R. (1973). The ancient concept of progress. In *The ancient concept of progress and other essays on Greek literature and belief* (pp. 1–25). Clarendon Press.

Dougherty, C. (2006). *Prometheus*. Routledge.

Eco, U. (1995). *The search for the perfect language*. (J. Fentress, Trans.). Blackwell.

Fuhrer, Th. & Erler, M. (2015). Introduction. In P. Derron (Ed.), *Cosmologies et cosmogonies dans la littérature antique* (pp. 1–14). Fondation Hardt. (*Entretiens sur l'antiquité classique*, vol. 61).

Garani, M. (2007). *Empedocles redivivus: Poetry and analogy in Lucretius*. Routledge.

Gould, S. J. (1987). *Time's arrow, time's cycle: Myth and metaphor in the discovery of geological time*. Harvard University Press.

Greenblatt, S. (2011). *The swerve. How the world became modern*. Norton.

Guthrie, W. K. C. (1957). *In the beginning: Some Greek views of the origins of life and the early state of man*. Cornell University Press.

Guthrie, W. K. C. (1969). *A history of Greek philosophy* (Vol. 3): *The fifth-century enlightenment*. Cambridge University Press.

Heath, J. (2005). *The talking Greeks. Speech, animals and other in Homer, Aeschylus and Plato*. Cambridge University Press.

Herren, M. (2017). *The anatomy of myth: The art of interpretation from the Presocratics to the Church Fathers*. Oxford University Press.

Hladký, V. (2017). Empedocles' Sphairos. *Rhizomata, 5*(1), 1–24.

Hladký, V. (2025). *Rejoicing Sphairos, wandering daimon, and other living beings: Studies on Empedocles' philosophy of life (and death)*. Karl Alber.

Hladký, V., Kočandrle, R., & Kratochvíl, Z. (2012). *Evoluce před Darwinem: Nejstarší vývojová stadia evoluční nauky [Evolution before Darwin: The earliest developmental stages of evolutionary thought]*. Pavel Mervart.

Kahn, C. H. (1981). The origins of social contract theory. In G. B. Kerferd (Ed.), *The sophists and their legacy* (pp. 92–108). Franz Steiner.

Kerferd, G. B. (1981). *The sophistic movement*. Cambridge University Press.

Laks, A., & Most, G. W. (2016). *Early Greek philosophy* (Vol. 1): *Later Ionian and Athenian thinkers* (Vol. 1). Harvard University Press (Loeb classical library).

Lovejoy, A. O., & Boas, G. (1935). *Primitivism and related ideas in antiquity*. The John Hopkins Press.
Luhanová [Fulínová], E. (2014). Blessed life without philosophy: Plato and Hesiod on prehistory of man and world. *Aither*, *4*(8): *International Issue*, *2*, 51–101.
Mattéi, J.-F. (1996). *Platon et le miroir du mythe: De l'âge d'or à l'Atlantide*. Presses Universitaires de France.
Most, G. W. (1998). Hesiod's myth of the five (or three or four) races. *The Cambridge Classical Journal, 43*, 104–127.
Nelson, S. (1998). *God and the land: The metaphysics of farming in Hesiod and Vergil*. Oxford University Press.
Rashed, M. (2018). *La jeune fille et la sphère: Études sur Empédocle*. Presses de l'Université Paris-Sorbonne.
Rechenauer, G. (2013). Archelaos aus Athen. In D. Bremer, H. Flashar, & G. Rechenauer (Eds.), *Die Philosophie der Antike* (Vol. 1): *Frügriechische Philosophie* (Vol. 2, pp. 797–810). In F. Ueberweg & H. Holzhey (Eds.), *Grundriss der Geschichte der Philosophie*. Schwabe.
Rossi, P. (1970). *Philosophy, technology, and the arts in the early modern era* (S. Attanasio, Trans.). New York. Original Italian edition: *I filosofi e le macchine: 1400–1700*. Feltrinelli, 1962.
Rossi, P. (1984). *The dark abyss of time. The history of the earth and the history of nations from Hooke to Vico* (L. G. Cochrane, Trans.). Chicago/London: University of Chicago Press. Originial Italian edition: *I segni del tempo: Storia della terra e storia delle nazioni da Hooke a Vico*. Feltrinelli, 1979.
Schmid, K. (2015). Von der Gegenwelt zur Lebenswelt: Evolutionäre Kosmologie und Theologie in Buch Genesis. In P. Derron (Ed.), *Cosmologies et cosmogonies dans la littérature antique* (pp. 51–103). Fondation Hardt. (*Entretiens sur l'Antiquité Classique*, vol. 61.)
Schrijvers, P. H. (1999). *Lucrèce et les sciences de la vie*. Brill.
Scully, S. (2015). *Hesiod's Theogony: From near eastern creation to paradise lost*. Oxford University Press.
Sedley, D. (2007). *Creationism and its critics in antiquity*. University of California Press.
Sonesson, G. (2016). Epilogue: Cultural evolution: Human history as the continuation of evolution by (partially) other means. In D. Dunér & G. Sonesson (Eds.), *Human lifeworlds: The cognitive semiotics of cultural evolution* (pp. 301–335). Peter Lang.
Sonesson, G., & Dunér, D. (2016). Introduction: The cognitive semiotics of cultural evolution. In D. Dunér & G. Sonesson (Eds.), *Human lifeworlds: The cognitive semiotics of cultural evolution* (pp. 7–21). Peter Lang.
Taylor, C. C. W. (2020). The sophists. In E. N. Zalta & U. Nodelman (Eds.), *Stanford encyclopedia of philosophy*. Retrieved October 12, 2021, from https://plato.stanford.edu/entries/sophists
Toman, J., & Flegr, J. (2017). Stability-based sorting: The forgotten process behind (not only) biological evolution. *Journal of Theoretical Biology, 435*, 29–41.
van Heusden, B. (2004). A bandwidth model of semiotic evolution. In M. Bax, B. van Heusden, & W. Widgen (Eds.), *Semiotic evolution and the dynamics of culture* (pp. 3–33). Peter Lang.
Vernant, J.-P. (1960). Le Mythe hésiodique des races: Essai d'analyse structurale. *Revue de l'histoire des religions, 157*(1), 21–54. Reprinted in *Œuvres: Religions, rationalités, politique* (Vol 1, pp. 255–280). Éditions du Seuil, 2007. English translation: Hesiod's myth of the races: An essay in structural analysis. In *Myth and thought among the Greeks* (J. Lloyd & J. Fort, Trans., pp. 25–51). Zone Books, 1996.
Vernant, J.-P. (1966). Le Mythe hésiodique des races. Sur un essai de mise au point. *Revue de philologie, 40*, 247–276. Reprinted in *Œuvres: Religions, rationalités, politique* (Vol. 1, pp. 281–313). Paris: Éditions du Seuil, 2007. English translation: Hesiod's myth of the races: A reassessment. In *Myth and thought among the Greeks* (J. Lloyd & J. Fort, Trans., pp. 53–87). Zone Books, 1996.
Vernant, J.-P. (1974). Le mythe prométhéen chez Hésiode. In *Mythe et société en Grèce ancienne* (pp. 177–194). Maspero. Reprinted in *Œuvres: Religions, rationalités, politique* (Vol. 1,

pp. 751–764). Éditions du Seuil, 2007. English translation: The myth of Promtheus in Hesiod. In *Myth and society among the Greeks* (J. Lloyd, Trans., pp. 183–201). Zone Books, 1990.

Vernant, J.-P. (1985). Méthode structurale et mythe des races. In J. Brunschwig, C. Imbert, & A. Roger (Eds.). *Histoire et structure: À la mémoire de Victor Goldschmidt* (pp. 43–60). Vrin. Reprinted in *Œuvres: Religions, Rationalités, Politique* (Vol. 1, pp. 314–334). Paris: Éditions du Seuil, 2007. English translation: Structural Method and the Myth of the Races. In *Myth and Thought among the Greeks* (J. Lloyd & J. Fort, Trans., pp. 89–112). Zone Books, 1996.

Vidal-Naquet, P. (1981a). Le Mythe platonicien du *Politique*, les ambiguïtés de l'âge d'or et de l'histoires. In *Le chasseur noir* (pp. 361–380). Maspero.

Vidal-Naquet, P. (1981b). Valeurs religieuses et mythiques de la terre et du sacrifice dans l'*Odyssée*. In *Le chasseur noir* (pp. 39–68). Maspero.

West M. L., (1966; Ed. & com.). Hesiod, *Theogony*. Clarendon Press.

West M. L. (1978; Ed. & com.). Hesiod, *Works and days*. Clarendon Press.

Wright, M. R. (1995a). *Cosmology in antiquity*. Routledge.

Wright, M.R. (1995b; Ed., trans. & com.). *Empedocles, the extant fragments* (2nd ed.). Hackett Publishing.

Zimmer, J. R. (2004). Semiotic questions raised by the evolution of talk and the emergence of complex society. In M. Bax, B. van Heusden, & W. Widgen (Eds.), *Semiotic evolution and the dynamics of culture* (pp. 106–116). Peter Lang.

Zlatev, J. (2016). Mimesis: The role of bodily mimesis for the evolution of human culture and language. In D. Dunér & G. Sonesson (Eds.), *Human lifeworlds: The cognitive semiotics of cultural evolution* (pp. 63–82). Peter Lang.

Chapter 2
Breaking Through the Boundaries of Time: Changing Modern Notions About the Age of Nature and Culture

Tomáš Hermann, Vojtěch Hladký, and Lucie Strnadová

Abstract This chapter traces the development of the idea of deep time since the Renaissance and the Enlightenment all the way to the beginning of modern historicism and evolutionism. It describes the experiences and theoretical steps that enabled the conceptualisation of a long historical development in four areas: 1. Earth's geological past and fossils (the archives of nature); 2. The ancient past against the backdrop of literary legacy and tradition (the archives of written history); 3. Cultural prehistory based on material artefacts (the archives of unwritten history); 4. Multilingualism and the history of human speech (the archives of language). It is shown how and why the notion of deep time based on revealed temporal layers and epochs has been of crucial importance to modern historical and biological thought.

Keywords Deep time · Geohistory · Biblical chronology · Prehistory and archaeology · History of science

2.1 Introduction

Sigmund Freud in his *Introductory Lectures on Psychoanalysis* described three milestones in the development of human knowledge. Starting with the Renaissance, these three turning points gradually transformed how people perceive their place in the world:

> In the course of centuries the naïve self-love of men has had to submit to two major blows at the hands of science. The first was when they learnt that our earth was not the centre of

T. Hermann (✉) · V. Hladký · L. Strnadová
Department of Philosophy and History of Sciences, Faculty of Science, Charles University, Prague, Czech Republic
e-mail: tomas.hermann@natur.cuni.cz; vojtech.hladky@natur.cuni.cz; lucie.cermakova@natur.cuni.cz

J. Toman et al. (eds.), *Culture Evolving*, Biosemiotics 10,
https://doi.org/10.1007/978-3-032-16124-6_2

> the universe but only a tiny fragment of a cosmic system of scarcely imaginable vastness. This is associated in our minds with the name of Copernicus [...]. The second blow fell when biological research destroyed man's supposedly privileged place in creation and proved his descent from the animal kingdom and his ineradicable animal nature. This revaluation has been accomplished in our own days by Darwin, Wallace and their predecessors, though not without the most violent contemporary opposition. But human megalomania will have suffered its third and most wounding blow from the psychological research of the present time which seeks to prove to the ego that it is not even master in its own house but must content itself with scanty information of what is going on unconsciously in its mind. (Freud, 1963: 284–285)

The revolutionary theories of all three of the abovementioned thinkers—Copernicus (1473–1543), Darwin (1809–1882), and Freud (1856–1939)—led to discussions that spread far beyond the confines of the science within which they were formulated and had a far-reaching impact on contemporary culture as a whole. In all three cases, the newly proposed theories met tremendous resistance and led to controversies. Heliocentrism, evolution, and psychoanalysis each formed the next step in weakening the exclusive status of humanity, which was according to traditional notions located both symbolically and literally at the centre of the universe.

The development of science thus followed the path of a gradual de-anthropocentring of human knowledge and culture in general. Both Copernicus's and Darwin's theories were based on a substantive broadening of the known world: in the former case, it was an expansion in space, in the latter an expansion in time. Modern science thus gradually arrived at the idea of an *infinite universe* and of *deep time*, but the processes that led to this shift in scale were far from straightforward or unequivocal.

The first person to postulate an infinite universe was Giordano Bruno (1548–1600), a passionate promoter of Copernicanism. He claimed that our Sun was just one of many stars in the universe. Until his time, stars were believed to be mere shining points in the sphere of fixed planets enclosing our world. Earlier, Bernardino Telesio had replaced the Aristotelian expression for 'place' (*locus*), which is defined by bodies present in it, with the concept of 'space' (*spatium*), which exists independently of bodies. The notion of infinite space thus formed gradually until it rose to prominence in Modern science (Koyré, 1957; Kristeller, 1964: 103–105). Its rise, however, was for various reasons far from straightforward: in fact, even Kepler spoke about the 'hidden terror' (*horror occultus*) this notion evoked (Kepler, 1606: 105, KGW I,253.15; cf. Koyré, 1957: 61).

What one can trace in the Early Modern Era was not only a shift 'from the closed world to the infinite universe' but also a transition 'from the world of "more-or-less" to the universe of precision' (Koyré, 1948, 1957). New scientific instruments enabled a more accurate measuring of reality, including observation of the night sky, navigation, and land surveying. Mathematisation of the world went hand in hand with its 'mechanisation', so that, for instance, if the heavenly bodies were to behave under all circumstances in accordance with exact mathematical laws, one had to deny them the ability to freely alter their movements. They had to become pieces of inert matter. Our vision of the world shifted away from viewing it as a gigantic living organism and towards seeing it as an ingeniously constructed machine. Initially, this transformation affected sciences such as astronomy or physics, while other sciences remained under a strong influence of the Platonic and Aristotelian tradition. Although there eventually emerged a chasm between the sciences that dealt with living bodies and those that investigated non-living bodies, even in the seventeenth century physics was sometimes viewed as dealing with *physis*, life as a whole, including for instance the areas now covered by biology (Dijksterhuis, 1961; Horský, 1966, 1967; cf. also Dunér, 2016: 247–255, in the context of biosemiotic discussion of conceptual metaphors of Early Modern science).

While this new conception of space was taking shape, new ways of thinking about time had also started to emerge, although that process was markedly slower. Between the Renaissance and the Age of Enlightenment, a gradual but profound change in the conceptualisation of nature and humans took place and, over time, this shift enabled a novel approach to the evolution of nature and culture. In the minds of scientists, the gradually dawning emptiness of the universe went hand in hand with its increasingly precise mathematisation and mechanisation. But when it came to boundlessly deep time, the notion was at least as terrifying and gapingly limitless as the idea of an empty infinite space used to be. Despite these sentiments, however, various people searched for tangible clues that set the vision of ancient times in an increasingly accurate and concrete context of particular layers, epochs, and events.

In Greek and Roman antiquity, the world was thought to go through cycles of roughly tens of thousands of years (see Chap. 1). The world of the Bible, on the other hand, had a beginning and an end: it evolved directly and over a relatively short period of time. Ever since the time of the Church Fathers, the Bible was believed to present a model of universal history with a uniform, linear time that started either with the creation of the world or with Adam. The only relevant framework was the sequence dictated by the scheme of Christian history of salvation. In the Middle Ages, particular Biblical or Church histories had to incorporate both the historical narratives of particular nations and the entire *historia naturalis*.[1] The monogenesis of humankind was another key part of Biblical exegesis. To some extent, the Old Testament allowed for a deeper temporal dimension, because although Christ's birth was supposed to be preceded by just 140 generations (already according to, e.g., Augustine), scholars assigned to many of them, i.e., not only the proverbially long-lived Methuselah but also others, markedly longer lifespans than those common in contemporary humans. According to some scholars, the

[1] Pliny's *Naturalis historia* described various *naturalia* and this sense of the term survived into the Modern Era. For Francis Bacon, *historia naturalis* still meant just a description: in his view, an investigation of causes belonged to the theoretical science (physics). The humanist Justus Lipsius (1547–1606) separated *historia naturalis* from *historia narrativa*. G. W. Leibniz then around 1694 included in the *histoire humaine* also universal history, geography, antiquities, philology, and literary history, as well as the history of customs, laws, and religions (Malina & Vašíček, 1990: 2).

exceptional age of some figures from the Old Testament was matched by their size, so that, for instance the French Academy of Sciences had in the eighteenth century determined that Adam must have been 40 meters tall and Eve just a meter and a half shorter. Such conclusions were in part inspired by the fact that the skeletal remains of palaeolithic animals were at that time still believed to have belonged to ancient humans.[2] In a similar way, the total number of people who would live between the creation of humanity and the Last Judgement was estimated in the range of hundreds of millions because the number was based on the number of Adam's sperm.

For a long time, the Bible and its stories furnished the tools for a periodisation within which natural and cultural finds and artefacts were situated. Long into the Modern Era, scholars spoke about the time before and after the flood (antediluvial and postdiluvial era), the time before and after the confusion of languages at the Tower of Babel, before and after Christ, or before and after the Greeks and Romans.[3] The mythical flood also served as the basic periodisation divider for human prehistory: for instance, the historian and topographer William Camden (1551–1623) divided the most ancient history into a period from creation of the world until the flood (the 'period of obscurity') and from the flood until the first Olympic games (the 'fabulous period'). Such periods were not long and could be interpreted using biblical history. Even the French *Encyclopaedia* takes the great flood, dated to 2262 or 1556 before the birth of Christ, to be the key point of periodisation of history. While history described in the Old Testament did inspire thinking in terms of numbers and sizes that exceeded the everyday experience, the timescale was nevertheless one of the obstacles that hindered a conceptualisation of deep time. The various estimates of the age of the world based on the Bible never exceeded 7000 years. The Irish bishop James Ussher (1581–1656), whose calculations were authoritative for almost two centuries, concluded—based on his own historical studies but also Luther's and Kepler's calculations—that the world was created at sunset on Sunday, 23rd of October 4004 before the birth of Christ. Just 2228 years were thus supposed to have elapsed between the creation of the world and the beginning of Greek history.[4] Even the French creators of the *Encyclopaedia* only shifted the beginning of

---

[2] Interpretation of palaeolithic skeletal remains as belonging to giant humans is an archaic practice but on the fringes of Europe, specifically in the Balkans, we encounter it even in the nineteenth century (Mayor, 2000: 3–10, 210–211).

[3] After a complicated and often unclear chronological translation of the entire literary corpus onto a single temporal axis, all we have retained is the putative year of Christ's birth, which coincides with the beginning of the Christian calendar.

[4] At around the same time, Johannes Kepler dated the creation of the world to 3977 years before Christ and Isaac Newton to 4000 years before Christ. A traditional Byzantine date for the creation of the world is 1 September 5509 before Christ, which concurs with the dating proposed by some Church Fathers. Ussher's calculations were made further 'more precise' by Bishop John Lightfoot (1602–1675) and the theologian, historian, and mathematician William Whiston (1667–1752), who in his *New Theory of the Earth* (1696)—which met with positive reactions from both Newton and Locke—dated the flood to 18 November 2349 before Christ.

creation of the world and creation of the first humans about two thousand years deeper in time.

Nevertheless, it is difficult for us to grasp why the discovery of the dizzying depth of time was so delayed compared to the discovery of infinite space. In fact, for nature and human society the depth of time was clearly conceptualised only in the nineteenth century. A satisfactory answer to this conundrum still eludes the investigations of historians and philosophers of science. It is a much tougher nut to crack than an explanation of the genesis of knowledge in physics, astronomy, or cosmology. Our goal here is therefore to outline some of the experiential and theoretical steps that enabled people to transcend the boundaries of human imagination so that the idea of a truly long historical development could become acceptable. In order to appreciate how complicated and varied the process of opening the gates of time was, we shall take a closer look at four model areas:[5]

1. The geological past of the Earth, examined through fossils and related phenomena (metaphorically speaking, opening the archives of nature).
2. The ancient past against the backdrop of literary legacy and tradition (the archives of written history).
3. Cultural prehistory, investigated based on material artefacts and non-literary sources (the archives of non-written history).
4. The multitude of languages and the history of human language (the archives of language).

2.2 The Archives of Nature: The Variety of Life, Fossils, and the History of the Earth

The Earth had lost its privileged position at the centre of the universe and became more accessible to study not only because of the application of mechanical principles but eventually also thanks to historical methods (Rudwick, 2005: 134; Rudwick, 2014). The process of 'deepening' the geological time was not only about its calculated length but crucially also about a shift in the assumed ratio between the hypothetical length of development of the Earth and human history. As the depth of time increased, human history went from broadly coinciding with the history of the Earth to becoming its ever smaller part. This shift inevitably called into question the belief that the Earth was created (or even created specifically for people).

Famous European collections, such as the Kircherian Museum in Rome or Ferrante Imperato's collections in Florence, which assembled an impressive range of *naturalia* from both the Old and the New World, including various fossils,

[5]We have been inspired by a number of basic monographs on the subject, especially by Rossi, 1987, Malina & Vašíček, 1990, and Eco, 1995. Of pivotal importance for the subject are also the now classical works of Toulmin & Goodfield, 1965 and Gould, 1987.

enabled a comparative approach (Findlen, 1994; Cooper, 2007: 9). The study of fossils facilitated certain crucial steps towards deep time, although initially this did not seem to be a likely consequence because, for a long time, the prevalent opinion was that fossils emerged *in situ* as 'freaks of nature' (*lusus naturae*) either by self-procreation or due to the action of a natural 'plastic force' (*vis plastica*). Athanasius Kircher (1602–1680) even tried to grow them in his laboratory to empirically test this explanation of their origin (Cutler, 2009: 144). But even after Kircher's time, scholars proposed theories according to which fossils were minerals (*lapides sui generis*) created in animal forms by crystallisation, either by accident or due to unknown causes, without achieving the perfection of organic shapes (Fig. 2.1).

Comparisons of fossils with current flora and fauna eventually inspired some naturalists' speculations about their organic origin. For instance, based on his observations, Niels Stensen (1638–1686) described similarities between fossilised *glossopterae* and the teeth of current sharks, while Robert Hooke (1635–1703) pointed to similarities between the forms of fossilised shells and current molluscs (Kusukawa, 2013; Thomsen, 2009). Hooke also investigated fossils under the microscope, compared fossilised and living wood, and based on various experiments concluded that fossilisation is a physical process that requires water with sufficient mineral content. In short, he supported his theory of an organic origin of

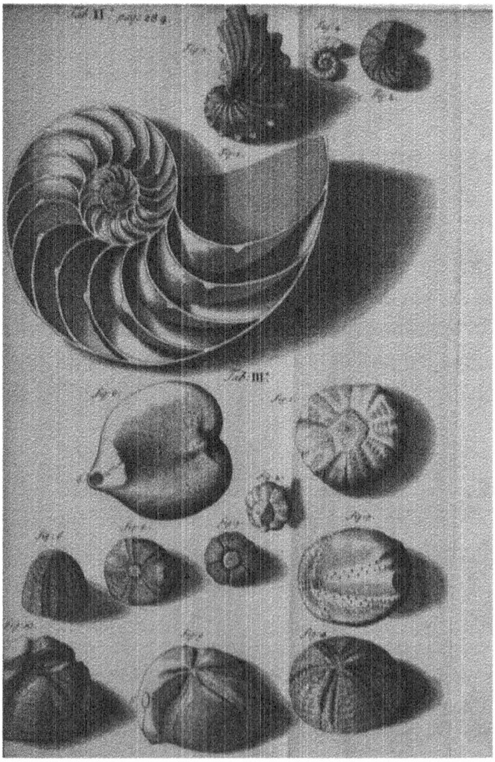

Fig. 2.1 A comparison of petrified echinoids 'Fig. 10' and contemporary echinoids 'Fig. 9' in Robert Hooke, *Posthumous Work*, 1705: tab. II, p. 248. Wellcome Collection (https://wellcome collection.org/works/xxggvyj3/items?canvas=278). See also Kusukawa, 2013

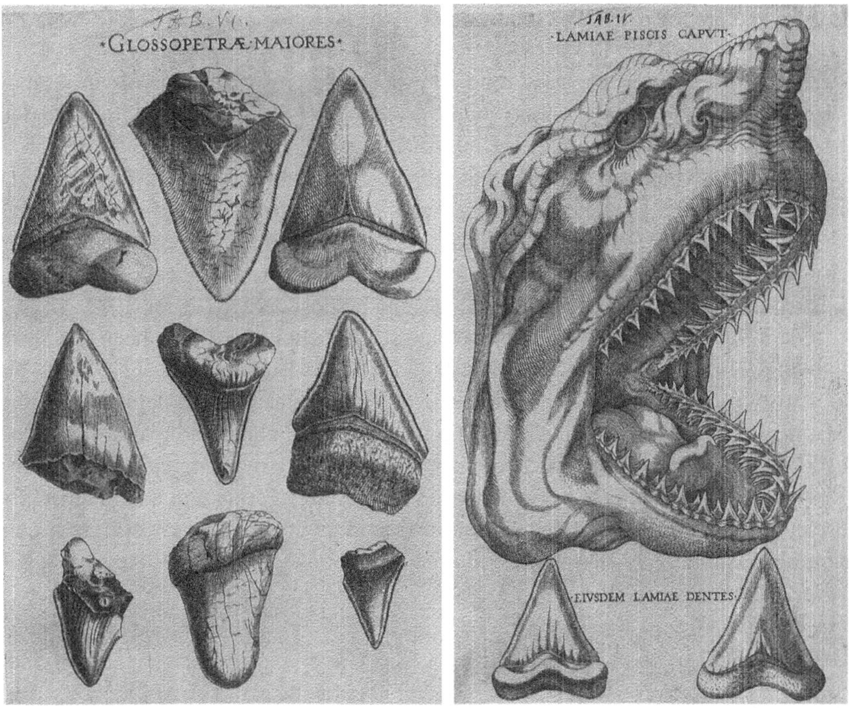

Fig. 2.2 Glossopterae, tongue-shaped stony objects similar to sharks' teeth. Niels Stensen used "Glossopterae"to argue that at least some fossils had been parts of living organisms. Niels Stensen, *Elementorum mytologiae specimen*, 1667; see Rudwick, 2014. e-rara (https://www.e-rara.ch/zut/content/zoom/4621718)

fossils by pointing to analogies between fossilised and living samples (Fig. 2.2 (a, b)).[6]

Yet even after scholars have accepted that fossils are of an organic origin, they did not necessarily infer their great age, and, even more importantly, it did not imply that the fossils came from extinct organisms. It was still possible to speculate that they belonged to organisms that still lived somewhere in the depth of the ocean, which is what for instance John Ray (1627–1705) claimed. Nevertheless, systematic research was finding fossils all over the place (in deep geological strata, at mountaintops, and even in the building material of ancient Etruscan temples or pyramids; Cutler, 2009: 145), which implied that they must be at least as old as the oldest human constructions. Through various estimates of the age of fossils, the

[6] Hooke was a pioneer of observation under a microscope. A microscope, which magnified tiny objects, expanded the boundaries of both the senses and human imagination much like the telescope, which enabled people to see distant things as if they were close (for more on the first observations with a microscope, see Freedberg, 2003). Importantly, Hooke also admitted that some species may have become entirely extinct (Rossi, 1987: 3, 15).

historical dimension had thus finally entered the *historia naturalis* (Rossi, 1987: 17; Toulmin & Goodfield, 1965: 281).

One of the first people who tried to explain the presence of fossils in various geological layers was again Niels Stensen. According to him, rocks initially had to be liquid, so that fossils could get into them. Over time, they solidified, which is why the lower strata are older than the top layers. In other words, rocks are organised in a temporal sequence. Stensen's aim was to prove that geological changes were common, but he refrained from specifying the rate of geological change (Rudwick, 2005: 97; Cutler, 2009: 146). Even he, however, believed that the most important motor of geological change was the biblical deluge (Rossi, 1987: 18).

For a very long time, the biblical flood was believed to have been the most dynamic geological event ever. For instance, Thomas Burnet (1635–1715) built his explanation of changes in the appearance of the Earth around the deluge, claiming that geological strata had emerged from the primordial chaos and over time settled according to their weight. The flood then caused a drop of geological strata and the formation of mountain ranges and continents. Burnet incorporated into his theory a direct timeline: he claimed that erosion was unidirectional and the world in its current form is in effect one vast postdiluvial ruin (Gould, 1987: 47; Rossi, 1987: 37) (Fig. 2.3).

Fig. 2.3 Burnet's linear history of the Earth. Depiction of the six stages (clockwise): 1. The initial chaos; 2. Paradisal antediluvial world; 3. Global flood; 4. The present world; 5. Future global volcanic 'conflagration' that will destroy the present world; 6. Perfection of the Earth during Christ's millennial reign; 7. The end, when the Earth will be transformed into a star. Frontpiece of Thomas Burnet, *Telluris theoria sacra*, 1699; see Rudwick, 2014.)

Both Burnet and Stensen managed to fit their theories into the biblical timescale. Stensen even found the time since the creation of the world (about 6000 years) to be too long for the petrified organisms in geological layers to survive to our times (Cutler, 2009: 145). Hooke proposed a longer timescale but he did not dare to define it more exactly. He believed that the Earth had undergone various dynamic natural events, such as earthquakes, floods, eruptions, and severe rains, which changed the Earth and life on it. Places that used to be underwater are now on dry land and vice versa, mountains have turned into plains and plains into mountains. The Earth was first fluid and then slowly and gradually solidified. Over time, entire regions (such as Atlantis) disappeared while others emerged, and even the species that inhabit the Earth had changed—and human cultures changed with them (Rossi, 1987: 14).

Fossils thus became a key element in the discussions about the history of the Earth. Later, they were even considered an accurate indicator of the age of individual geological layers because further observations had shown that each geological layer contained different kinds of fossils. Assuming that they were indeed of an organic origin, their accumulation would have required sufficient time—which was why they came to be viewed as a reliable indicator of the dating of individual strata. They were believed to provide a better foundation for estimating the relative position of geological layers, and thus also for estimating their relative age, than the composition of rocks (Toulmin & Goodfield, 1965: 162–163).

In the late eighteenth century, fossils started to lose their importance because scientists started to focus on primary rocks, which did not contain any. Based on his observations of emergence of a new island near Santorini, Anton Moro (1687–1764) speculated about gradual changes of the Earth's surface. He was convinced that such processes could explain the emergence of mountain ranges, the origins of Earth's structure, and the origin of fossils. He formulated a theory according to which geological layers are not sediments but rather the result of volcanic eruptions, which eject hot material that solidifies (Rossi, 1987: 82–83). Later, the observation of volcanic craters convinced William Hamilton (1730–1803) that they were of an age so remote as to be outside the reach of our history. In other words, if a volcano is formed by subsequent eruptions, each of which adds some material (and this is something known from actual observations), then the age of volcanos must exceed our understanding (Rudwick, 2005: 120).[7]

A breakthrough in understanding the transformations of both life and Earth's geological past came with the ideas and theories of Georges Louis Leclerc de Buffon (1707–1788), who explained the origin of the solar system and development of the Earth based on gradual rather than revolutionary changes. According to Buffon, the world did not change in a single catastrophic event, such as a great deluge, because the laws of nature are immutable and natural processes slow. We should therefore observe repeatedly occurring phenomena and daily events. Buffon was the first to divide the development of the Earth in six (and later seven)

[7] Hamilton had also undertaken archaeological excavations, observed the changes that came after the eruption of Vesuvius in 1767, and published his excellent observations of Etna and Vesuvius and the conclusions he reached based on these first-hand experiences in his *Campi Phlegraei* (1776).

geological eras. In doing so, he also tried to outline the temporal axis of development of the Earth. He assumed that the Earth was torn away from the Sun as a result of the impact of a comet. After this separation, it solidified for about 3000 years, cooled for about 35,000 years, after 74,382 years reached its current temperature, and about 45,000 years from now it will cool again and life will disappear from it. Buffon published these time estimates in his treatise *Les Époques de la nature* (1778),[8] but was not content with his calculations of the age of the Earth and in various editions of the book, he amended them. His manuscripts later showed that he felt this chronology was still too short: in fact, in private he speculated that the Earth may be about 3 million years old, which was a length of time hitherto felt to be utterly inconceivable. Buffon believed that longer time estimates would be closer to reality, but he was also convinced that one ought to keep the estimates short to accommodate the limited capacities of our reason (Rossi, 1987: 107–108). In his view, the greatest obstacle was thus the utter unimaginability of temporal spans of a higher order (Cutler, 2009; Rappaport, 1997; Rudwick, 1985, 2005). The notion of a truly long development was not only hard to grasp and unintuitive but also as terrifying as the idea of boundless space used to be.[9]

Georges Cuvier (1769–1832), the founder of modern palaeontology, viewed the conceptualisation of a long developmental period as an urgent problem already at the beginning of the nineteenth century. He struggled with the temporal scale because his geological theory of catastrophes and his rigid empirical methods not only required deep time but also opened the door to the biological history of the Earth (Toulmin & Goodfield, 1965: 164; Rudwick, 2005: 2). Cuvier did not fear deep time; on the contrary, he clearly perceived the need to think in such terms. When wondering whether one could find out about events that preceded the birth of humankind, and whether life sciences could 1 day also have their Newton, he anticipated something that took place in the following generation, when such a re-evaluation of the history of the world and life indeed took place.

Thanks to Cuvier, fossils once again came to play a key role in dating the periods of the world's development, and jointly with other geological phenomena, they gradually came to be seen as leads, hints, evidence, or sources which a naturalist must observe to uncover the history of the Earth. In fact, fossils were often compared to historical sources. Charles Lyell (1797–1875) on the title page of his foundational *Principles of Geology* (1832) symbolically visualised the close link between the history of civilisation and the history of nature with a picture of ruins

[8] Buffon made his calculations based on experiments with heated iron balls and cannonballs, where he followed their cooling and wrinkled surface reminiscent of Earth's surface. He also observed the development of life using a meat broth which over time started to team with microorganisms.

[9] In his extensive *Histoire naturelle* (1749–1789), which ran to many volumes, Buffon considered a gradual transformation of species by 'degeneration' in reaction to local conditions. His postulate of an 'inner form' (*moule intérieure*) shared by all organisms had influenced Lamarck's slightly later evolutionary speculations. Together with British empiricists, Buffon exerted a crucial influence on Diderot, Holbach, Rousseau, and other thinkers of the French Enlightenment and their pre-evolutionary thoughts about nature and culture.

Fig. 2.4 Pillars of the Temple of Serapis at Pozzuoli used by Lyell to show the difference between the original and current sea level. The frontispiece of Charles Lyell's *Principles of Geology* (1832). Wikimedia Commons (https://commons.wikimedia.org/wiki/File:Charles_Lyell_-_Pillars_of_Pozzuoli.jpg)

of the temple of Serapis in Puzzuoli in southern Italy, where one can clearly see the traces of marine molluscs imprinted in the stones, because although the temple was built in the Roman times, it was for several centuries completely inundated. For Lyell, it became a symbol of the changes and processes that can give rise to entire continents if allowed to continue over a long-enough period of time (Rudwick, 2008: 282–283) (Fig. 2.4).

2.3 The Archives of Written History: Nonbiblical History and the Pre-Adamite Theory

Human history was expanding its boundaries much like the history of nature: it had to be made much broader than what the biblical tradition had assumed. Discussions about the turning points of cultural development were as complicated as the discussions about fossils and the history of nature. The Bible-based image of history was challenged both by new discoveries about human culture and by novel interpretations of long-known historical events (Rossi, 1987: 145). New interest in ancient texts, which led to discoveries about ancient civilisations, was supplemented by findings acquired during various journeys of discovery.

During the Renaissance, interest in Greek and Roman antiquity acquired a new impulse from the writings that arrived in Europe after the conquest of the Byzantine Empire by the Ottomans. Print also played an important role: the preparation of manuscripts for print required detailed philological work, thorough revision of ancient texts, and comparisons of various versions, transcriptions, and translations. The humanists who studied ancient texts moreover discovered claims that were incompatible with the Bible and its description of the origins of humankind (Popkin, 1987: 31). Such texts inspired a feeling of respect for all things ancient, including the age-old traditions from the times when Hellenistic states and later the Roman Empire ruled over various nations and cultures, learned from them, and often also admired them.

Advanced culture and a high level of education were often ascribed to civilisations of the East: texts ascribed to the ancient Egyptians, Persians, and Chaldeans (inhabitants of ancient Mesopotamia) were viewed as a veritable well of ancient wisdom. According to Marsilio Ficino (1433–1499), each of the three then known continents had its own source of wisdom independent of the biblical revelation. In the case of Asia, it was the so-called Chaldean Oracles, which popular tradition ascribed to Zoroaster and his Magi. In Africa, this wisdom was supposed to be represented by the Egyptian Hermetic writings, and Europe was represented by poems whose authorship was ascribed to Orpheus, the mythical divine musician and initiator into mysteries. From him, a continuous tradition of divine philosophy was said to lead to Pythagoras and Plato.

The aim of these Renaissance thinkers was to integrate the newly discovered ancient religious and philosophical notions into the Christian worldview. Biblical revelations and ancient thought were either viewed as one tradition within which the two elements enrich one another (*prisca theologia*) or, alternatively, it was assumed that each of these traditions was capable of approaching the divine basis of reality using rational insight (*philosophia perennis*). Importantly, Renaissance thinkers tended to view the development of human knowledge as cyclical rather than linear (in the sense of discovery of new truths). They thus saw themselves as trying to reconstruct and re-learn the original wisdom that was given to humanity in its fullness after the creation of the world but later lost or forgotten. This approach to the past was endorsed not only by philosophers but also by some scientists, such as Kepler or Newton, who viewed their discoveries as rather just re-discoveries and actualisation of much earlier, mostly Pythagorean, knowledge (Hladký, 2014, 2022a, 2022b).

This gradually undermined the alleged uniqueness of Western Christian civilisation and opened space for appreciating other cultures, cultures that existed in a different space, at a different time, and which sometimes appeared rather exotic. A comparison of different cultures, especially Chaldean, Egyptian, Greek, and Hebrew, showed that the Jews had adopted many Egyptian traditions and customs and to some extent preserved them. Some scholars also contrasted the Egyptian wisdom with traditions of the Jews, an originally nomadic nation without advanced laws and legal structures (Rossi, 1987: 125, 129). From this perspective, the Jews ceased to be perceived as the chosen nation and came to be viewed as just one nation among many, a nation whose history was merely part of a much older story.

No less unsettling was the encounter with the New World, which, among other things, inspired ideas about the innocence and nobility of its inhabitants (Černá, 2016: 96–97). Ever since the time of Columbus and Vespucci, various scholars have wondered who the inhabitants of the New World really were. Spanish and Portuguese philosophers started investigating the origins of American natives not just out of curiosity but also in order to determine the status of native Americans and to see whether they are related to the newly incoming colonists. Scholars thus started to study their origins and their relationship to Europeans in the early sixteenth century (Popkin, 1987: 32–33; Almond, 2006: 165; Grafton, 1992).

More generally, the new discoveries and the realisation that both ancient and new nations and cultures were greatly diverse led to a reawakening of the pre-Adamite theory, which claimed that there were generations of humans in the world even before the first biblical man (Rossi, 1987: 134).[10] For many scholars, however, it was difficult to accept speculations about the great age of some cultures. They disputed such claims and argued that the newly discovered writings and stories of both ancient and more recent cultures were incomplete and unclear. Most contemporary chronicles of the New World tried to reconcile the new findings with the traditional biblical notions. For instance, they often claimed that native Americans are the descendants of Noah who came to the American continent after the deluge.[11] The Spanish humanist Juan Luis Vives (1493–1540) in his commentary on Augustin's *De civitate Dei* (1522) argued against the pre-Adamite theory. He was openly sceptical regarding arguments aimed at proving the pre-biblical age of the Chaldean or Egyptian culture and believed the alleged great age of these cultures is exaggerated (Popkin, 1987: 32). Other scholars pointed out that nations often fabulate regarding their origin and history to prove their venerable antiquity (Popkin, 1987: 30; Rossi, 1987: 158, 161–163). Even Isaac Newton, who among other things intensively—and quite unconventionally—studied ancient, especially Egyptian, chronologies, was similarly sceptical and he, too, spoke of the bad habit of moving one's roots deeper into the past in old literary sources (Rossi, 1987: 163; Schilt, 2021). Both Vives and Newton thus tried to defend the traditional biblical calculations and their position can be viewed as an expression of epistemological stubbornness similar to what we saw when tracing the discovery of geological time. Newton was driven by rational scepticism, which did not prevent him from respecting the calculations of biblical creation in his cosmology and physics and defending ancient chronologies

[10] In the Islamic environment, a pre-Adamite claim appeared already in the tenth century, when al-Maqdisi deduced from the Quran that some people lived on the Earth even before Adam (Eco, 1995: 89 lists also further references).

[11] This was also the view of, for instance, Oviedo (1478–1557), Hernández (1514–1587), or Acosta (1539–1600) (Černá, 2013).

against new challenges. Generally speaking, the persistence, entrenched status, and long survival of the traditional chronological framework was driven by a fear of the unimaginable chasm of time, a fear of loss of a rational coherence of unity of the world, humanity, and its history.

Efforts to preserve traditional certainties were nevertheless repeatedly challenged by new forms of the pre-Adamite theory. Already Paracelsus (1493–1541) in his *Astronomia magna* openly doubted that Adam's descendants could have travelled to the distant American islands. If, therefore, their inhabitants were truly human, they must be descended from a different line than Adam (Almond, 2006: 165; Popkin, 1987: 34). Giordano Bruno, too, discussed subjects connected to the existence of people before Adam[12] and, like Paracelsus, suggested that inhabitants of the New World were perhaps not human at all, although in their appearance and intelligence they did seem to be. If, he said, they were people, they must belong to a different 'developmental line' (like Pygmies and other African ethnicities).

Bruno wrote down these thoughts during his sojourn in England (in 1583–1585), at a time when several other people who explored the pre-Adamite notions were also active here. The adventurer and writer Walter Raleigh (1554–1618) or the mathematician Thomas Harriot (1560–1621) had personally visited the New World and, although no records of their experiences survive, they were often accused of holding pre-Adamite views (Popkin, 1987: 36; Almond, 2006: 166). Another impetus for such ideas came with the discovery of Australia in the early seventeenth century. Gabriel de Foigny (*La Terre australe connue*, 1676) claimed that the aboriginal Australians are hermaphrodites unburdened by Adam's original sin, who live in perfect harmony with the virginal nature of their homeland (Almond, 2006: 163).

Arguments pertaining to both the ancient and newly discovered cultures were systematically summarised and a theory about people before Adam was presented by the French theologian Isaac de La Peyrère (1596–1676). He laid out his thoughts in the *Theological system upon the presupposition that men were before Adam* (*Systema theologicum ex Praeadamitarum hypothesi*, 1655), which was subsequently published either separately or jointly with an even longer treatise, the *PraeAdamitae* (1656). La Peyrère started by pointing to some problematic passages in the Bible, such as Paul's letter to the Romans (5,12–14), which implied that if the moral law had entered the world with Adam, the world, albeit lawless, must have been already inhabited. He also speculated about Cain's wife, who was allegedly from a distant family, which implied that she could not have been a descendant of Adam and Eve. La Peyrère supplemented these indications from the Bible with new arguments based on the study of the history of both ancient and newly discovered pagan cultures (Popkin, 1987: 43; Almond, 2006: 167). His ideas put the biblical account of history into perspective: he concluded that Adam was the forefather of only the Jewish nation and that both the flood and the building of the Tower of Babel

[12] Especially in his treatise *Spaccio della Bestia trionfante* (3,2). Bruno discusses people of different cultures also in his texts *Cabala del cavallo pegaseo* (2,1) or *De immenso et innumeralibus* (VII,18).

were merely local episodes that touched upon the lives of only a part of the world's population (Rossi, 1987: 140). The Greeks, Egyptians, Babylonians, but also the inhabitants of Mexico or China all had a history that was older than the history of the biblical nation (Popkin, 1987: 42–43). According to La Peyrère, the first inhabitants of the Earth were wild and barbaric. He personally was more fascinated by advanced civilisations and claimed that Chaldean astronomy, Egyptian chronology, and Chinese history and philosophy can all be explained only by assuming a long-enough period of development.[13]

La Peyrère was accused of heresy, tried, and sent to prison, but he did not revoke his claim that the pre-Adamite theory was the best way of reconciling the Bible with the pagan history. He was convinced that the Bible contains unclear passages that make sense only when interpreted this way. He compared his views to the theory of Copernicus: it did not change the facts, merely viewed them from a different angle (Popkin, 1987: 44). He condemned the unwillingness of other thinkers to deal with the times that preceded the history described in the Bible and believed that during these thousands of unknown centuries, what took place was a pluralistic history, created by various nations and various cultures (Rossi, 1987: 136). Paracelsus, Bruno, and La Peyrère thus represent the earliest form of the polygenetic theory. The pre-Adamite theory can be, in this context, viewed as a predecessor of the racial theories and polygenism of the nineteenth and twentieth centuries (see Chap. 3), but it has also played an important role in deepening cultural time and establishing linguistic polygenism (see Chap. 12).

2.4 The Archives of Non-Written History: Stone Artefacts and the Present Past

The understanding of cultural evolution was crucially influenced by the gradual discovery of the prehistoric dimension of humans in the depth of time, in a time for which there are no testimonies in written sources. It opened a new space between the geological past and ancient cultural history that is captured in writings. The main testaments that guided this journey into the past were various objects in the landscape, found or excavated artefacts, but also various comparative ethnographic or ethnological observations, which were viewed as a living testament to the life of ancient nations and cultures (see Chaps. 5 and 7). Metaphorically, one could think of them as being our guide to the 'present past' (see Chap. 18). In what follows, we shall see both some differences and direct connections with the two journeys into the depth of time described above.

[13] La Peyrère quotes Scaliger (for more on him, see below) who claimed that year 1594 corresponds to the year 880,703 of Chinese chronology. He also claimed that American and Australian civilisations go back many centuries, and the Chaldeans, too, have a rich history, which they however measured not in years but in intervals of 60, 600, or 6000 years (Rossi, 1987: 136).

Biblical history enabled people to talk about various further undifferentiated 'ancients' and pagans but the monuments ascribed to them, especially the tumuli and megaliths, so prominent in the landscape of northern Europe, tended to be attributed either to a recent pre-Christian past or to a mythical time of giants (typically by Saxo Grammaticus). Medieval Christian nations or aristocratic families were keen on deriving their descent from the Romans or even Trojans. Nevertheless, curiosities of all kinds (preserved either in the literary encyclopaedic tradition or coming to light as more or less accidental finds) represented anomalies in the harmonically organised, finite universe. They could take the form of a message fallen from the sky, freaks of nature, or either divine or satanic creations (Malina & Vašíček, 1990: 14). Renaissance scholars with a better insight accepted Classical tales of the Golden Age and the sequence of epochs where the use of stone had preceded bronze and iron (see Chap. 1). They were also acquainted with old stories regarding ancient times, which the Greeks had recorded from the mythologies of neighbouring 'barbarians'.[14]

Nevertheless, the Greek and Roman ideas were for a long time not directly linked to contemporary finds and excavations. Typical examples of this are stone tools, which were well represented in Renaissance collections because they were considered to be closely linked to fossils. Stone axes were called 'thunder—bolts' (*cerauniae*, i.e., related to thunder, derived from Greek *keraunos*, 'thunder'), because they were supposed to be the bearers of the effective cutting force of lightning. They were also believed to possess magical qualities and, sometimes, they were ascribed some other natural or celestial origin. Such views were described in detail by Ole Worm (1588–1654), a physician and pioneer of Scandinavian archaeology, in his catalogue of the Museum Wormium (1654), that is, his 'cabinet of naturalia' in Copenhagen. In this work, he recorded various widespread views regarding the origin of stone tools, such as wedges, hammers, or axes. He noted that some people believed them to be meteorites, others took seriously their similarity to iron tools and believed them to be their fossils, and some people even claimed to have found them in places where a house or a tree was struck by lightning (Malina & Vašíček, 1990: 26) (Fig. 2.5).

Like in the case of fossils, such hypotheses were gradually integrated into scholarly speculations[15] and while for the fossils, the basic factor that helped open a new temporal dimension was the recognition of their organic origin, in the case of these artefacts the turning point was the realisation that they were made by humans. Once again, such views were rarely voiced and the road to their acceptance was slow.

[14] For instance, Thucydides (app. 460–395 BCE) in the introduction to his work (I,6,6) observed, and documented using concrete examples, that the old world of the Greeks must have resembled the lives of contemporary barbarians. In particular, he noted that '...one could show that the early Hellens had many other customs similar to those of the Barbarians of the present day'.

[15] Many scholars, including Paracelsus, Johannes Mathesius (1504–1565) in his *Sarepta oder Bergpostil* (1562), Athanasius Kircher, or, in Bohemia, Bohuslav Balbín (1621–1688) nevertheless still believed that not only fossils but also stone and metal tools, and even ancient pottery and vessels, were of a supernatural origin or the products of the 'creative force of nature'.

Fig. 2.5 A frontispiece showing the interior of Ole Worm's museum (Ole Worm, *Museum Wormianum*, 1655). Getty Research Institute (https://www.biodiversitylibrary.org/bibliography/63844.)

Nevertheless, already Georgius Agricola (1494–1555), the 'father of mineralogy', rejected the idea of a natural or celestial origin of these stone or metal tools and expressed his conviction that they were the product of human labour.[16] A little later, Ulisse Aldrovandi (1522–1605) claimed that people used stone tools before they learned about metals, and he supported this argument by quotations from Pliny (Malina & Vašíček, 1990: 20–21).

Classical ideas about ancient times, ethnographic observations of missionaries, and prehistoric findings from Europe were methodically and systematically brought together by Aldrovandi's compatriot Michele Mercati (1541–1593), naturalist and medical advisor to Pope Clement VIII. Mercati was also in charge of the Vatican botanical gardens and the Vatican collection of fossils and minerals. During his field excursions, he collected various curious objects, including fossils, and crucially also 'thunderstones', that is, prehistoric stone tools, on whose origin he then speculated. He was thoroughly acquainted with the Classical tradition of subsequent epochs, but he also drew on the Old Testament, which was no less rich in references to stone and bronze tools and even mentioned the introduction of iron by the Philistines. Another source of information he could access were the growing collections of Asian and American artefacts sent to the Vatican by scholars, missionaries, and travellers. Mercati was thus the first person we know of whose interpretation of

[16] In his treatise *De veteribus et novis metallis* (1546), known also as *De natura fossilium*.

prehistory was systematically based on three kinds of sources: his own field research, the Classical tradition, and contemporary ethnography. His significance for archaeology and the study of prehistory has therefore been compared to what Cardano did for mathematics, Vesalius for anatomy, Galileo for physics, or Copernicus for astronomy (Malina & Vašíček, 1990: 21–22). Mercati's large treatise on minerals, fossils, and metals, the *Metallotheca* (1540), remained unpublished in the Vatican library but some scholars were familiar with it. For instance, the French historian Lancelot de La Popelinière (1541–1608) under Mercati's influence emphasised the importance of studying contemporary native Americans for a reconstruction of the European past.

In the seventeenth century, intellectuals interested in prehistory increasingly started to take into consideration the specific development of each country and especially in England, northern Germany, and Scandinavia, there was much interest in the tumuli, megaliths, rune stones, and other remnants of pre-Roman past.[17] In some cases, the research of 'antiquities' was also politically driven: on the one hand, this was a continuation of the nationalist and political tendencies known since the Middle Ages but, on the other hand, it had also contributed to an increasing awareness of cultural and historical diversity (see Chap. 5).

Nevertheless, scholars were very slow to accept that megalithic and other constructions in the landscape could be pre-Roman and that stone tools were not just freaks of nature. The idea that they could be in fact the oldest tools gained general acceptance only in the course of the eighteenth century—once again, we see here a parallel with methods used to investigate the geological past. The process was significantly aided by the publication of Mercati's work, which in 1717 finally appeared in print—and once it became accessible in print, Mercati's views could spread and be further developed. The physician and botanist Antoine de Jussieu (1686–1758) in his lecture for the Royal Academy of Sciences in 1721 highlighted the similarities between the tools of Native Americans and prehistoric tools: based on this comparison, he then postulated the Stone Age. Also under Mercati's influence, Joseph-François Lafitau (1681–1746), a missionary and pioneer of an ethnographic study of North American natives, in his *Customs of the American Indians Compared with the Customs of Primitive Times* (*Mœurs des sauvages amériquains comparées aux mœurs des premiers temps*, 1724) systematically compared American natives with ancient European ethnics. Among other things, he described parallels between the customs of the Iroquois and the culture of ancient Greeks and Romans. Lafitau believed the two to be equal and assumed an identical (but not simultaneous) development, which then became the subject of further comparative ethnological research (Malina & Vašíček, 1990: 26). As the concept of three subsequent developmental epochs became, against the background of Mercati's revived legacy, ever more

[17] Ethnographic comparisons in interpreting stone tools and megalithic constructions, including monumental ones (such as Stonehenge), were used by John Aubrey (1626–1697), a natural philosopher and pioneer of British archaeology.

Fig. 2.6 A comparison of the Persian and Roman culture with Native American culture of Louisiana. Lafitau used this picture to illustrate his argument that sacred fire is common to all people, which—as he claimed—proves the universality of belief in God (Joseph-François Lafitau, *Moeurs des savages*, 1724; see Grafton, 1992). Bibliothèque nationale de France (https://gallica.bnf.fr/ark:/12148/btv1b8602942m?lang=FR)

influential, research profited from further prehistoric finds and ethnographic material, which resulted in large, richly illustrated Enlightenment synthetic works.[18] On top of that, the history of European antiquity started including various sources attesting to the preliterary development in southern and western Europe (Fig. 2.6).

Christian J. Thomsen (1788–1865), the Danish father of modern archaeology, drew mostly on the French environment when he adopted the three-period system of the Stone, Bronze, and Iron Age. He then used it to organise all prehistoric artefacts: the system he used was thus based on the material from which characteristic tools were made during a given era. Thomsen's anonymous publication of his three-period theory in 1836 is viewed as the turning point in the modern study of prehistory (see Chap. 5). This 'Darwin of archaeology' thus completed a universal system

[18] Nicolas Mahudel (1673–1747), under the influence of Mercati but also the classics such as Lucretius, newly formulated the idea of three subsequent epochs and dated a stone hand axe to the sixth century before Christ (1730). A little later, Antoine-Yves Goguet (1716–1758) pointed out that stone, bone, or reeds were used, often for the same purposes, before iron and bronze (1758). Large synthetic works with thousands of wonderful engravings were published for instance by Bernard de Montfaucon (1655–1741) or Anne-Claude de Caylus (1692–1765).

that described how human cultures developed at a time for which we have no 'speaking witnesses.' This took place at a time of emerging classical historicism, when Lyell had definitively broken through the depth of geological time and Darwin, after returning from his journey around the world, already had a clear idea of biological evolution in his mind.

2.5 The Archives of Language: The Confusion of Babel, Polygenesis, and the Origin of Language

For every culture, the birth of an articulate language is a constitutive point (see Chap. 12). European culture has, since its earliest beginnings, faced linguistic diversity, and the very existence of numerous languages called for an explanation. Efforts to explain the existence of different languages appear in the mythologies and theogonies of various peoples. On the other hand, it is something altogether different to believe that multilingualism is a shortcoming that should be set right. As Umberto Eco (1995) describes in detail, various such attempts at rectifying this situation were, for several centuries, driven by the desire to find the 'perfect language'. In the previous section, we have indicated a link between our understanding of geological history and of human prehistory. In this section, we shall see that the exploration of the history of languages became linked to ancient literary history: this hitherto unsuspected gate of time was thus opened by old (dead)—and later also living—languages.

According to Eco, attempts at a 'rectification' of multilingualism were driven either by a forward-looking desire to create a language of reason that would acquire a lost or sought-after perfection or, alternatively, by a backward-looking turn to history driven by a desire to reconstruct the original language. Initially, the perfect language which scholars tried to find or reconstruct was identified—along biblical lines—with the 'divine' language, initially imparted to Adam and humanity. For a long time, scholars wondered whether this language reflected the inner substance of things and expressed their nature (so-called mimological or naturalistic hypotheses) or whether it was just Adam's creation and thus in effect an expression of convention.[19] In any case, it was believed that the Adamic language was forgotten after the biblical fall of the Tower of Babel and the subsequent 'confusion of languages' (*Genesis* 11,1–9), which gave rise to the linguistic variety of (about seventy) individual nations. This well-known explanation was always complicated by the preceding passages in the Bible, which spoke about how Noah's descendants, the sons of Japheth, Ham, and Shem, left for different corners of the world and multiplied in different lands and islands, thus founding nations 'everyone after his tongue, after their families, in their nations' (*Genesis* 10, 5; 20; 31). Logical inconsistencies

[19] In ancient philosophy, this dilemma was exemplarily formulated in Plato's *Cratylus*.

between the various stories of the Old Testament meant that the combination of Adam, the flood, and the Babel confusion hid explosive tensions, which were then reflected in mutually incompatible claims regarding the origin of languages and nations (Eco, 1995: 7–10).

Since the time of the Church Fathers and throughout the Middle Ages, there was a general consensus that the original language was Hebrew. Gradually, in part also due to interest in the Kabbalah, Christian Europe had discovered the idea of a perfect language. Ideas from Late Antiquity were revived in the humanist environment and Renaissance culture, which aside from Classical education also rediscovered Hebrew and encouraged its study. Guillaume Postel (1510–1581), a French polyhistor, predecessor of comparative linguistics, and a religious visionary, was a fervent advocate of the reinstatement of Hebrew as the initial language which, apparently already among Noah's descendants, evolved into Arabic, Chaldean, Hindi, and Greek, from which further languages later evolved. Postel believed Hebrew to be the universal language which people initially used to speak directly to God. Its restitution as the language of universal concord was thus supposed to restore a peaceful coexistence among nations (Eco, 1995: 75–80). In more refined versions of this idea, Hebrew was not necessarily considered to be the original divine language, and medieval and Renaissance traditions were further developed by etymologising, but Hebrew retained its pre-eminent status, which was said to be based on the fact that other languages were derived from it.[20] Various findings of travellers and missionaries and the discovery of civilisations older than the Hebrew one nevertheless eventually led to hypotheses about an original pre-Hebrew language. The fall of Hebrew from its privileged position was expressed in very concrete terms by the English architect and scholar John Webb (1611–1672), who—upon reading reports from Jesuit missionaries—advocated the idea that the original Noahic language was Chinese.

In the course of the eighteenth century, efforts to find the original language came to be driven by comparative methods. The numerous studies conducted in this direction culminated in the synthetic work of Antoine Court de Gébelin (1725–1784), whose aim was to construct a universal grammar, reconstruct the original language, and prove the biological and social origin of language (Eco, 1995: 93–95).

Theories that viewed language as a human creation that emerges by a convention whose form in turn depends on the specific conditions of each nation were becoming more prominent. According to these conceptions, language, as the form of human speech, has no creator and emerges gradually.[21] Such views found support in the Classical tradition, especially in Lucretius and Epicurus, who were 'rediscovered' in the fifteenth century. Epicurus seemed to indicate that the formation of

[20] In this rich tradition, which also included for instance Franciscus Mercurius van Helmont (1614–1699) or Athanasius Kircher, universal language was supposed to provide the means for overcoming religious and political differences, unification of nations, and lead ultimately to world peace.

[21] For a comparison with current biosemiotic accounts of the origin of language, see for instance van Heusden (2004), Zimmer (2004), Carling (2016), and Zlatev (2016).

language depends on the tribe, climate, and location, and he anticipated the emergence of different families of languages in different ways and at different times. The French humanist and philologist Joseph Justus Scaliger (1540–1609) did not focus on the original language but rather on a group of original languages (*linguae matrices*). He distinguished between eleven European language families whose individual members were genetically related but, importantly, there was no relation among the language families. According to scholars who championed such views, the Bible did not expressly say anything about the nature of the original language; they therefore believed that languages became differentiated much earlier than under the Tower of Babel and the *confusio* phenomenon was in fact a natural process (Eco, 1995: 85–88). Hobbes thus emphasised the symbolism of Adam's 'free' adding of new names according to need, while John Locke (1632–1704) spoke about the emergence of language based on a freely motivated decision to use it as an instrument of human interaction. This convention-based view of the origins of language in conjunction with the variety of cultures of exotic nations discovered by travellers was also reflected in La Peyrère's abovementioned pre-Adamite theory, which assumed that individual tribes evolved in different conditions and their linguistic ability adjusted to the environment. Languages were thus believed to have emerged together with nations in a process of polygenesis (Eco, 1995: 89).

Against this background of polygenism, one can trace the constitution of (pre) nationalistic hypotheses. Ever since the Renaissance, there had been efforts to prove that, after Babel, the initial perfection was preserved by one particular branch of languages, for instance Etruscan or Celtic (according to the Irish grammarians). In the seventeenth century, emphasis on the nobility of one's own national language had proliferated, and scholars proposed various idiosyncratic 'proofs' aimed at demonstrating that their language is a true descendant of the original perfect language due to its unique character. Such claims appeared variously for Dutch, Flemish, so-called 'Teutonic,' Swedish, German, but also Hungarian, Breton, or Polish. In the British environment, there emerged a defence of the Celtic as opposed to the Germanic tradition, but by that time, efforts to find the origin of a language were driven mostly not by a desire to achieve universal peace and religious concord but—like the contemporary re-discovery of ancient monuments—by partisan state interest (Eco, 1995: 95–103).

Gradually, this led to an unexpected side-effect: as the search for etymologies driven by the desire to prove the origin of living languages led to increasingly detailed comparisons, the idea of one original language melted away and there emerged a need to create typologies of the basic linguistic trees. The search for roots thus changed the goal of the undertaking. The new aim of these efforts was to identify language families and to approximate a model of the original language as the mother of the particular languages and dialects. In other words, the search for one original language resulted in efforts to create a typology of existing languages that would sort them into families and generations and assign them a place in family trees (see Chap. 12).

A groundbreaking result of this research was the scientifically fascinating hypothesis about a shared origin of Indo-European languages, which was formulated and gradually elaborated—lexicographically and later also grammatically—in the mid-eighteenth to early nineteenth century. The crucial first step in postulating the existence of Indo-European languages was taken by William Jones (1746–1794), whose contribution can be viewed as analogical to Buffon's importance for the geological past of the Earth. Jones had expanded the existing interest in word roots to the study of grammatical structures, and that led to the discovery of relatedness of a wider group of languages (and the originality of Sanskrit). Based on these observations, he became convinced that all these languages had 'sprung from some common source, which, perhaps, no longer exists' (*On the Hindus*, 1786; Eco, 1995: 103–106). Based on further sources, a connection between Indo-European languages was proven in 1808 by Friedrich Schlegel (1772–1829), who helped to demonstrate that this entire family evolved from an extinct ancient language that could be ideally reconstructed.[22] This went hand in hand with a sharp criticism of monogenetic theories and of the notion of a perfect language associated with thinkers such as Jean-Jacques Rousseau, Étienne Bonnot de Condillac, Antoine Destutt de Tracy, and Johann Gottfried Herder, a passionate champion of cultural pluralism who celebrated the Indo-European language as deeply poetic. From the theory of linguistic polygenism, further developments then led to ideas such as the spirit of various languages, the inner form of language, and to the linguistic relativism of Wilhelm von Humboldt (1767–1835).

The Indo-European hypothesis also assumed an original ancient language but one substantively different from the Adamic language. For one thing, Indo-European was not assumed to be the original language of all humanity but only of one family of languages. The theorists of perfect languages therefore claimed that it has features characteristic of languages after the confusion of Babel. Moreover, it was thought to be just a model, an ideal, something akin to an archaeological discovery that is yet to be fully excavated (Eco, 1995: 104). Once again, this required a profound departure from the temporal and spatial range determined by the biblical tradition, but this investigation led to another myth, a modern anti-biblical myth of the primacy of culture and race (see Chap. 3).

The abovementioned developments did not lead to a disappearance of monogenetic theories of humanity and language but—and this is important here—these theories had to adjust to the basic findings of the polygenetic theories (pre-Adamic and linguistic ones). As a result, they, too, had to assume a depth of time that eventually had nothing in common with the biblical context and traditional views. If the results of racial theories led to criticism and fundamental revisions of polygenism, then the new, modern monogenism with its 'Eve hypothesis' and the model of a

[22] Aside from his brother August Wilhelm, Friedrich Schlegel's thoughts were also developed by Franz Bopp (1791–1867) and Jacob Grimm (1785–1863), who defined scientific criteria of this investigation, taking into account among other things also vowel shifts. Aside from that, together with his brother Wilhelm, he started to pay attention to fairy tales and myths as another historical source.

Pre-Nostratic language had to go all the way to the biological roots of the evolutionary development of human groups, to the origin of language, to a point where biological evolution, anthropogenesis, and evolution of culture are so closely intertwined that they become in effect indistinguishable.

2.6 Conclusions: Deep Time as a Precondition of Evolution

In this chapter, we saw that the conceptual approaches to the four 'gates' leading to the depth of time were often intertwined in terms of subject matter, methods, and even on a personal level. We believe that our choice of the four perspectives has been vindicated: it helped us highlight the methods but also challenges which had to be overcome in order to access the deep temporal dimension of the development of the Earth, of written history and prehistory, and of language. The four routes to the discovery of deep time had some notably similar characteristics. First of all, they all had to come to terms with limitations given by the biblical world mediated by the Christian tradition. They all had to react to the new experiences and findings made in an ever more expanding geographical space. In all instances, we also saw a nonlinear development, where numerous alternative approaches were pitted against each other and an articulation of a deeper temporal dimension repeatedly appeared either as a minority alternative or as the result of methodical steps initially not intended to address the issue of the temporal depth of development. In these processes, the self-understanding of the Christian world—and thus also of Western European culture, especially regarding human society and nature—was fundamentally transformed.

A deepening of time (which in science continues to this day) was in all these areas a secondary, often initially unexpected, and even undesirable consequence of more important conceptual and methodical steps which disrupted the prevailing ideas about the unity of the world and the place of humans within it. In four distinct areas—geology, ancient history, prehistory, and language—we were thus able to see a shared foundation: an investigation of a certain set of facts led to a novel recognition of 'layers' or 'epochs' (the two terms are often interchangeable, for instance in Buffon's geology), their mutual relations, contexts, distances, and origins. Such new realisations would have been impossible without a shift in the interpretation of key sources, as we saw in the case of fossils, stone artefacts, or linguistic structures, which were later supplemented by myths, fairy tales, and folklore, but also for instance certain organic structures and functions. The newly discovered 'contemporaneity' of certain artefacts or phenomena, which may have initially had a different dating in absolute chronology, also led to a heightened awareness of their uniqueness, context-dependence, and irreproducibility.

We have started our investigation with Freud's diagnosis of a gradual de-anthropocentring of human knowledge in the theories of heliocentrism, biological evolution, and the subconscious as the subject of psychoanalysis. Heliocentrism marked the very beginning of Modern science in the sixteenth century, and it anticipated the idea of unlimited space. Evolution and historicism completed this process: they found support in the idea of deep time. And the birth of psychoanalysis, which methodically transcended the hitherto conceivable boundaries of human consciousness, belonged to a time of extensive critical revisions of the traditional form of modern sciences.

In the above, instead of investigating the 'grand' evolution theories of the nineteenth century, we focused on the less eye-catching preceding period, a time that lacks a symbolic epoch-making turning point. We wanted to approach the theories of evolution in effect indirectly, via the preconditions and processes that enabled them, through a space where the history of nature and culture still shared their foundations and background. We have thus presented the phenomenon of discovery of deep time as one of the defining preconditions of the Modern idea of evolution of nature and culture.

References

Almond, P. (2006). Adam, pre-Adamites, and extra-terrestrial beings in early modern Europe. *Journal of Religious History, 30*(2), 163–174.

Carling, G. (2016). Language: The role of culture and environment in proto-vocabularies. In D. Dunér & G. Sonesson (Eds.), *Human lifeworlds: The cognitive semiotics of cultural evolution* (pp. 83–96). Peter Lang.

Černá, J. (2013). *Očitá svědectví [Eyewitness testimonies]*. Pavel Mervart.

Černá, J. (2016). *Dál a dál za Herkulovy sloupy: Přírodní tajemství Nového světa a španělská renesanční filosofie a věda [Onwards and further beyond the pillars of Hercules: Natural secrets of the New World and Spanish Renaissance philosophy and science]*. Triton.

Cooper, A. (2007). *Inventing the indigenous: Local knowledge and natural history in early modern Europe*. Cambridge University Press.

Cutler, A. H. (2009). Nicolaus steno and the problem of deep time. *The Geological Society of America Memoirs, 203*, 143–148.

Dijksterhuis, E. J. (1961). *The Mechanization of the world picture* (C. Dikshoorn, Trans.). London–Oxford University Press. Original Dutch edition: *De mechanisering van het wereldbeeld*. Meulenhoff, 1950.

Dunér, D. (2016). Science: The structure of scientific evolutions. In D. Dunér & G. Sonesson (Eds.), *Human lifeworlds: The cognitive semiotics of cultural evolution* (pp. 229–266). Peter Lang.

Eco, U. (1995). *The search for the perfect language* (J. Fentress, Trans.). Blackwell. Original Italian edition: *La ricerca della lingua perfetta nella cultura europea*. Laterza, 1993.

Findlen, P. (1994). *Possessing nature: Museums, collecting, and scientific culture in early modern Italy*. University of California Press.

Freedberg, D. (2003). *The eye of the lynx: Galileo, his friends, and the beginnings of modern natural history*. University of Chicago Press.

Freud, S. (1963). *Introductory lectures on psycho-analysis*, part 3, (J. Strachey Trans.). In *The standard edition of the complete psychological works of Sigmund Freud*, vol. 16. Hogarth Press. Original German edition: *Vorlesungen zur Einführung in die Psychoanalyse* (Part 3). Heller, 1917.

Gould, S. J. (1987). *Time's arrow, time's cycle. Myth and metaphor in the discovery of geological time*. Harvard University Press.

Grafton, A. (1992). *New worlds, ancient texts: The power of tradition and the shock of discovery*. Harvard University Press.

Hladký, V. (2014). From Byzantium to Italy: "Ancient Wisdom" in Plethon & Cusanus. In J. Matula & P. R. Blum (Eds.), *Georgios Gemistos Plethon: The byzantine and the Latin renaissance* (pp. 273–292). Palacký University Press.
Hladký, V. (2022a). Francesco Patrizi on Hermes Trismegistus and Orpheus: The use of *Hermetica* and *Orphica* in the *Nova de universis philosophia*. *Rinascimento 62*, 59–97.
Hladký, V. (2022b). Kepler on Patrizi, ancient wisdom, and astronomical hypotheses. *Acta Comeniana, 60*(36), 105–135.
Horský, Z. (1966). La Cosmologie de Marcile Ficin. *Acta Historiae Rerum Naturalium Nec Non Technicarum: Czechoslovak Studies in History of Science* (Special Issue) 2, 57–68.
Horský, Z. (1967). Le Rôle du platonisme dans l'origine de la cosmologie moderne. *Organon, 4*, 47–54.
Kepler, J. (1606). *De stella nova in pede Serpentarii.* Praga: Sessius. Standard modern edition (KGW): *Gesammelte Werke*, vol. 1: *Mysterium cosmographicum, De stella nova* (ed. M. Caspar). C. H. Beck, 1937.
Koyré, A. (1948). Du monde de l'"à-peu-près" à l'univers de la précision. *Critique, 28*(4), 806–823.
Koyré, A. (1957). *From the closed world to the infinite universe*. Johns Hopkins Press.
Kristeller, P. O. (1964). *Eight philosophers of the Italian renaissance*. Stanford University Press.
Kusukawa, S. (2013). Drawings of fossils by Robert Hooke & Richard Waller. *Notes and Records of the Royal Society, 67*(2), 123–138.
Lyell, C. (1832). *Principles of geology: Being an inquiry how far the former changes of the earth's surface are referable to causes now in operation*. John Murray.
Malina, J., & Vašíček, Z. (1990). *Archaeology yesterday and today: The development of archaeology in the sciences and humanities*. Cambridge University Press.
Mayor, A. (2000). *The first fossil hunters: Paleontology in Greek and Roman times*. Princeton University Press.
Popkin, R. H. (1987). *Isaac La Peyrère (1596–1676): His life, work, and influence*. Brill.
Rappaport, R. (1997). *When geologists were historians, 1665–1750*. Cornell University Press.
Rossi, P. (1987). *The dark abyss of time: The history of the earth and the history of nations from Hooke to Vico*. University of Chicago Press. Original Italian edition: *I segni del tempo: Storia della Terra e storia delle nazioni da Hooke a Vico*. Feltrinelli, 1979.
Rudwick, M. J. (1985). *The great Devonian controversy: The shaping of scientific knowledge among gentlemanly specialists*. University of Chicago Press.
Rudwick, M. J. (2005). *Bursting the limits of time: The reconstruction of geohistory in the age of revolution*. University of Chicago Press.
Rudwick, M. J. (2008). *Worlds before Adam: The reconstruction of geohistory in the age of reform*. University of Chicago Press.
Rudwick, M. J. (2014). *Earth's deep history: How it was discovered and why it matters*. The University of Chicago Press.
Schilt, C. J. (2021). *Isaac Newton and the study of chronology: Prophecy, history, and method*. Amsterdam University Press.
Thomsen, E. (2009). Niels Stensen–steno, in the world of collections and museums. In G. Rosenberg (Ed.), *The revolution in geology from the renaissance to the enlightenment* (pp. 75–91). Geological Society of America.
Toulmin, S., & Goodfield J. (1965). The discovery of time. Harper & Row.
van Heusden, B. (2004). A bandwidth model of semiotic evolution. In M. Bax, B. van Heusden, & W. Widgen (Eds.), *Semiotic evolution and the dynamics of culture* (pp. 3–33). Peter Lang.
Zimmer, J. R. (2004). Semiotic questions raised by the evolution of talk and the emergence of complex society. In M. Bax, B. van Heusden, & W. Widgen (Eds.), *Semiotic evolution and the dynamics of culture* (pp. 106–116). Peter Lang.
Zlatev, J. (2016). Mimesis: The role of bodily mimesis for the evolution of human culture and language. In D. Dunér & G. Sonesson (Eds.), *Human lifeworlds: The cognitive semiotics of cultural evolution* (pp. 63–82). Peter Lang.

Chapter 3
The Evolutionary Utopia and a Confusion of Discourses: Projecting Eugenic Para-Religious Visions onto Human Future

Tomáš Hermann, Michal V. Šimůnek, and Jan Musil

Abstract This chapter presents modern evolutionary 'new utopias' with their para--religious aspects that had disastrous real-world consequences, which in turn made our thinking about cultural evolution problematic. We focus on *eugenics* as the key context for interpreting the concept of the *gene* and its politicisation. In particular, we outline the evolutionist motivations and scientific tools of eugenics, as well as the changes brought about by the arrival of *new genetics*, using two concrete literary examples of a utopian extrapolation of these ideas (Francis Galton and Alexis Carrel). This form of 'secular religions' sprang from various interpretations of genetics and specific modernising tendencies which took the idea of evolution to the extreme and aimed at explaining and reorganising human society. These striking examples of modern scientific over-confidence were accompanied by a blurring of science, power, and the discourses of knowledge of nature and culture. Moreover, it was applied to humanity as the object of controlled manipulation: the repercussions were so disastrous that they form an internal corrective for future efforts aimed at applying experimental research to humans to this day.

Keywords Eugenics · History of genetics · Evolutionary utopias · Secular religion · Biopolitics

T. Hermann (✉)
Department of Philosophy and History of Sciences, Faculty of Science, Charles University, Prague, Czech Republic

Institute of Contemporary History, Czech Academy of Sciences, Prague, Czech Republic
e-mail: tomas.hermann@natur.cuni.cz

M. V. Šimůnek
Institute of Contemporary History, Czech Academy of Sciences, Prague, Czech Republic

J. Musil
Department of Philosophy and History of Sciences, Faculty of Science, Charles University, Prague, Czech Republic

J. Toman et al. (eds.), *Culture Evolving*, Biosemiotics 10,
https://doi.org/10.1007/978-3-032-16124-6_3

3.1 Introduction

Max Brod in his autobiography recalls the seminar of the economist and sociologist Alfred Weber, which he attended at the university in Prague in 1904/1905 as follows (Brod, 1960, p. 324):

> Weber confidently steered our study of Darwinism, Lamarck's 'direct adaptation', Mendel's investigations into the race theory [i.e. of heredity], August Weismann's 'germ plasm continuity' and his rejection of the 'inheritance of acquired characteristics' (which became a matter of dispute again with the racial teachings of the Third Reich and later, in the opposite direction, with the theories of the Soviet scholar Lysenko). We talked about the national community and its generative health – about problems that have so painfully and unscientifically affected all of our lives several decades later. (…) I participated in one of the first presentations on Schallmayer's book Inheritance and Selection in the Life Course of Peoples. Later, another student thoroughly analysed Tönnies' book on Community and Civil Society, then there was Semon's Mneme, Hugo de Vries' mutation theory; in the second year of the ongoing debate I spoke once more, on Kropotkin's Mutual Aid. I put forward the argument that "mutual aid" does not constitute a principle opposed to the "struggle for life", as Kropotkin claimed. (…) The "inheritance of acquired characteristics" kept returning as the central theme around which our analyses revolved. Looking back, I feel that neither the teacher nor the students were really aware of the urgency of the issues we were discussing. We were petting the hounds of hell who were already rattling their chains.

The author sketches a fascinating picture of the lively foment of early twentieth--century debates about evolution and heredity, their relationships, and implications for understanding humans and their cultural development.[1] Brod hints that already in the first decade of the twentieth century, one could witness the formation of two ideologies which would later, with devastating force, take the form of *secular political religions*: namely Hitler's Nazism and Stalin's version of Communism. These ideologies were based on opposing conceptions of heredity. The Nazi racial theory, which culminated in a massive state-organised genocide, emphasised Weismann's and Mendel's principles in an effort to purify and protect the material basis of heredity. The Stalinist doctrine of Lysenkoism and Michurinian biology, on the other hand, relentlessly suppressed Mendelian genetics and, in an idiosyncratic reinterpretation of Lamarck's (and Darwin's) theory of inheritance of acquired characteristics, enforced the idea of an active transformation of nature for the benefit of the 'New Soviet Man.'

Despite their differences, the two doctrines were deeply related. They both used a theory of heredity to underpin their conception of human identity and they both distorted the initial theory for the purpose of controlling people, including their consciousness, thinking, values, and behaviour. Both used the language of heredity to move towards a new society, but each appropriated it in its own way: while the Nazis were obsessed with the permanence of identity and immutability of heredity, the Communist theorists accentuated the malleability of heredity and possibilities of

[1] For more on the authors and their works mentioned by Brod, cf. Weismann (1892), Schallmayer (1903), Tönnies (1887), Semon (1904), or Kropotkin (1902).

its manipulation. For both, however, science was supposed to aid *selection*, a 'purification' of society. The concept of *genes* thus emerged as a key player in culture and politics and, as it turned out, became one of the most dangerous concepts in history (Mukherjee, 2017).

It was at the time of Weber's seminars that young Adolf Hitler arrived in Vienna (1906) and began to assemble the foundations of his nationalist, racist, and anti--Semitic worldview. This was also the time when Lenin formulated the theory of his revolutionary party (1902) and young Stalin after the Russian Revolution of 1905 published a series of articles called *Anarchism or Socialism* (1906–1907), in which he among other things embraced Lamarckism. The hounds of hell were already rattling their chains, but when Brod referred to discussions about 'the national community and its generative health', he meant debates about the conceptions of science and society that found an expression in the theory and practice of *eugenics*. It was eugenics that became the key arbiter of what the *gene* was to mean and eugenics that was responsible for its politicisation.

In the following, we present this intertwining of the discourses of science and politics in several steps. First, we place eugenics within a broad context of paradigmatic changes of modern thought with emphasis on the motivation and scientific tools of eugenics, its development, and its reformist visions. This should help us grasp the underlying motivation and para-religious anchoring of the eugenic program, which we illustrate using two utopias, namely Galton's and Carrel's, as our examples. In the conclusion, we suggest that this part of history is a striking example of modern scientific over-confidence characterised by a conflation of science and power and the discourse of nature and culture in relation to humans as the object of manipulation. The memory of what it led to remains a dark memento and an internal corrective for all further considerations about the relations between nature and culture or the application of experimental research to humans. This has numerous practical consequences, including striking changes of discourse in post-war anthropology (see Chaps. 5 and 7), or a continuing reluctance towards biologisation in humanities (see Chaps. 10, 11 and 13).

3.2 Historicisation and Biologisation of the Modern Mind

With the development of natural knowledge and the revolutionary social changes since the Age of Enlightenment, the meaning of 'revolution' and 'evolution' has radically changed. 'Revolution' ceased to be understood as a cyclical return to the original state (from Late Latin *re-volvere*) and 'evolution' as the unfolding of something given and pre-formed (from Latin *e[x]-volvere*). Both terms came to mean the very opposite: the arrival of a uniquely new condition, irreversible and transformative.

This semantic shift captures a radical change in the understanding of human society and life, a process we can think of in terms of *historicisation* and *biologisation* of nature and culture. Reflected historical consciousness, i.e., the awareness of existing at a particular point of history that is unique and almost all 'givens' are therefore also the product of the time (and as such inherently temporary) can be expressed by the concepts of *historicism* and *evolutionism*. Deep, linear, and unique time became a self-evident given, something 'natural', an *a priori* assumption of historical and biological thinking without which the past could not be investigated. Significantly, it was only in the nineteenth century that articulated historical consciousness (in the understanding of human society) and the evolutionary scenario (in the interpretation of life) came to be viewed as symptoms of the same process. Modern science responded to this change in a way that preserved its core principles: by postulating a more elementary unchanging basis that would guarantee repeatability, timeless regularity, rational coherence and intelligibility, and the possibility of exegesis or axiomatisation of knowledge.

Already in the Enlightenment, syntheses of natural science and history sometimes took the form of ingenious, radical, and often modern-sounding historicising and pre-evolutionary conceptions (Hume, Vico, Buffon, Rousseau, Diderot, Winckelmann, Herder, etc.). The grand theories of the nineteenth century, too, grappled with a rational approach to the irreplaceable historical singularity, contingency, or relativity—be it Comte's techno-progressivist universal development, Lamarck's evolutionary scheme, Hegel's deeper meaning of the world spirit, or Spencer's harmonious organic development. Later epochs kept returning to the legacy of speculative idealist systems such as *Naturphilosophie* or Romantic historiography, including their methods of comparison and analogy, for which the historicity of culture and nature was an obvious starting point: after all, various forms of holism, organicism, and structuralism all more or less consciously draw on the Romantic conceptions to this day. One can also trace an unbroken continuity in the positivist reaction against Romanticism after Hegel's death (1830), which took the form of philological criticism, materialism, and a tendency to mechanicism.

It was only in the second half of the nineteenth century that the modern evolutionary approach championed by Darwin, Haeckel, and others, and the methodological approach to historical consciousness (Droysen, Dilthey, etc.) were fully defined and completed. Classical Darwinism with its mechanism of natural selection and classical historicism with its hermeneutics and demythologisation were real attempts to anchor scientific thought not in timeless structures but in the dynamic changeability of historical reality. Reactions to the crises of early twentieth century (the crisis of historicism, crisis of Darwinism, the 'philosophy of life') and a search for new methodological foundations inspired a universal use of statistical methods, logicisation, and algorithmisation. Efforts to underpin Darwinian evolutionism by a synthesis with the newly established science of genetics continued in parallel with the later stages of debates about historicism, which permeated even various previously normative disciplines such as law, economics, or theology, and with debates about the multi-layered structuring of historical time (so-called 'new history').

This transformation of modern thought went hand in hand with two interrelated general phenomena: (a) a *belief in progress*, rooted in the Enlightenment faith in the power of reason and human ability to control nature (manifested by the development of science and technology) and (b) the continuing 'disenchantment' of the world, that is, *secularisation* (see Chap. 18). The connection between the two is clearly demonstrated in numerous nineteenth-century visions where religion is explicitly replaced by science: religion thus takes the form of a 'worship of Nature' and the authority of the Church is supposed to be replaced by the authority of science.[2] Eschatology of the traditional religion was replaced by a *concrete future* towards which the human race and civilisation were said to be heading. Nevertheless, this optimism was challenged by a pessimistic awareness of catastrophes and decline threatening a collapse of the existing order whose reconstruction was increasingly viewed as unfeasible. The historicisation and biologisation of the Western thought was paralleled by advances in the natural sciences and humanities, which were nevertheless also falling prey to new ideologies and a new para-religious utopian universalism. In short, the demythologisation of traditional values soon led to secularised 'new mythologies'.

This resulted in a *confusion of discourses*. In the ferment of thought in the second half of the nineteenth century, scientific vocabulary became part of political ideologies (e.g., in social Darwinism). The most striking example of this fusion of the vocabularies of science, medicine, and politics was until mid-twentieth century the eugenic movement, which combined experimental research, theory, and practical application. A fear of 'demographic degeneration' inspired experts from various areas of biology and medicine to propose various 'eugenic reforms' aimed at a 'regeneration' of society. Men of science saw themselves as guardians of the future who should use their knowledge for the common good. Radical improvement of population health, public hygiene, social policy, and a biologisation of identity were all phenomena in which scientific knowledge and medical care intersected with power—and the state was empowered to manage the living through a rational and systematic 'biopolitics' (in Foucault's sense) (Fig. 3.1).

Eugenicists often used a pseudo-religious vocabulary, presenting themselves as apostles of the future or as creators, *demiurges of new life* (Turda, 2010, p. 14). Eugenics was fuelled by a fear of 'degeneration' and propelled by scientific self-confidence, which took the form of belief that it was not only possible to find the causes of various pathologies but also to change society through the application of knowledge. In the name of a healthy future, eugenicists sought to intervene in public and private life. In many countries, they persuaded politicians to introduce both 'negative' and 'positive' measures and to invest considerable resources into the creation of a homogeneous and healthy society.

The movement flourished during the interwar period, when proposals of eugenic reforms drew legitimisation from the disastrous consequences of the Great War (WWI). Depopulation, differential fertility, and staggering 'counterselection'[3] were the most frequently cited problems. National eugenic movements varied in the

[2] Classics on the subject include Chadwick (1975), Burrow (2000), McLeod (2000), or Loewenstein (2009).

[3] A contemporary term for 'war's depletion of the best generations'.

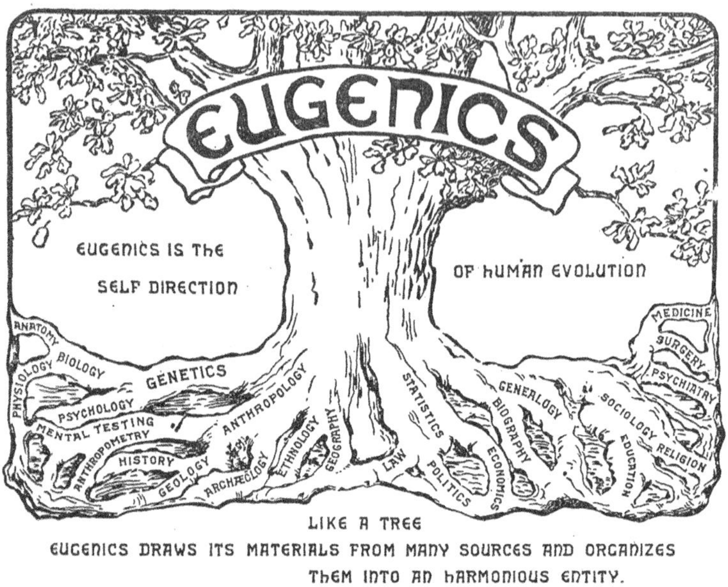

Fig. 3.1 *The Eugenic Tree: The Relation of Eugenics to Other Sciences*. Announcement of the Third International Eugenics Congress (New York City, 21–23 August 1932). Prof Joe Cain website (https://profjoecain.net/eugenics-tree-source-famous-image/)

methods and instruments they recommended: their choice was shaped by views regarding the heredity of acquired characteristics, the power of selection, or the importance of hygiene. To combat the weakening of society, proponents of a eugenic reform often advocated the use of invasive instruments to restrict the reproduction of medically undesirable individuals—or individuals who were deemed racially undesirable.

This was due to the fact that in many parts of the Western world, there prevailed in the second half of the nineteenth century an interpretation of culture as a product of human activity dependent on race. The racial imagery inferred the level of cultural maturity from various biological characteristics of members of the given 'race': much attention was therefore paid to investigating natural inequality between people. The tradition of literary racism (Arthur de Gobineau, H. S. Chamberlain, etc.) had evolved into an ideology aimed at 'renewing' a *racially defined population* in order to protect a certain 'cultural potential' defined by race and blood. In many cases, eugenics then went hand in hand with the rise of racial protectionism and racial hygiene. A parallel scientific program of national regeneration was endorsed by *constitutional medicine and biotypology*, which used updated humoral concepts to classify people into body types with the aim of placing them into the functional structure of a new, technocratically and rationally hierarchical, society.

Despite proclaiming they were based on a holistic perspective, both eugenics and constitutional medicine were based on a generalisation of specific (individual) characteristics, reduction of the complexity of actual personalities and their classification along standardised linear scales, a mathematical typification of diversity, and on deriving function from form. This approach spread in the scientific and medical discourse with the development of anthropology, anthropometry, and statistical models. In the first three decades of the twentieth century, various combinations of, or parallels between, eugenics (and racial hygiene) and constitutional medicine (and biotypology) emerged throughout Europe, the Americas, and East Asia.

As we are about to see, these movements were not innocent: since their inception, the new paradigms of evolution and inheritance had also practical ambitions. In particular, their aim was to shape the future of humanity by assigning important new roles to the modern state based on *bioinformatic management*, whose potential was gradually developed into 'new utopias'.

3.3 Body and Calculations: The Evolutionary and Statistical Foundations of Eugenics

Francis Galton (1822–1911) derived the term eugenics from the ancient Greek phrase *eu-genes* (well-born). In his *Inquiries into Human Faculty*, he noted that eugenics deals

> ... with questions bearing on what is termed in Greek eugenes, namely, good in stock, hereditarily endowed with noble qualities. This and the allied words, eugeneia, etc., are equally applicable to men, brutes and plants. We greatly want a brief word to express the science of improving stock (...) The word eugenics would sufficiently express the idea; it is at least a neater word and a generalised one than viriculture, which I once ventured to use. (Galton, 1883, pp. 24–25)

Two decades later, in 1904, when he petitioned for the establishment of national eugenics as a separate discipline at the University of London, he specified its scope as follows (Pearson, 1930, p. 222; Galton, 1904):

> The term National Eugenics is here defined as the study of the agencies under social control that may improve or impair the racial qualities of future generations either physically or mentally.

Galton began his scientific career shortly after the publication of Darwin's *On the Origin of Species* (Darwin, 1859) and died at a time of crisis of classical Darwinism and the constitution of modern (classical) genetics. He inspired several lines of biological research, most notably in biometrics and biostatistics (Stubbe, 1972, p. 178; Mayr, 1984, p. 558). In 1886, writing in *The Times*, he described how reading Darwin's work over three decades earlier was a turning point in his life (Pearson, 1924, p. 201):

> I have myself attempted some of the very inquiries to which he (Darwin) had drawn attention, in complete unconsciousness that he had done so. It was owing to this hereditary bent of mind that I was well prepared to assimilate the theories of Charles Darwin when they first appeared in his 'Origins of Species'. Few can have been more profoundly influenced than I

> was by his publications. (...) I rarely approached his genial presence without an almost overwhelming sense of devotion and reverence, and I valued his encouragement and approbation more, perhaps, than that of the whole world besides. This is the simple outline of my scientific history.

Galton was a strong supporter of the evolution theory and the centrality of natural selection. He has even been credited with playing a key role in transforming Darwin's hypotheses into Darwinism by further developing Darwin's conceptual network (Hertwig, 1921). He decided to focus on one part of the theory of evolution which Darwin also considered problematic: the issue of how the properties of living organisms are transmitted between generations and what changes they can undergo. This complex of relationships, which in many ways coincided with intuitive experience, came to be known as *inheritance* or *heredity*. Galton started to investigate heredity in the mid-1860s, around the time Mendel first summarised his rules of inheritance based on experiments with plant hybrids, but the two men most likely did not influence each other in any way (Stubbe, 1972, pp. 175–178; Orel, 1984; Olby, 1996, p. 523).

From the beginning, Galton was interested in the transmission of not only particular qualities but entire sets of 'psychic' and 'moral' qualities in their full complexity, that is, the transmission of what he called 'mental aptitudes' or 'general intellectual power'. The hierarchy of these qualities was naturally determined by the contemporary ranking of values (Pearson, 1924, p. 82). Galton was convinced that a better understanding of how such qualities are inherited would be a promising starting point of a *betterment of the mankind* by purposeful manipulation. The rules of human heredity, which he hoped to find, could thus be put to 'practical' use. Starting with his paper *Hereditary Talent and Character* (Galton, 1865), Galton dedicated himself to this research until the end of his life, publishing dozens of papers in anthropology, psychometrics, applied statistics, and biometrics. His student Charles Pearson (1857–1936) then placed his findings in the context of eugenic efforts (Pearson, 1901).[4]

In his theoretical considerations, Galton drew on Darwin's tentative theory of heredity, the so-called *theory of pangenesis* (Darwin, 1868), which centred around the idea that heredity is based on small, barely visible material particles, *gemulae*, which are produced by all cells of a living organism, circulate in them, vary in number in each individual, and cause variation in individual characteristics. Although the *gemulae* are present in all cells, only some are concentrated in the sex cells. Variability of the offspring was thought to be due to the mixing of different types of *gemulae* from both parents, whereby some *gemulae* were thought to be latent (Johannsen, 1909, pp. 147–148). Both the cells and the *gemulae* were moreover assumed to be subject to external influences.

[4] Regarding eugenics in the narrow sense, Galton's most important works were *Hereditary Genius* (1869), *English Man of Science, Their Nature and Nurture* (1874), *Inquiries Into Human Faculty* (1883), *Natural Inheritance* (1889) and one of his later works, the collection *Essays in Eugenics* (1909).

> Gemmules are supposed to be thrown off by every unit, not only during the adult state, but during each stage of development of every organism ... Lastly, I assume that the gemmules in their dormant state have a mutual affinity for each other, leading to their aggregation into buds or into the sexual elements. Hence, it is not the reproductive organs or buds which generate new organisms, but the units of which each individual is composed. These assumptions constitute the provisional hypothesis which I have called Pangenesis. (Darwin, 1868, pp. 370–371)

Although his theory was later proven wrong, it was thus Darwin who brought to life a concept that later led to the term 'gene'. More importantly, however, he had to deal with the fundamental objection that his theory was incompatible with the practical experience of heredity because a continuous dilution of variation in successive generations contradicted the evolutionary function of natural selection. The principle was not new: some scholars had noted that Darwin had, in effect, updated the ancient Hippocratic idea of 'mixed inheritance', which Aristotle had already rightly disputed (e.g. Bodenheimer, 1957, p. 72).

Galton tested the validity of Darwin's hypothesis experimentally by transfusing blood from different species of rabbits (Galton, 1871a, b; Dunn, 1965, pp 37–38). These experiments were not successful, so he began to work on his own hypothesis. From Darwin, he adopted the idea of separate material particles as the carriers of heredity. He introduced the term *stirp* or *stock* for the total potential of hereditary particles: this roughly corresponded to the later theory of *germ plasm* proposed by the German experimental biologist August Weismann (1834–1914) or the *idioplasm* of the Swiss biologist Carl Wilhelm von Nägeli (1817–1891). Galton sometimes referred to this precursor of eugenics using the contemporary American term *stirpiculture* (Kern, 1981). The abovementioned materialisation of heredity (paralleled by many contemporaries such as Haeckel, de Vries, etc.) led Galton, and later many other scholars, to reject the possibility of organisms acquiring heritable traits during life. In other words, it motivated a rejection of the hypothesis of *inheritance of acquired characteristics*.

Galton divided the observed characteristics in humans in normal (non-pathological) and abnormal (pathological) ones. During clinical research in a London hospital, he made correlations between the pathological manifestations of various diseases and, based on these correlations, wanted to investigate the genetic complexity of selected traits in a given population (Burdett, 2002; Ramsden, 2002). To describe and analyse selected sets of traits, he used an innovative method by which he applied statistics based on the probability theory. This led to a re-conceptualisation of inheritance as 'a statistical relationship between populations of successive generations' (Olby, 1996, p. 523; also Gutiérrez et al., 2002).[5]

[5] This was already noted by Galton's contemporary, the Danish geneticist Wilhelm L. Johannsen (1909, p. 5). For pathological conditions, Galton—like Darwin—could rely on the school of medical genealogy developed after 1810 by mainly French physiologists (Esquirol, Fodéré, Piorry, Lévy), which culminated in Prosper Lucas's treatise *De l'Hérédité Naturelle* (1847; López-Beltrán, 2004, p. 39; Olby, 1996, p. 522). In population statistics, he followed mainly the school of the Belgian mathematician and founder of modern statistics Adolphe L. J. Quetelet (1796–1874; Hankins, 1908).

Closely related to this method were Galton's attempts to identify people based on their biological characteristics, such as papillary lines or composite photography. Galton tried to reconstruct the process of inheritance in the intergenerational continuum by quantifying kinship: figuratively speaking, the life processes of each generation were compressed through mathematical formulas into a number (Olby, 1996, p. 521). This led to Galton's *law of ancestral heredi*ty (Galton, 1874, p. 326):

> The share a man retains in the constitution of his remote descendants is inconceivably small. The father transmits, on an average, one-half of his nature, the grandfather one-fourth, the great-grandfather one-eighth; the share decreasing step by step in a geometrical ratio, with great rapidity.

Another method Galton developed in relation to eugenics was based on genealogy, the study of ancestral and family relationships of individuals. For his work *English Men of Science* (Galton, 1874), he collected over 600 family trees of prominent contemporaries. The problem was that he usually acquired data only on the parents and children, so the family trees were very short and, moreover, the information he obtained from the questionnaires was mostly unverifiable.

Forty years after its initial formulation, eugenics was becoming increasingly accepted, mainly due to the supposedly ever-increasing potential of practical application. Its advocates in the early twentieth century agreed on its primarily expert nature, 'they were united in a belief that the biological expertise they commanded should determine the essential human issues of the new urban, industrial order' (Kevles, 1995, p 76). The eugenic perspective, which *de facto* amounted to applying Darwin's concept of artificial selection to humans, seemed clear and its ideals worth following, as Haycraft summarised shortly before the birth of classical genetics (Haycraft, 1895):

> But the end towards which we have to aim is the production in each generation of children from the best and healthiest of the population alone, for it is surely only reasonable that we should as a community pay the same care and attention to our own race propagation that a gardener does to his roses or chrysanthemums, or a dog-fancier to his hounds or terriers, or a cattledealer to his southdowns or shorthorns. That there is no means of improving our race so efficaciously as by selection we may be certain, and that there is no other way is highly probable; our interest in the subject, and the value we place upon changes the effects of which we shall never live to see, will determine whether we are prepared to adapt our ideas and modes of action to the lines that reason and the knowledge of the times have clearly pointed out as the only ones which it is expedient to follow.

3.4 New Genetics and the Eugenic Reform of Society

The emergence of classical genetics in the early twentieth century was completed in the late 1930s with population genetics (neo-Darwinian synthesis) and in the early 1950s with the discovery of the structure of DNA. But scientists realised the importance of experimental research on heredity, such as that undertaken by Gregor Mendel (1822–1884), even before the First World War (Dunn, 1965, p. XI, XIII):

> As research in heredity gathered momentum, scientists in other fields, as well as physicians, scholars, laymen and even politicians, became conscious of the changed aspect which many practical problems assumed when affected by the spread of ideas from genetics. […] The break in the continuity of ideas about heredity around 1900 seemed almost improbably sharp, and that, of course, was a part of the fascination of this new field in which a radical new orderliness seemed to have sprung full-blown from the head of one man [i.e. Mendel's].

Still, Mendelism was not a clear turning point in the further development of eugenics and the position of eugenicists was at first ambivalent. Its rejection by the proponents of biometrics led to one of the most heated methodological disputes in modern biology (Dunn, 1965, pp. 62–77; Olby, 1966). Those eugenicists who assumed the possibility of inheritance of acquired traits and worked with *euthenics* as a theory of all external (phenotypic) factors that influence individual development were becoming increasingly isolated (Davenport, 1910). Over time, most eugenicists started to side with the Mendelians and develop their beliefs about the hereditary conditioning of human traits within the Mendelian framework (Dunn, 1965, p. 223):

> [They] have dealt first with the transmission mechanism of heredity and the physical constitution of the genetic material; second, with the heredity and variation (mutation) as these operate in populations, and thus with the causes and mechanism of evolution; and third, with the functioning of the genetic material in controlling the metabolism and development or morphogenesis of the individual.

Eugenicists liked Mendelian heredity, which reduced the organism to a set of heritable elements (Lubinsky, 1993). They were fascinated by the possibility of interpreting certain hereditary conditions, both pathological and non-pathological, as mechanical consequences of these elements (Kevles, 1995, pp. 43–44). This inspired the vision of monitoring and manipulating these conditions in humans: especially in the therapy of certain diseases, it promised to open a whole new dimension of medicine. That is why eugenicists were soon reluctant to consider phenomena such as mutability, polymerism, or the distribution of traits in the population, which were later expressed in the Hardy-Weinberg equilibrium. Such phenomena challenged their position. Germany is one case where the influence of eugenic axioms was so strong that it became integrated into an entire national school of genetics[6] and in the United States, Mendelism soon became the mainstream methodological framework of national eugenics (Ludmerer, 1972; Fig. 3.2).

Nevertheless, the term *pangen*, which later evolved into *gene* (from the Greek *gennao*, 'I beget', 'I give birth') and eventually also the term *genetics*, emerged from the often overlooked phase of theories of heredity inspired by Darwin and Galton. The connecting link was Hugo de Vries (1848–1935), who was the first to adopt Darwin's term 'pangen' to designate the hereditary particle. But like Galton with his theory of pangenesis, de Vries no longer assumed that hereditary units could travel through the organism, and because they did not leave the cell, de Vries spoke of 'intracellular pangenesis' (Galton, 1889). He believed that particular

[6] Cf. the work of Erwin Baur, Eugen Fischer, and Fritz Lenz *Grundriss der menschlichen Erblichkeitslehre und Rassenhygiene* (Baur et al., 1923); for more details, see Saller, (1961), pp. 20–72.

Fig. 3.2 'Yea, I have a goodly heritage' (Psalm 16.6). Fitter Families Medal awarded in the 1920s by the American Eugenics Society. Mendelianum, Moravian Museum in Brno

pangenes had different representations in the different cells of an individual, which explained differences between siblings: the offspring was more like the parent whose pangenes predominated in it. It is important to note that the idea that the number of pangenes determined whether a trait would be prominent diverged from the Mendelian view, according to which 'factors' (*Anlagen*) were combinatorial units where each trait was represented by a pair of these factors (as in de Vries's mutation theory, 1901–1903). De Vries's belief that the influence of pangenes on development and their multiplication were closely related was important for the later development of genetics.

The term *genetics* was first proposed by William Bateson (1906, 1908) in his address entitled *Science of Genetics* at a London conference on hybridisation and crossbreeding of plants. Bateson claimed that this new science was not only well suited to the study of heredity and variation related to the theoretical problems of evolutionists and systematists but also suitable for practical applications required by breeders. The Danish scientist Wilhelm Johannsen (1909, pp. 151–173) also adopted Darwin's original 'pangen' from de Vries's theory, but he shortened it to 'gene', which he used to denote a numerical unit used for the evaluation of the results of crossbreeding. Consequently, he called the sum of all genes in a cell 'genotype'. Johannsen noted that genotype could not be demonstrated as such but differences between genotypes could be made apparent by inference from their 'phenotypes'.

In the early twentieth century, both Mendelian heredity and Galton's idea of eugenics were rediscovered: at this point, they entered an atmosphere filled with apprehension about the crisis of civilisation and instability of values. The 'hereditary component' played an increasingly prominent role in eugenic efforts and it was made ever more concrete by experimental research conducted essentially in the original Galtonian spirit (Galton, 1865, p. 321):

> Our bodies, minds and capabilities of development have been derived from them [i.e. our ancestors]. Everything we possess at our birth is a heritage from our ancestors.

After 1900, eugenicists started to think more comprehensively about genealogy than Galton did in his pioneering beginnings. They realised that since it was not possible to conduct hybridisation experiments on humans, the focus should be on collecting and analysing data that could reveal the most about past sexual relationships (Pearson, 1924, p. 76). The study of family trees therefore had to take the centre stage and genealogy was to become a prominent 'auxiliary science' of eugenics. Genealogical information was viewed, by both eugenics and experimental biology, primarily as the record of a series of experiments that had a bearing on the health of parents and their offspring. To use Johannsen's terminology (Johannsen, 1909), the 'genotype' of an individual was to be determined on the basis of as many 'phenotypes' as possible.

The practical impact of this eugenic interpretation of family trees led to a kind of mania, a desire to gather as much genealogical information as possible, especially regarding the health of family members. A family tree was to be supported by so-called *family records*. An unnamed associate of Charles B. Davenport (1866–1944), one of the most influential American eugenicists, aptly remarked that his superior's office resembled 'a medical museum of morbid anatomy with its charnel odors and gruesome sights' (Kevles, 1995, p. 46). The initial results of these activities, which effectively constituted the foundation of eugenic information management, were summarised in Davenport's *Heredity in Relation to Eugenics* (Davenport, 1912). At this time, there also emerged so-called *ancestral statistics*, whose aim was to summarise, in the greatest possible detail, data on individual families without considering whether they belonged to the same generation and lineage or not. Its importance lay in sorting the material—as was the custom of the time—with regard to some physically apparent innate condition, a particular characteristic of the children and parents. Based on the results of these elementary experimental methods, the ancestral statistics then investigated the proportions in which the different types appear.

Since eugenics claimed to be the science of 'improvement of the human species', and the improving was to be done by the humans themselves, the question arose as to who would determine what was 'good', 'useful', and 'desirable'—or 'bad', 'harmful', and hence 'undesirable'—for the future generations of offspring and how they would determine this. In the early twentieth century, answers to these questions limited any project for which eugenicists wanted to develop a specific social strategy and seek social and political support.

In the eyes of the eugenicists, who were at that time recruited mainly from the ranks of general practitioners, the progressive promise of eugenics lay in its potential to function as a systemic prevention of lifestyle diseases. The profound societal changes of the time forced scientists to consider human health in relation to social and economic factors, such as intensive industrialisation, migration of the rural population into cities, mass poverty, inadequate hygiene, etc. The interdependence of population health, social status, and the political situation was coming to the fore and health was more than ever before becoming a matter of public concern and public debates (Adams, 1990; Pauly, 1993). Growing interest in the phenomenon of heredity among the general public and their seemingly more detailed knowledge of it led to the development of a new dimension of medicine at whose core was a biologising vision of a 'hereditarily healthy' individual free of all hereditary diseases. If that could be achieved, some diseases could even be prevented in the unborn. To many, this vision seemed consistent with the premises of eugenics.

Debates about eugenics thus contributed to pessimistic scenarios about the progressive 'degeneration' of the population and its likely impact. Eugenics presented itself as a practical instrument of this fight against 'degeneration' that supposedly threatened the fitness of both individuals and entire populations (nations, races).[7] Degeneration was understood as the result of ongoing deviation from the desired norm in terms of deterioration of mental and physical abilities, whereby mental degeneration (various illnesses such as schizophrenia, Huntington's disease, etc.) was considered worse than physical degeneration (deviations such as disproportionate body parts). Degeneration was identified through symptoms defined by eugenicists based on extensive statistical research, and statistics provided arguments for initiating a collective eugenic 'therapy' through state-directed measures (Pearson, 1909). For instance, the British parliament had in 1903 established a Commission on National Deterioration, which was to carry out eugenic activities (Searle, 1976). Immigration became a major issue for eugenicists in both Britain and the United States, where 'national eugenics' was institutionalised in 1904 through the efforts of Charles Davenport. Another issue that came to the fore was *birth control* and *differential birth rates* (Whetham, 1909).

In some countries, eugenics acquired a political aspect because it opposed the privileges of the traditional aristocracy, noting that there was no guarantee that the status of hereditary privilege reflected 'natural fitness' (Pearson, 1930, p. 85). The adoption of eugenic principles was supposed to aid the development of desirable 'qualities' in the current generation and their transmission to the next, whereby 'talented' and 'healthy' individuals were considered 'eugenically desirable'. Metaphorically speaking, eugenicists wanted to replace the rule of the old 'aristocracy of lineage' by a new 'aristocracy of nature'. By the same token, they did not regard all people as equal 'by nature': eugenics conceptualised 'natural inequality' (Kevles, 1995, p. 73), which essentially corresponded to Galton's earlier vision of a 'better man'.

Eugenics thus reflected the long-standing demand for a 'biologisation' of politics, in other words, for presenting political decisions as determined by natural factors. This led to a fusion of eugenic and racist thinking and to the introduction of a new hierarchy of the value of human lives which disregarded the existing principles of European civilisation. The repercussions of this shift were exposed in all their horror two decades later, when they took the form of the Nazi perversion of eugenics/racial hygiene.

---

[7] On the interpretation of the term 'degeneration' and its distinction from 'decadence', see esp. Weingart et al. (1992, pp. 23–73).

By investigating kinship relations, eugenicists approached human sexuality from a different perspective: their explicit goal was to transform it in accordance with the ideals of eugenics (Conrad-Martius, 1955). This took place at a time when the general sentiment of the *fin de siècle* was reflected in various reformist considerations. Prominent among the various proposals of concrete measures, were calls for a regeneration (*Erneuerung*) of society, which were closely related to the perception of crisis of social progress and reflected the pessimistic, even chiliastic, moods of the time (Sapp, 2003, pp. 95–112). In many instances, they were complemented by calls for a *sexual reform* that would involve a rejection of existing sexual morals.[8]

Eugenics of the early twentieth century differed from its beginnings mainly by coming up with concrete suggestions how the eugenic ideals should be achieved. In the subsequent decades, the movement focused on the possibility of *practical eugenic measures*, whose implementation was to take place in three phases (exploratory, analytical, and executive), which would thus correspond to the diagnosis and therapy of traditional medicine. To use the eugenic vocabulary, both 'inferior' and 'particularly valuable' individuals would be selected based on information about their health and pedigree 'evaluation', which would be recorded and entered into special registries. After a comparison with other records, a decision would be taken on 'positive' (selection of suitable sexual partners, promotion of reproduction) or 'negative' measures (institutionalisation, sterilisation, prevention of reproduction). Eugenicists therefore paid special attention to information management, meaning a systematic collection, cataloguing, and evaluation of data relevant to a person's hereditary status. Unsurprisingly, eugenicists came to play a key role in the practical implementation of the abovementioned statistical methods: eugenic information management was to intersect with health statistics and medical genealogy. Practical eugenics was also supposed to include the creation of eugenic organisations and active support of practical measures, which eugenicists themselves usually called *eugenic propaganda* (Fig. 3.3).[9]

[8] For more details, see e.g. Bassett (1995) or Conrad-Martius (1955). Representatives of this trend in academic circles include Christian von Ehrenfels (1859–1932), Austrian-born philosopher and psychologist at the German University of Prague (see Chap. 4), or Willibald Hentschel (1858–1947), a German biologist and an outspoken anti-Semite.

[9] In 1904, after some complications, Galton managed to establish the first specialised eugenics laboratory in London, which since 1907 published *Eugenics Laboratory Memoirs* and since 1909 the *Eugenics Laboratory Lecture Series*. In 1908, the Eugenics Education Society was formed, which a year later began publishing the quarterly *Eugenics Review*; more in Farrall (1985, pp. 285–310).

Fig. 3.3 Propaganda poster of the British Eugenics Society (1930s). Mendelianum, Moravian Museum in Brno

3.5 Francis Galton's Eugenia

Galton's last, little-known work, which he produced in the final years of his life (1910–1911), was supposed to crown his efforts to promote 'the science of the improvement of mankind'. It is a utopian novel with the working title *Extracts from the Journal of the late Professor I. Donoghue* (Pearson, 1930, p. 411; Smith, 1988, pp. 73–92). Galton's utopia was set in the present: for example, it envisaged immigration from Britain and gave figures on contemporary population of Britain or Russia. But only a torso of the work survives.[10]

Let us take a look at its plot. A British professor of vital statistics named 'I' arrives in Town, the capital of a country called Kantsaywhere, a kind of Eugenia. In Town, he meets a young lady, Augusta Allfancy, who is about to take the Honours Examination at the Eugenics College. The professor falls in love with her and, as an outsider, decides to go through the process with her. He gradually learns about, and comes to admire, the mechanism of eugenic selection on which the local society is

[10] Before his death, Galton destroyed the manuscript—allegedly because of its bizarre ending. The surviving parts had been described by Galton's biographer Pearson (1930, pp. 411–425). This pertains to chapters V (Pass and Honours Examinations), VI (The Calendar of Kantsaywhere), VII (Measures and Marks) and VIII (Marks Gained by Me—Society of the Place).

based. Kantsaywhere is an island with a population of just about 10,000, where life is governed by eugenic principles. The Eugenics College is one of the most important local institutions and its efforts to improve the local population's hereditary condition are generously supported from the estate of a Mr. Neverwas (Pearson, 1930, p. 414):

> The College was to grant diplomas for heritable gifts, physical and mental, to encourage the early marriages of highly diplomaed parents by the offer of appropriate awards of various social and material advantages to relieve the cost of nurturing their children, to keep a minute register of results, and to discuss those results from time to time...

The highest political authority in the land is the Council, whose main mission is a systematic implementation of a eugenic policy through measures ranging from the selection of marriage candidates to a targeted financial support of 'eugenically fit' families. Immigrants who do not have documents proving their hereditary value are viewed as examples of the 'eugenic backwardness' of the lands they came from.

The eugenic test is a turning point in the life of all locals: it decides their future fate, dividing it into 'before-life' and 'after-life.' The very existence of the test is the first stage of selection, because individuals who refuse to take it are considered eugenically inferior and penalised. The examination has two rounds. The first is the Pass Examination, taken over 4 days in the Examination Hall. It consists of four parts: anthropometric (reaction to external stimuli, spatial orientation, height and weight measurements), medical (general constitution, vaccination check), aesthetic/literary (loud recitation of a poem, singing, writing of four short essays on set subjects within 1 h), and hereditary. The Examiners are specially trained officials capable of examining about twenty candidates at a time.

The most important part of the test is the hereditary part, during which Ancestral Efficiency is investigated based on genealogical documents, personal identification (fingerprints), and other documents. The result is an assessment of the individual's genetic quality with the formula 'P.G.' (Passed in Genetics) and assignment of a class. The system assumes that 'all that is transmissible by heredity, whether it be of ancestral origin or a personal sport or mutation' (Pearson, 1930, p. 415). Female adepts are examined by female examiners along similar lines as the males. Particular emphasis is placed on naturalness (Pearson, 1930, p. 422):

> The girls have the same massive forms, short of heaviness, and seem promising mothers of the noble race... As for the men they are well built, practised both in military drill and in athletics, very courteous, but with a resolute look that suggests fighting qualities of high order. Both sexes are true to themselves, the women being thoroughly feminine, and I may add, mammalian, and the men being as thoroughly virile.

The final evaluation is done by the Board of Examiners. Resulting grades are then recorded on a Pass Certificate and an interim Schedule. A candidate receives three marks of at most forty each; the threshold for passing the examination is a total of at least seventy. Additionally, a star or—after consultation with fellow citizens—Additional Marks for Social Utility can be awarded in each of the individual sub-examinations. A successful completion of examinations is announced not only to the family but also to the public via special lists published in the newspapers.

Individuals with the highest grades for their hereditary status are praised the most (Fig. 3.4).

For those who obtain exceptionally high grades in the first round, there is a second round, the Honours Examination, held only once a year and administered over four days to eighty candidates at most. Its motto is 'Grace and Thoroughness.' This examination is essentially the same as the first one, only reduced in form and more detailed, with a maximum score of two hundred.

Only persons who pass at least the Pass Examination are allowed to marry. The choice of a partner is a purely 'technical' matter, decided based on information contained in the Calendar of Eugenics College, which is used for selection and matching. It takes the form of a written codex of binding eugenic rules and a central inventory (central database) administered by specially vetted officials under the direction of the chief Registrar. All individuals who score seventy or more points in the examinations are recorded and registered in it. Selection of a partner is determined by one's Eugenic Worth (defined by the marks obtained in the examination). For girls, the most desirable mates are males who scored above two hundred and at least two stars. Such union confers the right to a College Marriage, which guarantees the offspring care in institutions managed by the Eugenics College, free healthcare, and all conceivable support during their early childhood. Women who pass the exam with distinction are paired with men at least three years older; their ages range around 22. Officials responsible for the Calendar of Eugenics College thus in effect

Fig. 3.4 *Eugenic Family.* Promotional illustrations in the Art Nouveau style, first half of the twentieth century. Mendelianum, Moravian Museum in Brno

manage and optimise the population and reproductive policy. The central inventory also includes records of all facilities, institutions, and special financial funds available to the Eugenics College to care for the eugenically 'fit' population.

Individuals who fail the test receive negative marks are considered 'inferior', and their future fate is kept from the public. They are subject to a special system of population control combined with social restrictions (Pearson, 1930, p. 421):

> The restriction placed by public sentiment and, in extreme cases, by penalty, on the number of offspring that a couple may propagate in Kantsaywhere, is based on that of their joint marks. (...) It was necessary to take into account the need of the parents, the advantages of family life and the well-being of the children, as well as that of the race.

If the 'inferiors' exceed the permitted number of offspring, it is a 'crime' against the eugenic State and they face various punishments determined by their Civic Worth. The form of punishment is determined by a special Board of Penalties. Punishment can take the form of being ignored by the society, fined, boycotted or, in extreme cases, deported to a Labour Colony, where people work hard and live in celibacy. But they can also emigrate from Kantsaywhere.

If, despite all the preventive eugenic measures, an obviously disabled child is born, the responsibility falls on the Eugenics College. The Kantsaywhere community, however, must accept such a child and it is not to be discriminated against except for prohibition of any sexual intercourse. In such instances, the country's eugenics system functions automatically.

Heredity is the most important value for the people of Kantsaywhere; it forms the basis of their attitude to life. Locals know the lives and stories of their closest ancestors, relatives, and friends in great detail. They spend most of their free time talking and reflecting on this topic. A person is seen primarily as a mediator between generations whose worth is determined by his or her hereditary equipment (Pearson, 1930, p. 415):

> A person is therefore more important as a probable progenitor of many others more or less like to him in constitution than as a mere individual.

In this system, eugenics was supposed to promote the interests of the whole over individual interests and ultimately lead to greater social cohesion. The awareness of intergenerational genetic unity was constantly reinforced in order to establish faith in the future transformation into a collective, transcendentally understood identity, an embodiment of Galton's vision of eugenics as the religion of modern people (Pearson, 1930, pp. 423–424):

> Their creed, or rather, I would say, their superstition – for it has not yet crystallised into a dogmatic creed, is that living beings, and pre-eminently mankind, are the only executive agents of whom we have any certain knowledge... the people of Kantsaywhere have the strong belief that the spirits of all the beings who have ever lived are around us, and regard all their actions... Everyone here feels that they themselves will, after their life is over, join the spirit legion, and they look forward with eager hope that their descendants will then do what will be agreeable and not hateful to them... Their superstition certainly succeeds, even as it is, in giving a unity of endeavour and a seriousness of action to the whole population. They have no fear of death. Their funerals are not dismal functions as with us, but are made into occasions for short appreciative speeches dwelling lovingly on the life-work of the deceased.

3.6 The Best-Selling Eugenicist Alexis Carrel: The Utopia of Rule by the Medical Order

The famous French physician Alexis Carrel (1873–1944) is mostly known for developing the technique of arterial suturing and organ transplantation, for which he received the Nobel Prize in Physiology and Medicine in 1912. He spent the majority of his career at the Rockefeller Institute for Medical Research in New York, where he studied the organisation of living matter in collaboration with Simon Flexner and Frederick Gates. In the 1930s, Carrel became interested in the causes of the decline of Western civilisation, the occult sciences, spiritualism, and telepathy. He summarised his views in *Man, the Unknown* (French original version Carrel, 1935a), a reflection on the state of our civilisation, which he saw as mostly in a worrying state of decline. The book is an example of the widespread intellectual pessimism of the time, found also in the works of Oswald Spengler (1880–1936), Madison Grant (1865–1937), or Julius Evola (1898–1974). What distinguished Carrel from the other fatalistic pessimists was his belief—in keeping with the expectations of eugenics—in the possibility of saving society and civilisation.

Carrel's book was extremely successful: millions of copies were sold in many languages. Its popularity—somewhat surprising given the unoriginal themes, admiration for dictators and negative eugenic tools—was enhanced by Carrel's reputation as a Nobel Prize winner and war hero as well as by positive reviews from notable personages such as the biologist Raymond Pearl (1879–1940) or the philosopher Henri Bergson (1859–1941). Thanks to its spiritual accents and criticism of mechanism, it was even positively received in the religious circles. But above all, it gave voice to the middle class's fear of the future: people bought it to show off in their libraries (Reggiani, 2007, p. 76).

Carrel expressed a criticism of capitalism, industrialisation, and consumerism, but also proposed eugenic measures aimed at a rapid improvement of the situation. These were supposed to go hand in hand with a renaissance of organic holism and subordination of the individual to the order of the whole (Lawrence & Weisz, 1998, p. 20). In a mixture of conservatism and philosophical and spiritual speculation, Carrel's solution to the decline of civilisation rested on three pillars: a comprehensive conception of the individual that should be formulated by science, the creation of technocratic bio-elites, and an introduction of eugenic measures (Reggiani, 2007, p. 67).

The book contains various highly conservative views including a rejection of women's emancipation and racist proclamations of white supremacy. Carrel attributed the mental supremacy of the 'white race' to living in a temperate climate, while tropical environments allegedly led to 'degeneracy'. At the same time, the 'white race' was the most at risk of the mental illnesses caused by the modern way of life. Carrel found symptoms of decline in decreasing mental capacity, morality, and birth rates, along with increasing crime, prostitution, and hereditary diseases. He believed that a life of material prosperity enjoyed by increasing numbers of people led to laziness and profligacy. Scientists, he said, should therefore shift their attention from physiology and chemistry to the hitherto undervalued manifestations of human life, such as the intellect, moral and spiritual qualities, and telepathic abilities.

Carrel's position was both anti-mechanistic and vitalistic. In many places, he expressed a concern about the fact that the principle of natural selection had been suppressed and no longer regulated the population and guaranteed its viability (Carrel, 1935b, p. 31):

> Thus it appears that the environment, which science and technology have succeeded in developing for man, does not suit him, because it has been constructed at random, without regard for his true self.

Modern civilisation suppressed the adaptive mechanism of people, who thus became accustomed to comfort and lost the natural desire for power, knowledge, greed, ambition, and conquest, which used to drive them forward. In the modern world, he said, there are increasingly many lazy and feeble-minded people whose existence is 'the fatal result of physiological and moral degradation brought about by the industrial civilisation' (Carrel, 1935b, p. 166). Another danger according to Carrel is the illusion of democratic equality promoted by liberalism. He claimed that universal rights guaranteed to all members of society are a crime against the order of nature: since the worth of each individual is biologically determined, people are by nature unequal and cannot claim the same rights. With universal rights, democracy, and liberalism, civilisation will perish because individuality will cease to exist, and the natural elite will lose its privileges (Carrel, 1935b, p. 200):

> The equality of their rights is an illusion. The feeble-minded and the man of genius should not be equal before the law. The stupid, the unintelligent, those who are dispersed, incapable of attention, of effort, have no right to a higher education. It is absurd to give them the same electoral power as the fully developed individuals. Sexes are not equal. To disregard all these inequalities is very dangerous. The democratic principle has contributed to the collapse of civilization in opposing the development of an elite.

A return to a harmonious positive development must, according to Carrel, start with a synthetic knowledge of humans, that is, with a 'constitutional research' that would establish each person's true worth and thus determine their place in society.

The first theoretical step in this direction was a distinction between the 'external time' (a measurable external physical quantity) and the 'inner time' (memory of the body). Analogically, Carrel distinguished between physiological and psychological time: physiological age is the rate of wear and tear on the body, which occurs depending on hereditary predispositions, environmental influences, and lifestyle, while the psychological age corresponds to the degree of intellectual maturity of an individual. This distinction plays a role in considerations of prolongation of life: medical and hygienic measures so far fail not only in their attempts to prolong life but especially in their efforts to improve it, that is, to extend the period of physiological youth that would enable a high level of psychological maturity. This is where

Carrel suggested eugenic considerations that should be taken into account if the medical science is to prolong life[11]:

> Before attempting to prolong life, we must discover methods for conserving organic and mental activities to the eve of death. It is imperative that the number of the diseased, the paralyzed, the weak, and the insane should not be augmented. Besides, it would not be wise to give everybody a long existence. The danger of increasing the quantity of human beings without regard to their quality is well known. Why should more years be added to the life of persons who are unhappy, selfish, stupid, and useless?

Carrel's concrete eugenic tools for the recovery of civilisation were to be mainly positive: the strong should be made stronger, gifted children separated and given a better education, women returned to their natural social role, and the reproduction of fit individuals financially supported. Carrel also proposed some negative instruments: pre-marital checks, immigration quotas, and voluntary sterilisation of persons who, through awareness campaigns, would realise that they are unsuitable for reproduction. A passage about the disposal of particularly dangerous or weak members of society in institutions where chambers filled with lethal gas would be installed for this purpose became infamous in light of later events (Carrel, 1935b, p. 232).

> Those who have murdered, robbed while armed with automatic pistol or machine gun, kidnapped children, despoiled the poor of their savings, misled the public in important matters, should be humanely and economically disposed of in small euthanasic institutions supplied with proper gases. A similar treatment could be advantageously applied to the insane, guilty of criminal acts. Modern society should not hesitate to organize itself with reference to the normal individual. Philosophical systems and sentimental prejudices must give way before such a necessity. The development of human personality is the ultimate purpose of civilization.

Carrel claimed that eugenics, with general practitioners in the vanguard, should strive to create a new 'hereditary biological aristocracy' by the 'remaking of Man' according to 'natural laws' (Carrel, 1935b, p. 221), where the old 'titular heredity' would be replaced by 'genetic heredity' (Carrel, 1935b, pp. 217, 219):

> We must single out the children who are endowed with high potentialities, and develop them as completely as possible, and in this manner give to the nation a non-hereditary aristocracy. (…) It is imperative that social classes should be synonymous with biological classes.

This vision of renewal included the creation of a centre of science where all knowledge about humanity would be brought together (Carrel, 1935b, pp. 46–47):

> The syntheses needed for the progress of our knowledge of man should be elaborated in a single brain. It is impossible to make use of the mass of information accumulated by the specialists. For no one has undertaken to co-ordinate the data already obtained, and to consider the human being in his entirety. (…) The best way to increase the intelligence of scientists would be to decrease their number.

[11] Carrel (1935b, p. 137). Carrel considers the transfusion of young blood as an option for rejuvenation.

Carrel expanded the utopian vision of a single saviour genius and called for the creation of a small group of the best scientists, a kind of collective centre of knowledge. This sovereign institution, a kind of 'brain trust,' would be periodically renewed and charged with the creation of a programme of preservation and management of civilisation. Medicine would play a leading role because it had the best knowledge of the individual and would thus be best equipped to synthesise the knowledge of other disciplines. Carrel called it 'super science' and his own field, surgery, a 'triumph of biology.' The physician was literally like God in the breadth of his knowledge and the possibilities of intervening in the human body (Carrel, 1935b, p. 153):

> Owing to the extreme ingeniousness and audacity of its methods, [modern medicine] has surpassed the most ambitious hopes of medicine of former times. Its attainments are the purest triumph of biology. He who has completely mastered its techniques, who understands its spirit, who has acquired the knowledge of human beings and the science of their diseases, truly becomes like God.

Such doctors should therefore be in charge of the management of the corrective process (Carrel, 1935b, p. 209):

> Medicine has received from anatomy, physiology, psychology, and pathology the more essential elements of the knowledge of ourselves. It could easily enlarge its field, embrace, in addition to body and consciousness, their relations with the material and mental world, take in sociology and economics, and become the very science of the human being. Its aim, then, would be not only to cure or prevent diseases, but also to guide the development of all our organic, mental, and sociological activities. It would become capable of building the individual according to natural laws, and of inspiring those who will have the task of leading humanity to a true civilization. (...) Medicine (...) could supply modern society with engineers understanding the mechanisms of the body and the soul of the individual, and of his relations with the cosmic and social world.

In the conclusion, Carrel presented his idea of a medical contemplative order whose members would live in complete isolation from the outside world: at the top of the social hierarchy would thus stand not the Philosopher, as in Plato, but the Physician. The order, which would function like an immortal brain of the whole society, would be in charge of constant work on a programme of correction and management of society. Selected individuals would study for 25 years, gathering knowledge of anatomy, pathology, physiology, chemistry, psychology, health, as well as genetics, pedagogy, aesthetics, ethics, morality, religion, sociology, food chemistry, and economics. They would be—as in monastic contemplative orders—protected from the harmful influences of the civilisation they were to rebuild. At the age of 50, they would attain the necessary faculties to 'effectively direct the production of human beings and the construction of a civilization created truly for these beings' (Carrel, 1935b, p. 209). The precursor of such a collective medical brain could take the form of secret scientific societies that would attempt this on a small scale and whose activities would lead to the dismantling of the old industrial regime.

3.7 Conclusions: Evolutionary Utopias and the Limits of Discourse Confusion

The etymology of the word 'utopia' is ambiguous: it can be understood as a compound of *u-topos* (a non-existent place) or *eu-topos* (a good place). A combination of both would indicate a good place that does not exist in reality. An interpretation along these lines becomes the starting point of understanding utopias as naive dreams of a better life that provide an escape from a worrisome present state of affairs. But this escapist understanding does not capture the essence and purpose of utopian ideas.

In the Western tradition, utopianism can be viewed as a consequence of modernisation and 'disenchantment of the world' (see Chap. 18), where biblical eschatology is increasingly explicitly relocated into the real world. Since the beginning of the Modern Era, utopias took on various forms: the prophecies and egalitarian ideas of the sixteenth century, religiously radical visions of the seventeenth century, descriptions of virtues of the 'noble savage' contrasted with the moral degeneration of Western society and plans for experimental communities in the eighteenth century, all the way to the revolutionary utopias of the nineteenth and twentieth centuries. Utopianism is the core of communism and centralised state socialism, of urbanism and architecture, of visions of a racially pure society, as well as technocratic movements and programmes of population regeneration.

As a literary genre, utopias took shape during the Renaissance. The name was taken from Thomas More's eponymous work (1516) but various utopian visions were also outlined by other thinkers: Francis Bacon, Tommaso Campanella, James Harrington, and later Edward Bellamy or Robert Owen. These utopias each presented their vision of a virtuous life, with Modern emphasis on the importance of the public good. Later utopias incorporated utilitarian ideas, where morality is based on the greatest good for the greatest number of people. Utopias aim at investigating the existing reality and proposing ways of improving it. They are characterised by *rationality*, *criticality*, *normativity*, and the pursuit of improvement of the physical, social, economic, and psychological conditions of life. Their goal is thus a comprehensive change of the entire society. In a *serious* form, they present not an accurate description of the future, but rather an appeal for improving the present. They assume a correlation between a perfect organisation of material conditions and a perfect society. They paint a world at the end of history, where chaos has been replaced by order, individual hardship by collective wellbeing, and the irrationality of chance by the rule of reason.

Galton's and Carrel's utopias belong to this context of widely popular and serious visions, where the biological basis of the human being (evolution, heredity, and constitution) becomes the chief instrument of a comprehensive redefinition of human society *directed by science as the new secular religion*. They were not meant as metaphors: all of the instruments and measures cited by Galton were found already in his scholarly writings, experimental research, and advocacy of eugenics. Shortly before his death, Galton thus merely further elaborated on his belief in the legitimacy of directing the evolution of our own species, which he expressed already over forty years earlier (Galton, 1869, p. 1):

> I conclude that each generation has enormous power over the natural gifts of those that follow, and maintain that it is a duty that we owe to humanity to investigate the range of that power, and to exercise it in a way that, without being unwise towards ourselves, shall be most advantageous to future inhabitants of the earth.

The role of the clergy was to be assumed by 'a kind of scientific priesthood' (Galton, 1874; Turner, 1993, p. 170). In his public defence of eugenics, Galton explicitly described his vision as 'a new religion' (Galton, 1904). In Carrel's work, the perception of decline was united with a belief in biological determinism and led to the notion of a new aristocracy: the renewal of society would be directed by a medical elite and the physician would thus become God-like.

Despite their differences, Galton's and Carrel's evolutionary utopias shared certain features, such as an emphasis on the importance of selection and a eugenic foundation. Moreover, such pseudo-religious visions were not unique; in fact, they were rather typical of the times. The leading evolutionary morphologist Ernst Haeckel (1834–1919), one of the main sources of inspiration for eugenics, whose book *The Riddle of the Universe* (*Die Welträtsel*, Haeckel, 1899) was a bestseller, even tried to institutionalise his Darwinian monistic religion, to place it in direct competition with the traditional Church (see Chap. 4). In his self-appointed role as the new pope of the 'scientific worldview', Haeckel resembled the messianic zeal of Galton or Carrel. Their undisputed scientific authority provided support for legions of their followers among both the scientific elite and the general public. The idea of managing society based on anthropometric, biometric, and hereditary information was expressed in literature also by the well-known writer and biologist Herbert G. Wells (1866–1946) in Wells, 1905.

On the other hand, several well-known literary anti-utopias (Yevgeny Zamyatin, Karel Čapek, Aldous Huxley) clearly show that some people found such scientific universalist projections troubling already at the time of their publication. An opposition to organised eugenic activities came from the Church as a sort of continuation its opposition to Darwin's notions (Clark, 1999). A notable opponent of eugenics was the Catholic writer and historian Gilbert K. Chesterton (1874–1936) (cf. e.g., Chesterton, 1922), while the French engineer and sociologist Georges Friedmann (1902–1977) formulated a comprehensive criticism of biological and technocratic utopias, including Carrel's, in the name of authentic civilisational progress (Friedmann, 1936). But the danger of 'confusing' the discourses of biology and culture was anticipated already in the first ever synthetic history of classical Darwinism by Emanuel Rádl (1873–1942; Rádl, 1909):

> Even Darwin did not yet contradict these [liberal and socialist] doctrines, although his theory had already laid the foundations of a different conviction. His doctrine of the individual diversity of men, animals, and plants, and the doctrine of heredity, was a denial of the idea of the equality of men, albeit unconscious, replacing the old belief in external inequality by the notion of internal, innate, hereditary inequality.

Opposition to eugenics among biologists themselves often targeted the very core of evolutionary theory, as was the case of Oskar Hertwig (1849–1922) (Hertwig, 1916; Hertwig, 1921). Alfred R. Wallace (1823–1913) (Wallace 1913), the legendary co-creator and supporter of classical Darwinism, vehemently opposed the specific proposals of a eugenic programme, calling it quackery, warning against the dangers of medical tyranny, and rejecting all efforts to 'improve' society through selection (Durant, 1979, p. 31):

> The world does not want the eugenicist to set it straight. Give the people good conditions, improve their environment, and all will tend towards the highest type. Eugenics is simply the meddlesome interference of an arrogant, scientific priestcraft.

But the warning voices were a minority, drowned out by the widespread promotion and acceptance of eugenics. Galton's vision of a community reformed according to eugenic principles foreshadowed much of what was later implemented in the twentieth century with catastrophic consequences. This applies especially to his idea of a centralist state whose policies would be based on the control and management of the 'biological' characteristics of the population implemented through measures proposed by scholars (Bayertz, 1994, pp. 39–58). In the United States and Germany, sterilisation programmes received a legal footing based on eugenic reasoning.

Galton did not live to see the consequences of these measures, unlike Carrel, who with full support of the Vichy regime collaborated in the creation of the *Fondation française pour l'étude des problèmes humains*, a eugenic institution founded in 1941, in which Carrel held a senior position until August 1944, when he was suspended after the liberation of France.[12] Still, a full merger of science and politics took place only in Nazi Germany, where scientific activity was not only controlled and directed by the state but scientists and scientific institutions were forced to conform to the racist political ideology and become instruments of its enforcement in the transformation of society.

The modern biologisation and historicisation, a gradual conflation and merger of the discourses of biology, culture, and politics, have under special historical circumstances resulted in the aberration of 'murderous science', which rejected all 'non-scientific' worldviews and based its aims on the notion of the race. Haeckel allegedly said: '*Politik ist nichts anderes als angewandte Biologie*' (Politics is nothing but applied biology). The implementation of Nazi eugenics finally highlighted the dangers of this confusion, led to a reassessment of the underlying motivation, and to a phasing-out of eugenic programs worldwide.

Max Brod's recollection of the intellectual journey to a radical restriction of human nature and the main ideological controversies of the twentieth century was intended to show that these were not mere 'abuses' of an otherwise 'neutral'

---

[12] Carrel is nowadays considered a clearly controversial figure. It started in the 1990s: after Carrel's ideas were revived in France by the National Front, then led by Jean-Marie le Pen, his name disappeared from the public sphere and accusations of his complicity in murdering the sick during the Vichy regime began to emerge. For controversies regarding his activities, see Benveniste (2013), or Drouard (1990).

science. The point is that from the beginning, the modern life sciences, which investigate the principles and foundations of life (including human life) with their reductionist, experimental, and statistical methods, were associated with a universal ambition to apply their findings to human society.

Importantly, the numerous analogies between nature and culture did not lose their relevance after the tragic consequences they led to in the hands of the Nazis. In fact, they are becoming increasingly relevant in the digital age and the age of AI. By drawing attention to possible risks, they can still inspire alternative solutions and lead to a prudent consideration of the discourses among which we live, research, think, and act in an increasingly experimentally and theoretically elaborate evolutionary and historical paradigm.

References

Adams, M. B. (Ed.). (1990). *The wellborn science: Eugenics in Germany, France, Brazil, and Russia*. Oxford University Press.

Bassett, W. W. (1995). Eugenics and religious law: Christianity. In W. T. Reich (Ed.), *Encyclopedia of bioethics* (pp. 779–783). Simon & Schuster Macmillan.

Bateson, W. (1906). The progress of genetic research. In R. C. Punnet (Ed.), (1928) *Scientific papers of William Bateson (Vol. 2, pp. 142–151). Cambridge University Press.*

Bateson, W. (1908). *The methods and scope of genetics*. Cambridge University Press.

Baur, E., Fischer, E., & Lenz, F. (1923). *Grundriss der menschlichen Erblichkeitslehre und Rassenhygiene*. J. F. Lehmann.

Bayertz, K. (1994). *GenEthics: Technological intervention in human reproduction as a philosophical problem*. University Press.

Benveniste, G. L. (2013). Alexis Carrel: The good, the bad, the ugly. *ANZ Journal of Surgery, 83*(9), 609–611.

Bodenheimer, F. S. (1957). *Studies in biology and its history*. Biological Studies Publishers.

Brod, M. (1960). *Streitbares Leben. Autobiographie*. Kindler.

Burdett, C. (2002). From the new Werther to numbers and arguments: Karl Pearson's eugenics. In R. Luckhurts & J. McDonagh (Eds.), *Transactions and encounters science and culture in the nineteenth century* (pp. 204–232). Manchester University Press.

Burrow, J. V. (2000). *The crisis of reason: European thought 1848–1914*. Yale University Press.

Carrel, A. (1935a). *L'homme, cet inconnu*. Plon.

Carrel, A. (1935b). *Man, the unknown*. Harper and Brothers.

Chadwick, O. (1975). *The secularization of the European mind in the 19th century*. Cambridge University Press.

Chesterton, G. K. (1922). *Eugenics and other evils*. Casell & Co.

Clark, S. R. L. (1999). *The political animal: Biology, ethics, and politics*. Routledge.

Conrad-Martius, H. (1955). *Utopien der Menschenzüchtung: Der Sozialdarwinismus und seine Folgen*. Kösel.

Darwin, C. (1859). *On the origin of species; or the preservation of the favoured races in the struggle for life*. John Murray.

Darwin, C. (1868). *The variation of animals and plants under domestication*. John Murray.

Davenport, C. B. (1910). Euthenics and eugenics. *Popular Science Monthly, 78*, 16–20.

Davenport, C. B. (1912). *Heredity in relation to eugenics*. Williams & Norgate.

Drouard, A. (1990). La Fondation française pour l'étude des problèmes humains et l'organisation de la recherche en sciences sociales en France. *Cahiers pour l'histoire du CNRS* 9. http://www.histrecmed.fr/images/pdf/drouard.pdf. Accessed 31.08.22.

Dunn, L. C. (1965). *A short history of genetics: The development of some of the main lines of thought: 1864–1939*. McGraw–Hill Book Company.

Durant, J. R. (1979). Scientific naturalism and social reform in the thought of Alfred Russel Wallace. *The British Journal for the History of Science, 12*(1), 31–58.

Farrall, L. A. (1985). *The origins and growth of the English eugenics movement 1865–1925*. Garland.

Friedmann, G. (1936). *La Crise du progrès: esquisse d'histoire des idées, 1895–1935*. Gallimard.

Galton, F. (1865). Hereditary talent and character (Part I and Part II). *Macmillan's Magazine, 12*(68), 157–166, 318-327.

Galton, F. (1869). *Hereditary genius: An inquiry into its laws and consequences*. Macmillan & Co.

Galton, F. (1871a). Experiments in pangenesis, by breeding from rabbits of a pure variety, into whose circulation blood taken from other varieties had previously been largely transfused. *Proceedings of the Royal Society of London, 19*(123-129), 393–410.

Galton, F. (1871b). Pangenesis. *Nature, 4*(79), 5–6.

Galton, F. (1874). *English men of science: Their nature and nurture*. Macmillan & Co.

Galton, F. (1883). *Inquiries into human faculty and its development*. Macmillan & Co.

Galton, F. (1889). *Natural inheritance*. Macmillan & Co.

Galton, F. (1904). Eugenics: Its definition, scope, and aims. *American Journal of Sociology, 10*(1), 1–16.

Galton, F. (1909). *Essays in eugenics*. Eugenics Education Society.

Gutiérrez, R. R., Guazo, L., & Suárez, L. (2002). Eugenics, inheritance, selection, and biometry. *Llull, 25*, 85–107.

Haeckel, E. (1899). *Die Welträtsel: Gemeinverständliche Studien über monistische Philosophie*. Emil Strauss.

Hankins, F. H. (1908). *Adolphe Quetelet as statistician*. Columbia College.

Haycraft, J. B. (1895). *Darwinism and race progress*. Sonnenschein.

Hertwig, O. (1916). *Werden der Organismen: Eine Widerlegung von Darwin's Zufallstheorie*. Fischer.

Hertwig, O. (1921). *Zur Abwehr des ethischen, des sozialen, des politischen Darwinismus*. Fischer.

Johannsen, W. L. (1909). *Elemente der exakten Erblichkeitslehre*. Fischer.

Kern, J. L. (1981). *An ordered love: Sex roles and sexuality in Victorian utopias: The shakers, the Mormons, and the Oneida community*. University of North Carolina Press.

Kevles, D. J. (1995). *In the name of eugenics: Genetics and the uses of human heredity*. Harvard University Press.

Kropotkin, P. (1902). *Mutual aid: A factor of evolution*. McClure Phillips & Co.

Lawrence, C., & Weisz, G. (Eds.). (1998). *Greater than the parts. Holism in biomedicine 1920–1950*. Oxford University Press.

Loewenstein, B. (2009). *Der Fortschrittsglaube: Geschichte einer europäischen Idee*. V&R unipress.

López-Beltrán, C. (2004). In the cradle of heredity: French physicians and L'Hérédité Naturelle in the early 19th century. *Journal of History of Biology, 37*(1), 39–72.

Lubinsky, M. S. (1993). Scientific aspects of early eugenics. *Journal of Genetic Counseling, 2(2), 77–92.*

Ludmerer, K. M. (1972). *Genetics and American society: A historical appraisal*. John Hopkins University Press.

Mayr, E. (1984). *Die Entwicklung der biologischen Gedankenwelt: Vielfalt, Evolution und Vererbung*. Springer.

McLeod, H. (2000). *Secularisation in Western Europe, 1848–1914*. St. Martin's Press.

Mukherjee, S. (2017). *The gene. An intimate history*. Scribner.

Olby, R. C. (1966). *Origins of Mendelism*. Constable.

Olby, R. C. (1996). The emergence of genetics. In R. C. Olby (Ed.), *Companion to the history of modern science* (pp. 521–537). Routledge.
Orel, V. (1984). *Mendel*. Oxford University Press.
Pauly, P. J. (1993). Essay review: The eugenics industry—Growth or restructuring. *Journal of the History of Biology, 26*(1), 131–145.
Pearson, K. (1901). *National life from the standpoint of science*. A. & C. Black.
Pearson, K. (1909). The scope and importance to the state of national eugenics. *Eugenic Laboratory Memoirs, 1*, 12–76.
Pearson, K. (1924). *The life, letters and labours of Francis Galton* (Vol. II). Cambridge University Press.
Pearson, K. (1930). *The life, letters and labours of Francis Galton. Vol. IIIA: Correlation, personal identification and eugenics*. Cambridge University Press.
Rádl, E. (1909). *Geschichte der biologischen Theorien II. Geschichte der Entwicklungstheorien in der Biologie des XIX. Jahrhunderts*. Engelmann.
Ramsden, E. (2002). Carving up population science: Eugenics, demography and the controversy over the "biological law" of population growth. *Social Studies of Science, 32*, 857–899.
Reggiani, A. H. (2007). *God's eugenicist: Alexis Carrel and the sociobiology of decline*. Berghahn.
Saller, F. (1961). *Die Rassenlehre des Nationalsozialismus in Wissenschaft und Propaganda*. Progress-Verlag.
Sapp, J. (2003). *Genesis: The evolution of biology*. Oxford University Press.
Schallmayer, W. (1903). *Vererbung und Auslese im Lebenslauf der Völker: Eine staatswissenschaftliche Studie auf Grund der neueren Biologie. Gustav Fischer.*
Searle, G. R. (1976). *Eugenics and politics in Britain 1900–1914*. Noordhoff.
Semon, R. (1904). *Die Mneme als erhaltendes Prinzip im Wechsel des organischen Geschehens*. Engelmann.
Smith, J. M. (1988). Eugenics and utopia. *Daedalus, 117*(3), 73–92.
Stubbe, H. (1972). *History of genetics. From prehistoric times to the rediscovery of Mendel's laws*. MIT Press.
Tönnies, F. (1887). *Gemeinschaft und Gesellschaft. Abhandlung des Communismus und des Socialismus als empirischer Culturformen*. Fues.
Turda, M. (2010). *Modernism and eugenics*. Palgrave Macmillan.
Turner, F. M. (1993). *Contesting cultural authority. Essays in Victorian intellectual life*. Cambridge University Press.
Wallace, A. R. (1913). *Social environment and moral progress*. Cassell & Co.
Weingart, P., Kroll, J., & Bayertz, K. (1992). *Rasse, Blut und Gene: Geschichte der Eugenik und Rassenhygiene in Deutschland*. Suhrkamp.
Weismann, A. (1892). *Das Keimplasma: eine Theorie der Vererbung*. Fischer.
Wells, H. G. (1905). *A modern utopia*. Chapman and Hall.
Whetham, W. C. D. (1909). *Family and the nation: A study in natural inheritance and social responsibility*. Longmans.

Chapter 4
On the Monistic and Dualistic Origins and the Development of Nature and Culture: Haeckel and Ehrenfels

Lenka Ovčáčková

Abstract The aim of this chapter is to present a comparative analysis of the concept of the origin and development of nature and culture in the thinking of the German zoologist Ernst Haeckel (1834–1919) and the Austrian philosopher Christian von Ehrenfels (1859–1932). The thinking of both these scholars is anchored in Darwinism, but each followed his own path within it. Haeckel, who keenly promoted his revolutionary scientific concept, not merely adopted Darwin's theory of evolution but took it further, among other things by integrating ideas of the German *Naturphilosophie*, to the point of establishing a scientific monistic religion. Ehrenfels was likewise strongly influenced by Darwinism but concluded that its arguments cannot explain the origin and development of nature and culture. He therefore looked for a 'missing link' that would explain evolution not just by reference to chance but by a dualistically postulated *gestaltende Kraft* (shaping force) defined in contrast to chaos.

Keywords Scientific monistic worldview · Dynamic Darwinian dualism · Gestalt culture · Monistic culture · Organic derivation

4.1 Introduction

Ernst Haeckel (1834–1919) and Christian von Ehrenfels (1859–1932), two notable thinkers from the German-speaking milieu, were both strongly influenced by Darwinism. It had a deep impact on the scientific and cultural concepts they had developed although their starting positions were different. The physician and zoologist Ernst Haeckel anchored his work in biological research, especially morphology and embryology, and although it had significant overlaps to philosophy, his monistic

L. Ovčáčková (✉)
Department of Philosophy and History of Sciences, Faculty of Science, Charles University, Prague, Czech Republic
e-mail: lenka.ovcackova@natur.cuni.cz

J. Toman et al. (eds.), *Culture Evolving*, Biosemiotics 10,
https://doi.org/10.1007/978-3-032-16124-6_4

worldview was thus largely based on natural sciences. The philosopher Christian von Ehrenfels drew in formulating his theory of Gestalt qualities mainly on his background in philosophy, but when elaborating the concept of Gestalt and Gestaltung (shaping), he increasingly moved into the realm of natural sciences.

Both Haeckel and Ehrenfels tried to find an anticipation of their approaches in ancient Greek thought and investigated in detail the long historical development of philosophical, natural scientific and cosmological concepts (see Chaps. 1, 2 and 6). Both were also significantly inspired by *Naturphilosophie*, which was in the first half of the nineteenth century represented mainly by Johann Wolfgang Goethe, and they both reacted to and radically opposed various contemporary scientific and philosophical worldviews and emphasised the interconnectedness of nature and culture and the primacy of humans as the culmination of cultural development. Both Haeckel and Ehrenfels used their worldview as the foundation of a new religion, and they both based their arguments for the new religion on scientific knowledge. In interpreting and analysing the relationship between the origin and evolution of nature and culture, both relied heavily on cosmogonic and cosmological arguments. And finally, they both had a lasting impact on natural science and philosophy: Haeckel's scientific monistic worldview and Ehrenfel's theory of Gestalt qualities are still discussed by the scientific community (Fig. 4.1).

Ernst Haeckel[1] was born in Potsdam in 1834 into a family with a strong Protestant tradition. In childhood and youth, Haeckel developed his artistic talents and was intensely interested in botany, which he intended to study. In the end, though, he followed the wishes of his parents and studied medicine in Würzburg and Berlin. His most important teachers were the anatomist and physiologist Albert von Kölliker, the pathologist Rudolf Virchow, and the anatomist, physiologist, and zoologist Johannes Müller. Haeckel was an avid reader of texts on natural philosophy and natural science, especially those written by the writer and natural scientist Johann Wolfgang von Goethe and the natural scientist and traveller Alexander von Humboldt. During his studies, Haeckel felt challenged by the then extremely popular scientific materialism, represented mainly by the physician and philosopher Ludwig Büchner. A major turning point in Haeckel's life came in 1860, when he read Darwin's groundbreaking work *On the Origin of Species by Means of Natural Selection* (Darwin, 1859). Under its influence, Haeckel decided to dedicate his life to promoting Darwinism but he also did not abandon the Goethean approach to natural philosophy. In 1862, he published a notable scientific exploration, *Die Radiolarien* [The Radiolarians]. Three years later, he became a full professor of zoology at the University of Jena and contributed to the development of its institute of zoology.

[1] Haeckel's life and work is the subject of many biographies, (e.g. Bölsche, 1900, May, 1909, Uschmann, 1961, Krauße, 1987, and Di Gregorio, 2005). Haeckel's biographical data are also part of a number of studies published in recent decades by Ernst Haeckel Haus, Institute for the History and Philosophy of Science (former Institute for the History of Medicine, Science and Technology), Friedrich-Schiller-Universität Jena.

Fig. 4.1 Ernst Haeckel in 1905. Haeckel (1905). *Ernst Haeckels Wanderbilder*. Bonn: Strauss

Haeckel's professional and personal development was shaped not only by his wide-ranging theoretical explorations but also by a series of scientific journeys to places near and far. He documented his expeditions in writing but also by illustrations depicting various natural entities or landscapes. Haeckel worked at the university in Jena until 1909. He remained active in science and promoted his worldview almost until the end of his life: his last book was published in 1917, just 2 years before his death, under the title *Kristallseelen. Studien über das anorganische Leben* [Crystal souls. Studies on inorganic life] (Fig. 4.2).

Christian von Ehrenfels[2] was born in 1859 in Rodaun near Vienna but spent his childhood at the family's castle in Brunn am Wald in the Austrian Waldviertel. As the firstborn son, he was expected to manage the family's estates, but he decided to cede this right to his brother. From 1879, he studied philosophy, Germanic studics, and history at the University of Vienna, where his most important teacher was the German philosopher and psychologist Franz Brentano. Later, he transferred to Graz, where especially significant was his encounter with the philosopher and psychologist Alexius Meinong, Brentano's student. Ehrenfels had a lifelong love of music: he was a great admirer of Richard Wagner and composed many epic operas himself. In 1890, he published his famous paper *Über Gestaltqualitäten* [On Gestalt qualities], which made him the founding father of the Gestalt theory. In later years, he elaborated his theory in other contexts, for instance in connection with prime

[2] Ehrenfels's biographies are included in several books dealing with his work, for example (Fabian, 1986; Höfer & Valent, 2017; Weinhandl, 1960).

Fig. 4.2 Christian von Ehrenfels in his later years. Photography taken in Prague (Central European Photopress). Österreichische Nationalbibliothek

numbers, cosmogony, or Darwinism. In 1896, he was appointed extraordinary and in 1899 a full professor at the Faculty of Philosophy at the German part of the Charles-Ferdinand University (in 1920 renamed German University) in Prague. He continued to work there even after his retirement and until his death in 1932.

In 1929, Ehrenfels was instrumental in founding the Department of Natural Philosophy at the Faculty of Science of the German University of Prague, where he was the only lecturer.[3] The main focus of Ehrenfels's academic work was the theory of value ethics but his lectures often overlapped with other disciplines. Aside from value ethics, he lectured on psychology, aesthetics, logic, prime numbers, Gestalt qualities, but also cosmic morphology, Darwinism, vitalism, and social biology. His impact was potentiated by the fact that his lectures were attended by many members of the famous Prague Circle, including Franz Kafka, Max Brod, or Felix Weltsch.

4.2 Haeckel's Darwinian Nature-Philosophical Monism

For Haeckel, it was unthinkable that natural science should view nature as a soulless mechanism. He believed that all 'natural manifestations' and 'all living matter' are soulful (Haeckel, 1909, p. 50) and the difference between various natural entities is

[3] In 1932–1936, this department was headed by the famous philosopher and mathematician Rudolf Carnap.

merely a matter of the degree of formation of the soul, which ascends from crystals and cells to the conscious and rational human soul (Haeckel, 1878). The soulfulness of all nature played an important role in Haeckel's argument for his particular monistic worldview.[4] He was convinced that everything in the world is connected in one whole: organic and inorganic nature, matter and energy, God and nature, but also natural sciences and humanities. Haeckel believed in the existence of a single cosmological law, the 'Law of Substance', which included the law of conservation of matter and conservation of force, and later also the psychological law of conservation of perception (*psychom*). In Haeckel's view, matter, energy, and *psychom* were thus inseparably harmoniously linked—as were materialism, energetics, and psychomatics or panpsychism (Haeckel, 1914, pp. 34–36).

Haeckel's worldview is strictly anti-dualistic: it rejects the idea of an 'anthropomorphic creator' and is strongly opposed to any form of teleology, metaphysics, or vitalism. Events in the world are governed by natural causal laws. Haeckel appeals to the use of 'pure reason', which he contrasts with 'belief in traditional dogmas and alleged revelations' (Haeckel, 1914, p. 50). For Haeckel, the history of creation is a natural story that encompasses all spheres of life, including culture. We shall see below, that Haeckel speaks in this context of a 'monistic culture' (cf. Ovčáčková, 2017).

Already during his medical studies Haeckel was interested in the history of human development and the comparative history of development of the lower and higher animals based on descriptive embryology, but it was the encounter with Darwin's ideas that had a decisive impact on him. He later wrote: 'Only then did the veil of the great mystery of germ history fall and I clearly recognised that its true causes could only be found in its relationship to phylogeny. This is how I arrived at the establishment of the basic biogenetic law...' (Haeckel, 1912b, in Breitenbach, 1914, p. 165). Haeckel elaborated the theory of the basic biogenetic law in his two-volume *Generelle Morphologie der Organismen* [General morphology of organisms, Haeckel, 1866], where he outlined the outlines of his research, which he then elaborated throughout the rest of his life: based on the study of the general anatomy and morphology of organisms, he analysed the general developmental history of organisms with special emphasis on ontogeny and phylogeny. Significantly, Haeckel had in this book shifted his attention from microscopic examination and description of individual forms to a search for 'lawful connections in the multiplicity of individual phenomena'. In this quest, he was guided by his conviction that 'the art of describing forms gives rise to the knowledge of forms' (Haeckel, 1906a, pp. 4–5).

[4] In the historical context, Haeckel repeatedly points to the presence of monistic thinking in the natural philosophers Thales, Anaximenes and especially Anaximander: 'They already taught the unified law of nature as the primeval basis of diverse phenomena, the unity of all nature and the constant change of form' (Haeckel, 1889: 69, transl. Lenka Ovčáčková). According to Haeckel, however, the modern evolutionary ideas of the second half of the nineteenth century also find indications in Heracleitus, Empedocles and Democritus, and in poetic form in the Roman poet and philosopher Lucretius Carus in his work *De rerum natura* (see Chap. 1). The historical thought of Giordano Bruno, Spinoza, and the aforementioned Goethe was an important inspiration for Haeckel when formulating his pantheistic ideas (cf. Ovčáčková, 2017).

Based on the research of symmetrical structures, he also pointed to 'the simplest and most imperfect organisms *Moneren*' and the organic nature of crystals (Haeckel, 1906a, p. 79). The nature-philosophical underpinnings of the entire *Generelle Morphologie der Organismen* were highlighted by quotations from Geothe's work, which introduced each individual chapter. At the end of the second volume, Haeckel had for the first time presented a comprehensive outline of his views on the unity of God and nature—and thus his monistic religion (Haeckel, 1866, p. 456; Fig. 4.3).

Haeckel subsequently published several books, both academic and popularising, such as the *Natürliche Schöpfungsgeschichte* [The history of creation, Haeckel, 1868], *Anthropogenie oder Entwicklungsgeschichte des Menschen* [Anthropogeny or the history of human development, 1874], *Die Welträthsel* [The riddle of the universe, 1899], or the *Kunstformen der Natur* [Art forms in nature, Haeckel, 1904d], which brought him enormous popularity. Haeckel's most popular book was *Die Welträthsel*, which had a print run of 400,000 and was translated into 30 languages. It was divided in four parts: an anthropological, psychological, cosmological, and theological one.

One of the most concrete manifestations of the overlap between nature and culture in Haeckel's thought is his conception of ethics. The monistic categories of truth, the good, and beauty played a pivotal role in Haeckel's quest to promote a monistic worldview:

> Monistic natural science as knowledge of the true, monistic ethics as education for the good, monistic aesthetics as the cultivation of beauty – these are the three main areas of our monism; through their harmonious and coherent development we gain that truly happy bond between religion and science that is still sorely missed by so many today. (Haeckel, 1905, pp. 35–36).[5]

For Haeckel, the most important instrument he used to argue for his monistic worldview was the beauty and symmetry of natural forms.[6] His ultimate aim was to systematise the diversity of nature and describe the developmental order and regularity (cf. Breidbach, 1998, 2006; Fig. 4.4).

In 1904e, Haeckel published his *Thesen zur Organisation des Monismus* [Theses on the organisation of monism], which briefly summarised the foundations of both his theoretical monism, which was based on experience, reason, and science, and practical monism, which called for a rational conduct of life in accordance with the aforementioned monistic ethical ideals of truth, the good, and beauty (Haeckel, 1904c, in Breitenbach, 1914). In doing so, Haeckel took an important step towards founding a monistic organisation and one such organisation was indeed eventually established at the Zoological Institute of the University of Jena in 1906 under the

[5] 'Die monistische Naturforschung als Erkenntnis des Wahren, die monistische Ethik als Erziehung zum Guten, die monistische Ästhetik als Pflege des Schönen—das sind die drei Hauptgebiete unseres Monismus; durch ihre harmonische und zusammenhängende Ausbildung gewinnen wir jenes wahrhaft beglückende Band zwischen Religion und Wissenschaft, das heute noch von so Vielen schmerzlich vermisst wird.'

[6] Haeckel's discovery of 'myriads of beautiful forms' thus aroused a 'different aesthetic sense' that also gave a new direction to the visual arts (Haeckel, 1900, p. 393).

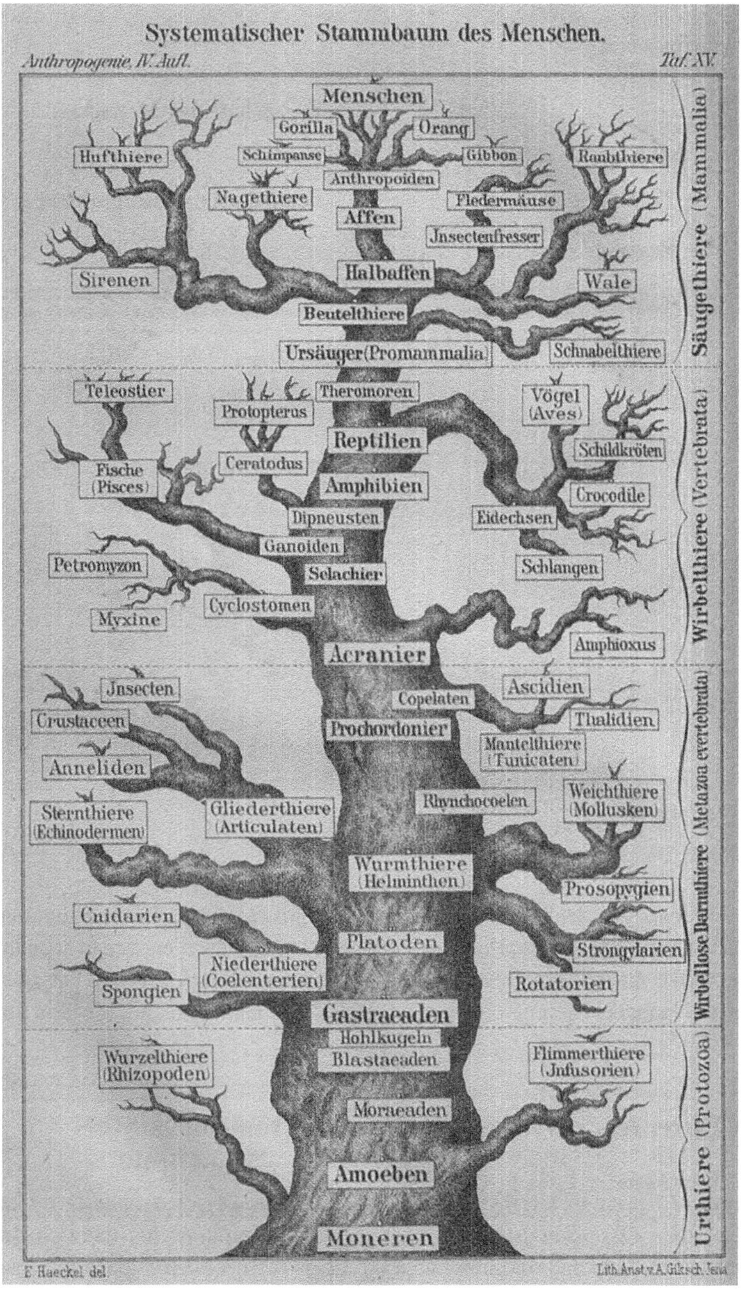

Fig. 4.3 Haeckel's human pedigree. Haeckel (1891). *Anthropogenie* (fourth edition). Leipzig: Engelmann

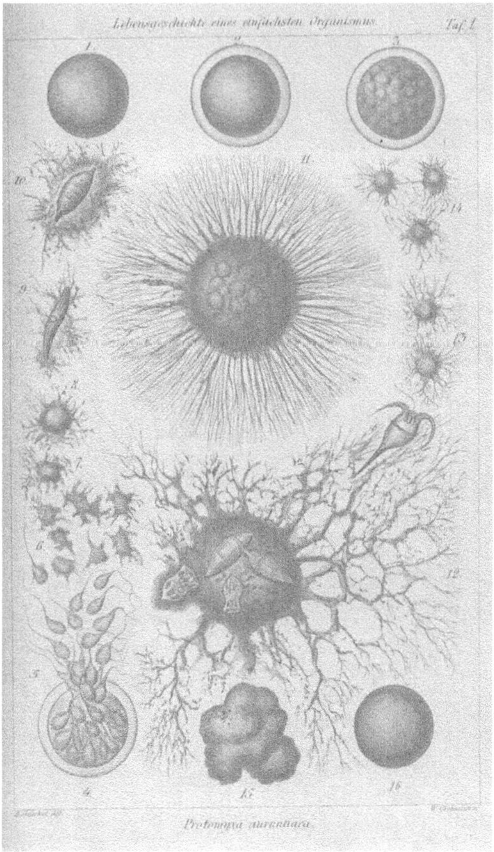

Fig. 4.4 Development of the simplest organism. Haeckel. (1889). *Natürliche Schöpfungsgeschichte* (first edition). Berlin: Reimer

name of *Deutscher Monistenbund* (German Monist League). In the late nineteenth and early twentieth century, Haeckel's monistic worldview then spread at an unprecedented pace not only in Germany and German-speaking countries but also elsewhere and Haeckel became one of the best-known natural scientists of his time.

4.3 Ehrenfels's Dynamic Darwinian Dualism

The core of Ehrenfels's reflections on Gestalt qualities is presented in his famous treatise *Über Gestaltqualitäten* (1890). Ehrenfels explains his theory using the example of melody and harmony, which reflects his deep experience of music. He characterises Gestalt qualities as the contents of ideas (*Vorstellungsinhalte*), which are linked to complexes of ideas in the consciousness (*Vorstellungskomplexen im Bewußtsein*) and consist of separable elements (Ehrenfels, 1890, in Weinhandl, 1960, p. 21). Memory and the emergence of new unified ideas play an important role in this process, as attested by Ehrenfels's characterisation of Gestalt qualities written in the last year of his life:

> The decisive step in the establishment of the doctrine of the Gestalt qualities was my claim: If the images of the memories of successive tones are available as a simultaneous complex of consciousness, then an idea of a new category can emerge in consciousness, namely, a unified idea that is linked in a characteristic way to the idea of the given complex of tones. (Ehrenfels, 1932, in Weinhandl, 1960, p. 61)[7]

For Ehrenfels, the relationality of the parts and the whole plays an important role: the whole is not merely the sum of the parts or elements, which are in some ways related. What is crucial is that the whole becomes a higher reality than the sum of its parts and their mutual relations: the whole acquires a new quality. This became the basis of all Ehrenfels's later Gestalt-oriented reflections and the essence of his dualistic worldview, which allows orientation in both the experiential and metaphysically conceived world. In fact, already in 1890 Ehrenfels suggested that his theory of Gestalt qualities could bridge the gap between 'different categories of the imaginable' and include various apparently disparate phenomena in a 'unified system' (Ehrenfels, 1890, in Weinhandl, 1960, p. 41).

In his essay of 1890, Ehrenfels characterised three qualities which in his view could become the basis of a new higher order: spatial qualities (the common features of spatial formations), non-temporal qualities (static momentary states), and temporal qualities (having a kinetic character). In later writings, he took further steps in this direction and—in line with his dualistic cosmogony—characterised also other 'types of formation' in the metaphysical and experiential world (Ehrenfels, 1916, p. 93).

His reflections upon the 'height and purity of the Gestalt' (Ehrenfels, 1916, p. 193) played an important role not only in the development of his cosmogony but also in his dualist Darwinian reflections. For each Gestalt, the *Höhe der Gestaltung* (height of shaping) is evident, whereby 'The higher shapes also differ from the lower ones in that the product of unity and diversity is greater in the latter than in the former' (Ehrenfels, 1916, p. 94). According to Ehrenfels, one can see the increase in Gestalt height for instance in the phylogenetic developmental series from the lowest organisms to humans. Higher Gestalten shape a unity from a higher diversity, so that for example a rose has a greater Gestalt height than a pile of sand. But Gestalt development can also lead to dead ends or decreased Gestalt height. In this context, Ehrenfels draws an analogy between Gestalt height and beauty (cf. Ovčáčková & Švorcová, 2024): beauty is equivalent to a Gestalt that has achieved full unity, while ugliness is a disharmony arising from competing Gestalt elements

[7] 'Der entscheidende Schritt für die Begründung der Lehre von der Gestaltqualität war nun die Behauptung meinerseits: Wenn die Erinnerungsbilder der aufeinanderfolgenden Töne als ein gleichzeitiger Bewusstseinskomplex vorliegen, so kann im Bewusstsein die Vorstellung einer neuen Kategorie auftauchen, und zwar eine einheitliche Vorstellung, welche auf eine eigentümliche Weise mit der Vorstellung des betreffenden Tonkomplexes verbunden ist.'

that move in incompatible directions (Ehrenfels, 1922b, in Weinhandl, 1960, p. 51).[8] In relation to Gestalt height, Ehrenfels stresses that it can increase indefinitely, while Gestalt purity can reach a maximum, which occurs at a relatively low Gestalt height. Pure and perfect Gestalten are represented by, for example, mathematical solids.[9] Ehrenfels further elucidated the connection between Gestalten and the mathematical domain in his treatise *Das Primzahlengesetz* [The law of prime numbers, 1922]. This work includes the 'Weiterführende Bemerkungen' [Explanatory notes], where Ehrenfels developed his theory of Gestalt qualities further, in part by reacting to his colleagues' comments, with a significant overlap into the theory of evolution and thus the origin and development of nature and culture.

Although Ehrenfels developed his worldview in line with the theory of evolution, he was critical of its excessive emphasis on chance as the sole principle of emergence and development (Ehrenfels, 1922a, in Fabian, 1990, p. 240) because in a world of blind chance—which Ehrenfels identified with chaos—nothing graspable or permanent can come into being. In reaction, Ehrenfels posits that chaos is not present in the world by itself. Instead, 'apart from it there is a second, the absolutely unified, instinctive, the source of all causality, order and shape', and thus also 'of all necessity and Gestalt' and all 'beauty, truth and good in the world' (Ehrenfels, 1929, in Fabian, 1990, p. 285; Ehrenfels, 1922a, in Fabian, 1990, p. 248). All unity therefore stems from a *gestaltende Kraft* (shaping force), which Ehrenfels identifies with God, while all multitude comes from chaos. According to Ehrenfels, Gestalten do not emerge from chaos—they are shaped by its resistance. 'Without chaos, God was just as incapable of creating the world as chaos was without God' (Ehrenfels, 1929, in Fabian, 1990, p. 285). This dynamic dualism is for Ehrenfels 'the missing link that completes Darwin's chain of thought' and ensures that the theory of evolution achieves 'a final elimination of teleological consciousness as a cosmic principle of explanation' and a way of explaining how uniform Gestalten can emerge from random stimuli (Ehrenfels, 1916, p. 76).

Ehrenfels viewed the principles of chaos and the unified shaping force reflecting the opposition of multiplicity and unity. He posits the origin and development of the first Gestalt into the bottomless, blind, and random chaos that existed in eternity before the world came into being (Ehrenfels, 1916, pp. 198–199). Then the first shaping and, in the absence of a teleological dimension, a random miracle took place, after which all other shapings followed. A Gestalt quality does not emerge necessarily: it depends on either the *gestaltende Kraft* itself[10] or on our creative

[8] For Ehrenfels, for example, an ugly animal is a long torso on short legs because the desired unity was not achieved in the shaping, i.e., the Gestalt of a long torso with long legs or a short torso with short legs did not assert itself.

[9] In this context, Ehrenfels asks a remarkable question whether height and purity of shaping are values 'for the psychoid source of all Gestalt' (Ehrenfels, 1916, p. 95). Although the nature of the unitary principle is psychic rather than physical, it cannot be clearly defined as psychic. This is why Ehrenfels uses the term psychoid (cf. Ehrenfels, 1916, p. 58).

[10] Like Haeckel, Ehrenfels found the sources of inspiration for his wealth of ideas in Presocratic philosophy, referring primarily to Anaxagoras. He was also deeply inspired by Johann Wolfgang

capacity. Moreover, it goes over and beyond connecting or combining the individual elements: something new is created (Ovčáčková, 2022). According to Ehrenfels, when we look at a genealogy, we should see its qualitative and quantitative ramifications:

> The history of the origin of the organic world is the greatest proof of this. The earlier the geological layers we descend into, the more similar the shapes appear – the smaller the range of variation they exhibit. From a few and relatively little differentiated Ur-beginnings, the diversity of shapes grows up towards us in a fan-like manner. (Ehrenfels, 1916, p. 11).[11]

For Ehrenfels, a family tree represents a type of Gestalt development in the organic world. It does not contradict the dualistic hypothesis: the stream of shaping springs from a primeval source, an Ur-source, 'dividing into hundreds, thousands and thousands of branches that weave their way through chaotic resistances' (Ehrenfels, 1922a, in Fabian, 1990, p. 249). According to Ehrenfels, while the inorganic nature contains various profound analogies with the world of living organisms (such as the Gestalt individuality of crystals), it lacks the shaping compulsion. The inorganic nature seems disconnected from the source of shaping and is, in a sense, left to itself (Ehrenfels, 1922a, in Fabian, 1990, p. 251). As will be shown below, the emergence of new Gestalt qualities, which is accompanied by considerable resistance, is evident in the emergence of not only new biological species but also of the new qualities of cultural phenomena, such as social systems or ethical ideals.

4.4 The Origin and Development of Culture

4.4.1 Monistic Culture

As mentioned above, Haeckel believed that morphology is the most important of all the biological sciences. This was because he recognised the genetic, that is, evolutionary aspect of the thousands of organic forms and derived their infinite variety from one simple source. Haeckel not only developed the field of comparative morphology but by incorporating elements from the theory of evolution and *Naturphilosophie* made it the basis of an entire monistic philosophy and monistic religion (Haeckel, 1914, pp. 17–18). Haeckel's argument for the development of a

---

Goethe's thought. Central for Ehrenfels was Goethe's anti-teleological statement that the only purpose of life is life itself. Life was not created to follow a purpose (Ehrenfels, 1922a, in Fabian, 1990, p. 243). Ehrenfels further developed the idea of the shaping force in relation to the vitalist conception of the German biologist Hans Driesch. He even lectured on Drieschian vitalism in Prague, in the winter term of 1920/1921 and the winter term of 1924/1925 (cf. *Ordnung der Vorlesungen an der Deutschen Universität in Prag*). It is interesting to note that Driesch was Haeckel's student but their paths diverged because of worldview differences.

[11] 'Das erweist am großartigsten die Abstammungsgeschichte der organischen Welt. In je frühere geologische Schichten wir hinabsteigen, desto ähnlicher sehen wir die Formen, – eine desto geringere Variationsbreite weisen sie auf. Aus wenigen und relativ wenig differenten Uranfängen wächst die Mannigfaltigkeit der Gestaltungen fächerförmig bis zu uns empor.'

new monistic philosophy, religion, and thus also culture, was closely intertwined with the development of monistic natural science at the turn of the nineteenth and twentieth centuries:

> The marvellous height of culture which man has scaled in the 19th century, the astonishing progress of natural science and its practical application in technology, industry, medicine, etc., allow us to hope for a mighty further upswing of culture in the twentieth century. However, this desired progress will only be possible if the familiar paths of traditional dogmas and clerical superstition are abandoned and rational monistic knowledge of nature gains the upper hand instead. (Haeckel, 1914, p. 49)[12]

In the nineteenth century—which Haeckel referred to as the 'century of natural science' (Haeckel, 1906b in Breitenbach, 1914, p. 51)—significant advances were made in natural science. According to Haeckel, the most important milestones was the discovery of cellular theory and the theory of evolution. Haeckel calls the subsequent twentieth century the 'monistic century' (Haeckel, 1912a in Breitenbach, 1914, p. 156) and claims that a critical historical comparison shows that 'every great advance in deep knowledge' (Haeckel, 1905, p. 13) leads to a move away from dualism and pluralism and towards monism.

On the other hand, Haeckel also repeatedly criticised the cultural inertia of his time, the inability to break out of a 'barbarian' state, noting that the 'uneducated cultured man' is, like the 'crude natural man', surrounded by the 'riddle of the universe', which he could solve only by pure scientific knowledge that is not available to him. This type of person has as little reason as his mammalian relatives, such as monkeys, dogs, and elephants (Haeckel, 1900, pp. 17–20). It is important to Haeckel that animals cannot be denied reason and religion: he believes that cultural life has over millennia led to the refinement of human soul but influenced also the development of domestic animals, such as dogs or horses. 'In constant coexistence with man and under the influence of his education, higher hereditary associations of ideas and a more perfect judgement have also gradually developed in their brain' (Haeckel, 1905, p. 11). In this context, Haeckel refers to the heredity of acquired characteristics. He believes that the 'differentiation and perfection of the brain'—and thus of spiritual life as 'the highest cerebral function' (Haeckel, 1889, p. 718)—drove the evolutionary transition from animal ancestors to humans, with upright posture and the development of speech being crucial moments on this evolutionary path.

Haeckel opposes the idea that only humans are endowed with language and argues that organisms acquire language through historical development. 'Ontogeny of speech has shown that the gradual development of speech in the child

[12] 'Die bewunderungswürdige Höhe der Kultur, welche der Mensch im 19. Jahrhundert erklommen hat, die erstaunlichen Fortschritte der Naturwissenschaft und ihrer praktischen Anwendung in Technik, Industrie, Medizin usw. gestatten die Hoffnung auf einen mächtigen weiteren Aufschwung der Kultur im zwanzigsten Jahrhundert. Dieser erwünschte Fortschritt wird aber nur dann möglich sein, wenn die gewohnten Bahnen der traditionellen Dogmen und des klerikalen Aberglaubens verlassen werden und die vernünftige monistische Naturerkenntnis statt ihrer die Herrschaft gewinnt.'

(corresponding to the basic biogenetic law) is a recapitulation of the phylogenetic process' (Haeckel, 1904a, p. 541). Haeckel also considers the origin of language and distinguishes between its monophyletic and polyphyletic origins. Higher linguistic forms derive from a single root language and are thus monophyletic; these include Germanic, Romance, Slavic, Celtic, Greek, Indian, that is, the Indo-European languages, but also Semitic and Mongolian languages. Haeckel refers to the languages of the 'lower' and 'imperfect language classes' as polyphyletic and strangely associates them with 'lower' human races (Haeckel, 1879, in Breitenbach, 1914, pp. 160–161) (cf. Chaps. 2, 5 and 12). In the same way, lower organisms have a polyphyletic origin while higher ones have a monophyletic one. These considerations are closely related to Haeckel's understanding of the multi-regional origin of individual human races, which he presented mainly in his *Anthropogenie oder Entwicklungsgeschichte des Menschen* [Anthropogeny or history of human development, 1874].[13]

For Haeckel, the humankind is a child of nature. Its history is therefore also part of the history of nature. The same eternal laws therefore

> …determine the historical development of the universe and the earth, the plant world and the animal world, human individuals and human society, their science and art, their morality and religion. Even the highest and most admirable achievements of the human spirit are not of supernatural origin; they are natural products of the cells that compose its phronema, the 'thinking organ' of its cerebral cortex. (Haeckel, 1910, in Breitenbach, 1914, p. 129)[14]

When arguing for a monistic culture, Haeckel stresses that the phylogenetic observations concern not only the body but also the soul, so that just as the brain developed in the course of evolution, so too did mental life as a function of brain cells (Haeckel, 1911, in Breitenbach, 1914, p. 113). This monistic interdependence enabled Haeckel to develop a natural explanation of the origin and development of many cultural phenomena. The most sublime performance of mental life (that is, reason and consciousness) is clearly in evidence in humans and higher animals but an interplay between nature and culture can be described even at a cellular level.

In his lecture *Zellseelen und Seelenzellen* [The souls of cells and the cells of souls], delivered in Vienna in 1878, Haeckel shifted his attention from human mental life to the inner life of cells. His starting position was based on the claim 'that each individual cell has a certain degree of individual independence, has its own shape and leads its own life' (Haeckel, 1909, p. 15). In this context, Haeckel speaks of a division of labour, as with ants or bees. In cells, it pertains to the need to adapt

---

[13] Haeckel's work in the field of racial theory development is also an important part of the overlap between knowledge of evolution and culture. This aspect of Haeckel's work has been intensively studied. For further details, cf. critical works on Haeckel's anthropology and social science: e.g. Hoßfeld (2005), or Preuß et al. (2006).

[14] '… bedingen die historische Entwicklung des Weltalls und der Erde, der Pflanzenwelt und Tierwelt, der menschlichen Individuen und der menschlichen Gesellschaft, ihrer Wissenschaft und Kunst, ihrer Moral und Religion. Auch die höchsten und bewunderungswürdigsten Leistungen des Menschengeistes sind nicht übernatürlichen Ursprungs; sie sind natürliche Produkte der Zellen, welche sein Phronema, das "Denkorgan"seiner Grosshirnrinde zusammensetzen.'

to changing situations and endure the struggle for life. Cells behave like 'the well-educated citizens of a well-established cultural state' (Haeckel, 1909, p. 20) (cf. Chap. 14). Haeckel's subsequent reflections on the hierarchy of the human body or the body of higher animals forming a 'state of cells' are also remarkable:

> The so-called 'tissues' of the body, muscle tissue, nerve tissue, glandular tissue, bone tissue, connective tissue, etc. correspond to the various states or departments of the state, or even more precisely to the hereditary castes as we find them in ancient Egypt and still today in India. The tissues are hereditary cellular castes in the cultural state of the multicellular organism. The organs, however, which are again composed of different tissues, can be compared to the various offices and institutes. At the head of all is the powerful central government, the nervous system, the brain. (Haeckel, 1909, p. 20)[15]

Within the framework of practical monism, Haeckel placed special emphasis on sociology (Haeckel, 1904c, in Breitenbach, 1914, pp. 45–50), noting that culture is most associated with humans. It is culture that 'has given [people] the rule over nature'. Human interaction, on the other hand, is based on a division of labour and mutual complementarity and support. We can thus see that human sociology is closely related to the social formations of evolutionarily close mammals and within this cultural interdependence, division of labour (which is also found in animals) is essential: 'The biological foundations of socialisation are already present in sociable animals (at first primates); their herds and states are held together by social instincts (inherited habits)' (Haeckel, 1904c, in Breitenbach, 1914, p. 45).

The monistic blending of the natural sciences and humanities, and thus of nature and culture, brings a freedom that can be projected onto the culture of the future: a free monistic culture. For Haeckel, this concept of freedom is associated especially with the idea of free science and free teaching (Haeckel, 1904b, in Breitenbach, 1914, p. 224). Haeckel and his monist contemporaries placed great emphasis on the need to reform education, hoping to create a solid foundation for a transfer of scientific knowledge into the realm of culture. At the turn of the nineteenth and twentieth centuries, scientific knowledge was not commonly taught in schools and played a relatively minor role (Weber, 2000, p. 4). Haeckel repeatedly argued for a monistic pedagogy, pointing out that 'the scientific adaptation of teaching to young people ranks among the important tasks of cultural humanity' (Haeckel, 1904a, p. 547). He was convinced that the reform of education should be based on a natural acceptance of evolutionary natural science: 'In the new school, nature must become the main object; man is to acquire a correct idea of the world in which he lives; he is not to stand apart from nature or to be opposed to it, but to appear as its highest and noblest creation.' (Haeckel, 1900, p. 418).

---

[15] 'Die sogenannten "Gewebe"des Körpers, Muskelgewebe, Nervengewebe, Drüsengewebe, Knochengewebe, Bindegewebe usw. entsprechen den verschiedenen Ständen oder Korporationen des Staates, oder noch genau den erblichen Kasten, wie wir sie im alten Ägypten und noch heute in Indien antreffen. Das Gewebe sind erbliche Zellenkasten im Kulturstaate des vielzelligen Organismus. Die Organe aber, die sich wieder aus verschiedenen Geweben zusammensetzen, sind den verschiedenen Ämtern und Instituten zu vergleichen. An der Spitze aller steht die mächtige Zentralregierung, das Nervensystem, das Gehirn.'

A reform of education was one of the important goals of the German Monist League, which was founded in 1906. Haeckel's aim was to bring together, especially on the platform of a religion, adherents of the monistic worldview and other evolutionary-minded people, be it free thinkers or others. For such audience, the Monist League published books, monist magazines, and informative pamphlets, organised lectures, monist art evenings, and celebrations of monist holidays. In the summer, it organised holiday courses where questions of theoretical and practical monism were discussed, and it set up monist libraries and reading groups (Weber, 2001, p. 125). On Sunday mornings, monistically minded people gathered for monist sermons.[16] As evidence of the tendency to spread monism as a religion, a monist catechism (Frei, 1908) was published 2 years after foundation of the monist organisation. Haeckel's worldview covered all areas of life, and he urged their reform. We have already mentioned monistic pedagogy, ethics (based on the three pillars of truth, good, and beauty) and monistic biology, but there also existed for instance monistic geology, chemistry, physics, mathematics, medicine, psychology, and psychiatry.

4.4.2 Gestalt Culture

The significant overlaps between the Gestalt theory and culture are closely related to Ehrenfels's tendency to view his theory as an explanatory principle applicable to many fields of research. The theory of Gestalt qualities had significantly inspired research in psychology, especially studies of the role of fantasy and its correlation with experience, where it was argued that one should not link fantasy to experience only in the domain of elementary ideas: imagination or fantasy can move much more freely and extensively in the domain of Gestalt qualities than in the domain of elementary ideas. Gestalten allow us to grasp and thus interpret many phenomena or events much better than for instance the law of association by resemblance, because the human spirit possesses a creative power that is not independent of experience but goes further. Ehrenfels's emphasis on gradual development is crucial: 'For here too, as everywhere in the cosmos, it turns out that the height of the Gestalt cannot be reached in a single throw, but can only be climbed by degrees' (Ehrenfels, 1922b, in Weinhandl, 1960, p. 52).

The Gestalt theory also plays an important role in the theory of knowledge, where one can distinguish two distinct kinds of knowledge: the first is achieved with the help of logic, while the second is based on intuition. According to Ehrenfels, intuitive cognition reveals a higher ordering or more encompassing Gestalt ideas, but only a small proportion of people have this spiritual ability. Importantly, even

[16] From 1911 onwards, the German Monist League was led by the German chemist and philosopher Wilhelm Ostwald, who even sought to establish a Monist monastery. In contrast to Haeckel's substance monism, Ostwald promoted 'energy monism', which coined the imperative 'Don't waste energy: use it!' (Weber, 2001, p. 127).

this mode of cognition operates according to a logic Ehrenfels, 1922b, in Weinhandl, 1960, p. 52). Most significant, however, is according to Ehrenfels the overlap of Gestalt theory with cosmogony. Ehrenfels distinguishes three important phases of cosmic development: the beginning of the world, the emergence of life and its psychic dimension, and finally the emergence of purposive volition in humans (Ehrenfels, 1916, p. 197). This third level is closely intertwined with a significant development of culture: 'With the realm of human artefacts, Gestalt convergence seems to want to elevate itself to a cosmic power factor.' (Ehrenfels, 1916, p. 194).

In considering the origins and development of human culture,[17] Ehrenfels builds on and further develops what happens in the organic world. In this context, he speaks of an 'organic derivation' (Ehrenfels, 1916, pp. 9–16), in which one can trace an even higher degree of qualitative branching into the future, because living humans produce not only living offspring but also material and immaterial cultural objects. This derivation from the organic world plays a crucial role in transferring the Gestalt aspect to cultural artefacts. When organisms create artefacts, a similar law of origin applies as in the organic world. On the other hand, Ehrenfels admits that in the human world, artefacts often take on a life of their own (cf. Chap. 12): he cites the example of the gramophone, which spread from America through the world of 'cultivated humanity' at an incredible speed. Such spread does not require any physical particles; it is based on a 'communication of the principles of construction through words and drawings' (Ehrenfels, 1916, p. 191). Human artefacts also include intangible goods, such as scientific knowledge, the art of the carpenter or tailor, various 'mythical and artistic fantasy images', but also social institutions, private property, or monogamous marriage (Ehrenfels, 1916, pp. 191–192). The dissemination or evolution of these Gestalt qualities follows the same dualistic principle that defines chaos and the unifying shaping force:

> The tendencies of reproduction and of qualitative and quantitative branching into the future [...] apply to incorporeal organic derivatives in the same way as to corporeal ones: namely – compared to living organisms – with a significant removal of the chaotic component. (Ehrenfels, 1916, p. 192).[18]

For Ehrenfels, not only organic individuals but also noncorporeal artefacts are subject to the law of dead ends. In the development of human cultural history, one can see 'a tendency towards shaping convergence, towards the shaping of uniform, higher shapes from multi-branched, lower origins' (Ehrenfels, 1916, p. 192). For instance, in the construction of a medieval cathedral, one can see how tens of thousands of artistic, scientific, technical, and social Gestalt impulses converged towards

[17] Gestalt cultural morphology, too, requires a comprehensive grasp of Gestalt qualities and we ought to discard the notion of working with just the sum of the parts. The problem with cultural morphology is that it is difficult to prove experimentally, unlike for instance Gestalt psychology, where particular claims can be experimentally demonstrated (cf. Anderle, 1960, p. 75)

[18] 'Von den unkörperlichen organischen Derivaten nun gelten die Tendenzen der Fortpflanzung und der qualitativen und quantitativen Verzweigung in die Zukunft [. . .] in analoger Weise wie von den körperlichen:—nämlich—verglichen mit den lebenden Organismen—mit einem starken Zurücktreten der chaotischen Komponente.'

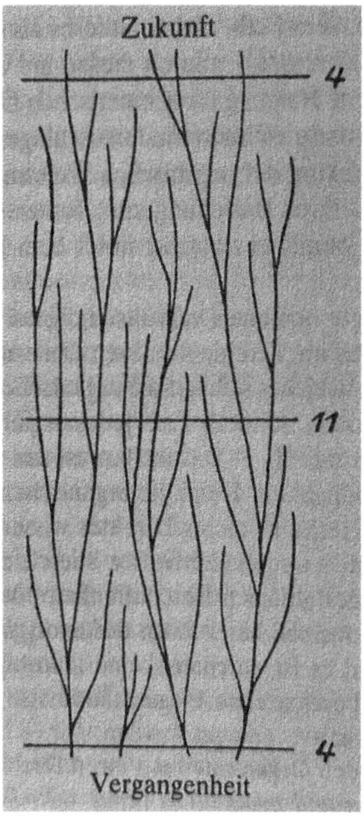

Fig. 4.5 Illustration from Ehrenfels' *Kosmogonie* (1916). A diagram of the quantitative and qualitative sequence of forms in the germplasm. In the lower half, the tendency to progressive branching leads to an increase in the number of branches. In the upper half, the tendency continues but due to increasing mortality (blind endings), the number of branches decreases. von Ehrenfels (1916, p. 12). *Kosmogonie*. Jena: Diederichs

a unified Gestalt. In nature, too, we find the convergence of common artefacts, for instance in beehives or bird nests, but it is in the artefacts of the human world that Gestalt convergence has a greatest power and significance (Ehrenfels, 1916, p. 193). With the participation of people, higher and therefore more valuable forms can be created (Fig. 4.5).

Intertwined with this process was Ehrenfels's tendency to anchor his extended Gestalt reasoning institutionally. In 1922, Ehrenfels (1922a) laid the foundations of his scientific religion and published the *Gedanken über die Religion der Zukunft* [Thoughts on the religion of the future], which was based on the theory of evolution. These were his long-standing convictions, as demonstrated by the fact that he outlined some features of his religion already in 1916, in his *Kosmogonie*, in a summary of principles of his dualistic worldview in the fourth part of chapter 'Die neue Dogmatik' [The new dogmatics]. According to these principles, humans with their creativity became a part of God's inner life and thus his collaborators. Only at this point can one even consider divine purposefulness: 'In and with man, God is looking for a leading idea that would be able to guide his hitherto instinctive shaping into the path of purposeful consciousness.' (Ehrenfels, 1916, p. 207). In doing so, humans also assume responsibility for the development of individual parts and the world as a whole.[19] Ultimately, though, Ehrenfels's reformist religion of natural

[19] The role of humans also played an important role in Ehrenfels's thoughts on a reform: under the influence of Darwinism, he developed reform concepts guided by the idea of 'regeneration of cultural humanity' (Ehrenfels, 1907). From 1902 onwards, he promoted his concept of sexual ethics in selected lectures, publications, and epic operas. He focused on criticising the 'cultural sexual

science on a dualistic basis, his *Real-Katholicismus* (real Catholicism), was not even founded. Ehrenfels presented his 'Religion of the Future' in 1929 in the form of an open letter to the then Czechoslovak President T. G. Masaryk, whom he knew from his studies in Vienna. Despite the friendly relationship between the two thinkers, the Czechoslovak president refused involvement in the new scientific religion, justifying his refusal not only by his old age but also by the need to focus mainly on political issues (cf. Ehrenfels, 1929, in Fabian, 1990).

4.4.3 Scientific Monism, Gestalt Qualities and Umwelt

Having compared some features of Haeckel's and Ehrenfel's theories of the origin and development of nature and culture, it may be instructive to relate them to another type of holistic approach, represented by Jakob von Uexküll (1864–1944). The wealth of ideas of this German biologist of Estonian origin, particularly those related to the theory of the Umwelt (cf. Brentari, 2011; Tønnessen et al., 2016), also enables us to relate Darwinism to holistic explanations of phenomena in nature and culture from a different perspective.

Like Haeckel and Ehrenfels, Uexküll was initially enthusiastic about Darwinism. Unlike the two thinkers, however, he soon became disappointed in it. His interest in zoology consequently shifted away from morphology and towards physiology, which was central to the formulation of his theory of Umwelt. According to Uexküll, every organism has its own Umwelt consisting of sensory perceptions, motor skills, and subjective experiences. An Umwelt cannot be understood in isolation: it can only be grasped as a part of a network of countless Umwelten. In this context, Uexküll emphasises that the human Umwelt does not have a privileged position over other Umwelten—it is just one among many (cf. Ovčáčková & Švorcová, 2024, p. 347).

Uexküll argues (2010, p. 151) that the form of plants and animals is determined by their ability to evaluate meaning factors coming from outside. A meaningful form (*bedeutungsvolle Form*) is thus always linked to the subject, not to the object. The key challenge, though, is to explain the origin and development of life from such a holistic point of view. As noted above, for Haeckel a Darwinian

morality', which—following the sublimation effect of sexual power—leads to a decline of the constitution of the Western society and thus also its degeneration. Ehrenfels called for a natural sexual morality whose adoption would lead to a re-invigoration of a constitutively degenerate society. A prerequisite of such improvement was a departure from the monogamous social structure. In general, Ehrenfels operated within a Darwinian framework and used the findings of the race theory, but his eugenic visions were characterised by criticism of the Aryan race and stress on the need to allow the mixing of races, especially Jews and Germans (cf. Ehrenfels, 1904). After the First World War, Slavs also played an important positive role in this context. It is also interesting to note that Ehrenfels corresponded with Sigmund Freud on subjects concerning sexual morality and even held a lecture on this subject in 1908 in the Vienna Psychoanalytic Society (for further detail, see Fabian, 1986, Rug & Mulligan, 1986, or Valent, 2016).

understanding of chance is essential. Ehrenfels rejects the principle of pure change and searches for a new explanatory principle of shaping in nature, the *missing link*, while trying to link it to a unified principle of shaping (cf. Ovčáčková & Švorcová, 2024, p. 353). Ehrenfels thus remains fundamentally faithful to Darwinism. He is moreover highly critical of Haeckel's monistic cosmogony, finding it incompatible with his dualistic position (cf. Ehrenfels, 1916, pp. 36–46). Uexküll's position is also incompatible with Haeckel's monism but, in this case, it is due to Uexküll's vitalistic worldview, which includes the assumption of a teleological principle and thus a plan underlying all developments in living nature. This basic setting is reflected also in Uexküll's terminology, in particular expressions such as *Bauplan, Entstehungsplan,* or *Planmäßigkeit* (Ovčáčková & Švorcová, 2024, p. 352).

Ehrenfels's, Haeckel's, and Uexküll's arguments are based on different worldviews, which are associated with different basic approaches to explaining living nature. As noted above, Ehrenfels's theory of Gestalt qualities is rooted in a cosmogonic dualistic dynamic principle, while the origin of nature is anchored in a synthesis of random chaotic diversity with formative unity. Haeckel's worldview, on the other hand, is strictly monistic and he rejects all forms of teleology, metaphysics, or vitalism. Uexküll's position, meanwhile, rests on a *purposive* understanding of the development of living beings. A comparison of Haeckel's evolutionary development, Ehrenfels's 'height of Gestalt', and Uexküll's 'height of Umwelt' nevertheless also shows that they were all keenly aware of the importance of development and increase in complexity that implies both diversity and unity (Ovčáčková & Švorcová, 2024, p. 349). In fact, this relationship between ontogeny and evolution is a cornerstone of biosemiotic research to this day (cf. Giorgi & Bruni, 2016), and this applies not only to Uexküll's biosemiotic interpretation but, more broadly, also to Haeckel's and Ehrenfels's reflections on the origin and development of living beings.

In their thoughts on the origin and development of nature and culture, these three thinkers also offer a basis for comparing their theories to environmental ethical narratives. What all these three approaches share is a strong emphasis on addressing the origins and evolution of natural entities and nature as a whole. On this foundation, they then develop cultural/scientific concepts that call for a holistic approach to nature. Haeckel's formulation and monistic elaboration of the concept of ecology (Haeckel, 1866), Uexküll's Umwelt theory (1910), and reverberations of Ehrenfels's Gestalt theory (Ehrenfels, 1890; Naess, 1989) continue to form the foundations of various approaches to environmental ethics, including biosemiotics (see, e.g. Wheeler, 2006). Despite the differences in the positions of these thinkers, the terms ecology and Umwelt (as in ecological vs. environmental ethics) are in contemporary environmental ethical discourse often used synonymously—and it does not diminish their conceptual significance (for contemporary biosemiotic ecological insights, see e.g. Wheeler, 2006).

4.5 Conclusions

In contemporary discussions of the theory of evolution, both Haeckel and Ehrenfels assumed a self-evident and close interconnection between nature and culture. From a monistic and dualistic cosmogonic perspective, respectively, they documented the evolution of not only individual natural entities but also natural wholes and related cultural manifestations. They both assigned the central position in nature and culture to humans. Haeckel was critical of the dualistically found elevation of humans above nature; his sufficiently modern cultured person, educated in natural sciences and thus monistically minded, occupied the top position thanks to the ability of rationally understanding reality. Ehrenfels, on the other hand, promoted the notion of a dynamic dualistic formation of the world. Although he speaks of a purposiveness (*Zweckbewusstsein*) linked to human activity, the human position lacks a teleological direction and alongside culture and everything else subject to the cosmogonic principle of a never-ending struggle between chaos and the unifying shaping force. Moreover, God does not act according to a plan but in response to stimuli and resistance coming from chaos (Ehrenfels, 1930, in Fabian, 1990, p. 300). Therefore, as with Haeckel, humans are inherently a part of nature. Both Haeckel and Ehrenfels were deeply affected and inspired by a combination of Darwinism and Goethe's *Naturphilosophie* and they both believed with Goethe that life was not created to follow a purpose.

Haeckel's direct response to Ehrenfels's dynamic dualism, which was closely linked to Darwinism, is not known. Ehrenfels, on the other hand, clearly felt the need to react to the monistic worldview, represented primarily by Haeckel, which he perceived as incompatible with his approach. While aware of the intellectual attractiveness of the monistic worldview, Ehrenfels believed it was built on shaky foundations. He was convinced that the origin and development of the world (and thus also the development of life) cannot be explained by a single principle, neither 'from the one primeval basis of all regularity, order and Gestalt', not 'from absolute chaos' (Ehrenfels, 1916, p. 36). Ehrenfels was convinced that some form of disguised dualism is inherent in all forms of monism. It has been interesting to explore, albeit briefly, various connections between Haeckel's view of monistic unity, Ehrenfels's theory of Gestalt qualities, and Uexküll's theory of Umwelt, which has also significantly contributed to the development of a holistic understanding on the origin and development of nature and culture.

Haeckel's monistic scientific religion had fallen on a more fertile ground than Ehrenfels's dynamic dualism. Haeckel's popularity went beyond the borders of the German-speaking countries and spread worldwide. Ehrenfels was immensely popular but relatively little-known outside Prague. At the university, his lectures were packed with students and observers but when it came to promoting his worldview and reformist positions, he was relatively isolated. As a result, few people are nowadays aware of the links between Ehrenfels's Gestalt qualities and the dynamic interplay between chaos and the unifying shaping power. In that respect, his work had a smaller impact than Haeckel's monism. Still, both Haeckel's and Ehrenfels's approach illustrate the specific tendencies of an era that sought to reform both humanity and society and looked for inspiration both to recent discoveries in natural sciences and to various interpretations of Darwin's theory.

References

Anderle, O. F. (1960). Christian von Ehrenfels und das Problem einer wissenschaftlichen Kulturmorphologie. In: F. Weinhandl (Ed.), *Gestalthaftes Sehen. Ergebnisse und Aufgaben der Morphologie* (pp. 65–78). Wissenschaftliche Buchgesellschaft.

Bölsche, W. (1900). *Ernst Haeckel*. Verlag von Carl Reißner.

Breidbach, O. (1998). Kurze Anleitung zum Bildgebrauch. In E. Haeckel (Ed.), *Kunstformen der Natur* (pp. 9–18). Prestel.

Breidbach, O. (2006). *Ernst Haeckel. Bildwelten der Natur*. Prestel.

Brentari, C. (2011). *Jakob von Uexküll. The discovery of the umwelt between biosemiotics and theoretical biology*. Springer.

Darwin, C. (1859). *On the origin of species by means of natural selection*. John Murray.

Di Gregorio, M. A. (2005). *From here to eternity. Ernst Haeckel and scientific faith*. Vandenhoeck & Ruprecht.

Ehrenfels, C. (1890). Über Gestaltqualitäten. *Vierteljahrsschrift für wissenschaftliche Philosophie, 14*, 249–292.

Ehrenfels, C. (1904). Über den Einfluß des Darwinismus auf die moderne Soziologie. In R. Fabian (Ed.), *(1988). Christian von Ehrenfels, Philosophische Schriften 3: Psychologie, Ethik, Erkenntnistheorie* (pp. 251–263). Philosophia Verlag.

Ehrenfels, C. (1907). *Sexualethik*. J. F. Bergmann.

Ehrenfels, C. (1916). *Kosmogonie*. Eugen Diederichs.

Ehrenfels, C. (1922a). Gedanken über die Religion der Zukunft. In R. Fabian (Ed.), *(1990), Christian von Ehrenfels, Philosophische Schriften 4: Metaphysik* (pp. 231–280). Philosophia Verlag.

Ehrenfels, C. (1922b). Weiterführende Bemerkungen. In F. Weinhandl (Ed.), *(1960). Gestalthaftes Sehen. Ergebnisse und Aufgaben der Morphologie* (pp. 47–60). Wissenschaftliche Buchgesellschaft.

Ehrenfels, C. (1929). Die Religion der Zukunft. In R. Fabian (Ed.), *(1990), Christian von Ehrenfels, Philosophische Schriften 4: Metaphysik* (pp. 281–288). Philosophia Verlag.

Ehrenfels, C. (1930). Das dualistische Glaubenbekenntnis. In R. Fabian (Ed.), *(1990), Christian von Ehrenfels, Philosophische Schriften 4: Metaphysik* (pp. 296–309). Philosophia Verlag.

Ehrenfels, C. (1932). Über Gestaltqualitäten. In F. Weinhandl (Ed.), *(1960). Gestalthaftes Sehen. Ergebnisse und Aufgaben der Morphologie* (pp. 61–63). Wissenschaftliche Buchgesellschaft.

Fabian, R. (Ed.). (1986). *Christian von Ehrenfels. Leben und Werk*. Rodopi.

Frei, L. (1908). Katechismus der monistischen Weltanschauung. Frei nach E. Haeckels´ Lehre. Verlag Neue Weltanschauung.

Giorgi, F., & Bruni, L. E. (2016). Germ cells are made semiotically competent during evolution. *Biosemiotics, 9*, 31–49.

Haeckel, E. (1862). *Die Radiolarien (Rhizopodia radiaria). Eine Monographie*. Georg Reimer.

Haeckel, E. (1866). *Generelle Morphologie der Organismen*. Georg Reimer.

Haeckel, E. (1868). *Natürliche Schöpfungsgeschichte*. Georg Reimer.

Haeckel, E. (1878). Zellseelen und Seelenzellen. In E. Haeckel (Ed.), *Gesammelte populäre Vorträge aus dem Gebiet der Entwicklungslehre* (pp. 143–181). Emil Strauss.

Haeckel, E. (1879). Einstämmiger und vielstämmiger Ursprung. In W. Breitenbach, *(1914), Monistische Bausteine II.* (pp. 125–162). Dr. W. Breitenbach.

Haeckel, E. (1889). *Natürliche Schöpfungsgeschichte II*. Verlag von Georg Reimer.

Haeckel, E. (1891). *Anthropogenie oder Entwicklungsgeschichte des Menschen*. Wilhelm Engelmann.
Haeckel, E. (1900). *Die Welträtsel. Gemeinverständliche Studien über monistische Philosophie*. Emil Strauß.
Haeckel, E. (1904a). *Die Lebenswunder. Gemeinverständliche Studien über biologische Philosophie*. Alfred Kröner.
Haeckel, E. (1904b). Ueber die Biologie in Jena während des 19. Jahrhunderts. In W. Breitenbach (Ed.), *(1914) Monistische Bausteine II* (pp. 201–227). Dr. W. Breitenbach.
Haeckel, E. (1904c). Thesen zur Organisation des Monismus. In W. Breitenbach (Ed.), *(1914) Monistische Bausteine I* (pp. 35–50). Dr. W. Breitenbach.
Haeckel, E. (1904d). *Kunstformen der Natur*. Verlag des Bibliographisches Instituts.
Haeckel, E. (1904e). Der Monistenbund. Thesen zur Organisation des Monismus. *Das freie Wort, 13*, 481–489.
Haeckel, E. (1905). *Der Monismus als Band zwischen Religion und Wissenschaft*. Alfred Kröner.
Haeckel, E. (1906a). *Prinzipien der Generellen Morphologie der Organismen. Wörtlicher Abdruck eines Teiles der 1866 erschienenen Generellen Morphologie*. Georg Reimer.
Haeckel, E. (1906b). Monismus und Naturgesetz. In W. Breitenbach (Ed.), *(1914), Monistische Bausteine I* (pp. 51–105). Dr. W. Breitenbach.
Haeckel, E. (1909). *Zellseelen und Seelenzellen*. Alfred Kröner.
Haeckel, E. (1910). Die Grenzen der Naturwissenschaft. In W. Breitenbach (Ed.), *(1914) Monistische Bausteine I* (pp. 123–138). Dr. W. Breitenbach.
Haeckel, E. (1911). Die Fundamente des Monismus. In W. Breitenbach, *(1914), Monistische Bausteine I.* (pp. 111–114). Dr. W. Breitenbach.
Haeckel, E. (1912a). Energetik und Substanzgesetz. In W. Breitenbach (Ed.), *(1914), Monistische Bausteine I* (pp. 147–156). Dr. W. Breitenbach.
Haeckel, E. (1912b). Zur Geschichte der Entwicklungslehre. In W. Breitenbach (Ed.), *(1914), Monistische Bausteine I* (pp. 159–176). Dr. W. Breitenbach.
Haeckel, E. (1914). *Gott-Natur (Theophysis). Studien über monistische Religion*. Alfred Kröner.
Haeckel, E. (1917). *Kristallseelen. Studien über das anorganische Leben*. Alfred Kröner.
Hoßfeld, U. (2005). *Geschichte der biologischen Anthropologie in Deutschland: Von den Anfängen bis in die Nachkriegszeit*. Franz Steiner.
Höfer, U., Valent, J. (Eds.). (2017). *Christian von Ehrenfels: Philosophie – Gestalttheorie – Kunst. Österreichische Ideengeschichte im Fin de Siècle*. De Gruyter.
Krauße, E. (1987). *Ernst Haeckel*. Teubner Verlagsgesellschaft.
May, W. (1909). *Ernst Haeckel. Versuch einer Chronik seines Lebens und Wirkens*. Johann Ambrosius Barth.
Naess, A. (1989). *Ecology, community and lifestyle*. Cambridge University Press.
Ordnung der Vorlesungen an der Deutschen Universität in Prag. Deutsche Universität Prag (1920–1932).
Ovčáčková, L. (2017). Od evolucionismu k monismu. Haeckelův evoluční příběh. *[From evolutionism to monism. Haeckel's evolutionary story]*. In L. Ovčáčková (Ed.), *O původu kultury. Biologické, antropologické a historické koncepce kulturní evoluce* (pp. 87–104). Academia.
Ovčáčková, L. (2022). Přírodní filozofie Christiana von Ehrenfelse. O gestaltové teorii a darwinismu. *[Philosophy of Nature of Christian von Ehrenfels. About Gestalt Theory and Darwinism]. Živa, 4*, 43–46.
Ovčáčková, L., & Švorcová, J. (2024). Metaphysics of the organic whole: Ehrenfels, Uexküll, and Merleau-Ponty. *Biosemiotics, 17*(2), 341–360.
Preuß, D., Hoßfeld, U., & Breidbach, O. (Eds.). (2006). *Anthroplogie nach Haeckel*. Franz Steiner.
Rug, R., & Mulligan, K. (1986). Theorie und Trieb – Bemerkungen zu Ehrenfels. In R. Fabian (Ed.), *Christian von Ehrenfels. Leben und Werk* (pp. 214–246). Rodopi.
Tønnessen, M., Magnus, R., & Brentari, C. (2016). The biosemiotic glossary project: Umwelt. *Biosemiotics, 9*(1), 129–149.
Uexküll, J. (1910). Die Umwelt. *Die Neue Rundschau, 5*, 638–648.

Uexküll, J. (2010). The theory of meaning. In J. Uexküll (Ed.), *A foray into the worlds of animals and humans; with a theory of meaning*. The University of Minnesota.

Uschmann, G. (1961). *Ernst Haeckel. Forscher, Künstler, Mensch*. Urania-Verlag.

Valent, J. (2016). Christian von Ehrenfels. Eine Intellektuelle Biographie: Neue Forschungsergebnisse Aus Dem Nachlass. In M. David & M. Antonelli (Eds.), *Existence, fiction, assumption: Meinongian themes and the history of Austrian philosophy* (pp. 175–206). De Gruyter.

Weber, H. (2000). *Monistische und antimonistische Weltanschauung*. Verlag für Wissenschaft und Bildung.

Weber, H. (2001). Der Deutsche Monistenbund. In K. Buchholz, R. Latocha, H. Peckmann, & K. Wolbert (Eds.), *Die Lebensreform. Entwürfe zur Neugestaltung von Leben und Kunst um 1900* (Vol. 2, pp. 125–127). Häusser.

Weinhandl, F. (Ed.). (1960). *Gestalthaftes Sehen. Ergebnisse und Aufgaben der Morphologie*. Wissenschaftliche Buchgesellschaft.

Wheeler, W. (2006). *The whole creature: Complexity, biosemiotics and the evolution of culture*. Lawrence & Wishart.

Part II
Cultural Evolution and the Human Sciences: Archaeological and Anthropological Perspectives

Radek Kundt

Two observations spring to mind when considering the rise of the concept of cultural evolution in the nineteenth century. First, in the closing decades of the nineteenth century this idea had taken the intellectual world by storm. Second, the concept of cultural evolution gave rise to two new human sciences: archaeology and anthropology. These observations are each correct and accurate in its own right, but when explored separately, one might lose sight of their interaction. Moreover, as our minds play tricks on us, simply placing one idea before the other could lead to incorrect assumptions about causality (cf. Chap. 2). *The first point* of this introduction is therefore a reminder that one ought to keep in mind their mutualistic relationship.

Before Darwin entered the scene with *On the Origin of Species* (1859), Herbert Spencer had captivated the minds of Victorians with the notion of universal evolution. In his view, a single law of progressive development (which he subsequently called the principle of evolution) permeates everything in the known and unknown universe, from the development of individual minds (1855) all the way to societal changes (1851). In the eyes of the broader public, Darwin thus introduced scientific rigor to one particular extension of this guiding principle, one specific domain of this universe – biology. When ideas of cultural evolution started to appear, they found a fertile ground because they fit well with the prevalent progressivist sentiments of the colonial era (cf. Chap. 14). Nevertheless, this relationship was bidirectional: in a feedback loop, as the concept of cultural evolution started to form in the work of early archaeologists and anthropologists, they started to enrich and reinforce various previously merely latently present ideas and sentiments in the society. All four chapters of this section show and discuss this bidirectional dimension of the relationship.

R. Kundt
LEVYNA Laboratory for the Experimental Research of Religion, Masaryk University, Brno, Czech Republic

The *second point* of this introduction is that the concept cultural evolution is not confined to its origins in the nineteenth century. After a period of dismissal (cf. Chap. 3), the concept of cultural evolution gradually in the course of the twentieth century distanced itself from the ethnocentric and supremacist ideologies it had inspired, and that allowed for several solid and valuable revivals. It has become a vibrant and successful field supported by a highly interdisciplinary scientific society, the *Cultural Evolution Society*, which makes significant progress in both disciplines. Historical developments, as well as modern advances, are mirrored in both the content of each chapter of this section and in its overall structure.

This section is broadly divided in two parts: an archaeological perspective and an anthropological perspective. Each perspective is represented by two texts. The first text provides an overview of how the concept of cultural evolution developed over time within the respective discipline, while also addressing key issues concerning its current use, while the second text presents an original in-depth case study.

The first chapter discusses the cultural identity of past societies, a subject that has traditionally been addressed through the concept of archaeological cultures. This subject is problematic because even after a century of investigations, the paradigm of archaeological cultures still cannot persuasively show how, if at all, the material cultures which are central to its argument correspond to any social entities. Nevertheless, this concept has over the past decade re-emerged in connection with the rise of *archaeogenetics* – and once again, although the state-of-the-art techniques for ancient DNA study allow us to identify biological kinship between individuals and track the genetic lineage of historical populations, they have not yet come up with new theories and instead reuse the outdated concept of archaeological cultures. This chapter evaluates current initiatives aimed at addressing this deficiency through the development of a unifying theory.

The second chapter illustrates certain problems related to archaeological cultures using the complicated story of interpretation of the Greek Bronze Age through the prism of cultural evolution as it developed from the late nineteenth century until the present. Our understanding of this period had shifted from relatively simplistic evaluative conclusions – such as a vision of the cultured Minoans versus the barbaric Mycenaeans, which was shaped largely by the conceptual frameworks of early cultural evolutionists – towards more nuanced, less definitive, but more accurate interpretations. This shift has been driven by the growing involvement of archaeological sciences, including paleopathology, osteoarchaeology, and ancient DNA analysis.

The third chapter, which is the first text presenting the anthropological perspective, follows the development of thinking about cultural evolution in anthropology. It shows that although this line of thought had inspired anthropology since its beginnings, the early Victorian anthropologists often applied evolutionary principles to cultural change rather crudely. Since that time, the evolutionary study of cultural and social phenomena has experienced several waves of popularity, eventually giving rise to fields such as sociobiology, memetics, evolutionary psychology, human behavioural ecology, and cultural evolution. By examining topics such as religion and ritual behaviour, this chapter illustrates how the interdisciplinary nature of the cognitive and evolutionary sciences of culture can shed new light on these complex subjects.

The fourth chapter is a case study that presents research on how cultural technologies, including religion, can help solve certain biological problems of our species. The authors explore the reproductive roles of religious systems in various socio-ecological contexts and demonstrate the effects of religious practices on mating strategies. To this end, the chapter examines two concrete cases: first, it investigates how religious systems influence the trade-off between offspring quality and quantity by promoting religious alloparenting, and secondly, it analyses how religious systems encourage greater paternal investment by regulating female sexuality.

References

Darwin, C. (1859). *On the origin of species by means of natural selection, or the preservation of favoured races in the struggle for life*. John Murray.

Spencer, H. (1851). *Social statics: Or the conditions essential to happiness specified, and the first of them developed*. John Chapman.

Spencer, H. (1855). *The principles of psychology*. Longman, Brown, Green & Longman.

Chapter 5
Genes, Identity, and Material Culture: In Search of a Unifying Theory

Petr Květina and Václav Hrnčíř

Abstract This chapter discusses the cultural identity of past societies, an issue that has traditionally been addressed through the concept of archaeological cultures - discrete evolutionary units defined by the spatiotemporal distribution of sets of characteristic material remains. Although the paradigm of archaeological cultures creates easily understandable historical narratives, after over a century of research it still fails to convincingly demonstrate how, if at all, such material cultures correlate with any social entities. The failure of even ethnoarchaeological studies of contemporary societies to clearly demonstrate such a link casts further doubt on the usefulness of applying this idea to past societies. The concept of archaeological culture has therefore been gradually deconstructed and rejected, especially in the British and Scandinavian archaeological communities. Nevertheless, the last decade has witnessed its return in connection with the progress of archaeogenetics. While groundbreaking methods of ancient DNA analysis enable us to trace the genetic ancestry of past populations and uncover biological kinship between individuals, they do not in themselves provide any new theories and tend to fall back on the outdated concept of archaeological cultures. This chapter concludes with an assessment of present efforts to remedy this shortcoming by formulating a unifying theory that might link the interpretation of biological, social, and material data.

Keywords Archaeological culture · Cultural identity and cohesion · aDNA · Ethnicity · Material culture

P. Květina (✉)
Institute of Archaeology of the Czech Academy of Sciences, Prague, Czech Republic
e-mail: kvetina@arup.cas.cz

V. Hrnčíř
Institute of Archaeology of the Czech Academy of Sciences, Prague, Czech Republic

Department of Linguistic and Cultural Evolution, Max Planck Institute for Evolutionary Anthropology, Leipzig, Germany
e-mail: vaclav_hrncir@eva.mpg.de

J. Toman et al. (eds.), *Culture Evolving*, Biosemiotics 10,
https://doi.org/10.1007/978-3-032-16124-6_5

5.1 Introduction

In the study of the past, there is a large difference between studying societies that left written records and are thus categorically 'historical' and those that left no written sources and are therefore considered 'non-historical' or 'prehistorical'. The term 'prehistorical' is somewhat misleading because it evokes an ancient past (Corfield, 2018) although in fact, it refers to a particular characteristic of a culture, not to a time period. In other words, prehistory may have taken place in distant past as well as recently. For instance, in Egypt, prehistory ended around the thirty-second century BC with the emergence of hieroglyphic writing, while in the Americas, prehistory ended much more recently, with Christopher Columbus's voyage of 1492 AD. In Europe, the 'historical' societies of ancient Greece and Rome in the south of the continent coexisted for several centuries with 'prehistoric' cultures in the north and east, where the first written records did not appear until Christianisation after 1000 AD. Historical societies, unlike prehistoric ones, create written records that allow us to place the information they contain into an intuitive narrative context. This is because written records are based on events and such events help us create historical narratives composed of stories that include names, dates, and places known from written records. Thanks to its narrative nature, written history can be quite complex and easily understood. The past of prehistoric societies is, in contrast, much more difficult to study because without knowledge of events, we cannot create stories (see Chap. 18). Material evidence, which is the only source of information available to us for such societies, survives primarily on its own, with no direct relation to its creators, users, or bearers. Isolated and fragmentary archaeological findings are thus imperfect proxies for past societies and our situation is not so different from studying for instance animal cultures (see Chap. 9).

While written history presents the past as dynamic and turbulent, the testimony of artefacts shows it as static, cyclic, or evolving but slowly. In this, ancient prehistoric societies resemble recent aliterate societies studied by anthropology and ethnology. Claude Lévi-Strauss allegorically called such societies 'cold' in the sense that they are like mechanical machines, such as clocks. Cold societies seem to resist change and try to continue operating in the same energy-conserving patterns as long as possible. In contrast, 'hot' societies seem to be constantly changing and have a clearly visible history (Charbonnier, 1969, p. 33). While historical societies appear to us—thanks to the written records—as 'hot', prehistoric ones make the impression of being 'cold'. In reality, though, the lives of people in both kinds of societies moved at the same speed: it is just that we cannot see this equivalence.

Archaeology, methodologically born in the first half of the nineteenth century, has throughout its existence attempted to meet the challenge of meaningfully structuring prehistoric findings. With some degree of generalisation, one could say that theoretical archaeology moves in only two main directions. According to the first, archaeology is part of historical sciences and its goal is to find a way of telling the story of prehistory in a way comparable with written past, that is, in the form of stories and grand narratives (Kristiansen, 1998). The second direction considers similarities between archaeologically studied societies that no longer exist and living or recent non-literate societies studied by anthropology. The two sciences, archaeology and anthropology, then share their epistemological approach based on structures or long-term constellations of cultural patterns. Understood in this way, archaeology becomes a specific subdiscipline of anthropology, one that studies the development of culture and societies over long periods of time (Barfield, 2008).

The main point of friction between the two paradigms is the issue of reflection of social identity and cohesion (e.g., ethnicity) in archaeological remains, which can be expressed using the concept of 'archaeological culture'. Since its first formulation in the early twentieth century, this concept had long seemed to be the ideal agent of the first of the abovementioned directions (the culture-historical approach to archaeology), because archaeological cultures have been thought of as spatiotemporal agents of grand prehistoric narratives. Like political entities or nations of the written history, archaeological cultures were encountering or annihilating one another, mixing and migrating. Since the 1960s, however, this concept of archaeological culture has come under increasing criticism, as the internal diversity of apparently uniform cultural groups of the past has been increasingly recognised. Since the turn of the millennium, though, the grand prehistoric narratives seem to be making a comeback, largely due to a new variable in the archaeological matrix. Alongside material evidence, which—as noted above—serves as a proxy for members of vanished societies of the past, data from archaeogenetics also started to contribute to our knowledge of the past (Kristiansen, 2014). The data produced by these new methods led to the creation of a completely new category of identities of prehistoric populations based on biological relatedness. The spatiotemporal spread of individual haplotypes and mixing of genetically distinct populations enabled the formulation of new stories of human migrations, but we should keep in mind that genes are just one of the many layers of human social world, alongside for instance material culture and social identity (Furholt, 2021).

The aim of this chapter is to review the more than a century-long history of research into the cultural identity of past societies. Beginning with the emergence of the concept of archaeological culture in the nineteenth century, through the post--World War II polythetic models and the development of ethnoarchaeology, and all the way to the bioarchaeological revolution of the early twenty-first century. We need all this to understand the current initiatives which aim at developing a unifying theory that would link the interpretation of biological, social, and material data.

5.2 Prehistoric 'Nations'

Prehistoric objects have long fascinated people by their clear difference from things of the present. Archaeological findings thus featured in various collections, but they were treated as mere curiosities. In fact, until the Age of Enlightenment no one gave much thought to the possibility that such artefacts might be evidence of earlier, and significantly different, human societies (see Chap. 2). Moreover, except for items from classical antiquity (see Chap. 6), no one was able to date these prehistoric finds. By this, we do not mean any absolute dating in terms of years, centuries, or millennia, which radiocarbon dating made possible in the second half of the twentieth century. We mean a relative dating that would show which artefacts are older and which are younger. Consequently, excavated antiquities were placed on the shelves without any system or mutual relation. While it is true that in the early nineteenth century scholars have introduced the principle of archaeological stratigraphy, according to which objects found vertically lower should be older than those above them, in Europe this was applicable only to deep Palaeolithic stratification in caves. This helpless vacillation of antiquarians ended only in 1819, when Christian Jürgensen Thomsen, a Danish museum curator, was asked to prepare an exhibition of prehistoric finds for the Museum of Northern Antiquities in Copenhagen (Harris, 1989, pp. 7–8; Trigger, 1996, pp. 125–127). He approached the task analytically and grouped the artefacts based on whether they appeared together in closed assemblages. The outcome was the three-age system, which classified prehistory into the Stone Age, Bronze Age, and Iron Age. Thomsen's classification may have been inspired by the classical authors, such as Hesiod, Homer, or Lucretius, or perhaps by another scholar (see Chaps. 1 and 2) but, in any case, the system turned out to be functional and proved to be a key methodological advance in archaeology (Sklenář, 1983, pp. 87–89). Thomsen published his method much later: in 1836 in Danish, in 1837 in German, and in 1848, it finally appeared also in English (Daniel, 1943, p. 5; Heizer, 1962). Although Thomsen himself did not think his system could have a universal validity, his followers tried to prove the opposite (Wilson, 1851; Morlot, 1861, pp. 336–337).

The Scandinavian school of archaeology was closely connected with the intellectual climate of the Enlightenment, which adopted the concept of evolutionism (see Chaps. 2, 3, 4 and 7). This was not the case in all of Europe, but this new direction of thought gained momentum and by 1860 evolutionism formed the archaeological mainstream.

As both evolutionism and Thomsen's periodisation spread, scholars no longer viewed written records as the sole source of evidence in their efforts to contextualise archaeological findings. Driven by the notion that all human societies on Earth evolve in a single direction at whose pinnacle is the current Western civilisation (cf. Chaps. 3 and 7), scholars started to turn to ethnological analogies (Trigger, 1996, p. 166). In practice, it meant they sought parallels for societies of European prehistory among the modern tribal societies. John Lubbock justified this by arguing that just as one can find living species in different parts of the world which are related to species extinct in Europe, 'in the same manner, if we wish clearly to understand the antiquities of Europe, we must compare them with the rude implements and weapons still, or until lately, used by the savage races in other parts of the world' (Lubbock, 1865, pp. 440–441).

This quote clearly shows that early evolutionary archaeology also had a dark side. It often went hand in hand with some form of racial theory which causally linked the different historical development of various civilisations, nations, and tribes with innate predispositions of particular races. Those who achieved the greatest 'progress' were considered more advanced and capable. Such races were then believed to have the right to show the way to other races, which naturally suited the imperial powers of Western Europe, who used such claims as a moral basis for their colonial policies. Racial theories also seemed to explain the 'natural' differences between populations of the superpowers of the day, that is, the French, the Germans, and the British.

The racial discourse was also reflected in the historical sciences, including archaeology, which tried to link finds to ancient 'nations', such as the Celts or the Germans, whose existence was known from written sources. On the other hand, the developmental lines of artefacts ordered in accordance with Thomsen's scheme also pointed to the existence of other, older inhabitants of Europe. Moreover, these finds indicated that the smooth development of certain forms of artefacts was repeatedly disrupted, they were replaced by completely new forms—and evolutionary archaeology was unable to explain why this happened because its main methods were typology and seriation (Flinders Petrie, 1899). The former consists in forming a developmental series of certain categories of artefacts, such as bronze axes, pottery, or iron swords, based on their morphological characteristics and design, while seriation in archaeology refers to the study of the distribution of certain types of objects over time. Both methods thus work with development over time—and both ignore the geographical dimension. The geographical aspect was integrated into the archaeological study of time-dependent grouping of particular types of artefacts only later, with the arrival of a new theoretical approach: diffusionism (see Chaps. 7 and 12).

Classical evolutionism assumed convergent development, that is, the idea that cultural innovations may have appeared independently in different parts of the world. Diffusionists, on the other hand, were convinced that particular discoveries and innovations occurred only once in human history—and once discovered, they spread forth by diffusion in the form of either acculturation of cultural elements or through actual migration of people. In anthropology, diffusionism was first proposed by two German ethnologists, Friedrich Ratzel and Franz Boas (Boas, 1940, pp. 251–252). Ratzel was an advocate of 'absolute diffusionism': he argued that even artefacts as basic as the bow and arrow must have originated in one place, and that place should be located (Trigger, 1996, p. 155). Boas was not sympathetic to such simple solutions. He codified the doctrines of cultural relativism and historical particularism. Cultural relativism rejected the existence of universal standards according to which one can compare the stage of development of distinct cultures, while historical particularism claims that each culture is the result of a unique development within which changes take place mainly via cultural diffusion. Boas was convinced that the only effective way to study societies with no written history was to determine the direction of particular episodes of diffusion (Redman, 1999, pp. 54–55). In other words, he believed that one ought to search for the origins of cultural innovations and determine their routes and time of their spread.

Anthropological diffusionism soon found its way also into archaeology. Leaving aside some hyper-diffusionist excesses, such as Grafton Elliot Smith, who sought the origin of all important ancient discoveries in Egypt (and explained even the fall of the Mayan civilisation by an interruption of trade links with Egypt), the most prominent proponent of this direction of thought was Oscar Montelius (Trigger, 1996, pp. 223–232). Montelius belonged to the generation of archaeologists who, due to increasing awareness of both spatial and chronological diversity, were gradually abandoning the evolutionist framework and started to focus on the causality of geographical distribution of different forms of artefacts. In accordance with the diffusionist theory, Montelius (1899) developed a typological method that helped him to bring together various previously unclassified categories of archaeological finds from a large part of Europe into a single chronological system. The key characteristic of archaeological diffusionism was its tendency to explain all changes by external influences, either on the level of transmission of ideas or in terms of migration of populations. Montelius advocated the *ex oriente lux* theory, which claimed that most innovations originated in the Eastern civilisations. From the perspective of central and northern Europe, this in effect meant looking for the foci of diffusion to the south and south-east. Unlike the previous evolutionary approach, diffusionism in archaeology had shown that artefacts evolve in time within a particular spatial context, and within that context, one can trace certain defined groups.

The message of both approaches found support in cultural anthropology. This was a time when the concept of 'culture' had for the first time entered the anthropological discourse (Tylor, 1888, pp. 7–9). In historical sciences, the concept of culture was probably first used in the work of Eduard Meyer, who wrote about the Egyptian, Greek, Trojan, and Mycenaean culture (cf. Chap. 6). It should be noted, however, that the idea of treating 'geographically and temporally restricted assemblages of formally similar prehistoric archaeological material as cultures or civilisations and identifying them as the remains of ethnic groups seem to have occurred independently to a number of archaeologists' (Trigger, 1996, p. 233).

Scholars from Central Europe played a leading role in this process, with Gustaf Kossinna, a German and originally a philologist, being one of the most prominent proponents. From our perspective, he was a hard-line nationalist who tried to identify characteristically German elements in particular archaeological finds, aiming at the earliest possible ones, in an effort to determine the original homeland of the Germans (Kossinna, 1911; Martens, 1989). For the northern half of Europe, he proposed a chronological system which implied that the superior German race had always been at the forefront of progress. From there, innovations spread to the inferior prehistoric 'nations' of western, southern, and eastern Europe. This doctrine easily lent itself to political misuse. For instance Franz Petri, a historian and member of the NSDAP, used his study of medieval local names and funerary monuments to justify German claims to the territory of Belgium and northern France (Petri, 1937). Nazi leaders, too, used Kossinna's theories in their propaganda, with Heinrich Himmler claiming that 'Prehistory is the doctrine of the eminence of the Germans at the dawn of civilization' (Daniel, 1981, p. 151). In all this, Kossinna was not without his portion of blame but it should also be noted that patriotic and even downright nationalist tendencies were not uncommon among archaeologists at this time. We encounter them—in the spirit of imperial colonialism—also in North

America, Africa, and Australia (Shennan, 1989, p. 9; Diaz-Andreu, 2007, pp. 278–313).

Kossinna's doctrine organised archaeological findings in Central Europe into a system of cultures (*Kulturen*, *Kultur-Gruppen*) whose localisation and content changed over time. But this methodological step was rather prelude to the main hypothesis, which claimed that archaeological cultures reflect ancient ethnicities. Kossinna's 'settlement archaeology' (*Siedlungsarchäologie*) therefore should not be understood as the study of settlement remains but rather as a proclamation of the importance of search for the origins of various ethnic groups and tracing their (pre) historic development (Sklenář, 1983, p. 148).

With Kossinna, the era of classical evolutionary archaeology had definitively ended. It was replaced by the paradigm of cultural history, which aimed at tracing the course of prehistoric past much like historians traced the development of written history. This included following the fortunes of the agents of development, especially at pivotal moments which affected the future of political or ethnic units of the past. But while for history, these agents were individuals, for culture-historical archaeology, the agents were nations, ethnicities, or races, albeit identified only via proxy data in the form of artefacts and cultural practices that left a material trace (such as burial customs). Groups of certain types of material culture (especially pottery) distinguished by their geographic location or chronology were identified as archaeological cultures and their variable presence across space and time was explained by the movements of ethnic groups, that is, invasions and migrations.

By the early twentieth century, archaeological cultures had thus been created and their existence accepted. Their systematic conception was, however, still missing. It came only in 1925 with *The Dawn of European Civilisation*, a work of the Australian archaeologist Vere Gordon Childe (1925). Childe accepted Kossinna's concept of archaeological cultures as spatially and temporally defined groups of finds associated with prehistoric peoples but rejected Kossinna's nationalist connotations. On top of that, he accepted Montelius's diffusionist claim about the eastern origin of most innovations, identifying the area from which they came as the Near East (Childe, 1925, p. xiii). He described his concept of archaeological culture as follows:

> We find certain types of remains – pots, implements, ornaments, burial rites, house forms – constantly recurring together. Such a complex of regularly associated traits we shall term a 'cultural group' or just a 'culture'. We assume that such a complex is the material expression of what would to-day be called a 'people' (Childe, 1929, pp. v–vi).

Childe defined archaeological cultures based on a small range of diagnostic types of artefacts but also considered their function within the context of all (hypothetical) cultural elements. He assumed that home-made pottery, decorations, and burials have a greater potential to reflect local traditions and are more likely to resist change than, for instance, artefacts determined primarily by their purpose, such as tools and weapons, which are often direct agents of technological diffusion or subjects of barter and exchange (see Chaps. 11, 13 and 15). In taking this step, Childe endorsed functionalism, a contemporary school of thought in anthropology which appeared on the scene in early 1920s (Trigger, 1980, p. 102). Childe did not believe that the main task of archaeology was to formulate clear chronological phases of cultural development. Instead, he tried to identify nameless prehistoric populations using the concept of archaeological cultures (Fig. 5.1).

The culture-historical paradigm, including the concept of archaeological culture, had remained the mainstay of archaeological theory for a very long time, and with respect to some current archaeological interpretations, one cannot claim that its influence has ever ended. But Childe was probably the last great theorist of archaeology who conceived of archaeological culture as equivalent to anthropological culture, which he also saw as a direct reflection of ethnicity:

> Culture is a social heritage; it corresponds to a community sharing common traditions, common institutions and a common way of life. Such a group may reasonably be called a

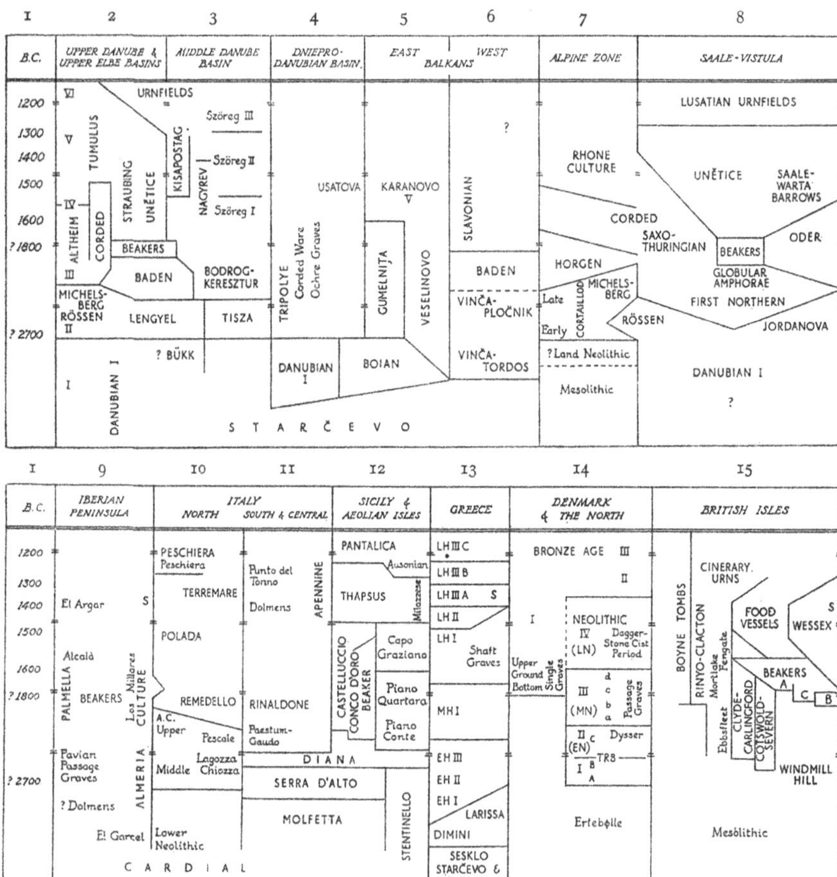

Fig. 5.1 Archaeological cultures in time and space according to Gordon Childe (1925). Childe (1925). *The Dawn of European Civilisation*. London: Routledge and Kegan Paul

people.... It is then a people to which the culture of an archaeologist must correspond. If ethnic be the adjective for people, we may say that prehistoric archaeology has a good hope of establishing an ethnic history of Europe, while a racial one seems hopelessly remote (Childe, 1933, pp. 198–199).

5.3 Cultures as Polythetic Units

After the Second World War, archaeology and anthropology had taken a different view of what 'culture' means. This was also a time of a conceptual split between European continental archaeology and American archaeology. While in Europe, prehistory followed the well-trodden path of cultural history, American archaeology (traditionally in conjunction with anthropology) periodically adopted new trends, which usually reached also Great Britain. It was here that Glyn Daniel in 1950 published a definition which emphasised the difference between 'culture' in anthropology and in archaeology:

In anthropology this word is used for the sum-total of man's material, moral, mental and spiritual inheritance, or for a particular pattern of inheritance; in archaeology the word is used for an assemblage of artefacts assumed to be the material manifestation of a particular pattern of culture inheritance (Daniel, 1950, p. 7).

On the other side of the Atlantic, Walter Taylor in 1948 had likewise clearly rejected any equivalence between archaeological and anthropological culture. He claimed that culture is unobservable and nonmaterial. Behaviour is observable but nonmaterial, and it is only through the consequential phenomena of behaviour that we come to the artefacts, architecture, and other concrete materials that make up the archaeological record: 'this [third] order consists only of objectifications of culture and does not constitute culture itself' (Watson, 1995, p. 685). This scepticism regarding the possibility of understanding culture through the archaeological record logically resulted in an even closer link between archaeology and anthropology, leading Philip Phillips to observe that 'New World archaeology is anthropology or it is nothing' (Phillips, 1955, pp. 246–247).

In the 1960s, a new approach to prehistoric archaeological finds was formulated in the USA by Lewis Binford, who rejected not only the traditional anthropological concept of culture but also the classical Childean normative paradigm (Binford, 1962). His criticism focused on two notions linked to archaeological culture. The first pertained to Childe's claim that 'distributional changes (in diagnostic types) should reflect displacements of population, the expansions, migrations, colonisations or conquests with which literary history is familiar' (Jones, 2002, p. 25). This meant that any transfer of particular forms or designs of material culture was generally viewed as evidence of an interaction between individuals and groups. A low variability of forms of material culture was considered a sign of high social cohesion, while discontinuity in the distribution of material culture was viewed as evidence of social and geographical distance. Binford criticised it, noting that

> [a]rchaeologists tacitly assume that artifacts, regardless of their functional context, can be treated as equal and comparable 'traits'. … I suggest that this undifferentiated and unstructured view is inadequate, that artifacts having their primary functional context in different operational sub-systems of the total cultural system will exhibit differences and similarities differentially, in terms of the structure of the cultural system of which they were a part (Binford, 1962, pp. 217–218).

Binford believed that cultural variability has numerous possible causes, which can moreover overlap and combine. For instance, a change in ceramic types can be linked to the vessel's function, manner of cooking, household size, status of the individual within the society, the method of production (specialised or household-based), transmission of production skills, etc. He was nevertheless convinced that leading among all these possible causes of change are environmental changes, which human culture, as a mode of adaptation, must cope with. Binford is therefore often credited with proposing a view according to which culture is primarily an extrasomatic human adaptation (Trigger, 1996, p. 394).

The other concept Binford criticised was the perception of archaeological cultures as precisely spatiotemporally defined entities. He was convinced that they cannot be exactly chronologically and geographically defined because cultures are internally variable, which makes it impossible to draw strict taxonomic boundaries. Archaeological cultures are usually defined based on just a few diagnostic elements of extant material evidence (such as pottery decoration, burial customs, dwelling forms), one of which is commonly perceived as primary and defining. Moreover, archaeological cultures often differ not only qualitatively (as in 'type A of ceramic decoration differs from type B') but also quantitatively (for instance, several archaeological cultures feature both types A and B in different frequencies). Binford claimed that this strict taxonomy of archaeological cultures is misguided because every human society is essentially a dynamic system with a certain variability of cultural norms (Fig. 5.2) and therefore also artefact domains (Shennan, 1989, pp. 12–13). This clearly shows that Binford not only rejected any identification of archaeological culture with any ethnic identity but even cast doubt on the normative concept underlying the definition of archaeological culture as such (Jones, 2002, pp. 27, 129).

In England, too, the prevailing paradigm of archaeological culture came under attack. This was mainly due to David Clarke, who criticised the 'intuitive' interpretation of archaeological finds by British archaeologists. He rejected Childe's approach to defining cultural complexes based on a small number of diagnostic types of artefacts and, instead, advocated a concept according to which 'cultural assemblage is regarded as a polythetic set of artefact-types' (Clarke, 1968, p. 246). Such polythetic definition of culture should then be supported by quantification and statistical methods (Roberts & Vander Linden, 2011, p. 8). Similar conclusions were reached at the same time behind the Iron Curtain by the Czech archaeologist Bohumil Soudský, who conducted large-scale excavations of a Neolithic settlement in Bylany. The total number of finds from this site was reaching hundreds of thousands, which was far more than traditional methods could deal with. To sort the vast amount of data (consisting primarily of pottery fragments) in a uniform way,

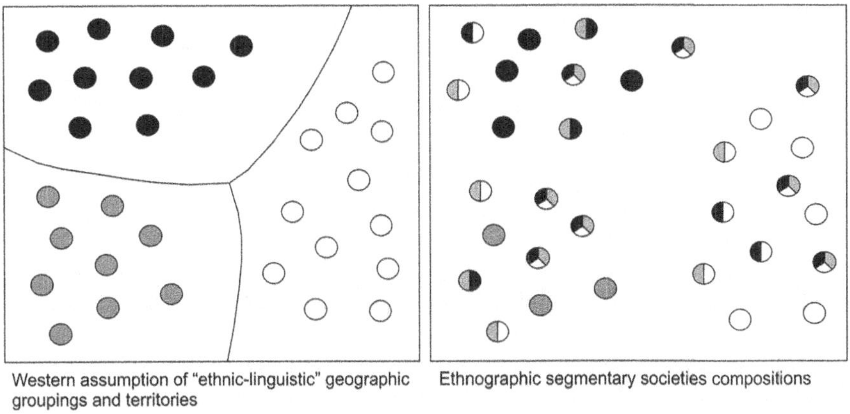

Fig. 5.2 Western perceptions of cultural groupings versus ethnologically known compositions and relations. Different shadings represent different linguistic or other cultural practices. Pre-state societies are usually not discretely bounded homogeneous social units; they typically form hybrid groups, for example due to intermarriage. Redrawn and adapted from Ensor (2021)

Soudský created a code-like system which enabled a semiautomatic analysis of data using punch cards in a unit-recording machine (Soudský, 1967). In this system, each observed quality of an artefact was treated as a trait that could be assigned a numerical code. Using this system, Soudský and his successors managed to create the relative chronology of a Neolithic settlement which remains a reference locality within Europe to this day (Pavlů et al., 1986). Proponents of this so-called 'analytical archaeology' did not reject the basic methodological framework of archaeological cultures but claimed that such characterisation of cultural groups is 'weak' and such schemes do not correspond to the actual cultural diversity of past populations (Gamble, 2007, p. 129).

Another important contribution to the ongoing debate about the nature of archaeological cultures came from Ian Hodder, who rather than to the traditional conception reacted to the discourse created by Binford (Hodder, 1982). It is interesting to note that the two proponents of different theoretical approaches came from the same methodological background of ethnoarchaeology: Binford studied the Nunamiut Inuit people of northern Alaska, while Hodder focused on several populations in eastern Africa. Perhaps relevant to their later disagreements is the fact that Binford studied a population of hunger-gatherers, while Hodder investigated pastoralist and agriculturalist groups.

Hodder's understanding of culture is based on a conception that emphasises its mental and symbolic nature, especially the active role of artefacts. Artefacts, their creation, use, and disposal, all these are 'symbols (i.e., culture) in (social) action' (Watson, 1995, p. 687). Material culture is thus not merely a reflection of environmental adaptation or socio-political organisation. It is also an active element which shapes the identity and mutual relations of various groups, whereby artefacts can equally well reveal and mask such relations. For instance, while successful and dominant populations tend to emphasise their distinctness, other groups may try to suppress their own identity through artefacts. In his ethnoarchaeological fields studies, Hodder had also shown that material culture does not symbolically reflect only one level of identity (e.g., the tribal one)—it can transcend identities and link members of otherwise distinct groups (Hodder, 1977). This hypothesis was based on findings pertaining to the style of spears and gourds in eastern Africa; they are stylistically equivalent to the age groups of men (spears) and women (gourds), whom they thus identify as a group regardless of their tribal affiliation (Larick, 1986). The entire conception is remarkably (bio)semiotic. Not only does it emphasise the constant, dynamic (re)interpretation of symbols, but we can also trace numerous analogies between the phenomena emphasised and biological phenomena. For instance, various aposematic coloration patterns, such as the yellow and black stripes characteristic of wasps, transcend species boundaries, either signalling danger or being mimetised by other, usually harmless, species.

In the 1980s, another important impulse for thinking about archaeological cultures emerged when archaeologists started wondering to what extent archaeological finds reflect human behaviour. Previously, it had been assumed that the relation between excavated artefacts and human activities is straightforward, so that for instance objects discovered in a defunct house relate to the activities that took place there. This assumption, however, turned out to be wrong, as explained by the famous 'Pompeii premise' (Binford, 1981, p. 197). In many cases, an assemblage of objects in extinct settlements is not the result of a single behaviour. Instead, it formed over a long period of time and is thus the outcome of factors belonging to cultural or natural transformations (Schiffer, 1987, p. 7). Complementarity of these two processes leads to the well-known 'black box effect' due to which we find in archaeological context objects whose properties differ from those they had in a living culture (Ascher, 1961, p. 324). Surprisingly, it turned out that the behaviour which contributes the most to the resulting state of archaeological finds from settlements is related to the treatment of waste (Květina & Řídký, 2016). Once this was taken into account, the only contexts which could still be considered 'closed' were burials, where however the material reflection of culture is deformed by a strongly ritualised behaviour. Given that all of this takes place in a semiosphere, that is, in a network of signs and meanings that transcend individual objects (Hoffmeyer, 1996; Herrmann-Pillath, 2021), which is difficult to capture archaeologically, can we at all claim that the remains of material culture can tell us anything about the group identity of ancient populations?

5.4 Pots Do Not Equal People

Childe and proponents of the culture-historical paradigm viewed diversity and changes in material culture as clear evidence of differences and changes in prehistoric group identity, or perhaps even ethnicity. Within their approach, artefacts were in their significance largely equal to people. For instance, the spread of a particular type of pottery in time and space was interpreted as equivalent to movements of peoples. Above, we saw that new theoretical approaches to archaeology had largely disqualified this kind of interpretation. These new concepts and impulses did not, however, form a continuous story, as our description might suggest; in fact, one should rather speak of episodes of longer or shorter duration and always of a limited geographical reach. This is how it happened that majority of continental Europe—and even more so countries behind the Iron Curtain—assumed deep into the 1990s that pots equal people.

The situation was different in Anglophone and Scandinavian countries. American archaeologists, unlike their European counterparts, tend to operate within an anthropological framework and readily incorporate into their developmental models both anthropological theory and ethnographic data (Bender, 1978, p. 206). This is why in the United States, researchers realised that archaeological culture does not reflect a living anthropological culture and started wondering whether one can link the diversity of archaeological finds with prehistoric identities at all. Impulses for asking this question, as well as the various answers that emerged, came from specifically focused research called ethnoarchaeology, or sometimes also 'living' or 'action' archaeology (Longacre, 1981, p. 49).

> Ethnoarchaeology is neither a theory nor a method, but a research strategy embodying a range of approaches to understanding the relationships of material culture to culture as a whole, both in the living context and as it enters the archaeological record, and to exploiting such understandings in order to inform archaeological concepts and to improve interpretation (David & Kramer, 2001, p. 2).

Within this line of thought, researchers started to focus on explaining the formal variability of material culture, on the 'style'. Stylistic variety of artefacts is something that already Binford had noted, believing that style is something that cuts across the technological, social, and conceptual functions of artefacts. He understood style as a set of formal qualities 'promoting group solidarity and serving as a basis for group awareness and identity' (Binford, 1962, p. 220). But what level of group integration does style actually relate to? Rather than a mechanism of integration, style may be a way of transmitting social information. Any part of material culture can contain meaning in the form of stylistic variability, which is open to interpretation. Different levels of material culture can be activated depending on the particular level of social interaction (e.g., the house architecture refers to a different level of identity than the style of dress). The nature of message expressed by style, as well as the medium in which it is expressed, is derived from their mutual understanding, usually the social distance between the 'sender' and the 'recipient'. This means that a person familiar with the symbolism of a particular style can read from it a range of detailed information about the group but also about the personal identity of the 'sender'. Persons not familiar with the language of the style may be able to read the message superficially, but for the most part they understand only that the object in question and its bearer are 'different' (Wobst, 1999).

The notion of such active and intentional manner of information transfer through style was revised by James Sackett (1977), who claimed that while style is potentially present in any artefact, instances where an object has a truly purposeful informational charge are rare. Most stylistic variability is basically silent (isochrestic), meaning that under normal circumstances, no message is exchanged between the sender and the recipient. But this can change with a more remote geographic or historical perspective, because information about social identity based on style becomes clearer and more comprehensible in confrontation with a different environment. Stylistic variety conceived in this way interestingly resonates with the definition of *habitus* in the theoretical conception of Pierre Bourdieu (1977, pp. 72–95). Isochrestic variability is not created intentionally, but that does not mean that differences in the form and shape of functionally identical artefacts would not have been perceived by their users. Numerous ethnoarchaeological works show that people can identify concrete traditions of production based on style, even down to the level of individual craftsmen (for pottery, see, e.g., Dietler & Herbich, 1989, p. 154; Graves, 1994; Stark, 1994, 1999).

In practice, implementation of the ethnoarchaeological approach to style led to efforts to identify the boundaries of ethnic or other social identities based on the material culture of living non-industrial populations. This was the focus of Polly Wiessner's work on the San hunter-gatherers from the Kalahari, where she showed that the style of material culture, especially arrows and bead bands, clearly differentiated the individual linguistic groups. This was quite naturally due to the communication barrier between the speakers of these languages (Wiessner, 1983; Emberling, 1997, p. 312). At around the same time, Hodder had conducted a detailed analysis of the spatial organisation of material culture in certain parts of Kenya, Zambia, and Sudan (Hodder, 1982, pp. 37–57). He highlighted the complex relations between artefacts and different levels of social identity. For instance, in the Lake Baringo area, gourd vessels and wooden stools that serve as headrests are specific to individual tribes, and it is possible to identify the distribution map of particular types with areas inhabited by the Njemp, Tugen, and Pokot tribes (Hodder, 1977). On the other hand, while the style of hand-forged spears is also an ethnic indicator, it is much more closely related to age groups (cohorts), which naturally cut across tribes. Here, too, researchers identified a certain flexibility of ethnic identity, whose eventual change is declared by the individual's respective spear style (Larick, 1986). In other words, people are part of a number of different, not necessarily hierarchical 'closures' (see Chap. 17).

Large-scale and long-term projects followed, promising to bring enough field data to solve the key question of correlation between material culture and ethnic boundaries. Within their research in the Kalinga region on the island of Luzon, researchers investigated a correlation between the style of pottery produced by local farmers and ethnic boundaries within the area. Their work has demonstrated that ethnic affiliation does not have the same meaning for locals as it does for Europeans. More important for self-identification are small kinship groups and, on a larger scale, groups based on regionality. This regional level of social structuring is evident in material culture, too, because for instance exchange of ceramic vessels takes place primarily within regions (Graves, 1994; Stark, 1994). A similarly conceived project took place also in northern Cameroon, where results likewise indicate that local populations do not perceive ethnicity as a key level of self-identification. Ethnicity could not even be considered a stable social unit here. In this region, any correlation between social identity and material culture exists on a scale either smaller or larger than ethnicity (MacEachern, 1992). Other ethnoarchaeological field projects investigated the geographical and social boundaries of style, and they likewise failed to confirm any clear link between material culture and ethnicity (David et al., 1988; Dietler & Herbich, 1989; Gosselain, 2000; Lemonnier, 2012; Welsch & Terrell, 1998).

The results of ethnoarchaeological research into the overlapping boundaries of material culture and social identity are thus highly ambiguous; the connection between the two is so complex that it does not allow for any straightforward generalisations. In most cases, the boundary of differences between artefacts runs parallel to some form of social categorisation, but this is usually a form with which no anthropological or archaeological theories operate. The match between the style of material culture and social identity usually transcends the boundaries of settlements (villages) but is significantly smaller than natural geographical regions, cultural areas, or ethnic groups (Stark, 1998, p. 10). Moreover, in cases where one can trace unique types of artefacts (or their characteristics) that are typical of easily interpretable types of social identity, this pertains to objects which leave no traces in archaeological record. This is partly because group identity is not only created through the mere sharing of artefacts, but more importantly through shared narratives about them (Rédei & Ranta, 2016).

To sum up: if one cannot clearly define the link between social/ethnic borders and material culture of living societies, it is naturally even less possible to identify them based on archaeological evidence. The concept of archaeological culture as a proxy for the social identification or categorisation of extinct societies has thus been refuted. Moreover, while elsewhere 'archaeological culture' has survived, at least as a methodological tool for taxonomy of archaeological findings, British scholars—and even more so their Scandinavian colleagues—have rejected the concept altogether.

5.5 Bioarchaeological Revolution

The direct evidence for mobility and migration in prehistory provided by recent bioarchaeological methods has brought about far-reaching changes in the concept of archaeological cultures and the whole dynamics of prehistoric development. The notion of geographical movements of individuals, groups, and even entire 'nations' was not new in archaeology or anthropology. The culture-historical direction of research, which was dominant from early twentieth century until about the 1960s, already regarded population changes, including large migration events, as agents of cultural change. Methodologically, however, the assumed population movements were attested only indirectly, via artefacts, for instance by the presence of pottery associated with one culture in a different artefact environment of another archaeological culture. But, as we already know, pots do not equal people.

With this in mind, the concepts of migration and mobility started to fade out of archaeological reasoning. It was David Anthony (1990), an American anthropologist, who pointed out that it would be a mistake to completely overlook population dynamics in archaeology, but it took some years before he was able to prove his migration hypotheses. This is because the ability to prove migration directly from human remains came only with certain methodical breakthroughs in bioarchaeology. The first was linked to the potential of strontium isotope ($^{87}Sr/^{86}Sr$) analysis, which can differentiate between individuals who are local ($^{87}Sr/^{86}Sr$ ratios in accordance with the local range) and those who are not (Bentley, 2006). Based on this method, it started to become clear that even in ancient prehistory, some individuals moved across thousands of miles in their lifetime. The iconic evidence for this came with the finding of the so-called 'Amesbury archer'- the burial of a man found near Stonehenge and dated to the period of its construction (ca. 2300 BC). The results of isotope analysis of his tooth enamel were most surprising: they showed that the 'archer' was not born in Britain but somewhere in central Europe (Parker Pearson et al., 2016).

Nevertheless, a real revolution in the study of past populations and overall conception of archaeological cultures came with the methodological progress in DNA sequencing and analysis of both modern and, especially, ancient people (Nielsen et al., 2017; Reich, 2018) (cf. Chap. 12). Although scientists had been trying to use molecular genetics to reconstruct prehistoric population dynamics since the 1970s (Ammerman & Cavalli-Sforza, 1984), most conclusions of these pioneering studies later turned out to be wrong. Early researchers analysed selected genetic traits (such as blood groups) in modern populations and assumed that the population history of a given geographical area is contained in the DNA of its current population, whose analysis can therefore indicate certain demographic processes in the past. Gradually, however, it turned out that results yielded by this approach are unreliable.

The problem was not only that people have intermixed with their neighbours for most of their history, which blurred the genetic signatures of past events. Even more crucial is the fact that the structure of current populations does not reflect the structure of past populations. In other words, people who live in a particular area are not descended only from people who lived there thousands of years ago, because migration and population mixture took place frequently even in later times (Pickrell & Reich, 2014).

A key breakthrough came in 2010–2015, when scientists managed to fine-tune the methodology of analysis of ancient DNA (aDNA). It started with the publication of the first five ancient human genomes, including archaic Neanderthal and Denisovan genomes (Reich, 2018, p. xviii). Since that time, thousands of aDNA samples had been analysed and many more come in each year. The development of archaeogenetics promised to overcome the limitations of traditional archaeological methods and bring definite answers to many hitherto unsolved questions regarding human past, including genetic relations between ancient peoples, whose solution Childe (1933, p. 199) believed to be 'hopelessly remote'. Traditional archaeological methods could map the spread of particular cultural elements (such as pottery decoration, burial customs, or dwellings forms) across time and space, but they could not tell us to what extent their spread was driven by migration, the movement of people, or by diffusion, the movement of ideas. Thanks to the potential of the new ways of identifying genetic relatedness, it is now much easier to find answers to these questions.

As an example, we can use the beginnings of agriculture in Europe during the Neolithic (in the Aegean and Greece roughly 6500 BC, in central Europe 5400 BC, in Scandinavia and Britain 4000 BC). Archaeogenetic studies were able to solve the long-standing debate about whether the spread of agriculture was due to the arrival of farmers from Anatolia or rather driven by local hunter-gatherers adopting agricultural technologies and settled lifestyle. It turned out that the former is correct, although in many areas the local hunter-gatherers were not completely replaced and what took place was rather a mixing of the two populations (e.g. Allentoft et al., 2024; Hofmanová et al., 2016; Rivollat et al., 2020). On top of that, archaeogenetics was able to confirm the spread of the Neolithic population both by the continental route, which led through the Balkans to central Europe, and by a Mediterranean route along the coastlines all the way to the Iberian Peninsula, which is what had been assumed based on archaeology. Further studies then showed that farmers from the Near East spread also to other regions, including east Africa, the Eurasian steppe, and southern Asia (Lazaridis et al., 2016). It should be noted that ancient DNA enables scientists to reconstruct to some extent even the appearance of ancient people. This is how we know that, for instance, the western European hunter-gatherers had the—from our perspective rare—combination of blue eyes, dark skin, and dark hair, while the first European farmers had dark hair but mostly brown eyes and light skin (Reich, 2018, p. 96).

Aside from tracing the origins of ancient populations, archaeogenetics can also investigate the kinship relations between individuals in relatively great detail (cf. Chap. 8). This was used for instance in an investigation of a Late Neolithic (2880–2776 BC) mass grave from the Koszyce in Poland. Archaeogenetic analyses had shown that this was one extended family and that close relatives were buried close to each other: mothers with their children, siblings side by side (Schroeder et al., 2019). Another example are the Avar cemeteries from Hungary dating back to the Early Medieval period. Researchers uncovered a strict patrilineal kinship system, in which patrilocality and female exogamy were the norm and multiple reproductive partnering, and levirate unions were common. The absence of consanguinity indicates that this society maintained a detailed memory of ancestry over generations (Gnecchi-Ruscone et al., 2024).

5.6 Kossinna's Ugly Smile

Such groundbreaking archaeogenetic studies had caused a commotion in the archaeological community (see Chap. 12). Kristian Kristiansen, a leading European archaeologist, had noted that 'There seems to be a general feeling that some fundamental changes are underway in archaeological thinking and have been for some time now.' He called the methodical shift in bioarchaeology a 'third scientific revolution in archaeology', thus implicitly referring to the formation of archaeology as an independent discipline in mid-nineteenth century and the discovery of radiocarbon dating a century later (Kristiansen, 2014).

But there also appeared numerous critical voices. A lively and still ongoing debate was sparked by two articles on the population history of prehistoric Europe at the end of the Neolithic (Allentoft et al., 2015; Haak et al., 2015). Based on genomic analyses of 101 and 69 individuals living between 8000–3000 years ago, these studies had shown that Europe had undergone massive demographic changes in the third millennium BC. Their results indicated that, at this time, a population with genetic origin in the Eurasian steppes had superimposed itself on the original population of Europe, which was a genetic mixture of European hunter-gatherers and Near Eastern farmers. This new group was linked to the archaeological Yamnaya culture, characterised by a nomadic-pastoralist lifestyle and burials in characteristic pit graves under a specific type of tumuli (kurgans). These demographic changes, where in some places up to 79% of earlier population seems to have been replaced, were explained by a massive colonisation of 'steppe raiders' (especially warbands of young men), literally 'eastern cowboys' who developed a new, more mobile way of life based on a combination of wagons, dairying, and horseback riding (Curry, 2021). A plague epidemic, inadvertently brought by these newcomers, then contributed to a decimation of the earlier population of Europe (Kristiansen et al., 2017).

One of the first critics of this narrative was Volker Heyd, who called his contribution—hinting on nationalism in archaeology—*Kossinna's Smile* (Heyd, 2017). Like other scholars (Frieman & Hofmann, 2019; Hakenbeck, 2019), he tried to point out that archaeogenetic studies anachronistically identify archaeological cultures with linguistically, culturally, and biologically defined populations, just as the infamous German researcher did a hundred years ago. In the interpretations of archaeogeneticists, genetic samples from a handful of individuals at one burial site came to represent entire archaeological cultures (defined by pottery typology and burial morphology) and subsequently, albeit often tacitly, even an entire ethnic group. Just like archaeological cultures were defined in the twentieth century based on 'type sites', archaeogenetics often reduced past genetic populations to 'type persons' (Feinman & Neitzel, 2020). However, using a small number of individuals to represent a social or cultural group is problematic, because it *a priori* assumes that such groups were genetically homogeneous. But to what extent, if any, there actually existed a link between genetic origin and construction of social identities remains to be carefully investigated.

The tendency to correlate gene flow with migrations of specific populations and to use this to explain technological and social changes in the third millennium BC had likewise become a target of criticism. Although stories of mass migrations make an attractive reading, they are inherently simplifying and thus potentially misleading. There are at least two reasons for this simplification of complex cultural developments. First, archaeogenetic data are perceived as 'hard' evidence and therefore as somehow more objective because they are the result of work conducted in a laboratory and not based on 'mere theorising' of scholars in humanities. This false allure of objectivity had allowed natural scientists to incorporate ill-considered and unverified assumptions in their models, leading to anachronistic interpretations of human groups. The results of archaeogenetic analysis can be perfectly precise but they still attest only to one of many aspects of past identities (Frieman & Hofmann, 2019). The other underlying cause of simplifying narratives is related to the functioning of prestigious journals such as *Nature*, where archaeogenetic studies often appear. These media force the authors to describe complex situations within the space of just a few pages (relegating the rest to supplements) and the likelihood of being published increases if the study has a proper 'wow effect'.

The results of archaeogenetic studies were enthusiastically covered by mainstream media because they brought popular information about the origin of Europeans. But the power of ethno-genetic-historical narratives was also recognised by the extreme right (Hakenbeck, 2019). These groups tend to be interested in genetic studies in general, because they incline towards the so-called objectivist racial interpretation of inequality of human cultures, both current and past ones (Hofmann et al., 2021). In this context, studies about the Yamnaya culture could be interpreted as evidence of the ancient origin (and superiority) of Aryan and Proto-Germanic peoples. The focus of scientific interpretations on 'raiding male warbands' which abducted local women (Kristiansen et al., 2017, p. 338) moreover resonated with the masculine self-conception and the ideal of heroic warriors adopted by many such individuals.

But what better interpretation of events and processes that took place in Europe in the third millennium BC do these critics offer? Drawing on Clarke's (1968) polythetic approach, Martin Furholt (2019, 2020) had suggested that the steppe-related ancestry was not linked to any clearly definable archaeological culture but rather just to specific burial practices. The traditional simplistic view that steppe pastoralists, represented by east European Yamnaya culture, migrated around 2900 BC into central Europe and southern Scandinavia, where they formed the Corded Ware culture, and after 2500 BC continued further westwards, where they transformed into the Bell Beaker culture, is most likely erroneous. Neither of these cultures, which spread over more than half of the European continent, was homogeneous (Fig. 5.3a). They were rather variable, both between particular geographical areas and even within regions.

This is especially apparent in the relation between burial rites and pottery styles. In contrast to the traditional notion according to which Corded Ware and Bell Beaker cultures practiced different burial customs, Furholt argues that the practices were rather similar and differed only in details. All burials in both of these

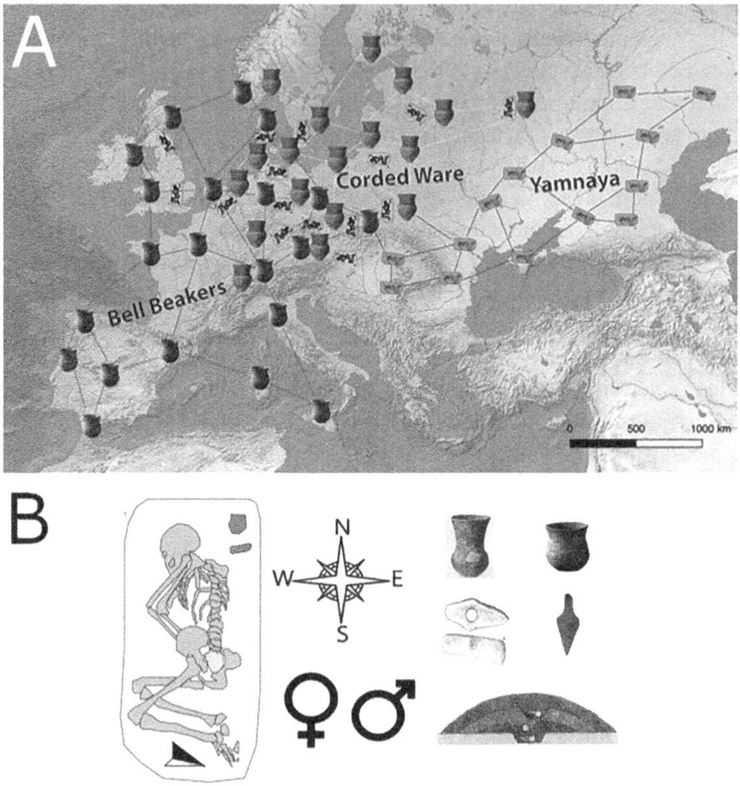

Fig. 5.3 (**a**) A map of Europe showing the spatial distribution of the three most prominent archaeological units of the third millennium BCE: Yamnaya, Corded Ware, and Bell Beakers. A newly emerging single burial complex intersects Corded Ware and Bell Beakers. (**b**) The main elements of the new, third-millennium BCE single burial complex in Europe: individual aggrandisement (a single burial in a crouched position, under a burial mound), gender differentiation (reflected by body orientation rules), male warriorhood (specialised weapons in male graves), and drinking vessels of different forms (Furholt, 2021)

archaeological cultures (as well as several later ones, which followed up on them) were in effect a variation on a common theme: individual burials, strict orientation rules, gender-related differentiation, and the association of males with specific weapons (Fig. 5.3b). But this specific type of burial customs, which appeared around 2900 BC and is known as Single Grave Burial Rituals (SGBR), was not the only type of burials practiced during this period. In its 'pure' form, it is found only in several regions associated with Corded Ware and Bell Beaker pottery. In other regions, it was rather different or even altogether absent. An even larger regional and local variability is found in other cultural elements, such as settlement patterns, types of dwellings, and subsistence strategies. Corded Ware pottery, for instance, appeared in hunter-gatherer societies as well as in societies based on various forms of agriculture and herding. Moreover, the pottery itself represented at least two distinct social spheres, so that while the style of fine ware was clearly superregional, coarse ware pottery was usually produced locally.

A comparison of aDNA results had shown that a vast majority of individuals with the new genetic steppe ancestry were found in SGBR burials, but also that these burials cannot be taken as synonymous with either Corded Ware or Bell Beaker culture, because material evidence associated with these cultures appears even in contexts where no such burials are found. The new pottery styles of the third millennium BC thus most likely formed independently in central and western Europe: they were not brought by groups of migrants from the east. Moreover, the proportion of steppe origin is probably overrepresented in aDNA of the Late Neolithic, because in many regions we lack sufficient information about the pre-existing local population. This can be due to other burial practices, which may be hard to identify by archaeological means, or simply due to local conditions leading to extremely poor bone preservation.

A recent archaeogenetic study confirmed the spread of steppe-related ancestry across western Eurasia around 3000 BC but also acknowledged that situations of interaction and admixture varied markedly between geographical regions (Allentoft et al., 2024). It also showed that steppe pastoralists had admixed with agricultural groups of the eastern European plains (e.g. the Globular Amphora culture) even before expanding further into Europe. Another study, regionally focused on Bohemia, showed that early SGBR graves contained a genetically diverse population, but one that gradually became homogeneous, perhaps restricted to a particular social segment, such as one a dominant lineage (Papac et al., 2021). Interestingly, researchers also suggested that several other demographic changes might have occurred several times before and after the appearance of steppe-related ancestry in Central Europe.

5.7 Conclusions: Towards a Unifying Theory of the Archaeological Universe

Methodological breakthroughs in bioarchaeology had forced archaeologists to reopen the debates about the identity and culture of past populations (Frieman & Hofmann, 2019). Although there are good reasons to abandon the concept of archaeological culture, rejection of any superregional identity of prehistoric populations is an untenable position. Ancient communities may not have been hierarchically nested within large-scale homogeneous social units with discrete boundaries but neither were they isolated units. The spread of a similar material culture (pottery production, burial practices, architecture) across large areas shows that most past societies were in some way interconnected and shared a similar worldview. In investigating the past, we therefore should not restrict ourselves just to the level of individual sites and small regions: we must be able to explain processes on a larger scale, to tell 'grand narratives'. The question remains, however, where the boundaries between different cultures and identities lay, and whether these boundaries were merely geographical.

To address these issues, it is important to bear in mind that people might relate to a numerous distinct social collectives or communities of practice and usually, they belong to a number of such groups that overlap in various ways (Furholt, 2020; Wenger, 1998) (see Chap. 17). In other words, social relationships and cultural expressions are not monothetic but polythetic, as Clarke (1968) emphasised over fifty years ago. From this perspective, the cultural boundaries of prehistoric societies were not sharp: they were rather blurry and multiple boundaries existed at different levels (Fig. 5.4). Different archaeological and anthropological data therefore need not be forcefully unified—each can be seen as a reflection of a different social world. Pottery making was doubtless linked to a different level of social life than burial rites, house construction, or the style of dress. In each social world, be it pottery making or house building, there existed different communities of practice, different groups of people who developed specific cultural standards for the area in question, which are reflected in the archaeological record by specific characteristics and enable archaeologists to create different typologies.

If people from different communities produced similar pottery and built similar houses, one can suppose that they shared a similar worldview and perhaps even had a similar origin. But was that enough for them to think about themselves as 'us', as members of the same social group? Enough to share the same semiotic lifeworld of experiences, meanings and interactions (Sonesson, 2016)? That we cannot tell. It is well possible that despite these shared elements, they perceived other members of such group as 'others', as people different from themselves, or even as 'strangers', people we do not know and with whom we do not share our cultural traditions and behaviour patterns (Belfer-Cohen & Hovers, 2020). In fact, the cultural elements that survived need not have been significant for their identity, and it is well possible that the elements that did define it will remain hidden and unknown to us. Finding that some cultural elements were of a local nature while others were superregional is likewise unlikely to help. The extent of spread perhaps attests more to the manner of production and distribution (exchange) of objects, to their social importance, or perhaps to the mobility and co-socialisation of producers across communities (Kolář, 2020). Explanation of homogeneity of some elements of material culture should not be sought only in massive migrations, but also in the high mobility of individuals who made them (Furholt, 2018).

Recent findings on the biological relatedness of ancient humans enriches our knowledge of past societies, but even this data should be viewed as merely another layer of identity (Fig. 5.4). Biology is a key element of kinship only to the extent to which it is contextualised via social relations (Brück, 2021). It should be borne in mind that in many cultural contexts, kinship does not emerge with birth. Instead, it is a product of social practices, including mythological cosmology, co-residence, food sharing, rites of passage, or exchange (cf. Chap. 8). Moreover, certain biological relations, such as those between father and son, can be emphasised differently in different societies and at different times. In some cultural contexts, they can be viewed as extremely important, while in other contexts, their significance can be

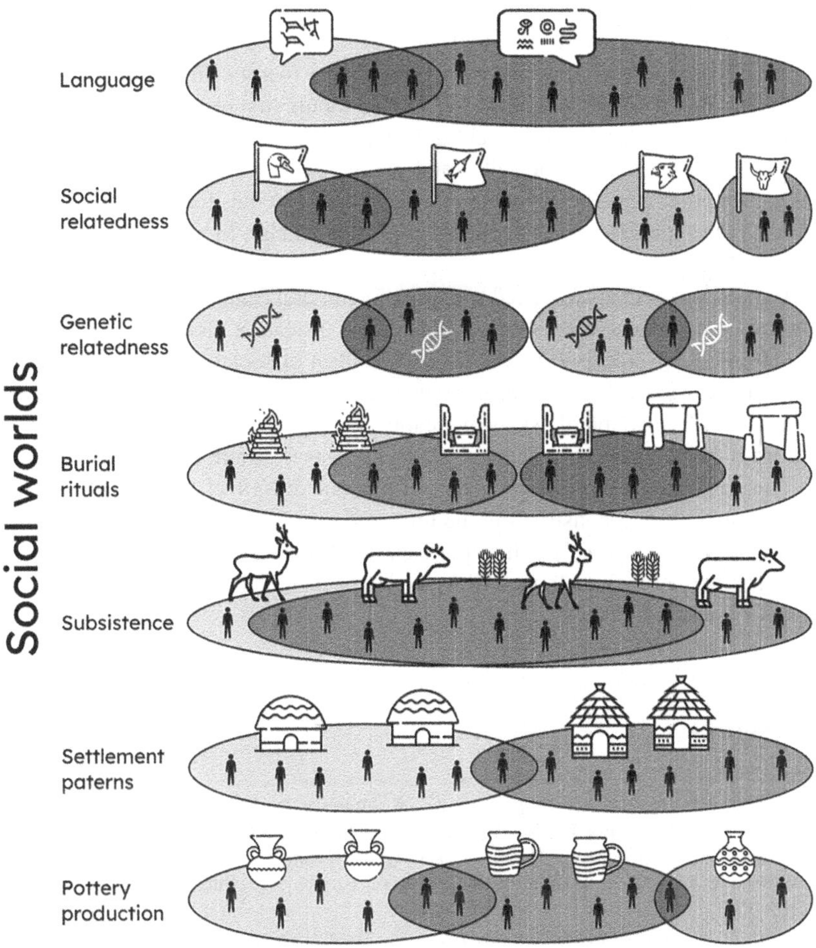

Fig. 5.4 Illustration of a polythetic culture model with different levels of sociocultural and biological identities and cultural practices that need not overlap. Author's drawing, inspired by Furholt (2020), icons designed by Freepik

marginal. Biosemiotics offers a promising framework for overcoming the divide between biology and culture, illustrating how human culture emerges as an extension of the semiotic processes found throughout nature (Herrmann-Pillath, 2021). From this perspective, an individual's biological and genetic makeup shapes their *Umwelt*, while their interactions with the world through semiotic processes reflect the intricate interplay of natural and cultural influences that contribute to identity formation (Cobley, 2016).

Another important point for understanding past identities is the need to abandon misleading notion that one story, or one set of mechanisms, can explain social processes that took place in different regions and communities. That is why we should move away from the 'top-down' studies that cover large areas and long periods of time, often involving entire continents and periods of thousands of years (cf. Chaps. 4, 7 and 12), to a more 'bottom-up' approach that focuses on smaller regions and particular communities (Veeramah, 2018). We should first resolve detailed relations between biological data, material culture, and identity on a local level, and only then formulate interpretations and tell grand narratives on a large (e.g., Europe-wide) scale. This process is already underway and the first studies focusing on local communities are appearing (e.g. Rivollat et al., 2023; Villalba-Mouco et al., 2022; Vyas et al., 2023).

Archaeology will always need some spatiotemporal frameworks for mutual communication and discussion among scholars. But instead of monolithic archaeological cultures, such frameworks could take the form of social worlds identified by a polythetic classification of archaeological and bioarchaeological data. Although these, too, will undoubtedly skew the reality of the past, because they would be created from the perspective of modern archaeologists (Furholt, 2020), in conjunction with for instance social network analysis (Feinman & Neitzel, 2020), they offer one of the best ways of reconstructing the form and change of past societies. Still, only the future will show whether this approach will bring the much-desired unification of conceptions of archaeological finds, archaeogenetic data, and cultural anthropological knowledge.

References

Allentoft, M. E., Sikora, M., Sjögren, K.-G., Rasmussen, S., Rasmussen, M., Stenderup, J., Damgaard, P. B., Schroeder, H., Ahlström, T., Vinner, L., Malaspinas, A.-S., Margaryan, A., Higham, T., Chivall, D., Lynnerup, N., Harvig, L., Baron, J., Casa, P. D., Dąbrowski, P., & Willerslev, E. (2015). Population genomics of Bronze Age Eurasia. *Nature, 522*(7555), 167–172.

Allentoft, M. E., Sikora, M., Refoyo-Martínez, A., Irving-Pease, E. K., Fischer, A., Barrie, W., Ingason, A., Stenderup, J., Sjögren, K.-G., Pearson, A., Sousa da Mota, B., Schulz Paulsson, B., Halgren, A., Macleod, R., Jørkov, M. L. S., Demeter, F., Sørensen, L., Nielsen, P. O., Henriksen, R. A., Willerslev, & E. (2024). Population genomics of post-glacial western Eurasia. *Nature, 625*(7994), 301–311.

Ammerman, A. J., & Cavalli-Sforza, L. L. (1984). *The Neolithic transition and the genetics of populations in Europe*. Princeton University Press.

Anthony, D. (1990). Migration in archeology: The baby and the bathwater. *American Anthropologist, 92*(4), 895–914.

Ascher, R. (1961). Analogy in archaeological interpretation. *Southwestern Journal of Anthropology, 17*(4), 317–325.

Barfield, T. (2008). Archaeology as anthropology of the long term. *Archeological Papers of the American Anthropological Association, 13*(1), 41–49.

Belfer-Cohen, A., & Hovers, E. (2020). Prehistoric perspectives on "others" and "strangers". *Frontiers in Psychology, 10*, 3063.

Bender, B. (1978). Gather-hunter to farmer: A social perspective. *World Archeology, 10*(2), 204–222.

Bentley, R. A. (2006). Strontium isotopes from the Earth to the archaeological skeleton: A review. *Journal of Archaeological Method and Theory, 13*(3), 135–187.

Binford, L. R. (1962). Archaeology as anthropology. *American Antiquity, 28*(2), 217–225.

Binford, L. R. (1981). Behavioral archaeology and the "Pompeii premise". *Journal of Anthropological Research, 37*(3), 195–208.

Boas, F. (1940). *Race, language, and culture*. Macmillan Publishers.

Bourdieu, P. (1977). *Outline of a theory of practice*. Cambridge University Press. Original French edition: *Esquisse d'une théorie de la pratique*. Librairie Droz, 1972.

Brück, J. (2021). Ancient DNA, kinship and relational identities in Bronze Age Britain. *Antiquity, 95*(379), 228–237.

Charbonnier, G. (Ed.). (1969). *Conversations with Claude Lévi-Strauss*. Jonathan Cape. Original French edition: *Entretiens avec Claude Lévi-Strauss*. Plon, 1961.

Childe, V. G. (1925). *The dawn of European civilisation*. Routlege and Kegan Paul.

Childe, V. G. (1929). *The Danube in Prehistory*. Clarendon.

Childe, V. G. (1933). Races, peoples and cultures in prehistoric Europe. *History, 18*, 193–203.

Clarke, D. L. (1968). *Analytical archaeology*. Methuen.

Cobley, P. (2016). *Cultural implications of biosemiotics*. Springer.

Corfield, P. J. (2018). Primevalism. Saluting a renamed prehistory. In A. Baysal, E. L. Baysal, & S. Souvatzi (Eds.), *Time and history in prehistory* (pp. 265–282). Routledge.

Curry, A. (2021). Milk fueled Bronze Age expansion of 'eastern cowboys' into Europe. *Science*.

Daniel, G. E. (1943). *The Three Ages: An essay on archaeological method*. Cambridge University Press.

Daniel, G. E. (1950). *A hundred years of archaeology*. Gerald Duckworth.

Daniel, G. E. (1981). *A short history of archaeology*. Thames and Hudson.

David, N., & Kramer, C. (2001). *Ethnoarchaeology in action*. Cambridge University Press.

David, N., Sterner, J., & Gavua, K. (1988). Why pots are decorated. *Current Anthropology, 29*(3), 365–389.

Diaz-Andreu, M. (2007). *A world history of nineteenth-century archaeology: Nationalism, colonialism, and the past*. Oxford University Press.

Dietler, M., & Herbich, I. (1989). Tich Matek: The technology of Luo pottery production and the definition of ceramic style. *World Archaeology, 21*(1), 148–164.

Emberling, G. (1997). Ethnicity in complex societies: Archaeological perspectives. *Journal of Archaeological Research, 5*(4), 295–344.

Ensor, B. E. (2021). *The not very patrilocal European Neolithic. Strontium, aDNA, and archaeological kinship analyses*. Archaeopress.

Feinman, G. M., & Neitzel, J. E. (2020). Excising culture history from contemporary archaeology. *Journal of Anthropological Archaeology, 60*, 101230.

Flinders Petrie, W. M. (1899). Sequences in prehistoric remains. *The Journal of the Anthropological Institute of Great Britain and Ireland, 29*(3/4), 295–301.

Frieman, C. J., & Hofmann, D. (2019). Present pasts in the archaeology of genetics, identity, and migration in Europe: A critical essay. *World Archaeology, 51*(4), 528–545.

Furholt, M. (2018). Translocal Communities – Exploring mobility and migration in sedentary societies of the European Neolithic and Early Bronze Age. *Praehistorische Zeitschrift, 92*(2), 304–321.

Furholt, M. (2019). Re-integrating archaeology: A contribution to aDNA studies and the migration discourse on the 3rd millennium BC in Europe. *Proceedings of the Prehistoric Society, 85*, 115–129.

Furholt, M. (2020). Social worlds and communities of practice: A polythetic culture model for 3rd millennium BC Europe in the light of current migration debates. *Préhistoires Méditerranéennes, 8*, 1–24.

Furholt, M. (2021). Mobility and social change: Understanding the European Neolithic period after the archaeogenetic revolution. *Journal of Archaeological Research, 29*(4), 481–535.
Gamble, C. (2007). *Archaeology: The basics*. Routledge.
Gnecchi-Ruscone, G. A., Rácz, Z., Samu, L., Szeniczey, T., Faragó, N., Knipper, C., Friedrich, R., Zlámalová, D., Traverso, L., Liccardo, S., Wabnitz, S., Popli, D., Wang, K., Radzeviciute, R., Gulyás, B., Koncz, I., Balogh, C., Lezsák, G. M., Mácsai, V., & Hofmanová, Z. (2024). Network of large pedigrees reveals social practices of Avar communities. *Nature, 629*(8011), 376–383.
Gosselain, O. P. (2000). Materializing identities: An African perspective. *Journal of Archaeological Method and Theory, 7*(3), 187–217.
Graves, M. W. (1994). Kalinga social and material culture boundaries. A case of spatial convergence. In W. Longacre & J. M. Skibo (Eds.), *Kalinga ethnoarchaeology. Expanding archaeological method and theory* (pp. 13–49). Smithsonian Institution Press.
Haak, W., Lazaridis, I., Patterson, N., Rohland, N., Mallick, S., Llamas, B., Brandt, G., Nordenfelt, S., Harney, E., Stewardson, K., Fu, Q., Mittnik, A., Bánffy, E., Economou, C., Francken, M., Friederich, S., Pena, R. G., Hallgren, F., Khartanovich, V., & Reich, D. (2015). Massive migration from the steppe was a source for Indo-European languages in Europe. *Nature, 522*(7555), 207–211.
Hakenbeck, S. E. (2019). Genetics, archaeology and the far right: An unholy Trinity. *World Archaeology, 51*(4), 517–527.
Harris, E. C. (1989). *Principles of Archaeological Stratigraphy*. Academic.
Heizer, R. F. (1962). The background of Thomsen's Three-age system. *Technology and Culture, 3*(3), 259–266.
Herrmann-Pillath, C. (2021). Biosemiotic foundations of a Darwinian approach to cultural evolution. *Cultural Science Journal, 13*(1), 16–33.
Heyd, V. (2017). Kossinna's smile. *Antiquity, 91*(356), 348–359.
Hodder, I. (1977). The distribution of material culture items in the Baringo district, western Kenya. *Man. New Series, 12*(2), 239–269.
Hodder, I. (1982). *Symbols in action: Ethnoaerchaeological studies of material culture*. Cambridge University Press.
Hoffmeyer, J. (1996). *Signs of meaning in the universe*. Indiana University Press.
Hofmann, D., Hanscam, E., Furholt, M., Bača, M., Reiter, S. S., Vanzetti, A., Kotsakis, K., Petersson, H., Niklasson, E., Hølleland, H., & Frieman, C. J. (2021). Forum: Populism, identity politics, and the archaeology of Europe. *European Journal of Archaeology, 24*(4), 519–555.
Hofmanová, Z., Kreutzer, S., Hellenthal, G., Sell, C., Diekmann, Y., Díez-del-Molino, D., van Dorp, L., López, S., Kousathanas, A., Link, V., Kirsanow, K., Cassidy, L. M., Martiniano, R., Strobel, M., Scheu, A., Kotsakis, K., Halstead, P., Triantaphyllou, S., Kyparissi-Apostolika, N., & Burger, J. (2016). Early farmers from across Europe directly descended from Neolithic Aegeans. *Proceedings of the National Academy of Sciences of the United States of America, 113*(25), 6886–6891.
Jones, S. (2002). *The archaeology of ethnicity. Constructing identities in the past and present*. Routledge.
Kolář, J. (2020). Migrations or local interactions? Spheres of interaction in third-millennium BC Central Europe. *Antiquity, 94*(377), 1168–1185.
Kossinna, G. (1911). *Die Herkunft der Germanen*. Kabitzsch.
Kristiansen, K. (1998). *Europe before history*. Cambridge University Press.
Kristiansen, K. (2014). Towards a new paradigm? The third science revolution and its possible consequences in archaeology. *Current Swedish Archaeology, 22*(1), 11–71.
Kristiansen, K., Allentoft, M., Frei, K., Iversen, R., Johannsen, N., Kroonen, G., Pospieszny, L., Price, D. T., Rasmussen, S., Sjögren, K. G., Sikora, M., & Willerslev, E. (2017). Re-theorising mobility and the formation of culture and language among the Corded Ware Culture in Europe. *Antiquity, 91*(356), 334–347.

Květina, P., & Řídký, J. (2016). Neolithic settlement space: Waste, deposition and identity. In D. Sosna & L. Brunclíková (Eds.), *Archaeologies of Waste. Encounters with the unwanted* (pp. 127–144). Oxbow.

Larick, R. (1986). Age grading and ethnicity in the style of Loikop (Samburu) spears. *World Archaeology, 18*(2), 269–283.

Lazaridis, I., Nadel, D., Rollefson, G., Merrett, D. C., Rohland, N., Mallick, S., Fernandes, D., Novak, M., Gamarra, B., Sirak, K., Connell, S., Stewardson, K., Harney, E., Fu, Q., Gonzalez-Fortes, G., Jones, E. R., Roodenberg, S. A., Lengyel, G., Bocquentin, F., & Reich, D. (2016). Genomic insights into the origin of farming in the ancient Near East. *Nature, 536*(7617), 419–424.

Lemonnier, P. (2012). *Mundane objects: Materiality and non-verbal communication*. University College.

Longacre, W. A. (1981). Kalinga pottery: An ethnoarchaeological study. In I. Hodder, G. Isaac, & N. Hammond (Eds.), *Pottery of the past: studies in honour of David Clark* (pp. 49–66). Cambridge University Press.

Lubbock, J. (1865). *Pre-historic times, as illustrated by ancient remains, and the manners and customs of modern savages*. Wiliams & Norgate.

MacEachern, A. S. (1992). Ethnicity and ceramic variation around Mayo Plata, Northern Cameroon. In J. Sterner & N. David (Eds.), *Papers in Honour of Peter Lewis Shinnie* (pp. 211–230). University of Calgary Press.

Martens, J. (1989). The Vandals: Myths and facts about a Germanic tribe of the first half of the 1st millenium AD. In S. Shennan (Ed.), *Archaeological approaches to cultural identity* (pp. 57–65). Unwin Hyman.

Montelius, O. (1899). *Der Orient und Europa. Königl.* Akademie der schönen Wissenschaften, Geschichte und Alterthumskunde.

Morlot, A. (1861). General views on archaeology. *Washington, DC, Annual Report of the Smithsonian Institution for 1860*, 284–343.

Nielsen, R., Akey, J. M., Jakobsson, M., Pritchard, J. K., Tishkoff, S., & Willerslev, E. (2017). Tracing the peopling of the world through genomics. *Nature, 541*(7637), 302–310.

Papac, L., Ernée, M., Dobeš, M., Langová, M., Rohrlach, A. B., Aron, F., Neumann, G. U., Spyrou, M. A., Rohland, N., Velemínský, P., Kuna, M., Brzobohatá, H., Culleton, B., Daněček, D., Danielisová, A., Dobisíková, M., Hložek, J., Kennett, D. J., Klementová, J., & Haak, W. (2021). Dynamic changes in genomic and social structures in third millennium BCE central Europe. *Science Advances, 7*(35), eabi6941.

Parker Pearson, M., Chamberlain, A., Jay, M., Richards, M., Sheridan, A., Curtis, N., Evans, J. A., Gibson, A., Hutchison, M., Mahoney, P., Marshall, P., Montgomery, J., Needham, S., O'Mahoney, S., Pellegrini, M., & Wilkin, N. (2016). Beaker people in Britain: Migration, mobility and diet. *Antiquity, 90*(351), 620–637.

Pavlů, I., Rulf, J., & Zápotocká, M. (1986). Theses on the Neolithic site of Bylany. *Památky archeologické, 77*(2), 288–412.

Petri, F. (1937). *Germanisches Volkserbe in Wallonien und Nordfrankreich. Die fränkische Landnahme in Frankreich und den Niederlanden und die Bildung der germanisch-romanischen Sprachgrenze*. Röhrscheid.

Phillips, P. (1955). American archaeology and general anthropological theory. *Southwestern Journal of Anthropology, 11*(3), 246–250.

Pickrell, J. K., & Reich, D. (2014). Toward a new history and geography of human genes informed by ancient DNA. *Trends in Genetics, 30*(9), 377–389.

Rédei, A. C., & Ranta, M. (2016). Narrativity: Individual and collective aspects of storytelling. In *Human lifeworlds: The cognitive semiotics of cultural evolution* (pp. 145–162). Peter Lang.

Redman, C. L. (1999). The development of archaeological theory: Explaining the past. In G. Barker (Ed.), *Companion encyklopedia of archaeology* (pp. 48–80). Routledge.

Reich, D. (2018). *Who we are and how we got here: Ancient DNA and the new science of the human past*. Oxford University Press.

Rivollat, M., Jeong, C., Schiffels, S., Küçükkalıpçı, I., Pemonge, M.-H., Rohrlach, A. B., Alt, K. W., Friederich, S., Ghesquière, E., Gronenborn, D., Laporte, L., Lefranc, P., Meller, H., Réveillas, H., Rosenstock, E., Rottier, S., Scarre, C., Soler, L., Wahl, J., & Haak, W. (2020). Ancient genome-wide DNA from France highlights the complexity of interactions between Mesolithic hunter-gatherers and Neolithic farmers. *Science Advances, 6*(22), eaaz5344.

Rivollat, M., Rohrlach, A. B., Ringbauer, H., Childebayeva, A., Mendisco, F., Barquera, R., Szolek, A., Le Roy, M., Colleran, H., Tuke, J., Aron, F., Pemonge, M.-H., Späth, E., Télouk, P., Rey, L., Goude, G., Balter, V., Krause, J., Rottier, S., & Haak, W. (2023). Extensive pedigrees reveal the social organization of a Neolithic community. *Nature, 620*(7974), 600–606.

Roberts, B. W., & Vander Linden, M. (2011). Investigating archaeological cultures: Material culture, variability, and transmission. In B. W. Roberts & M. Vander Linden (Eds.), *Investigating archaeological cultures: Material culture, variability, and transmission* (pp. 1–21). Springer.

Sackett, J. (1977). The meaning of style in archaeology: A general model. *American Antiquity, 42*(3), 369–380.

Schiffer, M. B. (1987). *Formation processes of the archaeological record*. University of New Mexiko Press.

Schroeder, H., Margaryan, A., Szmyt, M., Theulot, B., Włodarczak, P., Rasmussen, S., Gopalakrishnan, S., Szczepanek, A., Konopka, T., Jensen, T. Z. T., Witkowska, B., Wilk, S., Przybyła, M. M., Pospieszny, Ł., Sjögren, K.-G., Belka, Z., Olsen, J., Kristiansen, K., Willerslev, E., & Allentoft, M. E. (2019). Unraveling ancestry, kinship, and violence in a Late Neolithic mass grave. *Proceedings of the National Academy of Sciences of the United States of America, 116*(22), 10705–10710.

Shennan, S. (Ed.). (1989). *Archaeological approaches to cultural identity*. Unwin Hyman.

Sklenář, K. (1983). *Archaeology in central Europe: The first 500 years*. Martin's Press.

Sonesson, G. (2016). Phenomenological semiotics of iconicity and pictoriality. *Language and Semiotic Studies, 2*(2), 1–73.

Soudský, B. (1967). *Principles of automatic data treatment applied on Neolithic pottery*. Institute of Archaeology of the CAS. (the manuscript is deposited in the archives of the Institute of Archaeology of the CAS, Prague).

Stark, M. T. (1994). Pottery exchange and the regional system. In W. Longacre & J. M. Skibo (Eds.), *Kalinga ethnoarchaeology. Expanding archaeological method and theory* (pp. 169–197). Smithsonian Institution Press.

Stark, M. T. (Ed.). (1998). *The archaeology of social boundaries*. Smithsonian Institution Press.

Stark, M. T. (1999). Social dimensions of technical choice in Kalinga ceramic traditions. In E. Chilton (Ed.), *Material meanings. Critical approaches to the interpretation of material culture* (pp. 24–43). The University of Utah Press.

Trigger, B. G. (1980). *Gordon Childe, revolutions in archaeology*. Thames and Hudson.

Trigger, B. G. (1996). *A history of archaeological thought*. Cambridge University Press.

Tylor, E. B. (1888). *Primitive culture. Researches into the development of mythology, philosophy, religion, language, art and custom*. H. Holt and company.

Veeramah, K. R. (2018). The importance of fine-scale studies for integrating paleogenomics and archaeology. *Current Opinion in Genetics & Development, 53*, 83–89.

Villalba-Mouco, V., Oliart, C., Rihuete-Herrada, C., Rohrlach, A. B., Fregeiro, M. I., Childebayeva, A., Ringbauer, H., Olalde, I., Celdrán Beltrán, E., & Puello-Mora, C. (2022). Kinship practices in the early state El Argar society from Bronze Age Iberia. *Scientific Reports, 12*(1), 22415.

Vyas, D. N., Koncz, I., Modi, A., Mende, B. G., Tian, Y., Francalacci, P., Lari, M., Vai, S., Straub, P., & Gallina, Z. (2023). Fine-scale sampling uncovers the complexity of migrations in 5th–6th century Pannonia. *Current Biology, 33*(18), 3951–3961.

Watson, P. J. (1995). Archaeology, anthropology, and the culture concept. *American Anthropologist, 97*(4), 683–694.

Welsch, R. L., & Terrell, J. E. (1998). Material culture, social fields and social boundaries on the Sepik coast of New Guinea. In M. T. Stark (Ed.), *The archaeology of social boundaries* (pp. 50–77). Smithsonian Institution.

Wenger, E. (1998). *Communities of practice: Learning, meaning, and identity*. Cambridge University Press.

Wiessner, P. (1983). Style and social information in Kalahari San projectile points. *American Antiquity, 48*(2), 253–276.

Wilson, D. (1851). *The Archaeology and Prehistoric Annals of Scotland*. Sutherland and Knox.

Wobst, M. (1999). Style in archaeology or archaeology in style. In E. Chilton (Ed.), *Material meanings. Critical approches to the interpretation of material culture* (pp. 118–132).

Chapter 6
Shifting Perceptions of the Greek Bronze Age: From Simple Valuative Judgements Like 'Civilised' or 'Barbaric' to a Comprehensive Implementation of the Archaeological Science

Tomáš Alušík

Abstract Greek Bronze Age cultures have been gradually discovered since the end of the nineteenth century. At first, in the 1870s and 1880s, Schliemann uncovered parts of the main centres of the Mycenaean culture on the Greek mainland, while Kalokairinos conducted a small-scale excavation at Knossos (Crete), the most important site of the Minoan culture. As a sign of respect to the high level of advancement of these cultures and in accordance with the cultural-evolutionary concepts of the period, both were soon labelled civilisations. But due to their specific features, the overall character of these civilisations was interpreted differently: while the Minoans came to be viewed as peaceful and 'cultured', the Mycenaeans were presented as aggressive, warlike, and almost 'barbaric'. This chapter follows the development of interpreting the Greek Bronze Age from the late nineteenth century until the present. I argue that our understanding of this period evolved from relatively clear and simple, but wrong, valuative conclusions about both cultures, which were influenced by the dominant contemporary philosophy of history and by certain Classical works of literature, towards more complex and less definitive, yet more accurate, conclusions based on a critical appraisal of the literary sources, evidence from archaeological sources, and a growing increasing of science-based methods.

Keywords Aegean · Greece · Bronze Age · Archaeology · Civilisation

T. Alušík (✉)
Institute for History of Medicine and Foreign Languages, First Faculty of Medicine, Charles University, Prague, Czech Republic
e-mail: tomas.alusik@lf1.cuni.cz

J. Toman et al. (eds.), *Culture Evolving*, Biosemiotics 10,
https://doi.org/10.1007/978-3-032-16124-6_6

6.1 Introduction

The discovery of Bronze Age[1] civilisations in Greece in the second half of the nineteenth century is one of the most famous stories in the history of archaeology. The excavations conducted by Heinrich Schliemann, a German merchant, archaeological enthusiast, and 'self-made archaeologist', which started in Mycenae in 1876, met with well-deserved attention and public admiration, not least because of their spectacular presentation in the press, exhibitions, and lectures. Although his early excavations were a hunt for antiquities rather than a methodical research, Schliemann in the 1870s and 1880s uncovered parts of the main centres of the Late Helladic or Mycenaean culture on the Greek mainland. At the same time, Minos Kalokairinos, also a businessman and amateur archaeologist, conducted a small-scale excavation on the Kephala hill at Knossos, the most important site of the Minoan culture. He was the first person to uncover traces of a Bronze Age civilisation in Crete. A little later, from 1900 onwards, British archaeologist Sir Arthur Evans conducted large--scale excavations at Knossos. He uncovered a huge palace-like structure, which he later named 'Palace of Minos' after the legendary ruler of Crete. Their excavations have, so to speak, extended Greek history several centuries into the past and brought to light various insights into the Greek Bronze Age.

As a sign of respect to the high level of advancement of both prehistoric cultures and in accordance with contemporary concepts of cultural evolution, both of the cultures were soon labelled 'civilisations' (cf. Chap. 5; on the biosemiotic perspective of the term 'civilisation', see Wheeler, 2006, pp. 31–33). But due to certain specific features, the overall character of these civilisations was interpreted radically differently: while the Minoans were presented as peaceful and 'cultured', the Mycenaeans were portrayed as aggressive, warlike, and almost 'barbaric'. What played a crucial role in interpreting the general character of the two civilisations was the role of violence in the society (presence or absence of fortifications or weapons in tombs, evidence of violence in human skeletal remains, etc.), the use of writing, the main iconographical elements used in the art, and the origin of particular individuals buried in the specific chronological horizon of the Late Bronze Age.

Thanks to recent scientific and methodological progress, we can now also build on the findings of osteoarchaeology, palaeopathology, and archaeological sciences in general (e.g., aDNA and stable isotopic analyses) (cf. Chap. 5). These new advancements led to new discoveries, and new discoveries led to a rethinking of previous interpretations. For example, newly discovered elements of defensive architecture in Crete highlighted the presence of various forms of violence and warfare in the Minoan culture which, in turn, inspired a thorough reappraisal of the 'peacefulness' of that civilisation.

[1] In the territory of modern Greece, the Bronze Age dates to circa 3500/3000 to 1075/1050 BCE and is divided in three main periods: early (c. 3500/3000–2100/2050 BCE), middle (c. 2100/2050–1700/1675 BCE) and late (c. 1700/1675–1075/1050 BCE). Each of these periods can be further subdivided in shorter phases, labelled I, II, III, and horizons, labelled A, B, C, etc. (e.g., Manning, 2010). Slightly different regional chronologies exist for the mainland, the Cyclades, and Crete. In the text below, absolute dates are mentioned when referring to the individual phases/horizons.

In general, I will argue that our understanding of the Greek Bronze Age evolved from relatively straightforward, but wrong, valuative conclusions about both cultures, which were shaped by contemporary philosophy of history and cultural evolution and by the testimony of Classical works of literature, which have been studied for centuries, towards more complex and less definitive, but more accurate, conclusions based on a critical re-appraisal of the literary sources and archaeological evidence that profited from a growing integration of science-based methods.

In the following, I describe in detail certain relevant explorations and discoveries, summarise their interpretations, and analyse their point of view. I trace how the Greek Bronze Age was interpreted from late nineteenth century until the present, from the classical schematic explanations based on contemporary notions of cultural evolution, through theories of cultural exchange, and all the way to their reinterpretation and a formulation of more complex accounts since the end of the 1970s, including the biosemiotic perspective (Cobley, 2016; Dunér & Sonesson, 2016; Markoš, 2015; Wheeler, 2006). I will proceed chronologically and describe the discoveries and their changing interpretations from the time when the Greek Bronze Age was uncovered in the 1870s until the present.

6.2 Discovering the Greek Bronze Age: Calvert, Schliemann, Kalokairinos, Tsountas, Halbherr, and Evans

6.2.1 Expanding into Greek Prehistory: Calvert, Schliemann, and Others

As late as the mid-nineteenth century, it was believed that Greek ancient history began in 776 BC with the first recorded panhellenic games at Olympia. A vague awareness of ancient times was captured—not only for us, but also for Greeks of the Archaic and Classical period—by epic poems, especially Homer's Iliad and Odyssey and Hesiod's Theogony (cf. Chaps. 1 and 2). Homer recounts parts of the events described in the Trojan mythological cycle, which deals with a mythical campaign of a united Greek army[2] against Troy, a city on the coast of Asia Minor, and the harrowing return home from Troy of the hero Odysseus. In the nineteenth century, the authorship of the Homeric epics, including the historicity of Homer and the events described, became a prominent research topic in Classical philology.

Although before the Second World War, the Classical culture, Greek, and Latin had formed an essential component of education in the Western civilisation, serious scholars mostly doubted the historicity of events described by Homer. They were sceptical although it was known that some of the places where the heroes allegedly lived still existed, at least as local settlements and in some cases even with ancient ruins nearby (e.g., Cline, 2013).

Nevertheless, there were also some people who came to believe that there may be a grain of truth in the Homeric epics. One of them was the German merchant and

[2] Homer refers to the Greeks most often as Achaeans, Danaans, or Argives.

Fig. 6.1 The citadel of Mycenae (Argolid, Peloponnese): Detail of Late Helladic IIIA-B (c.1420/1410–1200/1190 BCE); fortifications with the Lion Gate. Photograph by the author

self-taught archaeologist Heinrich Schliemann (1822–1890), who was almost fanatically convinced about the historicity of places, persons, and events Homer had depicted (see e.g., Fitton, 1995, pp. 48–103). On his travels in northwestern Anatolia in 1868, he met with Frank Calvert (1828–1918), a native Briton who served as the American viceconsul in the Turkish sultanate. Calvert believed that the hill of Hisarlik was actually the site of ancient Troy.[3] Schliemann agreed to cooperate with him and in 1870 began an excavation of Hisarlik. During three excavation seasons (1870–1873), he uncovered a large part of the site with nine main settlement layers and numerous magnificent finds. Schliemann subsequently interpreted Hisarlik definitively as Troy, claiming that 'the town of Priam' was in the second level from the bottom (Troy II, now dated to about the middle or third quarter of the third millennium BCE; Cline, 2013).[4]

After his successes in Troy, Schliemann decided to explore residences of the Greek rulers who were supposed to have taken part in the Trojan War. First, he focused on Mycenae, where the remains of a citadel with massive enclosure walls were still visible on a low hill about 2 km above the modern village of Mykines (Fig. 6.1).

[3] The 'hill' was in fact a 'tell', that is, a mound formed by long-term settlement and gradual accumulation of settlement layers on top of each other.

[4] It was later shown that the Trojan War—if such conflict that may have inspired the Homeric epics took place at all—could only date to around 1200 BCE, which would correspond to the level of Troy VIIa in the Trojan stratigraphy. As was later demonstrated, the Trojan culture had influence on, extensive contacts with, and imported from areas and cultures of the Greek mainland and islands (especially during the Troy VI and VII phases).

Fig. 6.2 The silver Siege Rhyton from Mycenae (Grave Circle A, Shaft Grave IV), showing a fortified town in front of which a battle is taking place (see the drawing). Late Helladic I (c. seventeenth century BCE). National Archaeological Museum of Athens, inv. no. 481. Photograph by the author

Schliemann put in the first few test trenches in early 1874, selecting the trench sites based on information from Homer and other ancient authors. He was particularly keen on finding the final resting places of King Agamemnon, commander of the united Greek army at Troy, and his family members and retinue. In the summer of 1876, he thus launched full scale excavations—with success.

Not far behind the main 'Lion Gate' to the Mycenae citadel, Schliemann discovered a circle of rich shaft graves bordered by a double row of stone slabs, which came to be known as the Grave Circle A. In it, Schliemann uncovered five shaft graves with luxurious funerary equipment, including richly decorated weapons, vessels, and silver and gold objects weighing almost 15 kg in total (see Fig. 6.2). Among these were also several golden masks which had covered the faces of several deceased individuals.[5] Schliemann interpreted these finds as the burial site of King Agamemnon and his retinue because, according to ancient authors, they were buried in five graves (cf. Chap. 2 to see how this reflects the historical practice of ascribing archaeological findings to Biblical or historical figures). But just one year later, Panagiotis Stamatakis, the Greek Ephor of Antiquities who supervised Schliemann's explorations, discovered in the same grave circle a sixth grave (see Papazoglou--Manioudaki et al., 2009). Although it was convincingly proven later in the twentieth century that these graves and artefacts are several centuries older than the supposed

[5] These masks are now on display in the Prehistoric Hall of the National Archaeological Museum in Athens, where they are among the most important items in its collections. They remain the most famous artefacts of the entire Mycenaean civilisation of the Bronze Age.

time of the Trojan War,[6] Schliemann did discover a previously unknown Greek Bronze Age culture, which came to be known as the Mycenaean culture after the site where it was first found.[7] In 1884, Schliemann also carried out smaller excavations at the citadel of Tiryns, about 4 km from the town of Nafplion, with enormous, up to 17 m wide, fortifications constructed in 'Cyclopean' masonry, that is, made of massive stone blocks roughly fitted together without the use of mortar.

Because the Mycenaean period layers were well preserved and explored in the nineteenth century only at a few sites, the initial assessment of the Mycenaean culture was likewise largely based on the ancient texts, especially the Homeric epics. The Mycenaeans were thus portrayed as strong, aggressive warriors, and this view found support in the abundance of weapons found, massive fortifications, and numerous items with warlike iconography.[8] Although Schliemann and his successor in Mycenae, the Greek archaeologist Christos Tsountas (1857–1934) who worked in Mycenae in 1884–1902 (see e.g., Fitton, 1995, pp. 104–109), discovered during their excavations aside from weapons also an abundance of high-quality pottery and artistically highly accomplished objects, they failed to find all of the characteristics of an advanced civilisation (cf. Chaps. 1 and 5): specifically, they found no evidence of writing. As a result, the Mycenaean culture came to be viewed in the first decades after its discovery (and until about the 1920s) as rather crude, underdeveloped, and almost barbaric. This view was certainly influenced by progressivist cultural evolutionism, the prevailing anthropological model of the time (Kundt, 2015) (cf. Chaps. 3, 5, 7 and 12), which assumes certain stages which all cultures go through, such as savagery, barbarism, and civilisation. Given the state of knowledge about the Mycenaean culture in the first decades after its discovery and the absence of any major 'civilisational element' (see above), Mycenaean culture was thus at the time interpreted as rather 'barbaric'.

6.2.2 *Discovery of the Minoan Civilisation*

At the very end of the nineteenth century, in 1895–1899, the English archaeologist Arthur Evans (1851–1941; see e.g., Fitton, 1995, pp. 115–139; Branigan, 2000; Cadogan, 2000; Marinatos, 2015) travelled, sometimes together with another

[6] They are nowadays usually dated to the beginning of the Late Bronze Age (seventeenth century BCE).

[7] The general term 'Helladic' is currently used for the Bronze Age culture on the Greek mainland, while the term 'Mycenaean' sometimes refers specifically to its Late Bronze Age phase.

[8] Schliemann had discovered in one of the shaft graves of Grave Circle A (Shaft Grave IV) fragments of two silver vessels with representations of battle scenes. The 'Siege Rhyton' (National Archaeological Museum of Athens, inv. no. 481) shows a battle scene with over thirty armed figures taking place in front of a fortified town; see Fig. 6.2 (cf. e.g., Papadopoulos, 2019, with references). The 'Battle Krater' (National Archaeological Museum of Athens, inv. no. unknown) depicts two groups of four warriors engaged in combat in an axial composition of a continuous frieze; see, e.g., Blakolmer (2007) with references. Both vessels come from the same context and date to the Late Helladic I (ca. seventeenth century BC).

English archaeologist, John L. Myres (1869–1954; see e.g., Brown, 2000, pp. 9–12), through East Crete. They saw numerous strange remains of buildings made from the massive Cyclopean masonry (see e.g., Evans & Myres, 1895; for good reviews of these travels, see Brown, 2000, 2001). In his notes and diaries, Evans called them 'forts, strongholds, castles, guard stations, and acropoleis', eventually used the Greek term 'phrourion' (meaning 'castle'),[9] which shows that he assumed and emphasised their military function and importance. In accordance with contemporary knowledge and based on their construction technique, he assumed their Mycenaean origin (Evans & Myres, 1895).

Evans was a British gentleman who was educated at Oxford. In the early 1890s, he made his way to Crete to search for the origins of tiny artefacts known as 'milk stones', which were believed to help pregnant women and new mothers. These were engraved seals or tokens with special decorations, including symbols or characters that could be interpreted as writing. While searching in Crete, Evans came across the Kephala hill near the modern village of Knossos, about 6 km south of Herakleion, the current capital of Crete.[10] Evans realised the importance of the site and purchased a part of the hill. He started his excavations there in March 1900, and by 1903 uncovered most of the 'palace', as this large complex of several hundred rooms grouped around a large central courtyard came to be known (Brown, 1983). The abundant finds of utilitarian and artistic objects, elaborate architectural features, absence of any characteristic elements of fortification (or generally of defensive architecture), in conjunction with evidence of writing pointed to the existence of a hitherto unknown advanced civilisation that predated the Mycenaean culture. This impression was confirmed by the technically precise and artistically exquisite ceramics, traces of fresco decorations in the palace and, above all, by examples of several writing systems, mostly on clay tablets, including the Linear B script (see e.g., Palaima, 2010). Based on the Greek mythological tradition, Evans identified Knossos as the seat of the legendary King Minos, a ruler of Crete and later one of the mythical judges in the Greek Underworld (see Chap. 2). Consequently, he thus named the entire civilisation 'Minoan' after this important mythical figure (see e.g., Brown, 2000). He also worked out (with Duncan Mackenzie, 1861–1934) its first chronology (Evans, 1906a; Mackenzie, 1903). The 'Minoan' civilisation was an advanced Bronze Age civilisation; Evans soon divided it in the usual three phases (early, middle, and late) corresponding to contemporary ideas of the evolution of culture involving a phase of growth, peak, and decline (cf. Chap. 4). With some interruptions, Evans excavated at Knossos for nearly 30 years.

In the early twentieth century, excavations and research in Crete continued thanks to the efforts of Evans but also others, such as the Italian Frederico Halbherr (1857–1930), who started to uncover the second most important Minoan palace at Phaistos in south central Crete in 1900, the Americans Richard Seager, Harriet

[9] Nowadays the term 'guard house' or 'megalithic farmstead' is commonly used to describe buildings of this type (see below, also Chryssoulaki, 1999; Alušík, 2007, pp. 124–135).

[10] Minos Kalokairinos (1843–1907), an amateur archaeologist, had already previously conducted small-scale excavations there in the late 1870s (see e.g., Fitton, 1995, p. 123).

Boyd, and Edith Hall, or the Greeks Stefanos Xanthoudides and Joseph Hazzidakis (see e.g., Fitton, 1995, pp. 140–146; Branigan, 2000; McEnroe, 2002). Their findings had shown that until the beginning of the Late Bronze Age, the Minoan civilisation had a great influence on the Helladic (Mycenaean) civilisation and was thus more 'advanced'. This is evidenced, for instance, by various artefacts from Mycenaean citadels and tombs, including the famous Grave Circle A at Mycenae. Many of these objects are either of Cretan-Minoan provenance or they were influenced by Minoan styles. This conclusion probably reflected the growing popularity of diffusionism (see Chaps. 5, 7 and 12), which—in reaction to the increasing amount of new historical and archaeological data—was gradually displacing the original evolutionistic paradigm of primacy of endogenous cultural change and brought new insights into the mutual relationships and nature of the cultures under discussion.

Even so, Evans had long believed that the Mycenaeans were not very advanced and their culture was merely derivative, driven by Minoan impulses (see e.g., Fitton, 1995, pp. 134–135; Galanakis, 2014). He was also convinced that the dominance of Minoan Crete was not only cultural but also economic and military, consequently ascribing the strong Minoan influence on the Greek mainland (and some islands) to their political control of these areas (see e.g., Wiener, 2016). Aside from various references made by Greek authors of the Classical period, it was the numerous Minoan features in Mycenaean culture that reinforced his conviction that Crete's economic and military superiority was mirrored in the material culture.[11]

Evans was impressed by artefacts of a high utilitarian and artistic value and by the elaborate architecture and furnishings of the palace, which included light wells, staircases, an elaborate drainage system, etc. Although the palace was not protected by massive fortifications in Cyclopean masonry, Evans did discover within its grounds sections of walls and several special structures that probably had a defensive purpose, and he interpreted them as such (Alušík, 2007, pp. 74–85). For example, in 1930 Evans discovered on the outer side of the Western Court of the palace a section of the exterior fortification of the Old Palace that came to be known as the Outer Enceinte Wall. He also identified several rooms located in entrances to the palace or near them as 'guard rooms, warder's lodge, and porter's lodge' and concluded that this was where the guards would have checked the incomers.

6.2.3 *Reconstructions at Knossos: A Re-creation of the 'Art Nouveau Minoans'?*

Since many of the artefacts discovered survived only in fragments, Evans tried to reconstruct them and invited artists to join him in this effort. The task of reconstructing the original composition of the frescoes (including the stucco reliefs) fell to the

[11] Only in the second half of the twentieth century did it become clear that the relationship between the Minoan and Mycenaean civilisations was much more dynamic and complex, with alternating mutual influences; see below.

Swiss painter Émile Gilliéron (1850–1924), who later worked with his son (Stürmer, 2004), and to the English artist and architect Piet de Jong (1887–1967; Papadopoulos, 2007). Gilliéron and de Jong approached the reconstructions in a bold and overly creative way, often using just a few fragments to paint entire figures or settings, eventually repeating the basic motif several times within the composition. Some of Gilliéron's reconstructions later proved to be fundamentally incorrect, such as the fresco known as Saffron Gatherer and even the famous relief fresco of the Priest-King (Fig. 6.3). Gilliéron also produced drawings depicting the ideal form of some parts of the palace and its supposed inhabitants, sort of genre pictures of Minoan everyday life. It should be noted that his work reflected the spirit of his times (see e.g. Blakolmer, 2008), which is why the original reconstructions of the large fresco panels from Knossos, exhibited in several museums, are reminiscent of early twentieth-century Art Nouveau style fashionable at that time (cf. Chap. 14). The reconstructed frescoes and other artefacts make a highly polished and subtle impression: they show well-groomed women and toned men watching or participating in various ritual acts, and sometimes also animals and the natural environment. A similar overall sublime and advanced character of Minoan art has been confirmed by artefacts found at the same time by other archaeologists in southern and eastern Crete (see above). In contrast to the Mycenaean art, however, war motifs or battle scenes were almost absent.

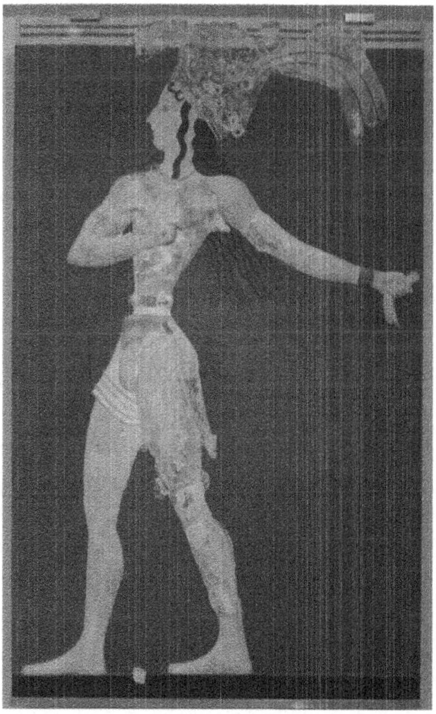

Fig. 6.3 Relief fresco of the Priest King (or Prince of the Lilies) from the Palace of Knossos, as reconstructed by Emile Gilliéron Jr. Middle Minoan IIIB to Late Minoan I (c. 1700–1470/1460 BCE). Heraklion Archaeological Museum, fresco room, on display. Photograph by the author

Fig. 6.4 The Palace of Knossos (central Crete): South Propylaeum, as reconstructed by Arthur Evans. Photograph by the author

Aside from the frescoes and some smaller artefacts made of various materials, Evans had also reconstructed some parts of the palace using modern materials, such as cement (Cadogan, 2000, pp. 20–24; Hitchcock & Koudounaris, 2002; Figs. 6.4 and 6.5). He drew on representations of various architectural elements in frescoes, seals, rings, and (terracotta) models of buildings. Unfortunately, in some cases he chose to reconstruct foundations that were poorly preserved and their exact form and original alignment uncertain—and some of these reconstructions were later proved wrong. One example is the so-called South Propylaeum in the southern part of the west wing of the palace, which was reconstructed as a propylon, that is, a monumental entrance with columns, behind which a large staircase leads to the upper floor (Fig. 6.4). In this case, the original foundations were poorly preserved in situ and the presence of columns in the entrance is highly uncertain.[12]

These reconstructions and re-creations of Minoan architecture and art became widely known in the Western world and sparked a wave of interest in both Minoan and Mycenaean art. In addition to Evans, the credit especially goes to the Gilliérons, both father and son,[13] and to de Jong. The scenes and reconstructions they created had influenced the art of the time, with some elements making their way into

[12] Moreover, it is likely that Evans was inspired in this reconstruction by a similar (clearly visible) feature in the palace at Phaistos, which was at that time excavated by Evans' friend, the aforementioned Frederico Halbherr (Hiller, 1980).

[13] He even set up a company that made replicas of Minoan artefacts and watercolour copies of frescoes and genre scenes from the Knossos palace.

Fig. 6.5 The Palace of Knossos (central Crete): North Entrance, as reconstructed by Arthur Evans. Photograph by the author

architecture.[14] Their creations shape the perception of Minoan archaeology to this day. The phenomenon of 'Cretomania' (Momigliano & Farnoux, 2017) is a good example of cultural elements being interpreted in a completely new context, helping to form the cultural milieu of the time, and retroactively affecting our thoughts about the culture in question in a sort of semiotic triangle.

6.2.4 *Minoan Thalassocracy and Pax Minoica*

The scenes depicted in Minoan art were generally peaceful and exuded a refined, sophisticated atmosphere. This applied especially to the reconstructed frescoes, which showed persons with distinguished expressions engaged in all kinds of

[14]A good example is the work of Jože Plečnik (1872–1957), a prominent Slovenian architect whose activity is linked to Prague, where he reconstructed and rebuilt some long unoccupied and decaying parts of Prague Castle. In the Bull Staircase, which connects the Third Castle courtyard and the Southern gardens, he used the characteristic Minoan columns in the form known from Minoan frescoes—with a pronounced rounded head and a downward tapering shank (see e.g., Alušík, 2004). For the phenomenon of 'Cretomania', see Momigliano and Farnoux (2017).

activities except for war. This, in conjunction with the absence—or rather neglect—of discovered elements of defensive architecture (at least not in the form of the massive Cyclopean fortifications known from the Mycenaean citadels) was responsible for the impression that the Minoans were a noble and peaceful people, at least within their island. Evans was inspired by the accounts of the Greek historians Herodotus (c. 484–425 BCE; especially his *Histories* III, 122.2) and Thucydides (460/455—c. 399 BCE, especially *The History of the Peloponnesian War* I, 4) which spoke of Minoan Thalassocracy, that is, naval supremacy. Based on these accounts but also evolutionary theories and nineteenth-century colonialism, Evans formulated a hypothesis of a 'Minoan peace', *Pax Minoica* (see e.g., Evans, 1928, pp. 571). According to this conception, clearly inspired by the Roman concept of *Pax Romana* and strongly defended until the end of the 1970s, Minoan towns and palaces did not have defensive walls because any possible threats would have been eliminated by a strong Minoan fleet at a sufficient distance from the Cretan shore (see below and Wiener, 1990; Borowka, 2018, pp. 7–9; cf. also Chap. 14). On the island itself, peace was clearly maintained, especially during the Neopalatial Period.[15] Evans did not completely rule out the existence of a Minoan warrior class or army, eventually the possibility of some military activities, but references to such possibilities and a handful of descriptions of elements of defensive architecture found on the site passed within his massive four-volume *Palace of Minos* almost unnoticed (Alušík, 2007, pp. 74–85). The overall impression of the Minoan civilisation, shaped by numerous beautiful artefacts, especially murals (mostly reconstructed or modified by Gilliéron and de Jong), seems to have been very positive. As a result, Evans's contemporaries viewed Minoans as a peace-loving nation, fond of art and flowers. This initial view had palpable consequences, such as the initial marginalisation of Evans's original interpretation of the site, failure to take into consideration the fortification walls and other elements of defensive architecture of the Knossos Palace, or the assumption that Minoan palaces were protected solely by palace guards who inspected incomers in 'guardrooms' or 'porter's lodges' (and thus served as a kind of security service; Alušík, 2007, pp. 81–82).

In 1918, the archaeologists Carl W. Blegen (1887–1971; see e.g., Fitton, 1995, pp. 155–165) and Alan J. B. Wace (1879–1957; see e.g., Fitton, 1995, pp. 150–155) published a basic tripartite chronological division of the Bronze Age civilisation of central and southern Greek mainland. They called this civilisation 'Helladic' and, reflecting Evans's scheme for Minoan Crete, divided it in three main periods (early, middle, and late) (Wace & Blegen, 1916–1918), whereby the Late Helladic period has been since the 1920s usually called 'Mycenaean'. In any case, from the interwar period onwards, the existence and nature of warfare, or more generally the relationship to organised violence, had gradually led to a redefinition of the overall character, historical role, and contributions of the two main Aegean civilisations.[16] In the case of Minoan Crete, this included the dispute over the absence of defensive architecture.

[15] On the other hand, Thucydides does not deny the warlike nature of the Minoans when he explicitly states that Minos settled the Cycladic Islands by expelling the original inhabitants, which of course cannot have been done without force.

[16] Crucial cultural/civilisational achievements such as the knowledge or use of writing were, of course, not overlooked.

6.3 Interpretation of Aegean Civilisations from the Interwar Period Until the End of the 1970s: A Slow Toppling of Traditional Ideas

6.3.1 The Interwar Period: Work in Crete and on the Mainland, Blegen at Troy and Pylos, and the First Physical Anthropology Studies

Although the notion of Minoan Thalassocracy and the *Pax Minoica* image of 'flower lovers' in various ways contradict each other, aggressive and either warlike or defensive features of the Minoan material culture have long been ignored. What was emphasised was this culture's sophistication, peacefulness, and 'civilised nature', especially in contrast to the Mycenaean culture. Based on Evans' authoritative influence (and to some extent also a misunderstanding of his original ideas), these images for a long time shaped the perception of the basic character of the Minoan civilisation and its relation to the Mycenaean culture. The conviction about a peaceful character of the Minoan civilisation was so entrenched that it persisted although, even before the Second World War, remains of defensive architecture and related artefacts associated with this civilisation were discovered both in Crete and on the Greek mainland.

Moreover, since the beginning of the twentieth century various Mycenaean artefacts, excavated and sometimes only briefly described by Schliemann, were subjected to a thorough study and the results were gradually re-published. Many of them, often despite their high artistic and craft quality, tended to reinforce the traditional view of the bearers of Mycenaean culture as aggressive warriors for whom war was an important cultural and civilisational feature.[17]

In 1932–1938, Blegen conducted thorough revision excavations of Troy. With his team, he recognised a total of 46 chronological phases within the nine main levels of Troy, and identified level VIIa, which had been destroyed by extensive fire, as the Homeric Troy (Blegen et al., 1950; Fitton, 1995, pp. 158–161). Over 70 years after the Calvert and Schliemann's excavations, the dispute over the existence of the 'Town of Priam' and historicity of the Trojan War thus finally had a solid archaeological basis (cf. Chap. 18) and no longer rested solely on the results of literary analysis and Classical philology. In 1939, Blegen even discovered a large complex on the hill of Ano Englianos in Messenia (southwestern Peloponnese), which soon came to be known as 'Nestor's Palace' (see e.g. Blegen & Rawson, 1966). Although this complex was not fortified with a massive wall like Mycenae or Tiryns, it was an important regional centre of Mycenaean civilisation in this part of the Peloponnese

[17] For example, in the early 1920s, there were published fragments of frescoes from the megaron in Mycenae (dated to Late Helladic IIIB, c. 1330/1315–1200/1190 BCE), which showed parts of a larger multi-storied building but also a battle. Although only the lower part of the body of a warrior sinking to the ground and the lower belly of a horse survived, the scene was reconstructed as a war chariot with an armed crew (see Rodenwaldt, 1921).

(Blegen & Rawson, 1966; Fitton, 1995, pp. 161–165). Numerous wall paintings were discovered in the palace, including a scene of a battle between warriors with typical Mycenaean weaponry (dagger, short sword, spear, helmets decorated with boar's teeth) and men dressed in animal skins and short trousers or skirts, perhaps some 'barbarian' raiders or pirates, which attracted a lot of attention.[18] Among the most significant artefacts excavated here were about a thousand clay tablets inscribed with the Linear B script (see e.g., Bennett, 1951) known already from Evans' excavations at Knossos. The Mycenaean culture had thus met one of the important criteria of a 'full-fledged' civilisation, putting to rest the long-held opinion (espoused by many, including Evans) that it was not very advanced or even primitive (Galanakis, 2014).

In the late 1930s, the concept of *Pax Minoica* also started to come under attack. The first to question it was the British archaeologist John D. S. Pendlebury (1904–1941; see e.g., Fitton, 1995, pp. 166–170; Branigan, 2000), who confirmed the existence of many Evans's 'forts' and 'guard stations' but also discovered many new similar locations. In his article, Pendlebury discussed the prehistoric 'forts/guard stations' in the area, emphasised their military purpose and questioned the concept of *Pax Minoica* (Pendlebury, 1936/1937). Shortly later, he published his most important work where he, among other things, supported the idea of Minoan thalassocracy (Pendlebury, 1939).

The interwar period also witnessed a gradual incorporation of physical (biological) anthropology into Aegean prehistoric and Classical archaeology, although it began to play a more significant role only from the 1970s onwards.[19] In Greece, the first works that used physical anthropology to study human skeletal remains of prehistoric populations were published in the 1930s (see e.g., Fürst, 1930). In the late 1930s, J. L. Angel (1915–1986), a prominent American physical (biological) anthropologist who had a great impact on the development of the discipline in the US, started to work in Greece (see e.g., Jacobsen & Cullen, 1990). Nevertheless, before the early twenty-first century, physical anthropology or osteoarchaeology entered the overall discourse on the nature of Bronze Age Aegean civilisations only a few more times, chiefly in the 1970s and 1980s.

[18] Found in Hall 64 and dated to Late Helladic IIIB2 (c. thirteenth century BCE); see e.g., Chapin (2016).

[19] As late as the end of the nineteenth century, human remains uncovered during archaeological excavations were still barely taken into consideration, among other things because there was still almost no interdisciplinary cooperation between Classical archaeology (itself still in the process of emerging and forming), prehistory, and physical anthropology. By the 1930s, physical anthropologists were already trying to understand not only the biological composition of populations (including pathologies), but also their origins, movement, mixing, and race issues. Within the preserved skeletal material, most attention was paid to skulls and one of the most widely used methods was craniometry, which is a system of measuring the external dimensions and angles of the skull based on precisely defined anthropometric points (cf. Chap. 5).

6.3.2 1950s to 1970s: Deciphering Linear B and Beginning of the Debate on the Destruction of Minoan Palaces

After the Second World War and Greek Civil War, excavations resumed in some of the main centres of Aegean Bronze Age civilisations (e.g., Mycenae, Pylos, Knossos, and Troy) in the 1950s and the results of excavations conducted just before the war were gradually processed and published. This was also a time of one of the most important achievements of Aegean archaeology: the deciphering of Linear B script, which was important not only for interpreting the overall character of the Minoan and Mycenaean civilisations but especially for learning about their mutual relationship. This feat was accomplished by the British architect and amateur philologist Michael Ventris (1922–1956) in 1953 (Ventris & Chadwick, 1953; see also Fitton, 1995, pp. 175–178), who realised that the script recorded an early form of the Greek language. In the last phase of this effort, he was assisted by the Cambridge-based philologist John Chadwick (1920–1998).[20] The deciphering of Linear B not only established a new linguistic discipline, Mycenaeology, but also had far-reaching implications for Aegean Bronze Age archaeology.

Evans originally dated the tablets with this script found at Knossos to the Late Minoan II period (c. 1470/1460–1420/1420 BCE), which followed the destruction of the New Palace of Knossos (Evans, 1935; Fitton, 1995, p. 133; Hatzaki, 2000, pp. 151–155). Because the material culture, including works of art, suggested a strong cultural influence of Minoan Crete on Mycenaean Greece, especially at the beginning of the Late Bronze Age (Late Minoan/Helladic), Evans concluded this was due to Minoan political control or strong influence (see e.g., Fitton, 1995, pp. 133–135; Wiener, 2016). Presence of the same type of script in the main palace centres of both civilisations could be taken as evidence of their close relationship and Cretan dominance, but the Knossos tablets date to a period after the fall of the New Palace, that is, to a time after the main period of the flowering of Minoan civilisation. They thus indicate the opposite: an incorporation of Crete into the sphere of Mycenaean civilisation. For some time, these tablets were even taken as evidence of a conquest of Knossos by the Mycenaeans (or at least their presence). The Mycenaeans were thus held responsible for the fall of the advanced Minoan civilisation, which had to bow to a militarily stronger aggressive neighbour—as happened much later to the Hellenistic states in the neighbourhood of the Roman Empire (see e.g., Hood, 1985; Wiener, 2015). From the perspective of cultural and social anthropology, the warlike character of the Mycenaean civilisation and the aggressive features of its ruling class were thus reaffirmed—not only by much later, but now also by the contemporary literary sources. But Evans' dating of the Linear B tablets from Knossos to the Late Minoan II phase was since the 1950s increasingly questioned and attention drawn to identical epigraphic or palaeographic features of some of the

[20] Together, they prepared for publication a book called *Documents in Mycenaean Greek*, which was published in 1956 just a few weeks after Ventris' sudden death (Ventris & Chadwick, 1956; second edition in 1973).

tablets from Knossos, Pylos and Chania, which were conclusively dated to the Late Helladic/Minoan IIIB, that is, to c. 1330/1315–1200/1190 BCE (see e.g., Hallager et al., 1992). Although the dispute over the dating of the Knossos tablets (Late Minoan II/IIIA1 versus Late Minoan IIIA2/IIIB1) still has not been fully resolved, their presence strongly suggest an adoption of some essential elements of Mycenaean culture and an incorporation of Crete into the sphere of Mycenaean influence.

In general, post-war Minoan archaeology was still strongly influenced by the conception of *Pax Minoica* and Minoan Thalassocracy until the 1980s, although Starr (1955) had challenged the thalassocracy concept much earlier, calling it a 'myth' (see e.g., Wiener, 1990). Until the 1980s, the discoveries of defensive architecture were treated as exceptional or isolated cases (Alušík, 2007, p. 10). An important shift in the interpretation of Greek Bronze Age civilisations and research priorities in their archaeology took place in the 1970s due to developments in philosophy and human sciences, including biosemiotics (Cobley, 2016; Markoš, 2015; Wheeler, 2006). These influences led to far-reaching changes both in archaeological theory and archaeological methods. One of the most fundamental theoretical works of Aegean prehistoric archaeology was published in the early 1970s (Renfrew, 1972) but although this was a time of a theoretical departure from the classical cultural evolutionism of the nineteenth century, many of the basic ideas persisted. For example, the transition between the Neolithic and the Bronze Age in Greece was, based on Service's conceptualisation, interpreted as a transition from tribes to chiefdoms or chiefdom states (Service, 1962; see also Hamilakis, 2002a, pp. 7–10).

In 1973, Angel published the results of his research on the skeletal remains of individuals buried in both grave circles at Mycenae. These individuals show a significantly better overall level of health than individuals from nearby contemporary sites such as Lerna or Asine. Angel emphasised their robustness, good physical condition, and skeletal evidence for involvement in strenuous activities (musculoskeletal markers of physical exertion and load). Moreover, most of the male individuals were perfectly capable of fighting and warfare and their engagement in such activities is further supported by their relative youth: only one individual was about 35 at the time of death, while another ten died before reaching 30 years of age (Angel, 1973).

In 1979, there was uncovered a large building, probably a sanctuary, at the site of Archanes-Anemospilia, on the northern slope of Mount Juktas in north-central Crete, about 7 km south of Knossos (see e.g., Sakellarakis & Sapouna-Sakellaraki, 1997). The finds included the skeletal remains of several people who were buried by the same earthquake that destroyed the building at the turn of the Middle Minoan IIB/IIIA periods (c. 1750/1700 BCE). In the so-called western room, there were discovered the skeletal remains of three individuals, including a woman about 28 years old and a strong tall man about 37 years old. Most interesting, however, were the remains of an approximately 18-year-old male who was lying on a rectangular platform, interpreted as an altar. These are some of the most controversial human skeletal remains from prehistoric Greece. The position of the body on its (right) side seems to indicate that he was tied with his hands and feet backwards. Placed on his body was a bronze weapon. Moreover, the right half of his remains is significantly darker than the left, which may indicate that he bled to death. According to the excavators, this young man was probably killed as a human sacrifice on the platform-altar: his throat was cut and he bled to death (Sakellarakis & Sapouna-Sakellaraki, 1997, pp. 305–11; see also Hughes, 1991, pp. 13–17; Warren, 2015). This interpretation, of course, caused a shock within Minoan archaeology, because it was the first piece of potential (osteo)archaeological evidence of human sacrifice from prehistoric and ancient Greece. Until that time, none of the archaeologists or historians could even imagine that these culturally advanced and more or less peaceful 'flower-lovers'—as the Minoans were still viewed—could have practiced such cruel rituals. While some archaeologists have accepted the interpretation of the debated remains as human sacrifice, others had reservations and dismissed it. Unfortunately, neither osteoarchaeology nor paleopathology can confirm or refute this unsettling issue.

The 1970s were also extremely important in drawing attention to Minoan defensive architecture. In 1976, Alexiou read his paper at the fourth International Cretological Congress in Heraklion (Alexiou, 1980), where he emphasised the number of sites either outright fortified or defensively situated (on high or rocky hills), which were built during all periods of Minoan history except for the Neopalatial period. In short, the discovery of a potential human sacrifice and both increased attention to and a growing number of examples of Minoan defensive architecture had finally significantly shaken the idea of 'cultural maturity' and non-violent nature of the Minoan civilisation.

6.4 New Discoveries, Methods, Interdisciplinarity, and Archaeological Science in the Past Forty Years: More Questions Than Answers?

6.4.1 1980s to 1990s: Extensive Fieldwork, New Methods, and a Reassessment of the Destruction of Minoan Palaces

The discoveries in Crete in the 1970s brought some disenchantment because they highlighted the fact that many aspects of the Minoan civilisation remain unknown and may even contradict certain long-held and deep-rooted ideas. This led to increased attention to fieldwork, both excavations and surface survey projects, throughout Greece.

In the 1980s, several research projects that started in Crete proved to be crucial for a better understanding of the defensive architecture and warfare in general. The Minoan Roads Research Programme, launched in 1984, led to the finding, description, and recording of several buildings of the 'guard house'[21] type in the region of

[21] These are freestanding closed structures built from massive masonry (often called 'megalithic' or 'Cyclopean'), which usually resemble a small fortress or stronghold (known also as 'phrourion'). For a basic overview of this architectural type, (see Chryssoulaki, 1999, pp. 78–83; Alušík, 2007, pp. 124–135). However, the function of these structures is often not quite clear and could

Sitia; these discoveries also resulted in the typology and chronological classification of this building type (see e.g., Tzedakis et al., 1989, pp. 60–74; Chryssoulaki, 1999; Alušík, 2016, pp. 56–59). The documentation of these buildings had for the first time drawn attention to the fact that, during the prehistoric period, the defensive character of a site could be architecturally expressed not only by the construction of massive walls around settlements and, moreover, prehistoric societies could develop effective site-specific defence strategies even without a large number of enclosed sites. Also in 1984, archaeologists have described sections of massive walls found in Palaikastro (east Crete), characterising them as town ramparts, that is, enclosure walls. Moreover, these walls would have protected the town even during the Neopalatial period, the golden age of the Minoan civilisation (MacGillivray & Sackett, 1984). Still in the 1980s, Rutkowski and Nowicki started their systematic research expeditions: they focused on the eastern part of Crete and the Lasithi area, where they recorded numerous defensively situated sites, often with defensive architecture, mainly from the Late Bronze and Early Iron Age (see e.g., Rutkowski & Nowicki, 1984). Later, Nowicki continued this part of the research alone and has been working on it since the 1990s. Since that time, he has extended his effort to the whole of Crete (see e.g., Nowicki, 1996) and the results of his work are described in detail in various publications, including three monographs (see e.g., Nowicki, 2000, 2014, 2025).

In the late 1990s, Driessen drew attention to modifications in the access systems into palatial buildings and villas that took place in Late Minoan I (c. 1700/1675–1470/1460 BCE). At this time, the number of entrances was in many cases reduced and the remaining entrance points were modified, which jointly led to a better defensibility of the building (Driessen, 1997). Driessen elaborated his hypothesis and, jointly with Macdonald, supported with by adducing many other examples (Driessen & Macdonald, 1997). In 1998, the first big conference on Aegean Bronze Age warfare took place in Liége: it focused on the military aspects of Aegean prehistoric civilisations (Laffineur, 1999). In the same year, Zielinski (1998) defended his doctoral dissertation on Cyclopean masonry in Minoan Crete, where he collected a large number of examples of its use in defensive contexts. In general, one can thus see that since the 1980s, the study of warfare became one of the important directions in Minoan archaeology (for a general overview of various aspects of Aegean Bronze Age warfare see Kvapil & Shelton, 2023). In particular, researchers examined the question of existence of Minoan warfare and interpersonal violence in general.

But the main research topics in Minoan archaeology of the 1980s, 1990s, and at the beginning of the new millennium, were broader issues of Minoan society during the Late Minoan IB (c. 1625/1600–1470/1460 BCE). Special attention was paid to the destruction (or destructions) at the end of the Late Minoan IB phase (c.

vary (including agricultural or economic use; see e.g., Beckmann, 2012). Although their real military value was in some cases rather symbolic, or psychosocial, GIS and Space Syntax analyses performed by Brabander (2012) prove that, although their defensibility varies, these structures had defensive qualities (the rate of defensibility differs per building and method).

1470/1460 BCE), which ended the Neopalatial period, and to the subsequent socio-cultural development during the Final Palatial period, including the precise dating of the 'final destruction' of Knossos Palace. Related to this are questions surrounding the Mycenaean invasion of Crete and a debate around the dating of the Knossian tablets, which started in the late 1950s (see above but also Fitton, 1995, pp. 175–178; for the debate from the 1980s onwards, see e.g., Niemeier, 1982; Driessen, 1990, 2000; Hatzaki, 2000). Although the dispute over the dating of the tablets has not yet been completely resolved, most scholars are now convinced that a vast majority of them—perhaps with the exception of those from the Room of the Chariot Tablets, which have been dated to the Late Minoan II/IIIA1 (c. 1470/1460–1390/1370 BCE; Driessen, 1990, 2000, 2008)—should be dated to the Late Minoan IIIB1 (c. 1330/1315–thirteenth century BCE), that is, to the Final Palatial period (Hallager et al., 1992; Palaima, 1993, 2010, pp. 361–362; Fitton, 1995, pp. 175–178).

Initially, the destruction that took place in the Late Minoan IB period was ascribed to natural factors,[22] but this hypothesis was soon abandoned and since about the 1980s the prevalent explanation lists anthropogenic factors or a combination of both (see e.g., Driessen, 2002). The most commonly considered factors have been intra-island warfare and a subjugation of the whole of Crete by Knossos (see e.g., Niemeier, 1983; Hallager, 1988; Rehak & Younger, 2001, pp. 440–441; Nafplioti, 2008, pp. 2307–2308), a Mycenaean invasion (see e.g., Hood, 1985; Wiener, 2015), a combination of the two (see e.g., Warren, 1991; Rehak & Younger, 2001, pp. 440–441; Wiener, 2015), or socioeconomic problems of the Minoan society, including social competition and rebellion against the elites (see e.g., Driessen, 2002; Driessen & Macdonald, 1997; Hamilakis, 2002a, b, c). In recent years, the last-named explanation is gaining ever more ground and fewer scholars believe that Mycenaeans played an important role in the fall of Knossos.

The hypothesis that linked the Late Minoan IB destructions to a Mycenaean invasion of Crete (and its eventual conquest) was based on the traditional sociocultural scheme that presented Mycenaeans as aggressive warriors and on a widespread—but methodologically incorrect—assumption that major changes in material culture must be caused by population movements (see e.g., Binford, 1965; Renfrew, 1972). Although the Late Minoan IB destruction marks a fundamental turning point in Minoan material and artistic culture, after which some previously characteristic elements (such as lustral basins) disappeared and were, from the Late Minoan II onwards, gradually replaced by new 'Mycenaean' elements (for instance in pottery), it need not have been the result of Mycenaean conquest or colonisation (cf. Chap. 5). The Mycenaeans in the Late Minoan IB may have been for example invited by the ruling class of Knossos to help it solve specific socioeconomic problems; some may have stayed and introduced new elements (see a summary of the discussion in Hatzaki, 2000, pp. 155–156, and in Rehak & Younger, 2001, pp. 440–441; Nafplioti, 2008, pp. 2307–2308). Alternative, the adoption of new

[22] See for instance Marinatos (1939), who linked these destructions with the eruption of the volcano in Thera (Santorini).

artistic and material elements of foreign origin could be explained as an attempt by new elites—which may have emerged with some Mycenaean help—to consolidate their power and status within the context of social competition and change after the Late Minoan IB destructions (Schallin, 1993). According to most scholars, there is ample evidence for some Mycenaean presence and, specifically, for a political and economic control of Knossos and other important centres, but only since the Late Minoan IIIA/B (c. second half of fourteenth century BCE). This is best illustrated by the introduction of a Mycenaean system of administration and written bureaucracy, that is, by the appearance of the Linear B script and tablets. On the other hand, there is no evidence of any mass Mycenaean invasion or active colonisation of Crete in the Late Minoan IB/II (c. 1470/1460 BCE; Rehak & Younger, 2001, pp. 440–441; for an opposing view, see Wiener, 2015) and we should therefore stop presenting the Mycenaeans as aggressive or 'barbaric' once and for all.

Some of the most important evidence of a physical presence of Mycenaeans at Knossos came from the so-called 'warrior graves' or 'burials with bronzes', found mainly in the area of Sellopoulo, Agios Ioannis, Zafer Papoura, Isopata, and Mavrospelio, and dated to the Late Minoan II-IIIA (c. 1470/1460–1330/1315 BCE; Evans, 1906b; Cadogan, 2000, pp. 22–23; Hatzaki, 2000, pp. 154–155). In these graves, the deceased were often buried with weapons, which is a custom almost undocumented in Minoan material and funerary culture until that time (see e.g., Alberti, 2004). Architecturally, they are shaft or shaft-niche graves and single-chamber tombs with a dromos (an entrance corridor). It can be expected that twenty-first century osteoarchaeology and archaeogenetics will weigh into the debates about the identity of these individuals.

In 1986, Wall and her team published the results of a study of several hundred bone fragments scattered on the floor of the basement of the North House at Knossos (Wall et al., 1986). They showed that most of the skeletal remains date to the Late Minoan IB (c. 1625/1600–1470/1460 BCE) and although some are animal, most are human. A vast majority of these remains belonged to three pre-adolescent individuals aged 8–12 years. What is most unusual about this assemblage is the presence of cutmarks on 79 of these bones. They were most likely made by a knife or a cleaver. In one of the skulls, the cuts are present on the underside, which may indicate a slitting of throat or decapitation. On another skull, there are incisions which must have been made during a removal of the brain. In addition to the skulls, cutmarks have also been found on the vertebrae, ribs, scapulae, bones of the free upper limb, pelvis, and bones of the free lower limb. X-ray imaging showed no other types of pathologies in this skeletal assemblage, which indicates that these deceased subadult individuals were, at the time of their death, healthy. The cutmarks in question, which occur on many different bones of the skeletons, indicate that soft tissues were deliberately removed from the bones, some parts of the bodies separated, and the brain, heart, and lungs removed. It is highly likely that at least some of these individuals were killed intentionally. According to the excavators, the physical anthropologists who studied these human skeletal remains, and some other experts, this points to human sacrifice or cannibalism. According to others, however, the cutmarks are related to burial rituals, including a possible secondary burial. This report had caused perhaps even more controversy than the remains of the potentially sacrificed young man from Archanes-Anemospilia, but its conclusions have unfortunately been somewhat overlooked. This is because only a few similar examples of unusual or deviant burials of Minoan subadults or young adults have been discovered to date and, moreover, many archaeologists are rather repulsed by the idea of Minoans sacrificing children, teenagers, or young adults.

The methods and approaches of physical anthropology were introduced into the study of prehistoric Aegean human skeletal remains, especially in Crete, since the 1980s by McGeorge (see e.g., Hallager & McGeorge, 1992; McGeorge, 2008). Likewise since the 1980s, researchers have used biochemical analyses of trace elements of the bone tissue to determine the predominant diet or geographical origin of deceased individuals.[23] Over the past two decades, these methods have had a large impact on Greek archaeology, including research into Greek prehistory. In the 1990s, there appeared hypotheses according to which, at the turn of the Final Neolithic and Early Bronze Age, there arrived in Crete a new population, probably from the Near East (most likely present-day Syria), that brought the cultural knowledge necessary for transition into the Bronze Age (Hood, 1990; Nowicki, 1999). These new arrivals would have played a role in the emergence and formation of the Minoan civilisation and their genetic signature should therefore be preserved in the Minoan genome (cf. Chap. 5).

In the 1990s, Aegean prehistoric archaeology began to be transformed by the introduction of new methods of archaeological science. Broad interdisciplinary cooperation enabled researchers to capture previously unexplored aspects of the material culture and details of the lives of specific individuals. This led to a deeper and more nuanced understanding of the civilisations of the time.

6.4.2 The New Millennium: Interdisciplinary and Science-Driven Research

The beginning of the new millennium brought several new important finds that further challenged the traditional image of peaceful and civilised Minoans versus bellicose Mycenaeans. In 2006, McGeorge presented a set of 16 serious cranial traumatic injuries—clear signs of interpersonal violence—from the Hagios Charalambos burial cave in East Crete.[24] This is the most numerous set of violent traumas occurring in a Greek Bronze Age site to date.[25]

[23] Since that time, there emerged a new and more accurate method based on an examination of the levels of stable isotopes of nitrogen (^{14}N and ^{15}N) and carbon (^{12}C and ^{13}C) to determine the diet, and strontium (^{87}Sr and ^{86}Sr) analysis, which is used to determine the geographical origin or migration of the individuals examined; (e.g., Richards & Hedges, 2008; Nafplioti, 2008; cf. Chap. 5).

[24] Prior to a later secondary burial, these individuals had been buried over a long period ranging from Final Neolithic until Middle Minoan IIB (especially to the Early Minoan III-Middle Minoan IIB), that is, between about, 3750–1700 BCE (McGeorge, 2008).

[25] Although, as noted above, these individuals died over a long period of time and their remains were deposited in the cave only secondarily.

In 2005, I have defended my doctoral dissertation on the defensive architecture of prehistoric Crete, where I collected almost 200 Minoan sites with at least probable examples of this architectural type, including both detached (exterior) and interior structures and construction elements (a revised English edition appeared as Alušík, 2007; further such research was published later as Alušík, 2016, 2023). While the actual military or defensive significance of some of these structures or sites may have been only secondary or questionable, this nevertheless shows that defence was also expressed in architecture during all major periods of the Minoan civilisation—which in turn clearly points to the presence of violence and a military nature of the Minoan civilisation.

In 2008, Nafplioti published the results of stable isotopic analyses of strontium from the bone tissue of individuals buried in several Late Minoan II-IIIA graves in Knossos, which were conducted to determine their geographic origins. These graves, known also as 'warrior graves', were long regarded as the final resting place of Mycenaeans who either came to Crete after conquering it or participated directly in the conquest, but this modern method had convincingly proven that the buried individuals were of a local, Cretan, and not mainland origin (Nafplioti, 2008). Archaeological science thus refuted the hypothesis of a Mycenaean conquest of Crete at the end of the Late Minoan IB or a massive influx of Mycenaeans to Crete after the Late Minoan IB/II destructions which ended the Neopalatial period. Unsurprisingly, the results of this research have been criticised and various scenarios suggested which still posit a Mycenaean control of Crete (or some similar involvement) at the beginning of the Late Minoan II and their dominant presence in Crete in the subsequent periods (see e.g., Wiener, 2015 but also below).

Just a little later came a thorough re-examination of the skeletal remains from Grave Circle A in Mycenae (see e.g., Papazoglou-Manioudaki et al., 2009). The research confirmed a physically robust and tall group of individuals who had a significantly better diet than other contemporary populations (as emphasised already by Angel in 1973). However, it also showed very limited skeletal evidence of traumatic injuries, especially those caused by interpersonal violence. Any actual involvement of these individuals in combat is thus uncertain. Around this time, scientists also investigated the fragmentary and dismembered remains of a young woman, 15 to 19 years old, who died in the Late Minoan IIIB1 (c. 1330/1315–thirteenth century BCE) (Andreadaki-Vlazaki, 2015; McGeorge, 2015). These remains were found in Chania (west Crete), scattered within concentrations of the animal bones. The overall context, with a traumatic injury to the back of the skull and characteristic cutmarks, strongly suggests the performance of human sacrifice. It would be the third documented case of this practice in Minoan Crete (Warren, 2015).

In recent years, there appeared the first results of ancient DNA (aDNA) studies of the genomic origins of the Minoans and Mycenaeans (Hughey et al., 2013; Lazaridis et al., 2017; Skourtanioti et al., 2023; cf. also Chap. 5). Their results support the idea of continuity, not isolation, in the history of Aegean populations both before and after the emergence of the earliest civilisations in the region. At the same time, these analyses revealed no measurable Levantine or African signature in the genomic structure of Minoans and Mycenaeans, which effectively ended speculation about the Aegean cultures being seeded by migrants from the old civilisations of those regions, as was suggested in the 1990s (see e.g., Hood, 1990) but also earlier.[26] A more recent genomic study confirmed an 'eastern' gene flow at the end of the Neolithic and during the Aegean Early Bronze Age; in Crete, this inflow came mainly from Anatolia (Skourtanioti et al., 2023). This research had also shown a probable demographic connection between the Greek mainland and Crete in the Late Bronze Age: in particular, it demonstrated that during the Late Minoan II–III, genetic signatures apparently typical for the mainland gradually started appearing also in Crete. Although this does not prove any mass invasion of the Mycenaeans to Crete—and especially not during the contested short period after the destructions at the end of Late Minoan IB—these genetic analyses show that at larger ports, the Cretan population biologically mixed with populations coming to the island over several centuries in the later phases of the Late Minoan period.

6.5 Conclusions: Aggressive Mycenaeans and Minoan Flower-Lovers... or How Was it Actually?

In the previous chronological overview, I have outlined the main points of development of the changing interpretations of Minoan and Mycenaean civilisations using as my examples the most important discoveries in the field, in museums, and in laboratories, and their interpretations. In general terms, one can trace over the past century or so radical changes in the range of available archaeological methods, including a gradual incorporation of 'scientific' approaches (such as aDNA analysis, stable isotopic analyses, physical anthropology, palaeopathology, or the Geographic Information System). This went hand in hand with a growing appreciation of their contribution and led to the formation of a multidisciplinary 'archaeological science', which can re-assess older hypotheses about both the overall character and specific features of the civilisations under study. Such re-examination of particular hypotheses then naturally leads to a more critical examination and progress in reassessing old theories (such as the 'peacefulness' of the Minoans) and to the formulation of new theories (about for instance the existence of defensive architecture and its specific types and functions or the role of warfare in culture and society). Of course, given that the new theories must be scientifically tested using the new methods, shifts in them creates feedback and lead to further progress in the methods used. Paradoxically, although these advances in archaeological theory and methods have brought a deeper knowledge of many aspects of both the Minoan and the Mycenaean civilisation, they also perhaps created more new questions than answers to old ones (cf. Chap. 18).

[26] Evans (1921, 1925) suggested that the Minoan civilisation was founded by refugees from the Nile Delta. Other archaeologists have argued for Cycladic migrations (Hutchinson, 1962) or, alternatively, for an autochthonous development of the Minoan civilisation from the original inhabitants of Crete (Branigan, 1970).

The early views of these Aegean civilisations were largely influenced by ancient sources, especially the Homeric epics and early historians such as Herodotus and Thucydides (cf. Chap. 1). Together with Evans, these are the most important figures who co-created their image as we know it today. Evans not only interpreted the Minoan culture based on the artefacts he discovered but, like the ancient authors, actually created it with his (and Gilliéron's and de Jong's) reconstructions (see e.g. Blakolmer, 2008). His basic ideological framework was shaped by classical cultural evolutionism, the prevailing nineteenth century philosophy of history, which—in analogy to a simplified model of biological evolution—viewed historical development as a linear and unilateral journey of progress, and by the colonialism of the British Empire (see e.g., McEnroe, 2002). For example, Evans' basic tripartite scheme of Minoan archaeology (Evans, 1906a; Mackenzie, 1903) reflects the evolutionist perspective of a gradual growth, maturity, and decline of civilisations. Evans then applied this scheme to the whole of Bronze Age Crete without allowing for regional differences and diverse regional developments (Hamilakis, 2002a, p. 6; cf. Chap. 5). In fact, evolutionist paradigms influence Minoan archaeology to this day.[27]

In general, archaeology of the Aegean Bronze Age had, during the first third of the twentieth century, gradually rejected the incorrect paradigms of classical Victorian evolutionism, especially its Eurocentric and often colonialist, even racist attitudes, which lead to various grave scientific errors that stemmed, at least in part, from a failure to fully appreciate some of the discoveries made in the field (cf. Chaps. 3 and 5). After the Second World War, there has been an effort to create new theories or models of cultural evolutionism that would not repeat the mistakes of the past. Since the end of the twentieth century, these theories have been part of an effort to synthesise the humanities and the natural sciences, demonstrated here by the increasing incorporation and appreciation of the contribution of archaeological science (cf. Chap. 5).

Since the beginning of formation of Aegean prehistoric archaeology in the second half of the nineteenth century, there existed two basic and rather contrasting images of the Minoan and Mycenaean civilisations. They were shaped by the cultural, anthropological, and sociological theories of the time and by the accounts of ancient authors. The Mycenaeans were supposedly skilled warriors, aggressive to the point of being 'barbaric'. The Minoans, on the other hand, were supposed to be culturally 'advanced', both economically and

[27] Hamilakis (2002a) has in his recent thought-provoking study briefly characterised the development of theoretical concepts in Minoan archaeology as different forms of evolutionism: cultural evolutionism expressed by Evans, neo-evolutionism that took into account the new approaches of New Archaeology represented by Renfrew (1972), and modified neo-evolutionism in the late twentieth century. He presents this idea in an introduction to a volume he edited (Hamilakis, 2002a, b, c), where the intellectual foundations of Minoan archaeology in its beginnings are examined, discussed, and critically evaluated within a broader framework of Eurocentric modernity (see also Barrett & Halstead, 2004).

politically strong, yet peaceful, that is, 'civilised'. These two initial simplistic conceptions correspond to L. H. Morgan's contrast between barbarism and civilisation (cf. Chap. 7). Aside from the accounts of ancient authors, these interpretations were in late nineteenth and early twentieth century based on two archaeological foundations, namely the (non)existence of fortifications and relationship to interpersonal violence. In terms of these markers, the two civilisations were said to have radically differed. The first starting point (fortifications) was based on the traditional notion of defence strategies of prehistoric populations that was nevertheless influenced by the appearance of medieval European or Crusader castles; not surprisingly, crucial importance was thus ascribed to high and massive fortifications, preferably made of stone. The second starting point was influenced by the large number of weapons in Mycenaean tombs and war scenes in Mycenaean iconography compared to their significantly smaller numbers in Minoan archaeology and art. In the 1950s, these ideas were boosted by the decipherment of Linear B, which resulted in the hypotheses about a mass Mycenaean invasion, conquest of Crete, and destruction of the Neopalatial palaces at the end of Late Minoan IB, which ended the main phase of bloom of the advanced Minoan civilisation. These elements have had an important role in interpreting the overall character of Minoan and Mycenaean civilisations throughout the twentieth century and until the present.

Until late 1980s and early 1990s, archaeological research was actually confirming the original interpretation of the character of Mycenaeans as aggressive barbarian warriors (in the 1950s, they were even called 'destroyers of Minoan palaces'). Eventually, though, new discoveries, a critical evaluation of the existing finds, and the gradual integration of archaeological science enabled not only a new dating of the Linear B tablets from Knossos but also led to new interpretations or models of the troubled period and destructions at the end of Late Minoan IB, which then retrospectively influenced the perception of Mycenaean civilisation. The most plausible current explanations of the destructions in Late Minoan IB, the arrival of the Mycenaeans to Crete, and the use of Linear B script point to internal processes in the Minoan society and possibly also to the effects of natural phenomena. While some admit a degree of Mycenaean involvement, they exclude mass immigration or 'colonisation' of Crete during Late Minoan II and absolve Mycenaeans from any direct blame for the destruction of Cretan palaces. Although a very recent genomic study has proven a gradual increase of typically Mycenaean genetic signatures in the Minoan genome in the course of several centuries in Late Minoan II–III, that is, some immigration to Crete from the mainland, it does not bring evidence of any Mycenaean mass invasion following the destruction of Minoan palaces at the end of Late Minoan IB (Skourtanioti et al., 2023). New elements in Minoan artistic and material culture are thus usually explained by intensified mutual contacts, including for instance travelling artists (see e.g., Niemeier, 1991) and some level of migration over several centuries (cf. Chaps. 11, 13 and 15), rather than by some large-scale population movements envisaged by the sociocultural schemes of the first half of the twentieth century (cf. Chap. 5). This also clearly shows that the specific nature of the Minoan and Mycenaean civilisation and their mutual interactions are complex and cannot be explained solely based on their genetic structure, that is, by reducing them to genes, as some neo-Darwinians or evolutionary psychologists tried to do (see e.g., Wheeler, 2006).

The perception of Mycenaean civilisation did not change significantly for a long time and only relatively recently did the Mycenaeans cease to be viewed as aggressive and 'barbaric' warriors and destroyers of civilisations (although warfare admittedly did have an important place in their culture). In contrast, the original image of an advanced, 'gentle', and peaceful Minoan civilisation was gradually modified with the discovery of new examples of defensive architecture and (probable) human sacrifice. In neither case is the traditional image tenable. Our understanding of the relationship between these two civilisations has also significantly changed. Gradually abandoning the original model of the economically, artistically, and politically more advanced Minoan civilisation influencing or—according to Evans—'teaching' the Mycenaeans, contemporary archaeologists have uncovered much more complex interactions and mutual influences between the two cultures (see e.g., Gorogianni et al., 2016), with an increasingly pronounced 'feedback' from the Mycenaean civilisation to the Minoans starting in the Late Minoan II period.

Our understanding of the Greek Bronze Age thus evolved from relatively clearcut and simple, but wrong, evaluative conclusions about both civilisations, considerably influenced by the prevalent philosophies of cultural evolution of the nineteenth century as well as the testimonies of Classical literature, to more complex and less definitive, yet more accurate, conclusions, which are based on a critical reading of the literary sources and archaeological evidence, with increasingly pronounced integration of science-based methods. Classical cultural evolutionary schemes have oversimplified the picture and provided only linear or superficial interpretations of both the Minoan and the Mycenaean civilisation. The development of theoretical concepts and archaeological methods since the 1960s and the introduction of archaeological science in the 1990s made it possible to capture certain more complex and multilayered, often seemingly contradictory, aspects of the two civilisations. These methods also provided a greater granularity of the data, sometimes allowing even for stories of particular individuals to emerge. Since the beginning of the new millennium, Aegean prehistorians have been developing new theoretical models which reflect contemporary theories of cultural anthropology (including the biosemiotic perspective) and are based on a greater integration of methods of archaeological science. As a result, some traditional misconceptions have been refuted but new discoveries, in conjunction with the expanding theoretical and methodological base of Aegean Bronze Age archaeology, not only provided answers to some long-standing issues but also, and perhaps even more importantly, raised many new questions.

References

Alberti, L. (2004). The late Minoan II–IIIA1 warrior graves at Knossos: The burial assemblages. In G. Cadogan, E. Hatzaki, & A. Vasilakis (Eds.), *Knossos: Palace, city, state. Proceedings of the conference in Herakleion organised by the British School at Athens and the 23rd Ephoreia of prehistoric and classical antiquities of Herakleion, in November 2000, for the centenary of sir Arthur Evans's excavations at Knossos* (pp. 127–136). British School at Athens.

Alexiou, S. (1980). Προανακτορικές ακροπόλεις της Κρήτης. In M. Χατζηδάκης (Ed.), *Πεπραγμένα του Δ΄ Διεθνούς Κρητολογικού Συνεδρίου, Herakleion, 29.8.–3.9.1976, A1 (Προϊστορικοί και αρχαίοι χρόνοι)* (pp. 9–22). Πανεπιστήμιο Κρήτης.

Alušík, T. (2004). Minoan-Mycenaean elements in the work of Josip Plećnik at the Prague Castle. *Studia Hercynia, 8*, 59–66.

Alušík, T. (2007). *Defensive architecture of prehistoric Crete*. Archaeopress.

Alušík, T. (2016). Fortifications of prehistoric Crete: The current state of research. In R. Frederiksen, S. Müth, P. I. Schneider, & M. Schnelle (Eds.), *Focus on fortifications: New research on fortifications in the ancient Mediterranean and the near east* (pp. 53–65). Oxbow Books.

Alušík, T. (2023). Fortifications and defensive architecture. In L. Kvapil & K. Shelton (Eds.), *Brill's companion to warfare in the bronze age Aegean* (pp. 25–88). Brill.

Andreadaki-Vlazaki, M. (2015). Sacrifices in LM IIIB: Early Kydonia palatial Centre. *Pasiphae, 9*, 27–42.

Angel, J. L. (1973). Human skeletons from grave circles at Mycenae. In G. E. Mylonas (Ed.), *O Taphikos Kyklos ton Mykenon* (pp. 379–397). Archaeological Society.

Barrett, J. C., & Halstead, P. (Eds.). (2004). *The emergence of civilisation revisited*. Oxbow Books.

Beckmann, S. (2012). *Domesticating mountains in Middle Bronze Age Crete: Minoan agricultural landscaping in the Agios Nikolaos region* (Ph.D. dissertation, University of Crete, Rethymno).

Bennett, E. L. (1951). *The Pylos tablets: A preliminary transcription*. Princeton University Press.

Binford, L. R. (1965). Archaeological systematics and the study of culture process. *American Antiquity, 31*(2), 203–210.

Blakolmer, F. (2007). The silver Battle krater from shaft grave IV at Mycenae: Evidence of fighting 'heroes' on Minoan palace walls at Knossos? In S. P. Morris & R. Laffineur (Eds.), *Epos: Reconsidering Greek epic and Aegean bronze age archaeology. Proceedings of the 11th international Aegean conference, Los Angeles, UCLA—The J. Paul Getty villa, 20–23 April 2006* (pp. 213–224). Université de Liège.

Blakolmer, F. (2008). Die 'Rekonstitution' des 'Palace of Minos' in Knossos im Spiegel des Art Déco. Und warum Evans recht hatte. *Studia Hercynia, 12*, 103–116.

Blegen, C. W., & Rawson, M. (1966). *The palace of Nestor at Pylos in Western Messenia* (Vol. I). Princeton University Press.

Blegen, C. W., Caskey, J. L., Rawson, M., & Sperling, J. (1950). *Troy I: General introduction. First and Second Settlement*. Princeton University Press.

Borowka, D. (2018). On the possibility of internal conflict on Crete between the proto- and Neopalatial period: The Pax Minoica re-examined. *Acta Archaeologica Lodziensia, 64*, 7–19.

Brabander, S. (2012). *Defensible or not defensible? Guard houses from Middle and Late Minoan Crete revisited, using GIS and space syntax* (M.A. thesis, University of Leiden).

Branigan, K. (1970). *The foundations of palatial Crete*. Praeger Publishers.

Branigan, K. (2000). Between the wars. In D. Huxley (Ed.), *Cretan quests: British explorers, excavators and historians* (pp. 28–33). British School at Athens.

Brown, A. (Ed.). (2001). *Arthur Evans's travels in Crete, 1894–1899*. Archaeopress.

Brown, A. (1983). *Arthur Evans and the palace of Minos*. Ashmolean Museum.

Brown, A. (2000). Evans in Crete before 1900. In D. Huxley (Ed.), *Cretan quests: British explorers, excavators and historians* (pp. 9–14). British School at Athens.

Cadogan, G. (2000). The pioneers: 1900–1914. In D. Huxley (Ed.), *Cretan quests: British explorers, excavators and historians* (pp. 15–27). British School at Athens.

Chapin, A. P. (2016). Mycenaean mythologies in the making: The frescoes of Pylos hall 64 and the Mycenae megaron. In E. Alram-Stern, F. Blakolmer, S. Deger-Jalkotzy, R. Laffineur, & J. Weilhartner (Eds.), *Metaphysis: Ritual, myth and symbolism in the Aegean bronze age. Proceedings of the 15th international Aegean conference, Vienna, 22–25 April 2014* (pp. 459–466). Peeters.

Chryssoulaki, S. (1999). Minoan roads and guard houses—War regained. In R. Laffineur (Ed.), *Polemos. Le contexte guerrier en Egée à l'âge du bronze. Actes de la 7e Rencontre égéenne internationale, Liège, 14–17 avril 1998* (pp. 75–84). Université de Liège.

Cline, E. H. (2013). *The Trojan war: A very short introduction*. Oxford University Press.

Cobley, P. (2016). *Cultural implications of biosemiotics*. Springer.

Driessen, J. (1990). *An early destruction in the Mycenaean palace at Knossos: A new interpretation of the excavation field-notes of the south-east area of the west wing*. Katholieke Universiteit Leuven.

Driessen, J. (1997). Observations on the modification of the access systems of Minoan palaces. *AEA, 2*, 67–85.

Driessen, J. (2000). *The scribes of the room of the chariot tablets at Knossos: Interdisciplinary approach to the study of a linear B deposit*. Ediciones Universidad de Salamanca.

Driessen, J. (2002). Towards an archaeology of crisis: Defining the long-term impact of the bronze age Santorini eruption. In R. Torrence & J. Grattan (Eds.), *Natural disasters and cultural change* (pp. 250–263). Routledge.

Driessen, J. (2008). Chronology of the linear B texts. In Y. Duhoux & A. Morpurgo Davies (Eds.), *A companion to linear B: Mycenaean Greek texts and their world* (Vol. 1, pp. 69–79). Peeters.

Driessen, J., & Macdonald, C. F. (1997). *The Troubled Island: Minoan Crete before and after the Santorini eruption*. Université de Liège.

Dunér, D., & Sonesson, G. (Eds.). (2016). *Human lifeworlds. The cognitive semiotics of cultural evolution*. Peter Lang.

Evans, A. J., & Myres, J. L. (1895). A Mycenaean military road in Crete. *The Academy, 47*, 469–470.

Evans, A. J. (1906a). *Essai de classification des époques de la civilisation minoenne: Résumé d'un discours fait au Congrès d'Archéologie à Athènes*. Quaritch.

Evans, A. J. (1906b). *The prehistoric tombs of Knossos*. Quaritch.

Evans, A. J. (1921). *The palace of Minos. A comparative account of the successive stages of the early Cretan civilisation as illustrated by the discoveries at Knossos* (Vol. I). Macmillan.

Evans, A. J. (1925). The early Nilotic, Libyan and Egyptian relations with Minoan Crete. *The Journal of the Royal Anthropological Institute of Great Britain and Ireland, 55*, 199–228.

Evans, A. J. (1928). *The palace of Minos. A comparative account of the successive Sstages of the early Cretan civilisation as illustrated by the discoveries at Knossos* (Vol. II). Macmillan.

Evans, A. J. (1935). The Palace of Minos. A Comparative Account of the Successive Stages of the Early Cretan Civilisation as Illustrated by the Discoveries at Knossos. Volume IV. Macmillan.

Fitton, J. L. (1995). *The discovery of the Greek bronze age*. British Museum Press.

Fürst, C. M. (1930). *Zur Anthropologie der prähistorischen Griechen in Argolis, nebst Beschreibungen einiger älteren Schädel aus historischen Zeit*. Lund Universitets årsskrift.

Galanakis, Y. (2014). Arthur Evans and the quest for the 'origins of Mycenaean culture'. In Y. Galanakis, T. Wilkinson, & J. Bennet (Eds.), *Αθύρματα: Critical essays on the archaeology of the eastern Mediterranean in honour of E. Susan Sherratt* (pp. 85–98). Archaeopress.

Gorogianni, E., Pavúk, P., & Girella, L. (Eds.). (2016). *Beyond Thalassocracies: Understanding processes of Minoanisation and Mycenaeanisation in the Aegean*. Oxbow Books.

Hallager, B. P., & McGeorge, P. J. (1992). *Late Minoan III burials at Khania*. Paul Åströms Förlag.

Hallager, E. (1988). Final palatial Crete: An essay in Minoan chronology. In A. Damsgaard-Madsen, E. Christiansen, & E. Hallager (Eds.), *Studies in ancient history and numismatics presented to Rudi Thomsen* (pp. 11–21). Aarhus University Press.

Hallager, E., Vlasaki, M., & Hallager, B. P. (1992). New linear B tablets from Khania. *Kadmos, 31*, 61–87.

Hamilakis, Y. (Ed.). (2002a). *Labyrinth revisited. Rethinking 'Minoan' archaeology*. Oxbow Books.
Hamilakis, Y. (2002b). Too many chiefs?: Factional competition in Neopalatial Crete. In J. Driessen, I. Schoep, & R. Laffineur (Eds.), *Monuments of Minos. Rethinking the Minoan palaces. Proceedings of the international workshop "Crete of the hundred palaces?" held at the Université Catholique de Louvain, Louvain-la-Neuve, 14–15 December 2001* (pp. 179–199). Université de Liège.
Hamilakis, Y. (2002c). What future for the 'Minoan' past? Re-thinking Minoan archaeology. In Y. Hamilakis (Ed.), *Labyrinth revisited. Rethinking 'Minoan' archaeology* (pp. 2–28). Oxbow Books.
Hatzaki, E. (2000). The Postpalatial period. In D. Huxley (Ed.), *Cretan quests: British explorers, excavators and historians* (pp. 151–161). British School at Athens.
Hiller, S. (1980). The south propylaeum of the palace of Knossos: Some reflections on its reconstruction. In M. Χατζηδάκης (Ed.), *Πεπραγμένα του Δ΄ Διεθνούς Κρητολογικού Συνεδρίου, Ηράκλειο, 29 Αυγούστου—3 Σεπτεμβρίου 1976, Α1 (Προϊστορικοί και αρχαίοι χρόνοι)* (pp. 216–232). Πανεπιστήμιο Κρήτης.
Hitchcock, L., & Koudounaris, P. (2002). Virtual discourse: Arthur Evans and the reconstructions of the Minoan palace at Knossos. In Y. Hamilakis (Ed.), *Labyrinth revisited. Rethinking 'Minoan' archaeology* (pp. 40–58). Oxbow Books.
Hood, S. (1985). Warlike destruction in Crete c. 1450 B.C. In Θ. Δετοράκης (Ed.), *Πεπραγμένα του Ε΄ Διεθνούς Κρητολογικού Συνεδρίου (Άγιος Νικόλαος, 25 Σεπτεμβρίου—1 Οκτωβρίου 1981), Α* (pp. 170–178). Εταιρία Κρητικών Ιστορικών Μελετών.
Hood, S. (1990). Settlers in Crete c. 3000 B.C. *Cretan Studies, 2*, 151–158.
Hughes, D. D. (1991). *Human sacrifice in ancient Greece*. Routledge.
Hughey, J. R., Paschou, P., Drineas, P., Mastropaolo, D., Lotakis, D. M., Navas, P. A., Michalodimitrakis, M., Stamatoyannopoulos, J. A., & Stamatoyannopoulos, G. (2013). A European population in Minoan bronze age Crete. *Nature Communications, 4*, 1861.
Hutchinson, R. W. (1962). *Prehistoric Crete*. Penguin Books.
Jacobsen, T. W., & Cullen, T. (1990). The work of J. L. Angel in the eastern Mediterranean. In J. E. Buikstra (Ed.), *A life in science: Papers in honor of J. Lawrence Angel* (pp. 38–51). Center for American Archaeology.
Kundt, R. (2015). *Contemporary evolutionary theories of culture and the study of religion*. Bloomsbury Academic.
Kvapil, L., & Shelton, K. (Eds.) (2023). Brill's companion to warfare in the Bronze Age Aegean. Brill.
Laffineur, R. (Ed.). (1999). *Polemos. Le contexte guerrier en Egée à l'âge du bronze. Actes de la 7e Rencontre égéenne internationale, Université de Liège, 14–17 avril 1998*. Université de Liège.
Lazaridis, I., Mittnik, A., Patterson, N., Mallick, S., Rohland, N., Pfrengle, S., Furtwängler, A., Peltzer, A., Posth, C., Vasilakis, A., McGeorge, P. J. P., Konsolaki-Yannopoulou, E., Korres, G., Martlew, H., Michalodimitrakis, M., Özsait, M., Özsait, N., Papathanasiou, A., Richards, M., Roodenberg, S. A., Tzedakis, Y., Arnott, R., Fernandes, D. M., Hughey, J. R., Lotakis, D. M., Navas, P. A., Maniatis, Y., Stamatoyannopoulos, J. A., Stewardson, K., Stockhammer, P., Pinhasi, R., Reich, D., Krause, J., & Stamatoyannopoulos, G. (2017). Genetic origins of the Minoans and Mycenaeans. *Nature, 548*(7666), 214–218.
MacGillivray, J. A., & Sackett, L. H. (1984). An archaeological survey of the Roussolakkos area at Palaikastro. *The Annual of the British School at Athens, 79*, 129–159.
Mackenzie, D. (1903). The pottery of Knossos. *The Journal of Hellenic Studies, 23*, 157–205.
Manning, S. W. (2010). Chronology and terminology. In E. H. Cline (Ed.), *The Oxford handbook of the bronze age Aegean (ca. 3000–1000 BC)* (pp. 11–28). Oxford University Press.
Marinatos, N. (2015). *Sir Arthur Evans and Minoan Crete: Creating the vision of Knossos*. I. B. Tauris and Co. Ltd.
Marinatos, S. (1939). The volcanic destruction of Minoan Crete. *Antiquity, 13*(52), 425–439.
Markoš, A. (2015). The birth and life of species–cultures. *Biosemiotics, 9*(1), 73–84.

McEnroe, J. C. (2002). Cretan questions: Politics and archaeology 1898–1913. In Y. Hamilakis (Ed.), *Labyrinth revisited. Rethinking 'Minoan' archaeology* (pp. 59–72). Oxbow Books.
McGeorge, P. J. P. (2008). The human remains. In P. P. Betancourt et al. (Eds.), Excavations in the Hagios Charalambos cave: A preliminary report. Hesperia, 77(4), 578–594.
McGeorge, P. J. P. (2015). The earliest archaeological evidence for a Mycenaean Greek ritual form of human sacrifice. *Pasiphae, 9*, 43–52.
Momigliano, N., & Farnoux, A. (Eds.). (2017). *Cretomania: Modern desires for the Minoan past*. Routledge.
Nafplioti, A. (2008). Mycenaean political domination of Knossos following the LM IB destructions on Crete: Negative evidence from strontium isotope ratio analysis (87Sr/86Sr). *Journal of Archaeological Science, 35*(8), 2307–2317.
Niemeier, W.-D. (1982). Mycenaean Knossos and the age of linear B. *SMEA, 23*, 219–287.
Niemeier, W.-D. (1983). The character of the Knossian palace society in the second half of the fifteenth century B.C.: Mycenaean or Minoan? In O. Krzyszkowska & L. Nixon (Eds.), *Minoan society. Proceedings of the Cambridge colloquium 1981* (pp. 217–236). Bristol Classical Press.
Niemeier, W.-D. (1991). Minoan artisans travelling overseas: The Alalakh frescoes and the painted plaster floor at Tel Kabri (Western Galilee). In R. Laffineur & L. Basch (Eds.), *Thalassa. L'Egée préhistorique et la mer* (pp. 189–201). Université de Liège.
Nowicki, K. (1996). Arvi Fortetsa and Loutraki Kandilioro: Two refuge settlements in Crete. *The Annual of the British School at Athens, 91*, 253–285.
Nowicki, K. (1999). Final Neolithic refugees or early bronze age newcomers? The problem of defensible sites in Crete in the late fourth millennium B.C. In P. P. Betancourt, V. Karageorghis, R. Laffineur, & W.-D. Niemeier (Eds.), *Meletemata: Studies in Aegean archaeology presented to Malcolm H. Wiener as he enters his 65th year. Vol. II* (pp. 575–591). Université de Liège, Histoire de l'art et archéologie de la Grèce antique; University of Texas at Austin, Programs in Aegean Scripts and Prehistory.
Nowicki, K. (2000). *Defensible sites in Crete c. 1200–800 B.C.: LM IIIB/IIIC through early geometric*. Université de Liège.
Nowicki, K. (2014). *Final Neolithic Crete and the Southeast Aegean*. Walter de Gruyter.
Nowicki, K. (2025). Settlement in Crete during the Bronze to Iron Age transition: The crisis, collapse and reconstruction, ca. 1230–900 BC. Walter de Gruyter.
Palaima, T. G. (1993). Ten reasons why KH 115 ≠ KN 115. *Minos, 27–28*, 261–281.
Palaima, T. G. (2010). Linear B. In E. H. Cline (Ed.), *The Oxford handbook of the Bronze Age Aegean (ca. 3000–1000 BC)* (pp. 356–372). Oxford University Press.
Papadopoulos, A. (2019). Mneme and propaganda in the Early Late Bronze Age Aegean: The case of the 'siege Rhyton'. In E. Borgna, I. Caloi, F. M. Carinci, & R. Laffineur (Eds.), *Μνήμη/Mneme. Past and memory in the Aegean Bronze Age. Proceedings of the 17th international Aegean conference, University of Udine, ca' Foscari University of Venice, 17–21 April 2018* (pp. 405–411). Peeters.
Papadopoulos, J. K. (Ed.). (2007). *The art of antiquity: Piet de Jong and the Athenian agora*. ASCSA.
Papazoglou-Manioudaki, L., Nafplioti, A., Musgrave, J. H., Neave, R. A. H., Smith, D., & Prag, A. J. N. W. (2009). Mycenae revisited part 1. The human remains from grave circle a: Stamatakis, Schliemann and two new faces from shaft grave VI. *BSA, 104*, 233–277.
Pendlebury, J. D. S. (1936–37). Lasithi in ancient times. *The Annual of the British School at Athens, 37*, 194–200.
Pendlebury, J. D. S. (1939). *The archaeology of Crete—An introduction*. Methuen.
Rehak, P., & Younger, J. G. (2001). Neopalatial, Final Palatial, and Postpalatial Crete. In T. Cullen (Ed.), *Aegean prehistory: A review* (pp. 383–465). Archaeological Institute of America.
Renfrew, C. (1972). *The emergence of civilisation. The Cyclades and the Aegean in the third millennium B.C.* Methuen.
Richards, M. P., & Hedges, R. E. M. (2008). Stable isotope evidence of past human diet at the sites of the Neolithic cave of Gerani; the Late Minoan III cemetery of Armenoi; grave circles A and B at the palace site of Mycenae; and Late Helladic chamber tombs. In Y. Tzedakis, H. Martlew,

& M. K. Jones (Eds.), *Archaeology meets science: Biomolecular investigations in Bronze Age Greece. The primary scientific evidence, 1997–2003* (pp. 220–230). Oxbow Books.

Rodenwaldt, G. (1921). *Der Fries des Megarons von Mykenai*. Niemeyer.

Rutkowski, B., & Nowicki, K. (1984). Report on investigations in Greece. II. Studies in 1983. *Archeologia Warsaw, 35*, 178–192.

Sakellarakis, J. A., & Sapouna-Sakellaraki, E. (1997). *Archanes: Minoan Crete in a new light*. Ammos Publications.

Schallin, A.-L. (1993). *Islands under influence. The Cyclades in the Late Bronze Age and the nature of the Mycenaean presence*. Paul Åströms Förlag.

Service, E.-R. (1962). *Primitive social organization: An evolutionary perspective*. Random House.

Skourtanioti, E., et al. (2023). Ancient DNA reveals admixture history and endogamy in the prehistoric Aegean. *Nature Ecology & Evolution, 7*(2), 290–303.

Starr, C. G. (1955). The myth of the Minoan thalassocracy. *Historia, 3*, 282–291.

Stürmer, V. (2004). Gilliéron als Vermittler der ägäischen Bronzezeit um 1900. *Studia Hercynia, 8*, 37–44.

Tzedakis, Y., Chryssoulaki, S., Voutsaki, S., & Venieri, Y. (1989). Les routes minoennes: Rapport préliminaire—Défense de la circulation ou circulation de la défense? *Bulletin de Correspondance Hellénique, 113*, 43–75.

Ventris, M., & Chadwick, J. (1953). Evidence for Greek dialect in the Mycenaean archives. *The Journal of Hellenic Studies, 73*, 84–103.

Ventris, M., & Chadwick, J. (1956). *Documents in Mycenaean Greek*. Cambridge University Press.

Wace, A. J. B., & Blegen, C. W. (1916–18). The pre-Mycenaean pottery of the mainland. *The Annual of the British School at Athens, 22*, 175–189.

Wall, S. M., Musgrave, J. H., & Warren, P. M. (1986). Human bones from a late Minoan IB house at Knossos. *The Annual of the British School at Athens, 81*, 333–388.

Warren, P. (1991). The Minoan civilization of Crete and the volcano of Thera. *Journal of the Ancient Chronology Forum, 4*, 29–39.

Warren, P. (2015). Human sacrifice in Minoan (Bronze Age) Crete. *Pasiphae, 9*, 171–180.

Wiener, M. H. (1990). The isles of Crete? The Minoan thalassocracy revisited. In D. A. Hardy, C. G. Doumas, J. A. Sakellarakis, & P. M. Warren (Eds.), *Thera and the Aegean world III. Vol. 1: Archaeology. Proceedings of the third international congress, Santorini, Greece, 3–9 September 1989* (pp. 128–161). The Thera Foundation.

Wiener, M. H. (2015). The Mycenaean conquest of Minoan Crete. In C. F. Macdonald, E. Hatzaki, & S. Andreou (Eds.), *The Great Islands: Studies of Crete and Cyprus presented to Gerald Cadogan* (pp. 131–142). Kapon Edition.

Wiener, M. H. (2016). Beyond the Versailles effect: Mycenaean Greece and Minoan Crete. In J. Driessen (Ed.), *Ra-pi-ne-u: Studies on the Mycenaean world offered to Robert Laffineur for his 70th birthday* (pp. 365–378). Presses universitaires de Louvain.

Wheeler, W. (2006). *The whole creature. Complexity, biosemiotics and the evolution of culture*. Lawrence & Wishart.

Zielinski, J. P. (1998). *Cyclopean architecture in Minoan Bronze Age Crete: A study in the social organization of a complex society* (Ph.D. dissertation, University of New York at Buffalo).

Chapter 7
The Evolutionary and Cognitive Perspectives in Research of Cultural and Social Phenomena

Danijela Jerotijević and Radek Kundt

Abstract The evolutionary perspective has been accompanying anthropology from its beginnings, but there is a significant difference between how the concept of evolution was used in the formative years of the discipline and later on. Victorian anthropologists, such as Edward Burnett Tylor or James George Frazer, were deeply inspired by the evolutionary theory, including its pre-Darwinian versions, but their application of the evolutionary principles to culture was oversimplified. While nowadays it may seem naive to use ideas such as monogenesis or unilinear evolutionism to explain the emergence of culture and cultural change, the evolutionary principles helped founders of anthropology to pioneer the notion of culture as a system of beliefs, moral rules, and customs transmitted from one generation to the next, and clarified essential anthropological concepts such as exogamy, totemism, matriliny, or kinship. Later research into cultural and social phenomena based on evolutionary principles had experienced several phases of varying popularity and led to the formation of approaches such as sociobiology, memetics, evolutionary psychology, human behavioural ecology, and gene-culture coevolution, which vary in the relative importance ascribed to culture in human evolution. This chapter guides readers through selected evolutionary approaches to human behaviour, paying special attention to cultural evolution, and exploring topics such as religion, ritual, or cooperation. In parallel, we show how the interdisciplinary nature of the cognitive and evolutionary approaches to culture can help us understand these complex subjects.

Keywords Ritual · Religion · Magic · Cognitive and evolutionary sciences of religion · Gene-culture coevolution

---

D. Jerotijević
Institute of Social Anthropology, Faculty of Social and Economic Sciences, Comenius University, Bratislava, Slovakia

R. Kundt (✉)
LEVYNA Laboratory for the Experimental Research of Religion, Masaryk University, Brno, Czech Republic
e-mail: radek.kundt@mail.muni.cz

J. Toman et al. (eds.), *Culture Evolving*, Biosemiotics 10,
https://doi.org/10.1007/978-3-032-16124-6_7

7.1 Introduction

Anthropology emerged during the turbulent times of late nineteenth and early twentieth century. It was a time of significant changes in both politics and social life: France was dealing with the aftermath of the French Revolution, Germany was coming to terms with the revolutionary year of 1848, which led to a rise of nationalism, and the Western world in general was becoming increasingly industrialised. The emergence of anthropology is also associated with colonialism, the spread of Western influence to other parts of the world, conquests of previously unknown territories, and ensuing encounters with other cultures. That does not mean that ethnographic writings about other cultures appeared only with the emergence of anthropology: such observations had been recorded ever since humanity discovered writing (McGee & Warms, 2008, pp. 5–6) but it is only relatively recently that they became the foundation of a systematic and scientific research.[1]

Just as the emergence of anthropology as a science reflects the era of colonialism, its further development reflects the political and social environment in which the various anthropological trends and theories appeared (cf. Chaps. 2, 3, 4, 5 and 18). While there has always been a substantial level of communication between the different anthropology traditions, the development of anthropology in the USA, Britain, and parts of continental Europe was still substantively shaped by local influences (cf. Barnard, 2004). Another factor that influenced the development of anthropology were the different starting points. For instance, in Victorian Britain, anthropology took a different path than an analogical research in Central Europe, where it was deeply influenced by the German interest in 'folk culture' that led to the formation of *Volkskunde*.[2] The meaning of the term 'anthropology' likewise varied in different cultural areas: in Great Britain and some other English-speaking countries, the study of cultures and societies was known as 'social anthropology', while in continental Europe and the United States the common term was 'ethnology' (Barnard, 2004, p. 4). Adding to the confusion, 'anthropology' was a little later also widely used as a synonym for biological anthropology.

The aim of this chapter is not to offer a comprehensive overview of all anthropological theories born out of these developments. Rather, the chapter, in the form of narrative review, aims at shedding some light on the main concepts that preceded (and resulted in) the current evolutionary approaches to cultural phenomena. The development of these ideas was not straightforward, and we want to show that the integration of evolutionary thinking into anthropology was neither inevitable not

[1] Interest in other cultures is apparent already in the texts of ancient authors. For example, written accounts of some parts of Africa date to the Classical Period in Greece and much information about other cultures can be found also in medieval Arabic sources or in the accounts of missionaries and travellers from the Early Modern Period (Barnard, 2004; Barnard & Spencer, 1996).

[2] In Germany and some countries of Central and Eastern Europe, *Volkskunde* was used to refer to the study of local customs and local manifestations of folklore. The term *Völkerkunde*, on the other hand, has a broader meaning: it refers to comparative social sciences and is used in Germany as the equivalent of *Ethnologie* (Barnard, 2004, p. 2).

linear: it was a process deeply influenced by broader scientific, philosophical, and social developments. At some points, the evolutionary ideas receded into the background only to vigorously reappear later, often in different settings and with a new meaning. Like many events in human history, the development of anthropological theories cannot be properly understood without its context and although the initial evolutionary influences appeared early, mainly in the form of analogies and broad patterns, it was not until the mid-twentieth century that anthropological theories and approaches started to incorporate the scientific principles derived from the theory of evolution (for a detailed discussion, see Roy D'Andrade, 1995).

In parallel to those anthropological evolutionary theories that drew directly on the methods and principles of evolutionary sciences such as population thinking and selectionist logic, another approach to the study of cultural phenomena, likewise rooted in evolutionary thinking, started to take shape in the mid-twentieth century: it is cognitive anthropology (Claidière & Sperber, 2007, pp. 89–111). It focuses on the cognitive mechanisms of the human mind and their role in shaping culture. While some strands of cognitive anthropology incorporate evolutionary perspectives, others are independent of evolutionary frameworks and grounded primarily in cognitive sciences and psychology.

In the following, we first introduce some early anthropological concepts inspired by evolutionary theories. Then we present some thinkers whose work was influenced by evolutionism through the prism of Marxism. Subsequently, we discuss the formation of cognitive anthropology and review the main developments since the 1950s. The last two sections deal with the current trends in the cognitive and evolutionary sciences of religion, which are the main fields of interest of both authors.

7.2 The Origins of the Evolutionary Approach in Anthropology

The natural place to start our expedition to the roots of anthropology is Britain, because it was there that the societal impact of evolutionary thought was the strongest. To understand the early anthropological concepts and arguments, one needs to consider the social context. According to Barth (2005, pp. 1–57), crucial to the emergence of anthropology was the Nonconformist movement and Quaker philanthropists who challenged the institution of slavery and its existence in the British colonies. After the abolition of slavery in 1833, these philanthropists founded the Aborigines' Protection Society with a motto '(of) one blood', whereby the term 'Aboriginal' was understood as referring broadly to the indigenous people of any continent (Barth, 2005, p. 4). Although various internal disagreements over the society's objectives eventually led to its dissolution, this initiative led to the establishment of a group focused on the study of the indigenous peoples, the Ethnological Society of London (1844). Its leading figures, two Quakers named Thomas Hodgkin and James Cowles, promoted the postulate of a 'unity of blood' and the view that cultural diversity is a consequence of environmental diversity. It should be noted, however, that these views were not shared by all members of this organisation. Equally popular was the concept of polygenesis (see Chap. 2), according to which differences between cultures are due to the existence of different races (Barth, 2005, p. 5), which should be understood as equivalent to different biological species (Barnard, 2004, pp. 24–25). Due to fundamental differences in core beliefs, divisions emerged within the early anthropological community. These ideological rifts reflected broader tensions between competing scientific and social perspectives of the time and revealed a growing incompatibility with the emerging evolutionary framework. As disagreements deepened, they prompted a series of institutional realignments aimed at redefining the discipline's theoretical foundations.

Until the nineteenth century, most people did not consider the indigenous people and their cultures to be equal to the British. From the Renaissance onwards, the dominant view was *degenerationism*, which relied on the Bible for its interpretation of human history and argued that after God destroyed the Tower of Babel (see Chap. 2), the previously unified empire where all peoples lived together broke up and led to the emergence of a multitude of languages and groups. Of these, some gradually degenerated compared to the pre-Babylonian state of civilisation. This claim was apparently supported by various ancient texts which showed that, in the past, the world was inhabited by cultures more advanced than some of the more recent ones (McGee & Warms, 2008, p. 5). Degenerationism was often exploited by politicians and Church leaders, who used it to justify the treatment of native peoples (Hodgen, 1931). The opposite of degenerationism, *progressivism*, argued that the development of humanity followed the opposite path: the ancient cultures did not gradually degenerate, they followed a path of constant progress and evolution from simpler to more complex forms. European countries were furthest along on this path and were thus viewed as the pinnacle of the development of humanity (McGee & Warms, 2008, pp. 5–6). The key manifestations of this highest stage of development were said to be writing, monogamy, and monotheism (cf. Chap. 8). In some otherwise advanced civilisations, some of these manifestations were absent, for instance belief in one God or monogamous marriage. This notion was popular among European intellectuals already in the seventeenth century: 'Their goal was generally to construct a universal history of humankind that moved from a primitive past to the development of European nations' (McGee & Warms, 2008, p. 6). On the other hand, the turbulences which Europe faced in the eighteenth and nineteenth century in many ways contradicted the idea of constant progress, since industrialisation led to technological progress but also went hand in hand with deepening poverty, increasing social disparities (especially in cities), conflicts between secularisation and religiosity, etc. On top of that, the fast-expanding stores of knowledge of biology, geology, and human history led to a rethinking of the previously dominant religious explanation of the origin of life on Earth, which was typically based on literal interpretations of sacred texts, such as the biblical account of creation in Genesis.

Evolutionary concepts started to emerge well before Darwin: his theory of evolution reacted to ideas proposed by thinkers such as Georges Louis Leclerc de Buffon,

Jean-Baptiste Lamarck,[3] or his own grandfather, Erasmus Darwin. In his first book, *On the Origin of Species* (Darwin, 1859), Darwin barely mentions humans, but in *The Descent of Man* (Darwin, 1871), a book he called 'ethnological', he compared humans extensively with other primates and mammals (Barnard, 2004, p. 29; Plotkin, 2004). Despite its growing popularity in scientific circles, the evolutionary theory had initially taken hold only in Great Britain and the United States. It was rejected by French intellectuals and had little impact in Germany until the popularisation efforts of Ernst Haeckel (McGee & Warms, 2008, p. 8) (see Chap. 4). Unlike some of his predecessors and contemporaries, Darwin did not emphasise the transmission of acquired traits from the parents to the offspring. Instead, he stressed that evolution proceeds by transmission of 'what we now call genetic traits'. Random mutations lead to a greater variation and individuals that are the best adapted to their environment will be the most successful in reproduction (Barnard, 2004, p. 29).

Darwin's ideas—or, more broadly, evolutionary ideas—strongly resonated in the emerging anthropology of late nineteenth century. Another idea that had a significant impact on anthropology was the notion of monogenesis, that is, a conviction about the common origin of all humans. While anthropologists may have described the indigenous communities as 'primitive', they did not—unlike many of their contemporaries—question the 'biological and intellectual similarity of all humans' (Barnard, 2004, p. 29). But just as Darwin's emphasis on similarities between humans and non-human animals was seen by many as shocking, the views advocating a common origin of humankind were perceived by many as radical.

The 'father' of anthropology, Edward Burnett Tylor (1832–1917), a man of a Quaker background, was part of this milieu, as were the founders of the Ethnological Society of London whose meetings he started to attend in 1862, after returning from his travels in Mexico.[4] The Ethnological Society at this time brought together ethnologists and archaeologists, who often discussed various similarities between the tools of prehistoric people and contemporary natives (cf. Chap. 2), which in turn spurred their interest in their own history (Barth, 2005, p. 6). Tylor's experience in Mexico, supported by discussions at the Ethnological Society, led to the publication of his most famous work, *Primitive Culture* (Tylor, 1871), which was published in the same year as Darwin's *Decent* (and Lewis Henry Morgan's *Systems of Consanguinity and Affinity*). Discussions with archaeologists as well as his own travels sparked Tylor's interest in 'survivals', whose study he considered crucial to understanding earlier evolutionary stages (McGee & Warms, 2008, p. 10). According to Tylor, many cultural elements that lost their original meaning and function are

[3] Lamarck's vision of evolution was markedly different because he also believed in the transmission of individually acquired characteristics (such as the ability to play the piano). But, as Barnard (2004, p. 28) points out, although Lamarck's theory was highly problematic in biology, it paradoxically better explained the transmission of cultural traits and cultural evolution.

[4] Tylor recorded his travels in Mexico in his first book *Anahuac: or Mexico and the Mexicans, Ancient and Modern* (1861), where he described the Aztec irrigation system, temples, stone tools, but also the contemporary ceremonies of the local people and their similarities with other cultures (Hodgen, 1931).

transmitted from one generation to the next thanks to the human tendency to conservativism and traditionalism (cf. Chaps. 10, 15, and 18). These elements thus have not disappeared—only their function did. Both Tylor and Morgan promoted the notion of a 'psychic unity of humanity', which, applied to this context, refers to a claim that the human mind is universal, persisting unchanged through time and space because the principles which guide the operation of the human mind are independent of culture (for a detailed overview, see Shore, 1996, pp. 15–41). Tylor's vision of evolution was unilinear: he believed that all societies undergo the same development, which happens in three phases: savagery, barbarism, and civilisation, and differences between societies therefore arise from the fact that they are at a different stage of development. This concept of a unilinear, progressive evolution allowed for an essentialist view of human nature (i.e. belief in universal traits shared by all humans) and explained cultural variations. It also facilitated comparisons between current indigenous cultures and those that were believed to have existed in early human history (Shore, 1996, p. 18). If indigenous cultures were viewed as 'living fossils' but 'psychic unity' was assumed, one could compare 'simple' and complex societies and, importantly, one could claim that development in different societies may have proceeded along similar lines (McGee & Warms, 2008, p. 10). Tylor applied this mode to religion, where he proposed that religious forms evolved from simple to complex ones, with monotheism as the final, most advanced stage. He started with the assumption that all cultures share a belief in a 'spiritual essence' that survives death; this belief, he claimed, is reflected in widespread ideas about death and associated rituals (Barnard, 2004, p. 36). Moreover, by identifying 'survivals' (artefacts, institutions, ideas) and tracing their evolution 'from its highest state in the civilized world till we reach its beginnings in the life of the rudest tribe of men' (Tylor, 1881, p. 16), we could come to understand how they changed over time. Ultimately, he saw religion as humanity's attempt to comprehend the surrounding world.

In North America, Lewis Henry Morgan (1818–1881), another proponent of unilinear evolutionism, applied a similar premise to marriage and kinship. Morgan and Tylor (as well as James George Frazer, on whom more below) drew on the work of Lorimer Fison, a missionary and journalist who was interested in anthropology and, while in Fiji, compiled a questionnaire about the kinship system. This survey found its way to Morgan, with whom Fison maintained lively correspondence, and Morgan distributed it worldwide to missionaries and travellers. As a result, early anthropologists relied on Fison's study of indigenous kinship systems that was based on communications with European settlers (McGee & Warms, 2008, p. 51).

Inspired by ethnographic data about various indigenous populations, James George Frazer (1854–1941), another pioneer of anthropology, focused on totemism, a subject that posed a challenge to several scholars at the time. Frazer (1910), Durkheim (1912), and Freud (1955 [1913]) all believed totemism to be the earliest form of religion. Although it is nowadays disputed whether totemism ever existed as a unified phenomenon (on totemism, see Jones, 2005), in the nineteenth and early twentieth century it was treated as such. Frazer collected a large corpus of ethnographic data that documented this phenomenon and concluded that the relationship between the individual or group and the totem is beneficial for both parties because it provides mutual protection. Members of the same totem clan are considered blood relatives, which regulates mutual relations within the group and implies various benefits as well as obligations. The relationship between totem and humans is essential because the totem not only represents certain religious beliefs but functions also as the foundation of a social system (Jones, 2005, p. 82). The religious aspect of totemism is manifested mainly through rituals, especially initiation rituals, which Frazer saw as a clear example of a link between the religious and social aspects of totemism (Jones, 2005, p. 82). Frazer was particularly interested in rituals related to death and rebirth, which he viewed as symbolically expressing the concept of 'immortality of the species' (Jones, 2005, p. 82).

Frazer's evolutionary inclination is especially notable in his explanation of the development of religious ideas. Frazer was an armchair anthropologist who became famous thanks to his book *The Golden Bough*, published in 1890 (Frazer, 2009). Despite its questionable scientific value due to Frazer's cherry-picking methodology, *The Golden Bough* makes for compelling reading, and it had a far-reaching impact, inspiring thinkers such as Bronislaw Malinowski or Claude Lévi-Strauss. In it, Frazer did not discuss a particular system of ideas he would consider the earliest religion; instead, he focused on magical beliefs. He was also a proponent of unilinear evolutionism and believed magic to be an early predecessor of science, a tendency that appears where ordinary knowledge fails—this notion later resonates also in Malinowski's approach. With the emergence of complex communities, magical ideas evolved first into polytheistic and later into monotheistic religions, which were in turn replaced by a further evolutionary stage of scientific explanations. In his writings, Frazer systematically applied this kind of evolutionism, which gradually fell out of favour with his younger colleagues, who increasingly explained the origin and spread of supernatural concepts through diffusion. They argued that, in the past, cultural innovations (including novel supernatural concepts) did not arise out of individual innovation but spread through contact, mainly exchange, trade, or war between groups (Veit, 2013, p. 190). Yet while the concept of diffusion is uncontroversial *per se*, the claim that cultural innovations always originated in one place and spread from there seems excessive (see Chaps. 5 and 12). Other opponents increasingly focused on the function of cultural innovations while emphasising cultural relativism and particularism. The work of both Tylor and Frazer was important for the emerging field of anthropology and stimulated interest in the study of human cultures even if their approach to cultural evolution found few successors. On the other hand, Tylor's definitions of culture and religion are familiar to every student of anthropology to this day, and the same holds of Frazer's conception of magic, which inspired several experimental studies on magical thinking (e.g., Paul Rozin et al., 1986). While interest in evolutionary explanations has in the subsequent decades waned,[5] interest in traditional cultures persisted and developed to include direct study of such groups, ethnographic methods, and field work.

7.3 The Second Wave of Evolutionism

A revival of the evolutionary approach took place in the 1930s, after a period when anthropology in Britain was dominated by functionalism. Evolutionary explanations were brought back to life thanks to the work of Vere Gordon Childe (1936) and Leslie White (1959) (cf. Chap. 5). Childe was an Australian archaeologist who spent most of his career in the United Kingdom. In his early work, he espoused the notion of spread of cultural phenomena through diffusion or migration (Trigger, 1987). He argued that a pivotal moment in human history came with the emergence and spread of agriculture. This was a conviction he shared with the ethnologist Grafton Elliot Smith and other archaeologists (Trigger, 1987, p. 2), who suggested that after the last Ice Age, there was a widespread drought that forced people to settle around available water supplies and begin to cultivate crops. Childe focused mainly on the economic development of humankind but explained technological progress within a broader political and economic context, where numerous factors contribute to cultural change. This allowed him to explain why societies with similar technologies can take different developmental paths (Trigger, 1987, p. 3).

Childe's perspective on the development of humanity changed after his visit to the Soviet Union in 1935, where he encountered Soviet archaeology that had been influenced by the Marxist philosophy. What intrigued Childe was that Russian archaeologists did not focus on technological development but on social organisation. They wanted archaeology to examine not only what people produced in prehistoric times but, even more importantly, how they lived (Trigger, 1987). According to Childe, cultural evolution is the process of development of human cultures, and he argued that one can understand the evolution and functioning of technology only when also considering the environment in which it emerged (Trigger, 1987). Nevertheless, Child never adopted unilinear evolutionism and argued that cultural evolution should always be understood as a complex and context-dependent process.

Initially, the evolutionary approach had more adherents in Europe than in the United States, which under the influence of Franz Boas, a German-born opponent of evolutionism, especially its forms that led to hierarchisation of cultures, remained strongly grounded in relativistic principles and later in the neo-Freudian approach of the Culture and Personality movement.

Despite some departures from his original ideas, majority of Boas' followers maintained his critical approach to evolutionism (McGee & Warms, 2008, p. 119). Boas tried to demonstrate how the human body plastically responds to various

[5] Classical cultural evolutionists of the nineteenth century often understood evolution teleologically and sometimes even viewed the reproduction of certain social groups (e.g., the poor, immigrants, etc.) as a threat that could slow or undermine progress (see Chap. 3).

environmental conditions. Boas's hypothesis of cranial plasticity posits that during growth, the cranium responds mainly to environmental pressures (Holloway, 2002). He also argued against the idea of unilinear evolution, especially the claim that similar cultural elements arise in different cultures because of certain universal evolutionary laws. Citing several examples, Boas showed that cultural elements can also spread through contact or trade and can be thus explained without the assumption of universal evolutionary laws (Boas, 1896). He explained the emergence of cultural elements by referring to various environmental, historical, and psychological factors but historical development seen as pivotal (McGee & Warms, 2008, pp. 121–125). Boas' strongly antiracist, pro-immigrant, and pro-minority stance was strongly reflected in his anthropological work. His rejection of racial determinism directly challenged the prevailing views of the Harvard and Smithsonian scholars (Washington Group), who promoted race-based hierarchies (McGee & Warms, 2008, p. 119).

Aside from Childe, another influential figure of the second wave of evolutionism was the American anthropologist Leslie White, whose views were initially influenced by Boas and cultural relativism. With Julian Steward and George Peter Murdock, White in the late 1940s inspired in American anthropology a resurgence of interest in evolutionism, which contributed to the formation of ecological anthropology and cultural materialism.[6]

After White obtained a position at the University of Buffalo, he became familiar with the work of the famous local native, Lewis Henry Morgan (1818–1881), which was the target of much criticism in Boas's circles. Having read Morgan's work, White identified with most of his views but also concluded that, despite all the correct observations, nineteenth-century evolutionists did not manage to come up with a non-ethnocentric scientific theory that would explain the origin of cultures (McGee & Warms, 2008, p. 227). White appreciated the dynamic, systemic, and symbolic dimensions of culture, but was convinced of the pivotal importance of technology. Moreover, inspired by Marx, he thought in terms of three spheres within culture, the ideological, sociological, and technological one, which are closely intertwined. Nevertheless, he was also convinced that social change takes place mainly due to changes in technology, which is thus the foundation on which cultural progress can

[6] Cultural materialism holds that material factors and environmental conditions play a crucial role in the formation of culture. According to Marvin Harris (1987) and others, perceptions, beliefs, ideas, etc. are secondary to technological factors and culture primarily reflects relations between things (Barnard, 2002, pp. 202–203), so that for instance the fact that Jewish people abstain from eating pork is due to ecological conditions that made pig farming virtually impossible. Harris argues that pigs were not rejected because they were considered unclean but because their breeding was made impossible once forests were cleared to make way to fields that could be used to grow crops needed to feed a rapidly growing population. By the time of the spread of Christianity, the rejection of pork took on symbolic significance and became an ethnic marker (Harris, 1987, pp. 67–79). The reciprocal relationship between humans and the environment is the subject of ecological anthropology, which investigates how people change their environment and how the environment they inhabit shapes their cultural practices, social relations, political organisation, or economics (Salzman & Attwood, 2002, p. 257).

be measured. A crucial concept in White's approach is 'energy', by which he meant that the evolution of human societies is driven by their ability to use energy more efficiently; in short, the control of energy is the driving force behind evolution (White, 1943; Barnard, 2004, p. 39). White was aware of maladaptive elements in cultures and towards the end of his life moderated some of his views, including his assertion that culture 'serves man' (White, 1975). His view of evolution was in a sense teleological: he believed that the driving force behind evolution is the (biological) craving for certainty.

Even this brief overview of evolutionary concepts suggests that the understanding of the term 'evolution' in cultural and social anthropology has been rather loose. Moreover, after having a significant impact in Victorian anthropology, the evolutionary approach became the dominant framework for both biological anthropology and archaeology (cf. Chap. 5) but in social and cultural anthropology, the situation was markedly different. In both of these scientific disciplines, both unilinear and universal evolutionism were the subject of robust criticism bordering on outright rejection (see, e.g., Laguens, 2014).

More promising seemed to be the concept of multilinear evolutionism developed by Julian H. Steward (1972 [1955]). Its main premise was rather straightforward: social and technological evolution is not uniform, it depends on specific ecological conditions, especially the historical development and natural environment. Steward himself called his approach 'cultural ecology' and in the 1950s, he started to use the term 'multilinear evolution' to distinguish it from White's notion of 'universal evolution' (Kerns, 2013). According to Steward, societies that exist in different ecological conditions thus can evolve in different directions.

The scope of this text does not allow us to chart the entire development of the evolutionary approach in anthropology. It should be noted, however, that although anthropologists and archaeologists have aspired to deliver evolutionary explanations of cultural phenomena, the notion of 'evolution' in their approach was historically considerably less anchored in evolutionary biology than it is today. Moreover, it was a time when little attention was paid to the proximate mechanisms underpinning cultural phenomena, such as the cognitive and other psychological functioning of the human mind, which influence the acquisition of cultural representations and thus also their spread (see Chaps. 9, 10, 11 and 13).

7.4 Culture and Cognition

From mid-twentieth century onwards, there started to emerge—in parallel to the evolutionary theories—calls for more focused research into the psychological processes which underpin cultural phenomena. On the one hand, almost all scholars who studied the spread or acquisition of cultural representations in the past have touched upon the mechanisms that make these processes possible. In fact, already Tylor's very first anthropological definition of culture was decidedly cognitive (Blount, 2011). On the other hand, such explanations worked largely with psychological intuitions, and it was felt that modern evidence-based psychology could provide much clearer and more scientific explanations. Such efforts were initially heavily influenced by the psychoanalysis and by behaviourism, which programmatically avoided references to the mind and whose representatives were convinced that only that which is observable can be studied (Kanovský, 2002). By the middle of the twentieth century, the number of various published ethnographies had boomed and while they dealt with various aspects of cultures all around the world, their theoretical foundations evolved hardly at all (D'Andrade, 1995, pp. 4–5). In psychology, on the other hand, the fall of behaviourism coincided with developments in computer science and gradually snowballed into a 'cognitive turn', which soon found its adherents in linguistics and then spread to other disciplines. Change was gradually coming both to Europe and the United States.

In Europe, this paradigm shift (D'Andrade, 1995) is associated with Claude Lévi-Strauss (1958) and the emergence of structuralism[7] (which eventually spread to numerous areas of science). Lévi-Strauss was influenced by linguistics and argued that the human mind is structured in a certain way, and this structuration is reflected in how people classify and think about the world. As Kanovský (1998, p. 124) argued 'structures are real because they are products of thought and can be revealed in an exact way in the products of culture, but at the same time they are ideal because they are models, classifications, and do not follow directly from empirical facts'. Although Lévi-Strauss's understanding of the mind and human reasoning was, from today's perspective, oversimplified (mind does not operate solely or mainly in binary oppositions, as he was suggesting), structuralism brought about a significant change where the attention shifted from what could be observed or deduced from the environmental conditions, towards the principles of reasoning, which cannot be directly observed.

Around the same time, there started to appear in the United States the first anthropological studies that used experimental methods supplementing qualitative research. This approach, which advocated more rigorous methodological standards in anthropology, was pioneered by Floyd Lounsbury (1956) and Ward Goodenough (1956), who published a series of articles on kinship terminology (D'Andrade, 1995, p. 14). Lounsbury, Goodenough, but also others have argued that the traditional ethnographic research yields data which are mutually incomparable and moreover based on Western conceptual categories. They wanted to use clearer ethnographic methods, including immersion in the studied culture—this approach came to be known as the 'new ethnography' (McGee, 2013, p. 232). These efforts were guided by their conviction that culture includes various conceptual categories (domains) and the task of ethnography is to map and describe the content of these 'folk categories' using componential analysis, a method for breaking down complex categories into their basic, contrasting features (Goodenough, 1956).

[7] It is important to note that even before this shift, there were some exceptions: one of the first scholars to discuss the psychological function of cultural institutions was Bronislaw Malinowski (1948).

In the following period, various anthropological analyses focused on classifying and categorising different domains (kinship, fauna, flora, medicine, colours, etc.) using structured interviews and elicitation techniques adopted from linguistics. In fact, the influence of linguistics on structuralism and ethnoscience went beyond the methods and all the way to the core assumption that reasoning and natural classification are anchored in language. This was partly due to the influence of the Sapir–Whorf hypothesis, popular in the 1930s (Sapir, 1924; Whorf, 1956), according to which language influences the reasoning, but another inspiration came from the Prague School of Linguistics. Yet while such studies provided a wealth of knowledge about taxonomies, they offered no psychological explanation of *how* these categorisations are established (Kanovský, 2002). Their premise was that the human mind categorises (without a clear explanation of *how*) and research was supposed to yield insights into how this translates into everyday life. In short, explanations of the cognitive mechanisms underpinning these processes were still missing (McGee, 2013, p. 232). Despite a more rigorous methodology, ethnoscientific research was still heavily context-dependent, which made cross-cultural comparisons difficult. Moreover, people cannot verbalise everything, that is, what people say need not be an unequivocal indication of what they know, much less of how their cognition is linked to their behaviour.

Nevertheless, ethnoscience—in conjunction with the emergence of brain sciences and advances in artificial intelligence—did lay the foundations of cognitive anthropology. In the second half of the twentieth century, several theoretical approaches within the cognitive anthropology aimed at explaining the relationship between mind and culture.[8] Of special importance for later development of cognitive and evolutionary approaches to cultural phenomena were the approaches that addressed the issue of diffusion and transmission of cultural representations and studied the features that promote their acquisition and transmission (see Chaps. 11, 12 and 13).[9]

7.5 Current Trends in Cognitive and Evolutionary Approaches to Cultural Phenomena

The 1970s brought a renewed interest in evolutionary explanations with several approaches that differed, among other things, by how they defined 'culture' and how important they believed culture to be to the overall evolution of humankind. Unlike previous evolutionary theories in anthropology, these approaches were

[8] For example, Eleanor Rosch's theory of prototypes (1975), the schema theory (D'Andrade, 1995; Mandler, 1985), connectionism (Strauss & Quinn, 1997), or cultural models (Holland & Quinn, 1987). As Naomi Quinn notes, the concept of cultural models permeated cultural anthropology, especially medical and ecological anthropology (focusing on the conceptualisation of disease and the environment, respectively), although often without the 'cognitive' part (Quinn, 2013, p. 122).

[9] Especially, Dan Sperber (1996), Pascal Boyer (2001), and Claidière and Sperber (2007).

actually based on Darwin's theory of evolution; they went beyond just a loose or metaphorical application of some of its ideas to human cultures.

In 1975, Edward Osborne Wilson, published a seminal book called *Sociobiology: The New Synthesis*, in which he defined sociobiology as 'the systematic study of the biological basis of all social behavior' (Wilson, 1975, p. 4). Wilson emphasised that since people are animals, the same biological laws apply to them as to any other living creatures. Human behaviour is equally influenced by nature and nurture, and behaviours which enhance the organism's ability to adapt to its environment will be favoured in transmission to the next generation. In contrast to the progressivist notions of earlier anthropological theories, evolution conceptualised in this way has no 'goal' or 'purpose': it is a mechanistic process (McGee & Warms, 2008, p. 406). In cultural anthropology, the sociobiological approach was viewed with much suspicion and resistance against it was especially strong among scholars influenced by the Boasian tradition.[10]

Another naturalistic concept rooted in evolutionary explanations that focus on human reasoning, specifically the ability to represent, is biosemantics (Millikan, 1989), which links insights from (evolutionary) biology with the philosophy of mind (L'Hôte, 2010). Naturalistic concepts explain reasoning processes by appealing to the evolutionary and biological functions of neural states. Based on teleosemantics, which explained meaning in terms of their evolutionary function (i.e., how mental states or symbols mean what they were selected to respond to; Kingsbury, 2004), biosemantics proposes that linguistic and cognitive representations acquire meaning through their role in adaptive systems much like biological traits evolve through natural selection. When it comes to cultural evolution, biosemantics aims at explaining how symbols, concepts, and linguistic structures emerge, stabilise, and persist over time.

There are currently three prominent evolutionary approaches to studying human culture: behavioural ecology (Winterhalder & Smith, 2000), cultural evolution (Boyd & Richerson, 1985), and evolutionary psychology (Cosmides & Tooby, 1987). Although they differ in focus and methodology, each has developed distinct theoretical frameworks to explain the interplay between human behaviour, cognition, and cultural dynamics.

The concept nowadays known as cultural evolution dates back to the 1970s and is closely associated with the dual inheritance theory (known also as the gene-culture coevolution theory) of Robert Boyd and Peter Richerson (Boyd & Richerson, 1985, 2005), which became historically more successful than its main competitor, memetics (Chvaja, 2020; cf. Chaps. 10 and 16). Boyd and Richerson have taken mathematical models from population genetics and adapted them to cultural transmission. Crucially, they assumed that information can be transmitted not only from

[10] For example, Marshall Sahlins argues that sociobiology implies genetic determinism, but culture makes humans so different from the rest of organisms that biology simply cannot explain cultural diversity. He believes that 'while [biology] is an absolutely necessary condition for culture, [it] is equally and absolutely insufficient: it is completely unable to specify the cultural properties of human behavior or their variations from one human group to another' (Sahlins, 1976, p. xi).

parents but also from peers, teachers, religious leaders, or individuals who gained prestige in the community. The flow of information can also be influenced by tendency to conformity, i.e., by people's desire to 'fit' in their community (Smith, 2013, p. 242). Explanations of cultural transmission were central to the study of culture not only in Boyd and Richerson's evolutionary approach, which stressed the importance of learning biases, but also in the cognitive approach of Atran (2002), Boyer (1994, 2001), Hirschfeld (1996), and Sperber (1996) who based their approach on an investigation of properties of mental representations.

Evolutionary theories mostly try to explain why individual species evolved the way they did and what adaptive mechanisms allow them to reproduce and survive. Cultural evolution attributes a greater role to nongenetic factors, especially the transmission of knowledge and its acquisition (Laland & Brown, 2002; Lewens, 2020). As cognitive scientists pointed out, transmission of cultural representations does not take place by 'copying' from mind to mind (Atran, 2002; Boyer, 2001; Sperber, 1996). In other words, the human mind does not transmit information in the form in which it received it: it can transfer only part of information (the part that is deemed relevant) or pass on information that is slightly changed or adjusted to a recipient. In the cognitive approach, representations that have a psychological advantage in intergroup, intragroup, or intergenerational transmission (i.e., they have some properties that make them more likely to be transmitted) are usually referred to as cognitive or cultural 'attractors' (Henrich et al., 2008; Sperber, 1996).

These cognitive explanations of transmission were further elaborated by Henrich et al. (2008), who introduced the perspective of cultural evolution and proposed that it is useful to consider models of transmission involving 'discrete replicators that evolve under the influence of natural-selection-like forces' (Henrich et al., 2008, p. 121). They argued that while individuals may share various representations, learning biases can lead to systematic patterns in how certain cultural variants are more likely to be retained and transmitted across generations.[11]

The process of acquisition or learning is influenced by two cognitive mechanisms, which work in tandem: inferential transformation and selective attention. While the former 'captures the manner in which cognitive processes of acquisition, storage, and retrieval alter mental representations in ways to favour some representations over others – cognitive attractors', the latter 'captures the tendency for individuals to pay particular attention to some individuals more than others', i.e., it describes strategic learning (Henrich et al., 2008, p. 122). Henrich argues that natural selection shaped cultural learning processes so that the acquisition of ideas, attitudes, beliefs, or behaviours would be as effective as possible. This led to two types of biases: content bias and context bias (Henrich, 2009; cf. also Henrich & Henrich, 2007). In the case of content bias, cultural learning is influenced by the content of cognition, meaning that attention is attracted especially to representations that convey information that is adaptive, i.e., relevant to fitness—and this information is

[11] For details of how Henrich and Boyd's (2002) model differs from Sperber's (1996) proposition, see Claidière and Sperber (2007).

then passed on (Henrich, 2009, p. 245). Simultaneously, there is also context bias at work, which means that people tend to accept and pass on information from individuals who are more successful, more skilled, older, better educated, or have a greater expertise (Henrich, 2009, p. 245). Other factors that influence context bias are the ethnicity or gender of the information bearer, where people are attracted to cultural models which are heuristically more likely to fit their future societal roles.

In other words, people choose, mostly unconsciously, whose attitudes and opinions they listen to, assimilate, and pass on. For example, persons who find themselves in an unfamiliar environment will eat what the locals eat and thus try to avoid potentially dangerous items. Although expertise is mostly domain-specific (a person is unlikely to be both a top surgeon and a top economist), prestige bias operates across domains, which is why we may be likely to accept an expert's statements even in domains where the person does not have expertise (and thus believe that, e.g., a successful tennis player's statements about vaccinations are relevant). This excessive copying is largely due to the fact that we cannot identify which factors and mechanisms are the cause of success and, as a result, we copy everything, even attitudes quite unrelated to the domain in which the person excels (see Chaps. 9, 10, and 11).

The concept of cultural evolution is thus a fairly complex. So far, it has been applied to various domains of culture, especially cooperation (Henrich & Henrich, 2007; Henrich & Muthukrishna, 2021), ethnic boundaries (Boyd & Richerson, 1987; Moya & Henrich, 2016), and religion (Norenzayan et al., 2016; Norenzayan & Gervais, 2012; Singh, 2021; Singh & Henrich, 2020; Winkelman, 2021) but, like other evolutionary and cognitive approaches to the study of culture, its impact on mainstream cultural anthropology has been so far limited. While contacts and constructive debates between the humanist and scientific approaches to cultural anthropology remain sparse, the efforts continue (Lang & Kundt, 2020; Slingerland & Collard, 2011).

7.6 Cognition and Evolution: Religious Beliefs and Behaviours

As noted above, one particularly fruitful area where cognitive and evolutionary theories of culture have taken hold is the study of religion (Kundt, 2015; Petersen et al., 2019), where our understanding of various religious phenomena was significantly enriched by combining traditional cultural anthropology with cognitive and evolutionary approaches (Lang & Kundt, 2020). In this sense, it is similar to, for instance, the study of morality (Moncrieff & Lienard, 2018; Purzycki et al., 2018). Moreover, these combining approaches contribute to an even deeper interdisciplinarity that is inherent in anthropology. Subjects repeatedly discussed within these approaches to religion include the transmission of religious ideas and behaviours, the relationship between religion and cooperation (see Fig. 7.1), and the reasons for

Fig. 7.1 Sharing the blessed milk. Communities of Tamil Hindus in Mauritius organise an annual ritual called *Thaipusam kavadi*. During the ritual, attendees participate in a slowly moving procession. They carry portable altars on their shoulders, while other participants witness their journey along the route. While men traditionally pierce their tongue, cheeks, and body with needles or hooks with attached chimes or bells, women carry special containers with sacrificial milk. Depicted is a combination of both, where the male attendee carries little vessels of sacrificial milk attached by hooks to his torso. Accompanied by family members, attendees reach the temple where priests remove their body piercings, and the milk, now with transformed qualities, is offered to the gods and the surrounding participants. After that, the milk is usually further distributed around the temple either by attendees themselves or by their family members. Phototgraph by Eva Kundtová Klocová. Reproduced with permission

the emergence of rituals and religion in our evolutionary history (e.g., Lang & Kundt, 2024).

A boom in the cognitive and evolutionary sciences of religion (CESR)[12] came with the new millennium, but some crucial works appeared earlier. The birth of

---

[12]Although evolutionary explanations were applied to religion in early anthropological theories, in the subsequent development of anthropology, they largely faded away. The 1990s witnessed a veritable boom in cognitive explanations of religious practices which led in 2006 to the creation of the International Association for the Cognitive Science of Religion (IACSR). The body of research that offers evolutionary explanations of religion has also been growing in recent decades (Kundt, 2019); this was reflected in the addition 'evolutionary' to the name of both the scientific approach (now abbreviated as CESR rather than CSR) and the organisation, which was renamed the International Association for the Cognitive and Evolutionary Sciences of Religion (IACESR) in 2021.

CESR is usually associated with the publication of Stewart Guthrie's 'Cognitive Science of Religion' (1980). Other pioneering works in the field of CESR are books by E. Thomas Lawson and Robert N. McCauley (Lawson & McCauley, 1993; McCauley & Lawson, 2002), who were inspired by Noam Chomsky's work in linguistics but also by Dan Sperber's work (1975). Lawson and McCauley argue that thinking about ritual activity is intuitive and people represent rituals in their minds like any other activity. Moreover, they suggest that intuitive reasoning is based on a cognitive mechanism, an 'action representation system' (Lawson & McCauley, 1993, p. 77). What is specific to religious rituals is that they involve a 'culturally postulated superhuman agent' (Lawson & McCauley, 1993, p. 77). In this, we can see the influence of Dan Sperber, who criticised the symbolic approaches to the explanation of cultural phenomena that ignore the cognitive mechanisms underlying symbolism, although all symbolic ideas emerge from the inner workings of the mind. Sperber's cognitive theory of symbols was for the development of CESR as groundbreaking as his epistemology of representations (Sperber, 1996) in which he summarised the cultural attractors thesis mentioned above (Jerotijević & Maňo, 2014).

This integration of scientific approaches into theoretical frameworks coming from the social sciences (Hoffmeyer, 1997; Kull, 1999; Barbieri, 2008; Brier, 2015) had important implications for biosemiotics (Kleisner & Tureček, 2017) both in terms of which phenomena are worth studying and in terms of the extent to which one can draw analogies between humans and other organisms (see Chap. 16). Biosemiotics views every living creature as a semiotic system and considers sign production as fundamental to life (Barbieri, 2008). Consequently, meaning-making is not unique to humans—it is present throughout the biological world. As the link between a sign and meaning can be established by a rule of code (Cobley, 2016; Wheeler & Westling, 2015), the social dimension of life forms an additional layer through the possibility of establishing conventions (Barbieri, 2009). Rituals are in this context of special interest because they can simultaneously establish convention and inform about that convention's acceptance by participants (Rappaport, 1999).

In the 1990s, there emerged another influential conception of rituals: its author, the British anthropologist Harvey Whitehouse (2000, 2004), argued that rituals come in two modes that differ in frequency and emotional response (Atkinson & Whitehouse, 2011). First, there are the rituals which are vivid, intense, accompanied by strong emotions, and some take place only once in a lifetime (e.g., initiation rituals). They function in what Whitehouse calls the 'imagistic mode'. Secondly, there are rituals that take place often, are less intense, and usually not accompanied by strong emotions. They function in what Whitehouse calls the 'doctrinal mode'. These different modes trigger different types of (epistemic and semantic) memory and have a different impact on participants. The painful and traumatic experiences produced by some kinds of rituals can have a powerful impact on participants and create a sense of belonging or togetherness. Whitehouse's conception initially aimed at explaining two distinct ritual forms, at least one of which is present in every culture. In his later work, he linked the emergence of doctrinal modes to the rise of agrarian societies (Atkinson & Whitehouse, 2011). According to this hypothesis, the adoption of a more sedentary lifestyle led to the emergence of doctrinal forms of religions characterised by frequent and less emotionally charged ritual forms, the establishment of priesthood, and a unification of regional religious systems and groups.

One of the seminal works of the CESR was Pascal Boyer's *Religion Explained* (Boyer, 2001), which analysed various aspects of religious beliefs and behaviours, including the question of what is specific about religious ideas. While religious beliefs of different cultures exhibit a vast diversity (supernatural agents differ in the appearance, manifestations, abilities, spheres of operation, etc.), the creativity of human mind is not boundless. Supernatural concepts can be classified into basic ontological categories (person, living thing, manmade object) but what makes them different from natural concepts is that they usually contain features that 'disrupt' the ontological expectations in their domain. As a result, they contain both an 'intuitive' and a 'counterintuitive' part. For instance, a ghost belongs to the category of persons but has properties which actual people do not have: it is invisible, it can affect people at a distance, walk through walls, etc. (Boyer, 2001, 2003). It is this counterintuitive information that makes supernatural concepts compelling and drives their diffusion within a group and from one generation to the next. Furthermore, Boyer suggested (2004) that 'successful' supernatural concepts, that is, those that persist over long periods of time, are those that do not contain too many counterintuitive elements (they must be memorable enough to be passed on by an individual human mind) and have some inferential potential (they must help explain some real-life situations).[13]

Boyer's counterintuitiveness thesis was further developed by Justin Barrett (2000), who noted that if a supernatural concept is to be successfully transmitted, it must be attention-grabbing and easy to remember. Moreover, one should bear in mind that certain kinds of concepts, for instance invisible furniture, are rarely the subject of religious beliefs. Most subjects of religious beliefs are intentional agents, that is, supernatural agents or agents with some supernatural properties (Barrett, 2000). The bias to detecting in the environment (intentional) agents that need not exist has been discussed already by Stewart Guthrie (1980). Developing Guthrie's thesis, Barrett argues that the human mind is probably equipped with a mechanism, a 'hyperactive agent detection device', which is highly sensitive to specific cues from the environment, and if it registers activity, it automatically tries to figure out who is behind it. If it detects no natural (human or animal) agent, it may attribute the activity to a supernatural agent. Whether this happens, and what kind of agent is assumed, depends on the cultural environment that moulds our expectations. Barrett suggests that such mental mechanism was favoured by natural selection because it boosted predator avoidance: the costs of assuming the presence of a potentially dangerous agent and adjusting one's behaviour accordingly outweigh the possibly fatal risk of ignoring potential cues even in absence of actual danger (Barrett, 2004)

[13] Boyer's approach was recently criticised by Purzycki and Willard (2016). See their text for a detailed analysis.

(see Chap. 10). van Elk, Karinen, et al. (2016a) thus noted that while Guthrie believed that the bias driving the detection of supernatural agents operates on perceptual inputs, Barrett shifted this bias to people's perceived need to explain intentionality.

The concept of a hyperactive agent detection device subsequently featured in many studies dedicated to representations of supernatural agents. It was attractively straightforward and intuitive, but experiments designed to test this hypothesis yielded ambiguous results. For example, van Elk, Rutjens, et al. (2016b) have shown that priming involving supernatural concepts does not automatically trigger agency detection (cf. also Norenzayan et al., 2008; Willard & Norenzayan, 2013). In their study, which focused on the perceptual level, the priming had activated agent search only in participants who declared themselves religious, which suggests the importance of specific cultural content (van Elk et al., 2016b).

Others, such as Andersen (2017) or Nenadalová et al. (2024), have even contested the idea that people in threatening situations have an increased tendency to detect agents in their environment. According to Andersen, early cognitive theories of religion were influenced by evolutionary psychology, which was at that time dominated by the domain-specificity model of mind (Cosmides & Tooby, 1987). In that model, the mind was likened to a Swiss Army knife that contains many tools designed to deal with different stimuli. Current models, on the other hand, assume common computational principles aimed at processing information from different domains, that is, 'procedural and domain-general models of perception and cognition' (Andersen, 2017, p. 6).

Such debates between representatives of different evolutionary paradigms have moved our understanding of the human mind, human behaviour, and cultural factors considerably forward. The CESR is a dynamic area of research. Over the past two decades, it has inspired a large number of both empirical and theoretical studies, which contributed to our understanding of religious phenomena, their psychological underpinnings, and their evolutionary function and significance. While the first wave of CESR was mostly theoretical (e.g., Boyer, 2001; Lawson & McCauley, 1993; Whitehouse, 2004), the next wave made significant strides in experimental research and some of the most recent studies even include cross-cultural research in ecologically valid settings. Cross-cultural research and mixed methods (behavioural experiments and ethnographic interviews) were used, for example, in the study by Purzycki et al. (2016), which continues the quest for an evolutionary explanation of moral behaviour. Purzycki and his team have tested in eight cultural settings[14] the hypothesis that belief in omniscient and punishing gods promotes parochial cooperation, prosociality, trust, and fairness, and their results support this hypothesis. Other researchers investigated specifically the influence of the environment (secular

---

[14] Their sample included communities of hunters and gatherers, horticulturalists, herders, farmers, and urban populations in Vanuatu, Fiji, Tanzania, Tuva, Mauritius, and Brazil, and different religious traditions (Christianity, Hinduism, Buddhism but also local traditions, such as animism and ancestor worship). This study was later replicated and extended and based on an even larger and more culturally diverse sample of 15 societies (Lang et al., 2019).

vs. religious) on prosocial behaviour in further cross-cultural studies (Lang et al., 2016; Nichols et al., 2020; Xygalatas et al., 2016).

Although one cannot map here the developments in the field of the study of religious phenomena in detail, this brief overview should suggest the potential of this interdisciplinary approach, which is similarly evident in other fields (especially morality and decision-making).

7.7 Conclusions

We have shown that in the history of research into cultural and social phenomena, the evolutionary perspective had fluctuated in prominence. Nowadays, supplemented by insights from cognitive science, it represents a significant—though not mainstream—trend in social and cultural anthropology. Cognitive research of culture has been flourishing since about the 1990s and while cognitive anthropology never became the dominant approach in social or cultural anthropology,[15] it produces a steady stream of significant findings and insights into how our psychological and behavioural mechanisms evolved in relation to specific ecological and cultural environments. Moreover, there has been a shift in the field towards linking cognitive explanations with evolutionary ones, which also helped fill some lacunae in our knowledge.

Cognitive and evolutionary approaches to cultural phenomena have been developing in several, not necessarily mutually exclusive, directions. Methodologically, these approaches are interdisciplinary, bringing together anthropologists and psychologists with behavioural scientists, neuroscientists, and biologists. In CESR, this interdisciplinary approach facilitated the integration of knowledge from different disciplines and, crucially, it enabled the testing of hypotheses, be it in laboratory or natural settings. Moreover, while in the early days of anthropology the evolutionary approach was but a metaphor or inspiration used to interpret cultural phenomena, the current concepts applied to cultural phenomena are fully anchored in the theory of evolution and are becoming its integral part (e.g., the gene-culture coevolution). The cognitive and evolutionary perspectives have greatly advanced our understanding of human culture, its evolution, the transmission of knowledge, the function of individual cultural phenomena, and the role of cumulative culture. The results of this research cannot be ignored by anyone who is trying to understand the role of culture in the evolution of humanity.

[15] In social or cultural anthropology, the symbolic or interpretative approach dominated in the 1960s and 1970s, at which point postmodern anthropology started to make inroads. Poststructuralism and feminist anthropology formed parallel distinct strands.

References

Andersen, M. (2017). Predictive coding in agency detection. *Religion, Brain & Behavior, 9*(1), 65–84.

Atkinson, Q. D., & Whitehouse, H. (2011). The cultural morphospace of ritual form: Examining modes of religiosity cross-culturally. *Evolution and Human Behavior, 32*(1), 50–62.

Atran, S. (2002). *In gods we trust: The evolutionary landscape of religion*. Oxford University Press.

Barbieri, M. (2008). Biosemiotics: A new understanding of life. *Naturwissenschaften, 95*(7), 577–599.

Barbieri, M. (2009). A short history of biosemiotics. *Biosemiotics, 2*(2), 221–245.

Barnard, A. (2002). Cultural materialism. In A. Barnard & J. Spencer (Eds.), *Encyclopedia of social and cultural anthropology* (pp. 202–203). Routledge.

Barnard, A. (2004). *History and theory in anthropology*. Cambridge University Press.

Barnard, A., & Spencer, J. (Eds.). (1996). *Encyclopedia of social and cultural anthropology*. Routledge.

Barrett, J. (2000). Exploring the natural foundations of religion. *Trends in Cognitive Sciences, 4*(1), 29–34.

Barrett, J. (2004). *Why would anyone believe in god?* AltaMira Press.

Barth, F. (2005). Britain and commonwealth. In F. Barth, A. Gingrich, R. Parkin, & S. Silverman (Eds.), *One discipline, four ways: British, German, French, and American anthropology* (pp. 1–57). University of Chicago Press.

Blount, B. G. (2011). A history of cognitive anthropology. In D. B. Kronenfeld, G. Bennardo, V. C. de Munck, & M. D. Fischer (Eds.), *A companion to cognitive anthropology* (pp. 11–29). WileyBlackwell.

Boas, F. (1896). The limitations of the comparative method in anthropology. *Science, 4*(103), 901–908.

Boyd, R., & Richerson, P. J. (1985). *Culture and the evolutionary process*. University of Chicago Press.

Boyd, R., & Richerson, P. J. (1987). The evolution of ethnic markers. *Cultural Anthropology, 2*(1), 65–79.

Boyer, P. (1994). *The naturalness of religious ideas: A cognitive theory of religion*. University of California Press.

Boyer, P. (2001). *Religion explained*. Basic Books.

Boyer, P. (2003). Religious thought and behaviour as by-products of brain function. *Trends in Cognitive Sciences, 7*(3), 119–124.

Boyer, P. (2004). Why is religion natural? *Skeptical Inquirer, 28*(2), 25–31.

Brier, S. (2015). Can biosemiotics be a science if its purpose is to be a bridge between the natural, social, and human sciences? *Progress in Biophysics and Molecular Biology, 119*(3), 576–587.

Childe, V. G. (1936). *Man makes himself*. Watts & Co.

Chvaja, R. (2020). Why did memetics fail? Comparative case study. *Perspectives on Science, 28*(4), 542–570.

Claidière, N., & Sperber, D. (2007). The role of attraction in cultural evolution. *Journal of Cognition and Culture, 7*(1–2), 89–111.

Cobley, P. (2016). *Cultural implications of biosemiotics*. Springer.

Cosmides, L., & Tooby, J. (1987). From evolution to behavior: Evolutionary psychology as the missing link. In J. Dupre (Ed.), *The latest on the best: Essays on evolution and optimality* (pp. 277–306). MIT Press.

D'Andrade, R. G. (1995). *The development of cognitive anthropology*. Cambridge University Press.

Darwin, C. (1859). *On the origin of species by means of natural selection, or the preservation of favoured races in the struggle for life*. John Murray.

Darwin, C. (1871). *The descent of man, and selection in relation to sex*. John Murray.

Durkheim, É. (1912). *The elementary forms of religious life*. G. Allen & Unwin.

Frazer, J. G. (1910). *Totemism and exogamy: A treatise on certain early forms of superstition and society*. Macmillan.
Frazer, J. G. (2009). *The golden bough: A study in magic and religion*. (Original work published 1890). Macmillan.
Freud, S. (1955). Totem and taboo: Some points of agreement between the mental lives of savages and neurotics. In *The standard edition of the complete psychological works of Sigmund Freud, Vol. XIII (1913–1914)* (pp. VII–162). (Original work published 1913).
Goodenough, W. (1956). Componential analysis and the study of meaning. *Language, 32*(1), 195–216.
Guthrie, S. (1980). The cognitive science of religion. *Current Anthropology, 21*(2), 181–203.
Harris, M. (1987). *The sacred cow and the abominable pig: Riddles of food and culture*. Simon & Schuster.
Henrich, J. (2009). The evolution of costly displays, cooperation, and religion: Credibility enhancing displays and their implications for cultural evolution. *Evolution and Human Behavior, 30*(4), 244–260.
Henrich, J., & Boyd, R. (2002). On modeling cognition and culture: Why cultural evolution does not require replication of representations. *Journal of Cognition and Culture, 2*(2), 87–112.
Henrich, N., & Henrich, J. (2007). *Why humans cooperate: A cultural and evolutionary explanation*. Oxford University Press.
Henrich, J., & Muthukrishna, M. (2021). The origins and psychology of human cooperation. *Annual Review of Psychology, 72*, 207–240.
Henrich, J., Boyd, R., & Richerson, P. J. (2008). Five misunderstandings about cultural evolution. *Human Nature, 19*(2), 119–137.
Hirschfeld, L. A. (1996). *Race in the making: Cognition, culture, and the child's construction of humankinds*. MIT Press.
Hodgen, M. T. (1931). The doctrine of survivals: The history of an idea. *American Anthropologist, 33*(3), 307–324.
Hoffmeyer, J. (1997). *Signs of meaning in the universe*. Indiana University Press.
Holland, D., & Quinn, N. (Eds.). (1987). *Cultural models in language and thought*. Cambridge University Press.
Holloway, R. L. (2002). Head-to-head with Boas: Did he err on the plasticity of head form? *Proceedings of the National Academy of Sciences of the United States of America, 99*(23), 14622–14623.
Jerotijević, D., & Maňo, P. (2014). Rituál a mysl. Kognitívní a evoluční teórie rituálu. [Ritual and the mind: Cognitive and evolutionary theories of ritual]. *Pantheon, 9*(1), 139–165.
Jones, R. A. (2005). *The secret of the totem: Religion and society from McLennan to Freud*. Columbia University Press.
Kanovský, M. (1998). Lévi-Strauss a štrukturálna antropológia. In C. Lévi-Strauss (Ed.), *Totemizmus* (p. 124). Chronos.
Kanovský, M. (2002). Kultúra a kognitívna myseľ. In J. Rybár, Ľ. Beňušková, & V. Kvasnička (Eds.), *Kognitívne vedy* (pp. 163–199). Kalligram.
Kerns, V. (2013). Julian Steward. In R. J. McGee & R. L. Warms (Eds.), *Theory in social and cultural anthropology: An encyclopedia* (pp. 812–816). Sage Publications.
Kingsbury, J. (2004). Biologising the mind. *Biology and Philosophy, 19*(3), 473–482.
Kleisner, K., & Tureček, P. (2017). Cultural and biological evolution: What is the difference? *Biosemiotics, 10*(1), 127–130.
Kull, K. (1999). Biosemiotics in the twentieth century: A view from biology. *Semiotica, 127*(1–4), 385–414.
Kundt, R. (2015). *Contemporary evolutionary theories of culture and the study of religion*. Bloomsbury Academic.
Kundt, R. (2019). Making evolutionary science of religion an integral part of cognitive science of religion. In A. K. Petersen, I. S. Gilhus, L. H. Martin, J. S. Jensen, & J. Sørensen (Eds.), *Evolution, cognition, and the history of religion: A new synthesis (pp. 141–158). Brill.*

L'Hôte, C. (2010). Biosemantics: An evolutionary theory of thought. *Evolution: Education and Outreach, 3*(2), 265–274.

Laguens, A. G. (2014). Unilinear evolution and lineal time: A critique. In *Encyclopedia of global archaeology* (pp. 7455–7463). Springer.

Laland, K. L., & Brown, G. R. (2002). *Sense and nonsense: Evolutionary perspectives on human behaviour*. Oxford University Press.

Lang, M., & Kundt, R. (2020). Evolutionary, cognitive, and contextual approaches to the study of religious systems. *Method & Theory in the Study of Religion, 32*(1), 1–46.

Lang, M., & Kundt, R. (2024). The evolution of human ritual behavior as a cooperative signaling platform. *Religion, Brain & Behavior, 14*(4), 377–399.

Lang, M., Mitkidis, P., Kundt, R., Nichols, A., Krajčíková, L., & Xygalatas, D. (2016). Music as a sacred cue? Effects of religious music on moral behavior. *Frontiers in Psychology, 7*, 814.

Lang, M., Purzycki, B. G., Apicella, C. L., Atkinson, Q. D., Bolyanatz, A., Cohen, E., & Henrich, J. (2019). Moralizing gods, impartiality, and religious parochialism across 15 societies. *Proceedings of the Royal Society B: Biological Sciences, 286*(1898), 20190202.

Lawson, E. T., & McCauley, R. N. (1993). *Rethinking religion: Connecting cognition and culture*. Cambridge University Press.

Lévi-Strauss, C. (1958). *Structural anthropology*. Basic Books.

Lewens, T. (2020). Cultural evolution. In E. N. Zalta (Ed.), *The Stanford encyclopedia of philosophy*. https://plato.stanford.edu/archives/sum2020/entries/cultural-evolution/

Lounsbury, F. (1956). A semantic analysis of the Pawnee kinship usage. *Language, 32*, 158–194.

Malinowski, B. (1948). *Magic, science and religion and other essays*. Beacon Press.

Mandler, G. (1985). *Cognitive psychology: An essay in cognitive science*. Erlbaum.

McCauley, R. N., & Lawson, E. T. (2002). *Bringing ritual to mind: Psychological foundations of cultural forms*. Cambridge University Press.

McGee, R. J. (2013). Ethnoscience/new ethnography. In R. J. McGee & R. L. Warms (Eds.), *Theory in social and cultural anthropology: An encyclopedia* (p. 232). Sage Publications.

McGee, R. J., & Warms, R. L. (2008). *Anthropological theory: An introductory history* (4th ed.). McGraw Hill.

Millikan, R. G. (1989). Biosemantics. *The Journal of Philosophy, 86*(6), 281–297.

Moncrieff, M. A., & Lienard, P. (2018). Moral judgments of in-group and out-group harm in post-conflict urban and rural Croatian communities. *Frontiers in Psychology, 9*, 212.

Moya, C., & Henrich, J. (2016). Culture–gene coevolutionary psychology: Cultural learning, language, and ethnic psychology. *Current Opinion in Psychology, 8*, 112–118.

Nenadalová, J., Szymanek, P., & Van Leeuwen, N. (2024). Revisiting feeling of threat and agency detection: A preregistered virtual reality study. *Psychology of Religion and Spirituality, 17*(3), 161–174.

Nichols, A. D., Lang, M., Kavanagh, C., Kundt, R., Yamada, J., Ariely, D., & Mitkidis, P. (2020). Replicating and extending the effects of auditory religious cues on dishonest behavior. *PLoS One, 15*(8), e0237007.

Norenzayan, A., & Gervais, W. M. (2012). The cultural evolution of religion. In E. Slingerland & M. Collard (Eds.), *Creating consilience: Integrating science and the humanities* (pp. 243–265). Oxford University Press.

Norenzayan, A., Hansen, I. G., & Cady, J. (2008). An angry volcano? Reminders of death and anthropomorphizing nature. *Social Cognition, 26*, 190–197.

Norenzayan, A., Shariff, A. F., Gervais, W. M., Willard, A. K., McNamara, R. A., Slingerland, E., & Henrich, J. (2016). The cultural evolution of prosocial religions. *Behavioral and Brain Sciences, 39*, e1.

Petersen, A. K., Gilhus, I. S., Martin, L. H., Jensen, J. S., & Sørensen, J. (Eds.). (2019). *Evolution, cognition, and the history of religion: A new synthesis. Brill.*

Plotkin, H. (2004). *Evolutionary thought in psychology: A brief history*. Blackwell.

Purzycki, B. G., & Willard, A. K. (2016). MCI theory: A critical discussion. *Religion, Brain & Behavior, 6*(3), 207–248.

Purzycki, B. G., Apicella, C., Atkinson, Q. D., Cohen, E., McNamara, R. A., Willard, A. K., Xygalatas, D., Norenzayan, A., & Henrich, J. (2016). Moralistic gods, supernatural punishment, and the expansion of human sociality. *Nature, 530*(7590), 327–330.
Purzycki, B. G., Pisor, A. C., Apicella, C., Atkinson, Q. D., Cohen, E., Henrich, J., McElreath, R., et al. (2018). The cognitive and cultural foundations of moral behavior. *Evolution and Human Behavior, 39*(5), 490–501.
Quinn, N. (2013). Cognitive anthropology. In R. J. McGee & R. L. Warms (Eds.), *Theory in social and cultural anthropology: An encyclopedia* (pp. 120–124). Sage Publications.
Rappaport, R. (1999). *Ritual and religion in the making of humanity*. Cambridge University Press.
Richerson, P. J., & Boyd, R. (2005). *Not by genes alone: How culture transformed human evolution*. University of Chicago Press.
Rosch, E. (1975). Cognitive representations of semantic categories. *Journal of Experimental Psychology, 104*(3), 192–233.
Rozin, P., Millman, L., & Nemeroff, C. (1986). Operation of the laws of sympathetic magic in disgust and other domains. *Journal of Personality and Social Psychology, 50*(4), 703–712.
Sahlins, M. (1976). *The use and abuse of biology: An anthropological critique of sociobiology*. University of Michigan Press.
Salzman, P. C., & Attwood, D. W. (2002). Ecological anthropology. In A. Barnard & J. Spencer (Eds.), *Encyclopedia of social and cultural anthropology* (pp. 256–261). Routledge.
Sapir, E. (1924). Culture, genuine and spurious. *American Journal of Sociology, 29*(4), 401–429.
Shore, B. (1996). *Culture in mind: Cognition, culture, and the problem of meaning*. Oxford University Press.
Singh, M. (2021). Magic, explanations, and evil: The origins and design of witches and sorcerers. *Current Anthropology, 62*(1), 2–29.
Singh, M., & Henrich, J. (2020). Why do religious leaders observe costly prohibitions? Examining taboos on Mentawai shamans. *Evolutionary Human Sciences, 2*, e32.
Slingerland, E., & Collard, M. (2011). Creating consilience: Toward a second wave. In E. Slingerland & M. Collard (Eds.), *Creating consilience: Integrating the sciences and the humanities* (pp. 3–42). Oxford University Press.
Smith, E. A. (2013). Evolutionary anthropology. In R. J. McGee & R. L. Warms (Eds.), *Theory in social and cultural anthropology: An encyclopedia* (pp. 241–245). Sage Publications.
Sperber, D. (1975). *Rethinking symbolism*. Cambridge University Press.
Sperber, D. (1996). *Explaining culture*. Blackwell.
Steward, J. H. (1972). *Theory of culture change: The methodology of multilinear evolution*. University of Illinois Press. (Original work published 1955).
Strauss, C., & Quinn, N. (1997). *A cognitive theory of cultural meaning*. Cambridge University Press.
Trigger, B. G. (1987). V. Gordon Childe: A Marxist archaeologist. In L. Manzanilla (Ed.), *Studies in the Neolithic and urban revolutions: The V. Gordon Childe colloquium, Mexico, 1986* (pp. 1–8) BAR International Series 349.
Tylor, E. B. (1861). *Anahuac: Or Mexico and the Mexicans, ancient and modern*. Longman, Green, Longman and Roberts.
Tylor, E. B. (1871). *Primitive culture: Researches into the development of mythology, philosophy, religion, art, and custom*. John Murray.
Tylor, E. B. (1881). *Anthropology: An introduction to the study of man and civilization*. Macmillan and Co.
van Elk, M., Karinen, A., Specker, E., Stamkou, E., & Baas, M. (2016a). Standing in awe: The effects of awe on body perception and the relation with absorption. *Collabra, 2*(1), 4.
van Elk, M., Rutjens, B. T., van der Pligt, J., & van Harreveld, F. (2016b). Priming of supernatural agent concepts and agency detection. *Religion, Brain & Behavior, 6*(1), 4–33.
Veit, U. (2013). Diffusionism. In R. J. McGee & R. L. Warms (Eds.), *Theory in social and cultural anthropology: An encyclopedia* (pp. 216–219). Sage Publications.
Wheeler, W., & Westling, L. (2015). Biosemiotics and culture: Introduction. *Green Letters, 19*(3), 215–226.

White, L. (1943). Energy and the evolution of culture. *American Anthropologist, 43*(3/1), 335–356.

White, L. (1959). *The evolution of culture: The development of civilization to the fall of Rome*. McGraw-Hill.

White, L. (1975). *The concept of cultural systems: A key to understanding tribes and nations*. Columbia University Press.

Whitehouse, H. (2000). *Arguments and icons: Divergent modes of religiosity*. Oxford University Press.

Whitehouse, H. (2004). *Modes of religiosity: A cognitive theory of religious transmission*. AltaMira Press.

Whorf, B. L. (1956). *Language, thought and reality*. MIT Press.

Willard, A. K., & Norenzayan, A. (2013). Cognitive biases explain religious belief, paranormal belief, and belief in life's purpose. *Cognition, 129*(2), 379–391.

Wilson, E. O. (1975). *Sociobiology: The new synthesis*. Harvard University Press.

Winkelman, M. J. (2021). A cross-cultural study of the elementary forms of religious life: Shamanistic healers, priests, and witches. *Religion, Brain & Behavior, 11*(1), 27–45.

Winterhalder, B., & Smith, E. A. (2000). Analyzing adaptive strategies: Human behavioral ecology at twenty-five. *Evolutionary Anthropology, 9*(2), 51–72.

Xygalatas, D., Kundtová Klocová, E., Cigán, J., Kundt, R., Maňo, P., Kotherová, S., Mitkidis, P., Wallot, S., & Kanovský, M. (2016). Location, location, location: Effects of cross-religious primes on prosocial behavior. *The International Journal for the Psychology of Religion, 26*(4), 304–319.

Chapter 8
Appreciating Culture in Human Mating and Parenting Research: The Reproductive Functions of Religious Systems

Terezie Dvořáčková, Eva Kundtová Klocová, and Radek Kundt

Abstract Thanks to the extraordinary cumulative character of human culture, our biology is no longer shaped solely by genetic inheritance. By incorporating cultural factors, the study of human reproductive behaviour had become increasingly complex. This chapter introduces research into the effects of cultural technologies that help address biological problems, with focus on religion as one of such technologies. The authors investigate the fitness benefits of various religious practices in relation to mating strategies (including childbearing and childrearing) and examine various reproductive functions of religious systems under different socioecological conditions. Existing research suggests that religious beliefs and associated behaviours promote reproductive success by facilitating partner choice, courtship, strategies for forming and maintaining a family, and by increasing the investment in children. To illustrate this point, the chapter analyses two examples: first, cases where the religious system influences the reproductive dilemma of offspring quality/quantity payoffs and promotes alloparenting by enhancing trust among members of a religious group and second, where the religious system increases paternity certainty by restricting female sexuality, thus motivating the fathers to invest more in the offspring.

Keywords Religion · Ritual · Religious alloparenting · Religious mate-guarding · Religion and paternal investment

RK acknowledges support by the European Regional Development Fund project 'Beyond Security: Role of Conflict in Resilience-Building' (reg. no.: CZ.02.01.01/00/22_008/0004595).

T. Dvořáčková · E. Kundtová Klocová · R. Kundt (✉)
LEVYNA Laboratory for the Experimental Research of Religion, Masaryk University, Brno, Czech Republic
e-mail: terezie.dvorackova@mail.muni.cz; radek.kundt@mail.muni.cz

J. Toman et al. (eds.), *Culture Evolving*, Biosemiotics 10,
https://doi.org/10.1007/978-3-032-16124-6_8

8.1 Introduction

Like all other sexually reproducing organisms, people need to choose, attract, and retain a mate. To address challenges pertaining to the timing of reproduction, relationship length, the number of sexual partners, and investment in the offspring, we have evolved adaptations that significantly affect our net fitness outcomes. Some of these adaptations take the form of mating strategies, that is, genetically based decision rules expressed by psychological mechanisms that drive particular behaviours. In humans, unlike in other species (see Chap. 9), acquired behaviours and practices rely heavily on social learning. Each society has developed certain norms regulating sexual behaviour, childbearing, and childrearing. Groups differ in their expectations, prescriptive rules, and prohibitions regarding mate choice (how to acquire a partner, whom to marry, and when), family form (how many partners, who is part of the household, where the family should settle), and child welfare (who cares, when, and how much). With time, the evolutionary study of human reproductive behaviour had increasingly come to appreciate the enormous impact of human culture.

We shall start with a brief reminder of the similarities and differences between three evolutionary approaches to studying human behaviour which currently dominate the scene: evolutionary psychology, human behavioural ecology, and cultural evolution. Next, we show that religion is the ideal candidate for studying normative cultural practices pertaining to mating and parenting and introduce two examples to demonstrate the impact of cultural technologies on solving biological problems such as offspring quality/quantity payoffs or paternity certainty. These religious practices can be seen as complex sign systems that facilitate communication and meaning-making in biological contexts and thus reinforce social norms and optimise reproductive success.

After the initial attempts of ethology and sociobiology, which either did not deal with culture (understood as information stored in brains through social learning) or their understanding of it was too simplistic, two other evolutionary approaches to studying human reproduction started to take shape in the late 1980s and early 1990s, and both took culture more seriously. One developed into a subdiscipline called evolutionary psychology (see Chap. 7), while the other gradually evolved into human behavioural ecology. Having emerged at around the same time, these two research directions share a deep appreciation of human evolution as the broad theoretical background which frames their research questions. But the two disciplines soon started to diverge, sometimes quietly agreeing to disagree, sometimes sharply clashing about their assumptions, goals, and conclusions. This mutual interaction has been fruitful, because it forced scholars on both sides to improve their arguments and, in the process, to realise that these are in fact two distinct approaches with different questions, levels of analysis, and methods.

Whenever evolutionary psychology studies cultural forms, it focuses on the psychological foundations which support and inform them. When confronted with the world's cultural variance, evolutionary psychology thus tends to focus on the underlying universality of cultural features rather than their differences. The added value of the evolutionary aspect of evolutionary psychology is its interest in the adaptations of the human mind, which emerged in the ancestral environment but are still active in creating, storing, and transmitting cultural forms. Evolutionary psychology is thus concerned with the inner workings of past psychological adaptations and their impact on current cultural outputs. Due to the adaptive lag, past adaptations need not be adaptive in the current environments, but investigation of their current adaptiveness is not what evolutionary psychology primarily focuses on. Typical research of human mating undertaken from the perspective of evolutionary psychology takes the form of questionnaires, distributed worldwide, where participants are asked to report their preferences when choosing a partner. Typically, the results highlight widespread differences in the male and female preferences, which points to cross-culturally shared sex differences in human mating strategies (Buss, 1989). Of the two abovementioned sub-disciplines, evolutionary psychology is much larger but its reputation sometimes struggles with oversimplification or sensationalism resulting from the work of a few problematic researchers or naive press coverage.

Human behavioural ecology, the other evolutionary approach associated with the growing interest in culture among researchers of reproductive strategies, focuses on the immense human adaptability to various environments. Couched in Richard Dawkins' terms (1989), cultural practices are seen as part of our extended phenotype that helps our species optimise behavioural strategies to the different adaptive challenges of particular ecologies. Human behavioural ecology thus tends to focus on the function and is less concerned with the mechanisms than evolutionary psychology is (although there are, naturally, many exceptions). This approach, sometimes referred to as phenotypic gambit, is deliberate. Given that humans are the most successful and, in terms of habitat, most widespread terrestrial mammal on the planet, human behavioural ecology brings cultural variance to the centre stage. Typical research of human reproductive behaviour from this perspective involves efforts to map the worldwide distribution of polygynous, monogamous, and polyandrous marriage systems and in-depth investigations of conditions under which these systems can be adaptive (e.g., polygyny threshold; Borgerhoff Mulder, 1990). The methods used by human behavioural ecology are usually similar to those used in anthropology (e.g., observation) or ethology (e.g., behavioural measurements). Of the two sub-disciplines, human behavioural ecology has a stronger tendency to view cultural elements as functional products which follow the logic of adaptationism (rather than as nonfunctional byproducts of other adaptations).

Human behavioural ecology and evolutionary psychology were later joined by a third approach, cultural evolution, which views cultural change as an intrinsically evolutionary process enabled by our ability to inherit cultural traits (just as genetic evolution is enabled by our ability to inherit genes) (Richerson & Boyd, 2008) (see Chaps. 7, 11, and 13). Cultural evolution shows that while genetic and cultural evolution need not affect each other, culture can also evolve so as to work with or go against our biological dispositions; for instance, while divorce and remarriage take advantage of our serial pair bonding psychology, our acquired preference for chili peppers must overcome the initial aversion to pain. Nevertheless, because that culture had biological carriers, cultural evolution assumes that most cultural elements are either neutral or adaptive[1] depending on whether they were selected for in the past or not (for instance, religious communities that strongly advocate celibacy, such as the Shakers usually do not thrive). Of the three approaches mentioned just above, cultural evolution places the greatest emphasis on investigating the complex interactions involved in the culture–gene coevolution[2] (cf. Chaps. 7, 10, 11, and 13).

Although evolutionary psychology and human behavioural ecology both aim at explaining cultural elements and practices (ask why they are the way they are), it is cultural evolution that attributes to culture the most prominent role in human evolution. It often not only shows how culture contributed to transforming human biology in the past but also argues that, ever since a specific point in our evolutionary history, cumulative culture had become the defining pressure that shapes human genetic evolution (for instance, the use of fire in food preparation led to smaller jaw muscles and shorter intestines). In the same vein, cultural evolution can incorporate the semiotic aspects of human culture. Later in this chapter, we will explore how this includes various, often costly, signalling methods used to demonstrate one's dedication to caring for the offspring (one's own or those of kin or group members), eventually indicating fidelity to a partner. These behaviours evolved and have been enforced at various levels of social organisation. Over time, they were reinterpreted in contexts sometimes unrelated to reproduction, thereby forming the substrate of various cultures. Combining methods from psychology, anthropology, and economy with historical analyses, cultural evolution research of human mating and parenting has for instance explained the recent spread of normative monogamy by pointing to the group adaptive benefits it confers in intergroup competition (for instance, it reduces crime rates by suppressing intrasexual competition and reduces intra-household conflict by increasing relatedness within the household) (Henrich et al., 2012).

[1] In most cases, adaptive benefits emerge through the process of unintentional 'inventions' and their fortunate combinations, which are retained by strategic learners in subsequent generations (Henrich, 2015). Otherwise, there would be no added value in explaining the process through the logic of natural selection (Kundt, 2015).

[2] As with cultural evolution, this name applies both to the process and the field that studies it.

8.2 Religion as a Subgroup of Cultural Practices

Religious practices are a specific subgroup of cultural practices that regulate mating and parenting. Compared to other cultural behaviours, religious practices stand out by how widespread they are, both geographically and historically, by introducing inner motivations based on supernatural beliefs, and by providing an interlocked package of beliefs and practices which permeate many, and often all, areas of an individual's life. This is why some authors view religion as a complex adaptive system (see also Chaps. 7 and 10) and argue that it is a structure of interrelated components that play specific roles within the system's overall function. Like any other system, religious practices require energy, an input which internal processes transform into an output. The system is complex, meaning that its internal components may interact in a nonlinear way, and it is adaptive, which refers to the assumption that it was, at a certain point in our evolutionary history, selected as a whole because it conferred fitness benefits (Lang & Kundt, 2020; Sosis, 2017, 2019a, b).

Compared to other adaptive systems, the component of belief in the supernatural is unique in its irrefutability, which lends it an aura of immutability and persistence. Interestingly, the spatiotemporal prevalence of religion suggests that this cultural adaptation can be beneficial in many environmental conditions, which indicates an extraordinary plasticity and adaptability of this system (cf. Chap. 15). Furthermore, the element of belief in the supernatural is partly responsible not only for the pervasiveness of religious norms in many areas of life but also for the functioning of other elements, such as costly signalling. While it has been shown that costly rituals increase trust and cooperation in religious groups (Power, 2017a, b; Xygalatas et al., 2013), this association does not seem to hold in secular groups such as university fraternities (Shaver et al., 2018a) or among for instance hikers to Santiago de Compostela (Chvaja et al., 2023).[3]

In view of all this, it is not surprising that religion actively regulates reproductive behaviour: in fact, there is hard to imagine a more direct way of influencing fitness than through expectations and norms pertaining to parenting, sexual behaviour, and marriage norms. By establishing clear guidelines and conventions, religious systems create a framework within which individuals navigate their reproductive strategies. These practices not only ensure the stability of social structures but also enhance biological fitness by promoting behaviours which increase the likelihood of successful offspring (Spake et al., 2022). From the biosemiotic perspective, religious symbols and rituals act as communicative tools that reinforce social norms and expectations and thus align individual reproductive behaviours with collective goals (Deacon, 1997; Lang & Kundt, 2024).

[3] Conversely, for research that shows how costly signals facilitate collective action through assortment of cooperators even in nonreligious contexts, see Lang et al. (2022).

8.3 Religion as a Cultural Scaffolding for Collective Childcare

8.3.1 Alloparenting

In contrast to other species, human babies are enormously costly: to survive and thrive, they require both intensive and long-term adult care (Royle et al., 2012, pp. 72–73). Aside from material support, babies also need emotional investment and adequate socialisation. One could thus wonder: with the limited resources of the preindustrial era, how was it even possible to bring up more than one or two children? Considering the harshness of many ecologies, how did our ancestors succeed in populating the whole world?

The difficulty with answering these questions stems from a long-standing bias in childcare research. Researchers rooted in the current WEIRD societies[4] tended to see the nuclear family (mother, father, and one or two children) as the 'natural' and 'universally optimal form' (Sear, 2021) and they focused mainly on the biological ontogeny and psychology of the offspring. In particular, they tended to investigate the children's emotional attachment with parents (Mercer, 2006) and internal parenting mechanisms, which they understood as generally applicable and universal. But such assumptions are the product of an ethnocentric view and, moreover, one based on specific recent development. In modern times, fertility has declined due to the increased cost of care which parents need to provide to raise a successful individual in a complex social environment. As a result, the effort required to ensure sufficient child quality has limited child quantity.

But once we step outside the WEIRD societies, we find a wide range of parenting behaviours. In general, if the offspring require additional care after birth (such as feeding and protection) to ensure the parents' reproductive success, such investment will start to be selected for. To better understand the features which evolved to enable us to interact with children, it is essential to look at childcare and possible variations in rearing behaviour across different cultural ecologies from the phylogenetic perspective.

As with any other human behaviour, childcare is shaped not only by internal biological conditions but also by ecological and cultural restraints. A growing body of evidence indicates that other people besides the mother (whom the WEIRD model views as the primary and sometimes even the sole caregiver) play a significant role in child development. If investment in the offspring is too costly for the mother to bear, the father and eventually other individuals will provide the

[4]The concept of WEIRD (Western, Educated, Industrialized, Rich, and Democratic) societies has been established by Henrich et al. (2010) to highlight that individuals from these societies tend to score as outliers in many psychological and behavioural domains, such as cooperation, visual perception, categorisation, moral and spatial reasoning, fairness, inferential induction, etc. The fact that WEIRD populations are unusual compared to other human populations makes all generalisations drawn from these samples problematic.

necessary care. Such collective care for the offspring is known as alloparenting. Aside from the support of fathers, the care of 'allomothers' (or alloparents, i.e., 'other parents') of children who are not their own can also play an essential role in the human species. Such care promotes overall fertility because, across a range of socioecological conditions, the mother alone could not provide the necessary care to the numerous children, which were needed for our species to achieve such reproductive success in the past. In a trope-shattering claim, Hrdy (2008, p. 412) therefore concluded that a hunter father and caregiver mother would not have been able to create an effective family unit, especially not in the Late Pleistocene, and only aid from other relatives and community members enabled them to rear a larger number of costly offspring. Alloparenting was the crucial component that boosted the rearing of costly and slowly developing offspring with large brains and led to the reproductive success and competitiveness of the human species.

Although alloparenting is flexible and takes different forms in different populations (Bentley & Mace, 2012), the role of allomothers is typically assumed by the mother's female non-reproducing relatives, such as younger siblings or older women during or after the menopause (Kramer, 2010, pp. 418, 422–423). The willingness to invest in someone else's infant is not automatic, alloparenting involves considerable costs in terms of time, energy, and other resources.

Like any other evolved behaviour, the propensity for alloparenting is shaped to optimise the trade-offs between possible costs and benefits. What profits have thus supported the evolution of this behaviour or offset the costs of this investment? Given that allomothers are usually biologically related to the children which they help care for, it seems that alloparenting has evolved to increase one's inclusive fitness, that is, the sum of own fitness and the fitness of one's kin. Hamilton's rule predicts that alloparenting evolved through kin selection (see Kramer, 2010, p. 425), whereby the amount of provisioning and caregiving strongly depends on relatedness (Emmott & Page, 2019, p. 7) and even the perception of relatedness can be crucial. Inherent paternal uncertainty, embodied in the well-known adage *mater semper certa est, pater incertus* ('mother is always certain, father is not'), explains the greater contribution of maternal relatives, because the paternal kin inherently cannot be certain enough about consanguinity to commit to allocare (Boyd & Silk, 2009, p. 462). Especially maternal grandmothers positively contribute to inclusive fitness by decreasing their grandchildren's mortality and shortening their daughters' interbirth intervals (Hrdy, 2006, pp. 13, 20). Moreover, the need for grandmothers living long enough to boost reproductive success may have led to human longevity and the existence of menopause (see Perls & Fretts, 2001). Carers in post-reproductive age not only contribute further calories for the offspring and free the mother's hands but may also be present when children are especially susceptible to cultural learning which grandmothers can provide.

Inclusive fitness is not the only possible driver of alloparental behaviour. Non-reproducing individuals cannot compete for reproductive opportunities, which is why they can provide allocare without any complicated dilemmas. But what about the motivation of reproductively active kin and other unrelated individuals who help with non-offspring? Wherever the pull of inclusive fitness seems insufficient, we should also consider the effects of direct reciprocity. Repeated interactions with the same individuals usually lead to mutual trust in the current investment being repaid in the future, and parents can thus create beneficial collaborative networks (Emmott & Page, 2019, p. 9). Another benefit can be found in cultural learning: given that raising an infant is not easy and must be learned, engagement in caregiving can be beneficial for younger allomothers, who can thereby improve their parenting skills and thus also their future prospects (Emmott & Page, 2019, p. 8). Finally, interest in children is facilitated by proximate mechanisms on a biological level: being in touch with infants leads to hormonal changes which lead to the perception of juvenile features as cute and attractive (Hrdy, 2006, pp. 21–27; Hrdy, 2009, pp. 212–231).

These principles explain a significant part of the presence and variability of alloparental behaviour, but other innate driving forces sometimes act in the opposite direction and lead to a disruption of allocare. For instance, conflicts between parents and their offspring reflect different interests of individual family members and, especially when resources are limited, siblings are prone to compete for the parents' investment to maximise their own survival chances (Emmott & Page, 2019, p. 11). Such antagonism is incompatible with the parents' interest, which aims at ensuring the survival of as many offspring as possible. Coercing the older offspring to participate in allocare can help resolve this conflict. Strassmann's research among the Dogon in Mali (Strassmann, 2011), which examined the expected altruistic cohabitation of relatives and actual kin conflict, can serve as an illustration. Strassmann found that Dogon siblings compete rather than provide care to each other and the presence of paternal grandmother is harmful to her grandchildren rather than a source of support. The author concludes that humans are a complex and flexible species that cannot be simply categorised as cooperative breeders. Not genetically determined to provide allocare, humans require additional rules to engage in collective care. Moreover, like any other cooperative behaviour, alloparenting probably struggles with the free-rider problem because even in kin-based communities, there will always be individuals who want to benefit from collective life while minimising their investment (Olson Jr., 1971).

To work sufficiently well, cooperative childrearing in humans depends on cultural conduct. Existing evidence shows that variability in successful childrearing (and thus also fertility rates) can be linked to cultural customs in different groups that coexist in the same environment. For instance, we can see differences in the population of New Zealand depending on ethnicity and the culture of origin: compared to New Zealanders of European descent, Māori and Pacific Islanders are likely to have more children, while persons of Asian descent have fewer children (Shaver et al., 2019, p. 320).

Human biology is plastic enough to co-evolve with culture. Like any other behaviour, alloparenting is shaped by our capacity for cultural learning and by our propensity to comply with social norms, which are, in turn, the product of cultural evolutionary processes such as cultural group selection. If intergroup competition is sufficiently strong and lasting, the most efficient package of institutionalised ideas and customs that lead to population growth will be selected because the more successful groups will outcompete the less successful ones (Henrich, 2015, pp. 167–168). The particular form of prescribed practice then influences our manners: for instance, older children may imitate alloparental behaviour seen in their cultural models and thus avoid being potentially penalised by their relatives for not following the norms. In contrast, complex societies with weakened social networks help primary caregivers with raising their children through various impersonal institutions (schools, nurseries, daycare, etc)—they thus require behaviours such as payment of child support. In this case, the penalty for not fulfilling the norm is enforced by another impersonal institution, namely the judicial system (Emmott & Page, 2019, p. 12).

Due to the spread of secularism since the Early Modern Period, it is possible to explore differences in fertility between secular and religious populations or between different religious groups. Recent research has been investigating the connection between the demographic transition in fertility rate and different levels of religiosity (Jenkins, 2020). In the globalised world, growing secularism leads to decreasing offspring numbers, which is probably due to an increase in women's autonomy, an anti-natalist perspective, and access to contraception even among religious women (Schnabel, 2021). Nevertheless, active church attendance is linked to having more children even in highly secularised Western countries such as Britain, France, the Netherlands (Peri-Rotem, 2016), or Canada (Dilmaghani, 2019). There are also significant differences in fertility rates between religious groups, reflecting different reproductive agendas. Distinct family planning practices and decisions regarding contraceptive practices lead to a significantly higher fertility of Muslim compared to Hindu women in India (Moulasha & Rao, 1999) and, despite high variation among the developing countries in Asia, Africa, and Latin America, Muslims have in general a higher fertility than Christians. For Christians, there is no substantial difference between Catholics and Protestants (Heaton, 2011). It should, moreover, be borne in mind that an effective pro-natalist agenda incorporated in religious traditions can work as a powerful tool for group expansion (see Chap. 10).

Like any other behaviour, alloparenting can be prescribed, monitored, and variously encouraged (by supernatural agents, family members, peers, etc.) and involvement in it can affect an individual's reputation, which is crucial for cooperative relationships and mating prospects. Religious norms regarding childcare contribute to overall cooperation among co-religionists. Shaver used the idea of a religious niche that promotes general cooperativeness to propose a religious alloparenting model, noting that 'religious cultures motivate alloparenting from large kin networks, as well as unrelated co-religionists, to enable high-quantity, high-quality reproductive strategies, and that shared parental care partially explains successful world religions' (Shaver, 2017, p. 325). As mentioned above, religiously supported alloparenting is one of the mechanisms that influence variability in the fertility rates of different religious groups (Chvaja et al., 2025; Shaver et al., 2024), and the efficiency of their alloparental support can determine the success of religious groups.

In connection with these life-enhancing and reproductive functions, one could take the argument further and speculate about the general origin of culture and, by extension, about the possible origin of religion as a mechanism that supports reproduction. For example, Blume (2015) assumes that religion evolved primarily to secure cooperation and alloparenting among women rather than as a way of signalling male fitness in mate choice or in order to promote male cooperation during hunting and fighting. Blume points out that women are more religious than men: globally, women form the majority of active participants in regular religious services and are thus more likely to directly benefit from cooperation among co-religionists. Still, while the evolution of religion may have been driven by one primary function, it is essential to acknowledge that religious signals are most likely multifunctional, meaning that during a ritual, there may be not only many signallers and many receivers but one signaller may be able to use one signal to simultaneously send different information to different receivers (for instance, prolonged dancing can signal good health to prospective mates and war party commitment to brothers in arms).

So how, specifically, can religion support cooperative childrearing? Religion contains a variety of beliefs and practices that may be useful in this respect. According to Palmer and Begley (2015), the abovementioned conflict of interests between siblings and parents can be resolved by fostering parental manipulation through supernatural ideas and traditionally prescribed altruism based on belief in the ancestors and their notions of proper conduct. The Old Testament commandment to 'honor your father and your mother' illustrates the need to solve the parent–offspring conflict well but while the biblical commandment is vague and offers room for interpretation, other religious systems present more concrete approaches. For example, the West African Beng practice intensive allocare, where mothers leave their infants with kin and strangers while working in the fields. This approach to childrearing is supported by a supernatural belief according to which each child is a reincarnation of some community member or ancestral being who has returned from the afterlife. To prevent them from wanting to go back to the afterlife, all community members embrace the baby and make themselves available to make it feel welcome and cared for. The baby is even formally introduced to any visitor to the village, who may then participate in caring for it. To increase the pool of potential babysitters, Beng mothers spend time decorating and adorning the baby to make it look especially attractive to caregivers (Gottlieb, 2009) (cf. Chap. 15).

Such examples illustrate the variability in religious ideas and customs concerning childcare. They also reflect the variations in alloparental norms, which can be linked to the diversity of socio-ecological demands. Alloparenting, as an important fitness-enhancing factor, should respond to this variability. In relation to ecological conditions, Martin et al. (2020) found that societies that live in harsh environments (with lower climate predictability, lower temperatures, and lower precipitation) provide more alloparental care to mitigate the potential impact of the hostile environment on child survival. Nevertheless, the prevalence of alloparenting decreases when the community faces the threat of starvation. On the one hand, prescribed altruism and reciprocity in childcare are crucial for reproductive success under challenging conditions but if the costs of alloparenting outweigh the benefits, the community will not force alloparents to invest.

The variability of culturally promoted alloparenting also reflects differences in social organisation and the marked differences in fertility rates, family structure, and forms of religion between modern large-scale populations after demographic transition and traditional small-scale societies present different challenges to alloparental behaviour. While a traditional society with large interconnected families who are in daily contact is more likely to promote assistance with direct care (to reduce mortality), in economically developed post-demographic transition societies which emphasise nuclear families the forms of alloparenting will be more indirect and focused on the quality of the offspring (for instance by providing financial assistance to ensure access to further education). The culturally promoted norms should thus reflect the needs of the society, be it through food sharing or help with organised schooling.

Trust among the members of a given society is yet another challenge. In kin-based small-scale societies, it is easier to promote willingness to engage in altruistic acts and cooperation because such behaviour boosts inclusive fitness, promotes direct reciprocity, and because members of such communities have a direct access to information about reputation. In contrast, in anonymous large-scale societies with virtually non-existent extended family networks, it is more difficult to earn trust. In both cases, however, religion can function as a trust-enhancing mechanism, whereby religiosity signals or directly fosters a commitment to group values (Shaver et al., 2018b). Moreover, religious environments provide various monitoring cues, such as religious music or statues which remind believers of the presence of omniscient moralising gods (Lang et al., 2019; Nichols et al., 2020; Xygalatas et al., 2015). Religious forms and their importance in a given culture also have an impact on the diversity of alloparental practices. In WEIRD societies, secularisation made religion visible and separate from other cultural domains, so that one can discriminate between secular and religious alloparenting (e.g., paid babysitting and public schools vs. aid from co-religionists and church camps), which is not the case in traditional cultures where religion is not so clearly separate. In this context, however, we should not concentrate on the differences between religious and secular alloparenting but rather on the various cultural institutions and norms pertaining to cooperative childrearing and the variability of religious commitment among co-religionists.

WEIRD populations provide a unique opportunity to compare alloparenting behaviour among believers and non-believers. Current research on religious alloparenting is represented by a project called the Evolutionary Dynamics of Religion, Family Size, and Child Success. Given the trade-off between offspring quantity and quality in modern high-complexity cultures, we find a surprising paradox in religious fertility: religious families have more children than non-religious families do but do not suffer from decreased offspring quality. Shaver et al. (2019) have proposed and tested a religious alloparenting hypothesis according to which the high-quantity, high-quality offspring production in religious families is enabled by increased alloparental help in religious communities. To support the claim that religious families have more offspring, the authors examined questionnaires completed by New Zealanders and found that religious identification and frequency of ritual attendance were, among people who had at least one child, associated with having more children. In terms of alloparental investment, their data showed that although the overall rates of alloparenting are low, members of religious communities do engage in more alloparental support than secular individuals. Nevertheless, these results may reflect a limited view of alloparental care, because the authors focused only on non-parents, meaning they discounted all people with children under 18 and children living in their home from the sample and defined alloparenting as the number of hours spent looking after children per week, which could hide potential alloparental networks among parents as well as other forms of alloparenting (such as material provisioning). Moreover, the data from the questionnaires could not indicate whether religious alloparenting is positively linked to children's quality.

These issues were fortunately addressed in another study (Shaver et al., 2020), which used data from the Avon Longitudinal Study of Parents and Children (ALSPAC database) on mothers living in the UK and their children during the first ten years of life. The results of this study show that the more frequently women attended church, the greater was both their (secular) social network support and the amount of assistance they received from co-religionists. But these two types of help had different consequences: while religious support was more long-term and linked to a higher fertility, secular support had a shorter duration and increased levels of this kind of help were associated with lower fertility, which suggests that mothers did not use it to have more children but rather to return to work. This supports the idea that ritual behaviour can promote cooperation and thus lead to a larger family size. As for the offspring quality, while in post-industrial societies the number of siblings has a negative effect, this study showed that both secular and religious support is associated with a mild improvement in cognitive abilities in the later stages of child development.

Regarding the number of allomothers, another study showed that religiosity is positively linked to the size of mothers' social networks (Lynch et al., 2022). Specifically, in rural Bangladesh, religiosity fosters bonds among female relatives: more religious mothers report having more helpers than less religious mothers, and while their networks contain a similar number of non-kin, they include more relatives than the social networks of less religious mothers. Higher religiosity seems to strengthen families and protect them against possible adverse economic outcomes, including uncertainties linked to labour migration and urbanisation (Lynch et al., 2022, p. 10). While it may be surprising that religiosity does not promote bonds between mothers and non-kin co-religionists, it is evident, from the perspective of behavioural ecology of religion, that religiosity sufficiently supports families in Bangladesh, a country where secular institutions are still not universally accessible and do not enable a transition from kin-based to non-kin-based society as in WEIRD societies (Lynch et al., 2022, p. 11).

This research shows how closely religion is linked to mating and parenting. Existing findings indicate that one of the factors that positively influence a given group's demography is cooperative childcare, which thus functions as a fertility-enhancing aspect of religion. It would be a mistake to omit religious norms in the study of the forces which shape human reproductive behaviour: by leaving them out, we would end up with a narrower and less comprehensive understanding of human nature. The following section is dedicated to the cultural factors which affect paternal investment, because a picture of alloparenting would not be complete without considering this crucial category. Moreover, fathers often have unique position in childcare.

8.3.2 Paternity Certainty

In modern high-income countries, whose societies are typically based on monogamous marriage and nuclear families, the role fathers play in providing and caring for children is usually taken for granted (Henrich et al., 2012), but across cultures the reality is much more varied. It is thus difficult to assess the importance of paternal investment from a global perspective while taking into account the various views, which are based on different findings. Some scholars view paternal care as rare and insufficient. They point out that since most alloparents are female relatives, paternal care is fluctuating and uncertain (Kramer, 2010). As a matter of fact, paternal care is not something a human mother can rely on: according to Hrdy (2008, p. 407), 'across cultures, there is more variation in paternal care in humans than in all 275 species of other primates put together'. Other scholars, however, point out that fathers provide at least some benefits for children in all societies: they create a beneficial environment for children by supplying food, supporting mothers, and providing social networks (Hewlett, 2017). This view emphasises that fathers' investment is more stable than the rest of alloparental care (Rosenbaum et al., 2021), a claim supported by the finding that paternal investment changes less in response to ecological challenges (such as extreme temperatures or starvation) than the amount of care provided by other alloparents (see the Supplementary Materials of Martin et al., 2020, pp. 6–7).

The parental investment theory claims that the parental investment of male relatives is much lower than the investment of female relatives. Due to the enormous reproductive potential of males and because they need not invest much more than gametes to create offspring, fathers face a steeper trade-off between mating and parenting efforts. While it is a risky strategy (because paternal investment increases offspring viability), it is tempting to abandon a pregnant mate in order to seek other partners (Geary, 2015, p. 484). Conversely, when other mating opportunities are limited and paternal certainty is high, fathers are more motivated to invest and secure the success of their progeny (Geary, 2015, p. 484).

Paternity, that is, genetic relatedness between a father and a child, is a strong driver of paternal behaviour but there is an inherent uncertainty, a risk of providing costly care in vain. As noted above, this affects the amount of allocare from patrilinear kin. While it is true that even unrelated men can provide care to the children of their current partner, it is a phenomenon commonly viewed as part of the mating effort rather than parenting (Boyd & Silk, 2009, pp. 459–462).

The inherent paternal uncertainty is given not only by the fact that men do not give birth but also due to the women's concealed ovulation, hiding the onset of pregnancy, and the threat of sexual infidelity. The existence of extrapair copulations shows that humans are not a strictly pair-bonding species and people can flexibly switch among different mating strategies. Moreover, women are prone to seeking additional genetic resources from multiple relationships, which establishes deep reproductive conflicts of interest between the sexes. On top of that, there is the danger of involuntary fertilisation due to sexual coercion by other males. The probability of being cuckolded is thus not null and paternity confidence does not necessarily match the actual genetic paternity: existing research shows that the average worldwide nonpaternity is about 9–10%, ranging between 1.7% and 29.8% (Anderson, 2006, p. 516). Actual variability across cultures might be even greater, as indicated by the study of Scelza et al. (2020b) which measured nonpaternity among Himba pastoralists. In that society, parallel relationships are permitted, which leads to 70% of couples having at least one unrelated child and a mean nonpaternity rate of 48%. Thanks to detailed genetic testing, this suggests that nonpaternity could be much more significant than previously thought. While the Himba culture responds to this high non-paternity rate by expecting male investment in their partner's children regardless of paternity, which equalises the net benefit across the group, this strategy is rather exceptional. Much more common are efforts to increase paternity certainty and reduce the risk of investing in someone else's offspring.

To increase paternal certainty, there evolved various mate-retention mechanisms aimed at preventing sexual infidelity. For instance, Buss (1988) listed 104 types of mate-retention behaviours oriented at one's mate (such as love and care, enhancement of physical appearance, monopolisation of time, emotional manipulation, violence) or at potential rivals (verbal signals of possession, possessive ornamentation, and intrasexual threats such as cold staring at the potential opponent, etc.). It should be noted that many of these mate-retention strategies are also used by women who are trying to secure their partners' investment in their offspring. In general, men use significantly more resource display to retain partners, while women emphasise their physical attractiveness. Buss (1988) further suggests that these individual and spontaneous behaviours are often driven by jealousy, an emotion that evolved to motivate us to actively guard our mates. Nevertheless, according to Scelza et al. (2020a), jealousy expressions are not universal. There are not only sex differences in jealous responses to sexual and emotional infidelity, which reflect specifically male and female concerns (paternal uncertainty vs. loss of resources), but the intensity of jealousy responses varies across cultures depending on the amount of expected paternal investment and norms regulating extramarital sex.

Individual, often spontaneous, mate-retention tactics can be further supported by coalitional actions: for instance, peers can assist with relationship problems, provide strategic information about the mate, or even directly guard the mate (Barbaro et al., 2015, pp. 67–68). Coalitional support in mate retention may be the precursor of various cultural forms that evolved to increase paternity certainty and prevent cuckolding, primarily by restraining women's sexuality. While coalitional support directly depends on the cooperative network of a given individual, cultural norms and practices provide solutions at the population level. The specific form of those norms and practices depends on the conditions and settings of the society in question, so that for instance where paternal care is essential and men occupy positions of power, cultural norms tend to reflect men's need to increase their paternity certainty and effectively guard their mates, which in turn leads to the creation of environment with barriers to potential female infidelity.

Grant and Montrose (2018) have suggested that patriarchy evolved as an efficient system for solving the male mating–parenting trade-off because it is a system that works even while the men are temporarily absent. Power over women's behaviour in patriarchal cultures increases paternal certainty by regulating women's sexuality, which in turn makes paternal investment a safer strategy. Historically, economic dependency on male resources or practices that restrict female bodies, such as Chinese footbinding or European tight-laced corsets, could restrict women's freedom—and even ability—to move (Grant & Montrose, 2018, p. 390). While these practices in their most extreme forms have largely disappeared, there still exit in many societies expectations and norms regarding female sexual purity and ways to punish its loss. The strictness of enforcement of such standards varies across cultures, ranging from gossip and status damage to honour killing or infanticide. These restrictive practices are often complemented by practices that increase in-pair fertility. Cultural conventions such as early marriage, preference for a high number of offspring, prohibition of contraception and abortion, and sex rules targeting the female fertile period promote having more offspring even in societies where women themselves might otherwise choose to postpone their first pregnancy and have longer interbirth intervals.

Grant and Montrose (2018, pp. 339–400) emphasise that religion often plays a significant role in enforcing sex-related behaviour in a direction that supports paternity certainty. Religious ideas are in turn often reflected in the laws and unwritten rules of conduct that permeate the society, thus creating an extensive mate-guarding system with a markedly pro-natalist function. A society driven by these norms can be demographically successful, which may suggest why the largest religions in the world are patriarchal.

The double standards for adolescent sexual behaviour, which exist worldwide, exemplify social norms that restrict girls much more than boys. Šaffa et al. (2022a) tested three theories aimed at explaining who benefits from restricting female premarital sexual behaviour, whether males, females, or parents. Their findings, which are based on cross-cultural data, support the male control theory: by limiting female sexuality, males benefit by a higher certainty of paternity. Moreover, stricter restrictions apply in societies which also prohibit female extramarital sex and where fathers provide wealth to their children. Another important factor is parental control, and this study had shown that in societies where mate choice is under parental—and especially paternal—control, strict premarital norms imposed on daughters may increase their mate value while increasing future certainty of paternity. Parents can thus arrange a marriage to a man of higher status and resources, who can provide for more offspring, which in turn benefits the biological fitness of the bride's parents.

Although there is a considerable amount of research on various forms of mate-guarding and practices that restrict women's sexuality, empirical studies of actual associations between those practices and paternity certainty are rare. One area that has been studied in this context are menstrual taboos. There are several reasons for their existence: aside from hygienic and psychological reasons or reasons related to power dynamics (Strassmann, 1992, pp. 91–93), adherence to prescribed forms of behaviour connected with menstruation provides invaluable information regarding a woman's fertility status. As a physiological phenomenon, menstruation indicates several aspects of female reproductive status, including whether a woman is healthy and of reproductive age or temporarily infertile due to pregnancy, eventually postpartum amenorrhea. But such information is available to the males only if menstruation is somehow signalled. In many cultures, there are various norms menstruating women are expected to follow; most such norms regulate sexual intercourse, female separation, appearance, participation in rituals, food preparation, and contact with male objects (Strassmann, 1992, p. 90).

Can men prevent the conception of an unrelated offspring by responding to information revealed by enacted menstrual taboos? Strassmann tested this assumption in her research among West African Dogons, a strongly patriarchal and patrilineal culture, where only men are allowed to own essential resources (land). Given that inheritance is passed along the male line, the husband and his relatives are concerned about paternity and want to prevent the passing of resources to someone else's offspring (Strassmann, 1992, p. 109). This concern is reflected by a strict requirement that women do not conceive illegitimate offspring; such cases lead to abortion attempts and even infanticide. On the other hand, it is difficult for Dogon men to prevent adultery by directly guarding the women, because men and women work and live mostly separately (Strassmann, 1996, p. 306). Therefore, they monitor the women's reproductive status differently, through a compulsory stay in special huts during menstruation. Given that the Dogon are considered a population with a high natural fertility—they do not use effective contraception and women are either pregnant or in postpartum amenorrhea for most of their childbearing years—it is important to note that menstruation is a rare event that is interpreted as a sign of opportunity for the next conception (Strassmann, 1992, p. 94). The public nature of this taboo enhances the quality of the signal, because the location and design of the menstrual huts allow the entire village, most notably the husband's family, to keep track of who is taking residence (Strassmann, 1992, pp. 111–112).

This publicly available information is used as an anti-cuckoldry tactic in three ways. First, it is customary to marry a woman right after she returns from the menstrual hut, thus increasing the likelihood of the bride not being pregnant. Second, the public nature of the taboo increases the risk of discovering adultery, because any information about the wife's contacts and the husband's presence or absence can be matched against the woman's reproductive status and thus efficiently used to assess paternity. Finally, the presence of a woman in the hut signals her future fertility, which allows the husband to increase his paternity certainty by paying his wife sexual attention specifically during the fertile period (Strassmann, 1992, pp. 120–125).

If, however, women did not properly follow menstrual taboos or even deliberately manipulated the signalling of their reproductive status, these practices would not provide the required information. Therefore, Strassmann (1996) tested the reliability of menstrual signalling using endocrinological measurements, which revealed that women did actually adhere to this taboo. Religious beliefs play a significant role in the internal motivation to honestly visit menstrual huts as required. The Dogons believe that transgression of the norm would lead to a desacralisation of magical objects which protect the community against famine and disease. The menstrual taboo is thus monitored and sanctioned not only by other group members but, in the minds of the women, also by supernatural forces that would punish the whole community for the actions of any female transgressors (Strassmann, 1996, p. 306).

In this case, the evidence indicates that the menstrual taboo reveals female reproductive status and that it is signalled honestly. But does this practice secure paternal certainty and limit the risk of misdirected paternal investments? The Dogon provide an especially good context for testing this hypothesis because they have one shared cultural background and live in one particular environment but vary in their religious affiliation and related menstrual taboos. A testing of fathers and their (putative) offspring for genetic relatedness revealed a link between specific forms of menstrual taboos and paternity by comparing particular religious groups within this ethnicity (Strassmann et al., 2012). It turned out that assumed paternity coincides the most with biological paternity among the Dogons who follow the indigenous religion, while among other Dogons (Catholics, Protestants, and Muslims), whose wives are not expected to visit menstrual huts, cuckoldry levels are higher. Given that all these groups share the same socio-ecological conditions, differences in religious beliefs and practices enable a comparison of the impact of menstrual taboos on cuckoldry levels. While Christians have no obligations concerning menstruation, Muslim women notify their husbands and must not pray or visit the mosque—and even this simple difference leads to lower cuckoldry levels in Muslims compared to Christians. Although all of the abovementioned religions share a patriarchal reproductive agenda, in this particular context the risk of cuckoldry is lowered by the use of menstrual huts, the community's traditional way of helping to secure paternity; it thus turns out to be the most efficient practice. It is possible that the cultural technologies brought by the Christian and Muslim tradition are not sufficiently successful here because they were imported from different socio-ecological contexts.

Another religion-based tradition that can serve mate-guarding purposes is female genital mutilation/cutting (FGM/C). Such practices are defined by the World Health Organisation as 'all procedures involving partial or total removal of the external female genitalia or other injury to the female genital organs whether for cultural or other non-therapeutic reasons' (World Health Organization, United Nations Population Fund, and United Nations Children's Fund (UNICEF), 1997, p. 3). Because these practices usually involve unprofessional interference with female reproductive organs, they often have a strong negative impact on a woman's psychological and physical health (World Health Organization, United Nations Population Fund, and United Nations Children's Fund (UNICEF), 1997, pp. 7–8). To explain the emergence of these apparently harmful practices and their persistence, often despite official prohibitions, is therefore a challenge for evolutionists.

One plausible explanation is that although FGM/C might be maladaptive, it can persist in a population as a side effect of evolved cultural learning capacities, such as the tendency to conform, imitate prestigious individuals, and other learning biases. On the other hand, the prevalence of FGM/C and the fact that it emerged independently in different places throughout human history suggest a convergent evolutionary process. Before concluding that it is a maladaptive practice that persists only as a byproduct of cultural evolution, we should therefore properly examine its potential hidden benefits and compare them with the costs it inflicts on individual and inclusive fitness. Where FGM/C benefits outweigh the costs, it could be a functional behavioural strategy further sustained by cultural learning.

A positive link between cultural learning and fitness is illustrated by Howard and Gibson's research (Howard & Gibson, 2017): in a sample of five West African countries, they found that FGM/C is frequency-dependent, i.e., the likelihood that an individual adopts this behaviour depends on how common it is in the population or group. But this kind of conformist bias is not the only mechanism that can help explain the persistence of FGM/C. Fitness benefits, too, are linked to FGM/C frequency: in ethnic groups where FGM/C is practiced, mutilated women have more surviving offspring than non-mutilated women do. In other words, women's reproductive success in such groups is lower if they go against the mainstream. But still, this only possibly explains the maintenance of the practice, not its origin or expansion.

A practice maintained as a by-product of learning processes need not be associated with reproductive success. Models by Ross et al. (2016) reveal differences between the emergence of FGM/C, its spread, and its maintenance in the population. The authors estimate that FGM/C emerges in stratified societies with differences in male mate value, that is, with intragroup differences in characteristics which increase attractiveness to potential mates (primarily social status and wealth). In such societies, FGM/C serves as a costly signal of virginity that may boost female marriage prospects. It can thus function as a cultural tool in contexts which encourage women's early entry into marriage, thus harnessing their full reproductive potential in a profitable union. Once it spreads to less stratified societies, the reproductive function of FGM/C loses its significance and the practice is likely to be maintained only through learning biases. It is possible that FGM/C acquires also some new functions, such as signalisation of group membership, commitment, willingness to cooperate, etc., which might affect reproductive success, albeit less directly.

Other studies suggest that female genital mutilation may have emerged as a practice that resolves the sexual conflict concerning male reproductive interests. A more detailed research into the origins of FGM/C shows that it is practiced mainly in polygynous, patrilineal, patrilocal, and bride-price societies where some type of scarification and male GM/C is already present (Šaffa et al., 2022b). This corresponds to patriarchal settings characterised by extensive control of female sexuality; in particular, FGM/C is closely associated with the separate residence of co-wives, which indicates that it serves mainly as a mate-guarding device aimed at increasing paternity certainty in contexts where it is difficult to guard women directly (Šaffa et al., 2022b, p. 638). Different types of FGM/C with varying degrees of physical restraint of women depend also on other socio-ecological conditions; for example, clitoridectomy, the most prevalent type of FGM/C, is strongly predicted by the presence of Islam, bride price, and restrictive extramarital norms. The connection with Islam is probably due to the fact that this religion spread easily to cultures that already practiced FGM/C, possibly due to a similar patriarchal ideology. As for the bride-price practice, FGM/C increases the price of brides and thus helps to differentiate women according to their mate value in the marriage market.

Nevertheless, infibulation, the most severe form of FGM/C, is associated with pastoralism (Šaffa et al., 2022b, pp. 637–638). Becker (2019) examined infibulation as a mate-guarding practice which persisted in originally pastoral cultures where care about herds often required male absence, so women could not be guarded directly by their husbands. At the same time, essential resources were held only by males, which led to a high paternal investment. In these cultures, the adoption of stricter norms and customs (such as infibulation as well restrictions on female promiscuity and mobility) could thus compensate for considerable paternity uncertainty.

The notion that FGM/C is a practice underpinned by patriarchal institutions that reflect male reproductive interests is further supported by the finding that where women have more freedom, they choose a different lifestyle. Women with access to education and economic independence tend to avoid this practice and adopt a slower life history strategy, that is, they usually marry later and focus on offspring quality rather than quantity (Reason, 2004).

Motivations for arranging an FGM/C ceremony for young girls can vary but it is interesting to note that they are often accompanied by religious justifications. For example, in central Guinea women report that the Qur'an prescribes that mothers educate their daughters, circumcise them, and get them married (Yoder et al., 2004, p. 13). This religious justification for maintaining the practice is complemented by beliefs such as that FGM/C 'ensures fertility', 'prevents maternal and child death', and serves additional hygienic functions (Amusan & Asekun-Olarinmoye, 2008, pp. 338, 346). Such ideas may help overcome the intuitive fear of being cut or arrange for one's daughter to be cut. In reality, though, FGM/C causes serious problems related to survival and sexuality, which makes this practice costly for the individual's fitness. On the other hand, the negative impact of FGM/C on female sexuality may be precisely why males prefer mutilated women in cultures where this practice exists. Women who have little desire to engage in sex, who cannot enjoy sexual intercourse and suffer from various other complications—which potentially leads to sex phobia (El-Defrawi et al., 2001)—are much less likely to seek additional sexual partners, which is how FGM/C increases paternity certainty.

Aside from monitoring women's reproductive status and interfering with their sexual organs, another potentially effective religiously sanctified practice of mate protection is to place restrictions on women's physical attractiveness. Appearance often plays a vital role during mate choice. It provides information relevant to probable future reproductive success, such as age, health, fertility, and genetic compatibility. Men's attention often focuses on the female silhouette (waist-to-hip ratio, breasts), because female body shape depends on estrogen levels, which are related to fertility (Singh et al., 2010). By concealing women's curves through strict dress codes, it is easy to avoid body assessment.

While the veiling of women is usually associated with Islam, similar practices have been adopted in many different societies and cultures for millennia. For instance, Christian nuns still mostly cover their bodies with loose habits and wear headscarves which hide their hair, while conservative female Christians usually cover their hair during worship in the church. Among members of the Bhil tribe in southern Rajasthan, it is customary for a married women to cover their hair and the entire face with a headscarf if there is any risk of being seen by men other than their husbands. Although the women view the face covering as impractical, they maintain the practice and if a woman were to violate this custom, others would shame her into compliance (Rojo, 2023, p. 180). Aside from the external motivation provided by peer pressure and potential shaming or other punishment, religion can also effectively internalise the motivation to follow a prescribed dress code. Although Muslim women wear different types of clothing according to the options available to them (see Fig. 8.1), their desire to adhere to a particular way of dressing is often religiously motivated; given that the Qur'an requires women to dress modestly, veiling can be an expression of devotion to Islam and submission to God (Ngwainmbi, 2015, p. 25; Carvalho, 2013).

Research into men's reactions has been conducted primarily in Muslim countries, where there is a considerable variability in religiously prescribed veiling. It had shown that the veiling of women has a significant impact on male mating perception and behaviour and, more generally, the covering of female bodies influences both the behaviour of men and women and their mutual interactions. Sufficiently loose clothing, such as a chador, reduces male interest and perceived female attractiveness (Pazhoohi & Hosseinchari, 2014). Men spend less time looking at the covered body parts and concentrate on what is exposed, which is often the face (Pazhoohi et al., 2017b). Men are also less likely to interact with more conservatively dressed women. For example, Pazhoohi and Burriss (2016) found in their experiment that 21.4% of drivers offered a ride to a liberally dressed female hitchhiker, who wore a headscarf and close-fitting garments, while only 3.9% of drivers stopped for a female hitchhiker who wore a conservative chador. Veiling may thus be a mate-guarding practice that limits intersex interactions and consequently also the likelihood of extra-pair copulations. Furthermore, veiling forces women to communicate a preference for long-term mating strategy by limiting the prospects of short-term mating or even of simply attracting the attention of a chosen man through their appearance. The particular practice of eye veiling (such as the burqa or the Indian ghunghat veil) prevents women from communicating their intentions even through eye contact (Pazhoohi, 2016).

Pazhoohi et al. (2017a) helped explain the differences in the strictness of female veiling in Muslim countries by showing how flexibility of mating strategies relates

Fig. 8.1 The variability of body coverings in current Mauritian society. Women waiting for public transport, July 2022. Notice the difference between the face and body and hair coverings which are contextually dependent cultural solutions to different biological challenges. First, a solution to the COVID-19 pandemic, where face masks, covering the nose and mouth, were used in public places to mitigate the risk of the spread of the virus. Second, a solution to the paternal uncertainty problem, where prescribed coverings of female bodies in public places restrict their sexuality, effectively thus functioning as mate-guarding instruments. Interestingly, both solutions allow for individual variance. Masks can be worn in a fashion that covers both nose and mouth, only the mouth, or neither. Headscarves can be worn in a fashion that covers the whole face and hair or leaves the whole face visible (with make-up applied). There is also a difference in the length of the garment, which can reach the knees, the calves, or end just above the ankle. Trousers worn under the garment differ in how loose they are, with one end of the spectrum occupied by a loose-fitting fabric and the other end by tight leggings (combined with high heels). Phototgraph by Eva Kundtová Klocová. Reproduced with permission

to parental investment and partner protection depending on the socio-ecological context. Harsher environments (i.e., countries with higher mortality rates, poorer health conditions, and lower GDP) are associated with higher religiosity and consequently also with a stronger emphasis on wearing the headscarf (hijab). Such adverse environmental conditions require extensive parental investment for reproductive success, and because more paternal care is needed, it is advantageous for men to limit female sexuality and increase their paternity certainty.

Still, unless a culture strictly prescribes veiling for women, women's willingness to veil themselves varies considerably. Blake et al. (2018) found that in Tunisia, where women can choose to move in the public either with hair uncovered or to wear different types of head coverings, the type of veiling is associated with the sex of their offspring. Mothers who have several sons have a stronger tendency to support veiling and wear a more conservative form of headscarf than mothers with multiple daughters; such mothers still wear some head covering but tend to opt for a more fashionable form such as a turban. It is possible that women in this way promote the reproductive interests of their offspring rather than their own. Perhaps, by signalling a more restricted sexuality through stricter covering, they indicate a certain mate preference for their sons, with an emphasis on chastity. In contrast, the mothers of daughters may encourage more fashionable clothing to enhance their daughters' attractiveness and marriage prospects. Both tactics should ultimately lead to higher maternal reproductive success.

Although veiling is prominent mainly in Muslim countries and in parts of India, an emphasis on chaste clothing can be observed in religious traditions across cultures, including even various religious communities in liberal WEIRD cultures. The rules of modest clothing vary; for example, among Christian communities, the rules range from the strict American Anabaptist requirements of anachronistic dress style and head covering for women and girls all the way to unwritten norms of avoidance of overly revealing clothes, such as short skirts, plunging necklines, overly tight clothing, or attention-grabbing colours and patterns in more liberal churches. The forms of enforcing and justifying modest dress also vary, but religion, individual religiosity, and devotion to God are often cited as the most important incentives for this behaviour (Taragin-Zeller, 2014).

In sum, recent research in cultural evolution suggests that the existence of menstrual taboos, FGM/C, and veiling can be at least partly explained by their reproductive function as mate-guarding practices which in turn promote higher paternal care. In addition to these specific practices, religious systems typically contain numerous other mate-guarding norms and practices, which by increasing paternity certainty lead to varying expectations of paternal investment.

8.4 Conclusion

This chapter shows that religious systems can influence alloparenting and paternal investment. In the first case, we have shown that religious systems influence the balance of the reproductive dilemma of offspring quality/quantity through promoting reproductive cooperation based on trust among co-religionists. In the second case, religious systems increase paternity certainty by restricting female sexuality, so that fathers are more motivated to invest in their offspring.

Still, religious systems influence reproductive success also in various other ways. For example, theories that explore the deep evolutionary pathway of human ritual behaviour in hominins (Lang & Kundt, 2024) suggest that religious systems evolved on top of a much older signalling system. Here, the authors argue that collective rituals—as systems that signal similarity, coalitions, and commitment—gradually evolved to facilitate mutualistic cooperation under socio-ecological pressures of the

Pleistocene and it was especially their ability to communicate cooperative intentions that drove their selection (Kundt & Lang, 2024). Collective rituals thus provide a platform where participants can express and receive information about the cooperative intentions of potential partners, be it partners in a future risky joint action (such as a bear hunt) or mating prospects (a suitable partner for a stable pair bond). The added value of ritual communication compared to other forms of communication is in the higher reliability of the signal. Many rituals require costly acts (such as body piercing) or allow for a substantial variation in the costliness of such acts (Xygalatas et al., 2021). Individuals who opt out of a ritual or choose a costlier behavioural version thus provide signal receivers with additional information about their commitment. Moreover, this kind of communication gives less space to deception and manipulation, tactics that pertain to other forms of symbolic communication (such as lying). Contrary to expectations, the added value of ritual communication thus does not rest in the use of purely symbolic signs (Deacon, 1997) but in its capacity to connect indexical communication (e.g., indices of painful body piercings) with symbolic communication (culturally specific meanings of these piercings as devotional instruments). In fact, a study of mate preferences among the Tamil Hindu in Mauritius by Xygalatas et al. (2022) had shown that parents perceive men who practice more costly rituals as better potential long-term mates for their daughters. Furthermore, Moon et al. (2018) have shown that sexuality and trust are surprisingly linked; people tend to trust religious people more because they associate information about religiosity with a restricted mating strategy; in short, religious individuals are viewed as more likely to be faithful in a long-term relationship. Atheists, on the other hand, are perceived as less trustworthy, less faithful, and more likely to cause conflicts in long-term romantic relationships (Brown, 2022; Brown & Neiswender, 2025).

Communication of commitment to the norms which regulate family life within a group is not the only signalling function through which religious behaviour contributes to the reproductive agenda. The diversity of religious activities enables the signalling of various information that assists mate choice. For example, Pyysiäinen (2008) suggests that ritual practices such as dancing, bringing of offerings, or rhythmic recitation can also indicate good genes by demonstrating good physical condition, skills for resource acquisition, or cognitive abilities.

Although mate quality signalling is one of the most common communicative functions of ritual behaviour in other species and it clearly has a biosemiotic dimension (Lewis, 2021), it remains relatively understudied in humans. It is thus a possible fruitful direction for further study (Slone, 2008; Slone & Van Slyke, 2015).

Human mating strategies and marriage systems are highly flexible, ranging from long-term monogamous lifestyles to serial one-night stands. Their variability is shaped by a combination of biological, ecological, and cultural factors. Regulation of sexual behaviour is central to religion (Moon, 2021) and religious beliefs, norms, and practices are potent factors in the promotion of fertility (religions incorporate various pro-natalist and childcare-centred rules). In some cases, however, such rules can also be a hindrance (for instance a postpartum sexual taboo exceeding the average human interbirth interval). To understand whether and how this interplay optimises reproductive success in a given context, it is crucial to harness the strengths of all evolutionary approaches at hand, that is, evolutionary psychology, behavioural ecology, and cultural evolution, and possibly also employ some novel methods, such as phylogenetic cross-cultural analysis and field-based genetic testing. Only by suitably combining such approaches can we hope to comprehensively address all four dimensions of the phenomenon (ontogeny, phylogeny, mechanisms, and survival value) as Tinbergen envisaged them (Tinbergen, 1963) and achieve a fully-fledged lineage explanation (Calcott, 2009). The complexity and importance of the topic deserve nothing less.

References

Amusan, O. A., & Asekun-Olarinmoye, E. O. (2008). Knowledge, beliefs, and attitudes to female genital mutilation (FGM) in shao Community of Kwara State, Nigeria. *International Quarterly of Community Health Education, 27*(4), 337–349.

Anderson, K. G. (2006). How well does paternity confidence match actual paternity? Evidence from worldwide nonpaternity rates. *Current Anthropology, 47*(3), 513–520.

Barbaro, N., Pham, M. N., & Shackelford, T. K. (2015). Solving the problem of partner infidelity: Individual mate retention, coalitional mate retention, and in-pair copulation frequency. *Personality and Individual Differences, 82*, 67–71.

Becker, A. (2019). *On the economic origins of restrictions of women's sexuality*. CESifo Working Paper 7770. Preprint retrieved from https://ssrn.com/abstract=3432818

Bentley, G., & Mace, R. (Eds.). (2012). *Substitute parents: Biological and social perspectives on alloparenting in human societies*. Berghahn Books.

Blake, K. R., Fourati, M., & Brooks, R. C. (2018). Who suppresses female sexuality? An examination of support for Islamic veiling in a secular Muslim democracy as a function of sex and offspring sex. *Evolution and Human Behavior, 39*(6), 632–638.

Blume, M. (2015). How it's with thy religion, pray? Selection of religiosity among individuals and groups. In D. J. Slone & J. A. Van Slyke (Eds.), *The attraction of religion: A new evolutionary psychology of religion* (pp. 63–71). Bloomsbury.

Borgerhoff Mulder, M. (1990). Kipsigis women's preferences for wealthy men: Evidence for female choice in mammals? *Behavioral Ecology and Sociobiology, 27*(10), 255–264.

Boyd, R., & Silk, J. B. (2009). *How humans evolved*. W. W. Norton & Company.

Brown, M. (2022). Preliminary evidence for an aversion to atheists in long-term mating domains in the southern United States. *Journal of Social and Personal Relationships, 39*(3), 711–733.

Brown, M., & Neiswender, P. R. (2025). Lay theories of mating interest and mate retention strategies for atheists and theists in the southern United States. *Psychology of Religion and Spirituality, 17*(1), 56–64.

Buss, D. M. (1988). From vigilance to violence: Tactics of mate retention in American undergraduates. *Ethology and Sociobiology, 9*(5), 291–317.

Buss, D. M. (1989). Sex differences in human mate preferences: Evolutionary hypotheses tested in 37 cultures. *Behavioral and Brain Sciences, 12*(1), 1–14.

Calcott, B. (2009). Lineage explanations: Explaining how biological mechanisms change. *The British Journal for the Philosophy of Science, 60*(1), 51–78.

Carvalho, J. P. (2013). Veiling. *The Quarterly Journal of Economics, 128*(1), 337–370.

Chvaja, R., Chinchilla, J., Goméz, Á., & Lang, M. (2023). Religious costly signal induces more trustworthiness than secular costly signal: A study of pilgrimage to Santiago de Compostela. *European Journal of Social Psychology, 53*(6), 1294–1308.

Chvaja, R., Spake, L., Hassan, A., Shenk, M. K., Sosis, R., Sear, R., & Shaver, J. H. (2025). Maternal religiosity and social support to mothers: Helpers' religious identity matters. *Religion, Brain & Behavior, 16*(1), 24–45.

Dawkins, R. (1989). *The extended phenotype*. Oxford University Press.

Deacon, T. W. (1997). *The symbolic species*. W.W. Norton.

Dilmaghani, M. (2019). Religiosity, secularity and fertility in Canada. *European Journal of Population, 35*(2), 403–428.

El-Defrawi, M. H., Lotfy, G., Dandash, K. F., Refaat, A. H., & Eyada, M. (2001). Female genital mutilation and its psychosexual impact. *Journal of Sex & Marital Therapy, 27*(5), 465–473.

Emmott, E. H., & Page, A. E. (2019). Alloparenting. In T. Shackelford & V. Weekes-Shackelford (Eds.), *Encyclopedia of evolutionary psychological science* (pp. 210–223). Springer.

Geary, D. C. (2015). Evolution of paternal investment. In D. M. Buss (Ed.), *The handbook of evolutionary psychology* (pp. 483–505). John Wiley & Sons.

Gottlieb, A. (2009). Who minds the baby? Beng perspectives on mothers, neighbours and strangers as caretakers. In G. Bentley & R. Mace (Eds.), *Substitute parents: Biological and social perspectives on alloparenting in human societies* (pp. 115–138). Berghahn Books.

Grant, R., & Montrose, V. T. (2018). It's a man's world: Mate guarding and the evolution of patriarchy. *Mankind Quarterly, 58*(3), 384–418.

Heaton, T. B. (2011). Does religion influence fertility in developing countries. *Population Research and Policy Review, 30*(3), 449–465.

Henrich, J. (2015). *The secret of our success: How culture is driving human evolution, domesticating our species, and making us smarter*. Princeton University Press.

Henrich, J., Heine, S. J., & Norenzayan, A. (2010). The weirdest people in the world? *Behavioral and Brain Sciences, 33*(2–3), 61–83.

Henrich, J., Boyd, R., & Richerson, P. J. (2012). The puzzle of monogamous marriage. *Philosophical Transactions: Biological Sciences, 367*(1589), 657–669.

Hewlett, B. S. (Ed.). (2017). *Father-child relations: Cultural and biosocial contexts*. Routledge.

Howard, J. A., & Gibson, M. A. (2017). Frequency-dependent female genital cutting behaviour confers evolutionary fitness benefits. *Nature Ecology & Evolution, 1*, Article 0049.

Hrdy, S. B. (2006). Evolutionary context of human development: The cooperative breeding model. In C. S. Carter et al. (Eds.), *Attachment and bonding: A new synthesis (the 92nd Dahlem workshop report)* (pp. 9–32). The MIT Press.

Hrdy, S. B. (2008). Cooperative breeding and the paradox of facultative fathering. In R. Bridges (Ed.), *The neurobiology of the parental brain* (pp. 405–415). Academic Press.

Hrdy, S. B. (2009). *Mothers and others: The evolutionary origins of mutual understanding*. Harvard University Press.

Jenkins, P. (2020). *Fertility and faith: The demographic revolution and the transformation of world religions*. Baylor University Press.

Kramer, K. L. (2010). Cooperative breeding and its significance to the demographic succes of humans. *Annual Review of Anthropology, 39*, 417–436.

Kundt, R. (2015). *Contemporary evolutionary theories of culture and the study of religion*. Bloomsbury Academic.

Kundt, R., & Lang, M. (2024). Communicating cooperative intentions drove the selection of collective ritual in hominins. *Religion, Brain & Behavior, 14*(4), 427–449.

Lang, M., & Kundt, R. (2020). Evolutionary, cognitive, and contextual approaches to the study of religious systems: A proposition of synthesis. *Method & Theory in the Study of Religion, 32*(1), 1–46.

Lang, M., & Kundt, R. (2024). The evolution of human ritual behavior as a cooperative signaling platform. *Religion, Brain & Behavior, 14*(4), 377–399.

Lang, M., Purzycki, B. G., Apicella, C. L., Atkinson, Q. D., Bolyanatz, A., Cohen, E., Handley, C., Kundtová Klocová, E., Lesorogol, C., Mathew, S., McNamara, R. A., Moya, C., Placek, C. D., Soler, M., Vardy, T., Weigel, J. L., Willard, A. K., Xygalatas, D., Norenzayan, A., &

Henrich, J. (2019). Moralizing gods, impartiality, and religious parochialism across 15 societies. *Proceedings of the Royal Society B: Biological Sciences, 286*(1898), 20190202.
Lang, M., Chvaja, R., Purzycki, B. G., Václavík, D., & Staněk, R. (2022). Advertising cooperative phenotype through costly signals facilitates collective action. *Royal Society Open Science, 9*, 202202.
Lewis, A. (2021). A semiotic modern synthesis: Conducting quantitative studies in zoosemiotics and interpreting existing ethological studies through a semiotic framework. *Biosemiotics, 14*(2), 295–327.
Lynch, R., Schaffnit, S., Sear, R., Sosis, R., Shaver, J., Alam, N., Blumenfield, M., Mattison, S. M., & Shenk, M. (2022). Religiosity is associated with greater size, kin density, and geographic dispersal of women's social networks in Bangladesh. *Scientific Reports, 12*(1), 18780.
Martin, J. S., Ringen, E. J., Duda, P., & Jaeggi, A. V. (2020). Harsh environments promote alloparental care across human societies. *Proceedings of the Royal Society B: Biological Sciences, 287*(1933), 20200758.
Mercer, J. (2006). *Understanding attachment: Parenting, child care, and emotional development*. Praeger Publishers, Greenwood Publishing Group.
Moon, J. W. (2021). Why are world religions so concerned with sexual behavior? *Current Opinion in Psychology, 40*, 15–19.
Moon, J. W., Krems, J. A., & Cohen, A. B. (2018). Religious people are trusted because they are viewed as slow life-history strategists. *Psychological Science, 29*(6), 947–960.
Moulasha, K., & Rao, G. R. (1999). Religion-specific differentials in fertility and family planning. *Economic and Political Weekly, 34*(42/43), 3047–3051.
Ngwainmbi, J. M. (2015). Feminism on religion: The veiling of Muslim women. *International Journal of Education and Social Science, 2*(12), 24–30.
Nichols, A. D., Lang, M., Kavanagh, C., Kundt, R., Yamada, J., Ariely, D., & Mitkidis, P. (2020). Replicating and extending the effects of auditory religious cues on dishonest behavior. *PLoS One, 15*(8), e0237007.
Olson, M., Jr. (1971). *The logic of collective action: Public goods and the theory of groups*. Harvard University Press.
Palmer, C. T., & Begley, R. O. (2015). Costly signaling theory, sexual selection, and the influence of ancestors on religious behavior. In D. J. Slone & J. A. Van Slyke (Eds.), *The attraction of religion: A new evolutionary psychology of religion* (pp. 93–109). Bloomsbury.
Pazhoohi, F. (2016). On the practice of cultural clothing practices that conceal the eyes: An evolutionary perspective. *Evolution, Mind and Behaviour, 14*(1), 55–64.
Pazhoohi, F., & Burriss, R. P. (2016). Hijab and "hitchhiking": A field study. *Evolutionary Psychological Science, 2*, 32–37.
Pazhoohi, F., & Hosseinchari, M. (2014). Effects of religious veiling on Muslim men's attractiveness ratings of Muslim women. *Archives of Sexual Behavior, 43*(6), 1083–1086.
Pazhoohi, F., Lang, M., Xygalatas, D., & Grammer, K. (2017a). Religious veiling as a mate-guarding strategy: Effects of environmental pressures on cultural practices. *Evolutionary Psychological Science, 3*(2), 118–124.
Pazhoohi, F., Macedo, A. F., & Arantes, J. (2017b). The effect of religious clothing on gaze behavior: An eye-tracking experiment. *Basic and Applied Social Psychology, 39*(3), 176–182.
Peri-Rotem, N. (2016). Religion and fertility in Western Europe: Trends across cohorts in Britain, France and The Netherlands. *European Journal of Population, 32*(2), 231–265.
Perls, T. T., & Fretts, R. C. (2001). The evolution of menopause and human life span. *Annals of Human Biology, 28*(3), 237–245.
Power, E. A. (2017a). Discerning devotion: Testing the signaling theory of religion. *Evolution and Human Behavior, 38*(1), 82–91.
Power, E. A. (2017b). Social support networks and religiosity in rural South India. *Nature Human Behaviour, 1*, 0057.

Pyysiäinen, I. (2008). Ritual, agency, and sexual selection. In J. Bulbulia, R. Sosis, E. Harris, R. Genet, & C. Genet (Eds.), *The evolution of religion: Studies, theories, and critiques* (pp. 175–180). Collins Foundation Press.

Reason, L. L. (2004). The behavioral ecology of female genital cutting in northern Ghana. In M. Alvard (Ed.), *Socioeconomic aspects of human behavioral ecology (Research in economic anthropology, 23)* (pp. 175–202). Emerald Group Publishing Limited.

Richerson, P. J., & Boyd, R. (2008). *Not by genes alone: How culture transformed human evolution*. University of Chicago Press.

Rojo, M. (2023). *Ženy, ktoré zostali. [Women who remained]*. Absynt.

Rosenbaum, S., Kuzawa, C. W., McDade, T. W., Avila, J., Bechayda, S. A., & Gettler, L. T. (2021). Fathers' care in context: 'Facultative,' flexible fathers respond to work demands and child age, but not to alloparental help, in Cebu, Philippines. *Evolution and Human Behavior, 42*(6), 534–546.

Ross, C. T., Strimling, P., Ericksen, K. P., Lindenfors, P., & Borgerhoff Mulder, M. (2016). The origins and maintenance of female genital modification across Africa: Bayesian phylogenetic modeling of cultural evolution under the influence of selection. *Human Nature, 27*(2), 173–200.

Royle, N. J., Smiseth, P. T., & Kölliker, M. (Eds.). (2012). *The evolution of parental care*. Oxford University Press.

Šaffa, G., Duda, P., & Zrzavý, J. (2022a). Paternity uncertainty and parent–offspring conflict explain restrictions on female premarital sex across societies. *Human Nature, 33*(2), 215–235.

Šaffa, G., Zrzavý, J., & Duda, P. (2022b). Global phylogenetic analysis reveals multiple origins and correlates of genital mutilation/cutting. *Nature Human Behaviour, 6*(5), 635–645.

Scelza, B. A., Prall, S. P., Blumenfield, T., Crittenden, A. N., Gurven, M., Kline, M., Koster, J., Kushnick, G., Mattison, S. M., Pillsworth, E., Shenk, M. K., Starkweather, K., Stieglitz, J., Sum, C., Yamaguchi, K., & McElreath, R. (2020a). Patterns of paternal investment predict cross-cultural variation in jealous response. *Nature Human Behaviour, 4*(1), 20–26.

Scelza, B. A., Prall, S. P., Swinford, N., Gopalan, S., Atkinson, E. G., McElreath, R., Sheehama, J., & Henn, B. M. (2020b). High rate of extrapair paternity in a human population demonstrates diversity in human reproductive strategies. *Science. Advances, 6*(8), eaay6195.

Schnabel, L. (2021). Secularism and fertility worldwide. *Socius, 7*, 23780231211031320.

Sear, R. (2021). The male breadwinner nuclear family is not the 'traditional' human family, and promotion of this myth may have adverse health consequences. *Philosophical Transactions of the Royal Society B, 376*(1827), 20200020.

Shaver, J. H. (2017). Why and how do religious individuals, and some religious groups, achieve higher relative fertility? *Religion, Brain & Behavior, 7*(4), 324–327.

Shaver, J. H., DiVietro, S., Lang, M., & Sosis, R. (2018a). Costs do not explain trust among secular groups. *Journal of Cognition and Culture, 18*(1–2), 180–204.

Shaver, J. H., Lang, M., Krátký, J., Kundtová Klocová, E., Kundt, R., & Xygalatas, D. (2018b). The boundaries of trust: Cross-religious and cross-ethnic field experiments in Mauritius. *Evolutionary Psychology, 16*(4), 1474704918817644.

Shaver, J. H., Sibely, C. G., Sosis, R., Galbraith, D., & Bulbulia, J. (2019). Alloparenting and religious fertility: A test of the religious alloparenting hypothesis. *Evolution and Human Behavior, 40*(3), 315–324.

Shaver, J. H., Power, E. A., Purzycki, B. G., Watts, J., Sear, R., Shenk, M. K., Sosis, R., & Bulbulia, J. A. (2020). Church attendance and alloparenting: An analysis of fertility, social support and child development among English mothers. *Philosophical Transactions of the Royal Society B, 375*(1805), 20190428.

Shaver, J. H., Chvaja, R., Spake, L., Hassan, A., Badjie, J., Prentice, A., Cerami, C., Sear, R., Shenk, M. K., & Sosis, R. (2024). Religious involvement is associated with higher fertility and lower maternal investment, but more alloparental support among Gambian mothers. *American Journal of Human Biology, 36*(12), e24144.

Singh, D., Dixson, B. J., Jessop, T. S., Morgan, B., & Dixson, A. F. (2010). Cross-cultural consensus for waist–hip ratio and women's attractiveness. *Evolution and Human Behavior, 31*(3), 176–181.

Slone, D. J. (2008). The attraction of religion: A sexual selectionist account. In J. Bulbulia, R. Sosis, E. Harris, R. Genet, & C. Genet (Eds.), *The evolution of religion: Studies, theories, and critiques* (pp. 181–187). Collins Foundation Press.

Slone, D. J., & Van Slyke, J. A. (Eds.). (2015). *The attraction of religion: A new evolutionary psychology of religion*. Bloomsbury Academic.

Sosis, R. (2017). Religions as complex adaptive systems. In N. Clements (Ed.), *Religion: Mental religion* (pp. 219–236). Macmillan Reference USA.

Sosis, R. (2019a). The building blocks of religious systems: Approaching religion as a complex adaptive system. In G. Gorgiev, J. M. Smart, C. L. Flores Martines, & M. Price (Eds.), *Evolution, development & complexity: Multiscale models of complex adaptive systems* (pp. 421–449). Springer.

Sosis, R. (2019b). Why cultural evolutionary models of religion need a systemic approach. In A. K. Petersen, I. S. Gilhus, L. H. Martin, J. S. Jensen, & J. Soerensen (Eds.), *Evolution, cognition, and the history of religion: A new synthesis* (pp. 45–61). Brill.

Spake, L., Hassan, A., Sear, R., Shenk, M. K., Sosis, R., & Shaver, J. H. (2022). Disentangling the relationships between religion and fertility. *Religion, Brain & Behavior, 12*(4), 343–346.

Strassmann, B. I. (1992). The function of menstrual taboos among the Dogon. *Human Nature, 3*(2), 89–131.

Strassmann, B. I. (1996). Menstrual hut visits by Dogon women: A hormonal test distinguishes deceit from honest signaling. *Behavioral Ecology, 7*(3), 304–315.

Strassmann, B. I. (2011). Cooperation and competition in a cliff-dwelling people. *Proceedings of the National Academy of Sciences of the United States of America, 108*(2), 10894–10901.

Strassmann, B. I., Kurapati, N. T., Hug, B. F., Burke, E. E., Gillespie, B. W., Karafet, T. M., & Hammer, M. F. (2012). Religion as a means to assure paternity. *Proceedings of the National Academy of Sciences of the United States of America, 109*(25), 9781–9785.

Taragin-Zeller, L. (2014). Modesty for heaven's sake: Authority and creativity among female ultra-orthodox teenagers in Israel. *Nashim: A Journal of Jewish Women's Studies & GenderIssues, 26*, 75–96.

Tinbergen, N. (1963). On aims and methods of ethology. *Zeitschrift für Tierpsychologie, 20*(4), 410–433.

World Health Organization, United Nations Population Fund & United Nations Children's Fund (UNICEF). (1997). *Female genital mutilation: A joint WHO/UNICEF/UNFPA statement*. World Health Organization.

Xygalatas, D., Mitkidis, P., Fischer, R., Reddish, P., Skewes, J., Geertz, A. W., Roepstorff, A., & Bulbulia, J. (2013). Extreme rituals promote prosociality. *Psychological Science, 24*(8), 1602–1605.

Xygalatas, D., Kundtová Klocová, E., Cigán, J., Kundt, R., Maňo, P., Kotherová, S., Mitkidis, P., Wallot, S., & Kanovsky, M. (2015). Location, location, location: Effects of cross-religious primes on prosocial behavior. *The International Journal for Psychology of Religion, 26*(4), 304–319.

Xygalatas, D., Maňo, P., Bahna, V., Kundtová Klocová, E., Kundt, R., Lang, M., & Shaver, J. H. (2021). Social inequality and signaling in a costly ritual. *Evolution and Human Behavior, 42*(6), 524–533.

Xygalatas, D., Maňo, P., Kundt, R., & Kundtová Klocová, E. (2022). Rituals as signals of mate quality. *Current Research in Ecological and Social Psychology, 3*, 100048.

Yoder, P. S., Abderrahim, N., & Zhuzhini, A. (2004). *Female genital cutting in the demographic and health surveys: A critical and comparative analysis*. DHS Comparative Reports 7. https://www.dhsprogram.com/pubs/pdf/CR7/CR7.pdf

Part III
Culture from the Perspective of Evolutionary Behavioural Science

Jan Havlíček

Evolutionary concepts had become deeply rooted in how contemporary biology approaches the study of various levels of biological entities, from molecules to ecosystems. For many biologists, it is challenging to even think beyond evolutionary concepts. This sharply contrasts with how culture is approached by social anthropology, ethnology, sociology, linguistics, and other related fields. Evolutionary concepts are not central to these disciplines, and many scholars even view their application as misplaced. Nevertheless, the last few decades have seen a vast increase in the study of cultural phenomena from an evolutionary perspective. This is by no means a unified movement; rather, it consists of several distinct schools of thought which oftentimes disagree even on the basic features of cultural evolution.

The aim of this section is to provide an overview of the main theoretical approaches and empirical insights into the current study of the evolution of culture. The individual chapters are far from offering just a full review of the current literature. Instead, they focus on some of the most studied aspects and hotly debated issues of cultural evolution. Because culture was traditionally attributed only to humans, we start with a chapter on traditions in non-human animals. This is followed by three chapters written from the perspective of different paradigms: memetics, attractionist theories, and phylogenetics. The section concludes with a chapter which explores the interface between social learning and demography in the diffusion of cultural characteristics.

As mentioned above, the first chapter of this section focuses on traditions in non-human animals. It presents a brief historical overview starting with the legendary diffusion of bottle-opening skills by great tits (Fisher & Hinde, 1949), the spread of potato washing among Japanese macaques (Kawai, 1965), and tool use by chimpanzees as first observed by Goodall (1964). These historical examples are complemented by more recent, intensively studied domains of cultural transmission such as

J. Havlíček
Department of Zoology, Faculty of Science, Charles University, Prague, Czech Republic

the learning of migratory routes, bird dialects, communication in cetaceans, and chimpanzee cultures. These behaviours are based on various forms of social learning, including local and object enhancement, emulation, imitation, and teaching. There is currently a consensus on the existence of cultural traditions in various species; what remains controversial is whether some of these traditions actually amount to a cumulative culture as seen in humans.

The second chapter introduces memetics, an approach to the study of culture which intensively draws on analogies between biological and cultural evolution. Similar to genes, memes are viewed as units transmitted between their bearers. The original memetics, as introduced by Dawkins (1976) and his followers, was heavily criticised for various oversimplifications and did not fully establish itself as an empirical field. Nevertheless, it produced various insights which can enrich our understanding of cultural evolution (for instance certain mechanisms, such as stability-based sorting or competition between memetic complexes). The authors also discuss the relationship between memes and semiotic signs, speculating whether memetics can be understood as a simplified but dynamic form of diachronic semiotics.

The third chapter introduces attractionist theories of culture, originally formulated by Sperber (1985, 1996) and subsequently elaborated by other scholars. The attractionist concept takes as its starting point transmission chains, which include an alternation of mental and public representations. During transmission, the content is transformed based on the individual's understanding and interests. But some cultural content is likely to be transformed in particular ways: such convergent transformations are then called attractors. They can emerge due to either ecological or psychological factors. Ecological factors include, for instance, the availability of local materials which may affect the shape of artefacts in a given culture. Psychological factors include biases, for instance the negativity bias, which suggests a higher chance of transmitting negatively valenced information. Interpretation is an important part of this process, leading authors to discuss the use of tools and narratives developed within the framework of semiotics and even a 'culturisation of biology'.

The fourth chapter explores how phylogenetic approaches can assist us in understanding cultural history. Phylogenetics is widely used in biology to study relatedness between species and the origin of various traits. Some cultural characteristics, too, share similar features due to their shared history. This is exemplified by phylogenies of languages, since languages provide large amounts of data, such as vocabulary corpora, which enable researchers to deduce their history. On the other hand, languages also incorporate terms from other languages through horizontal transfer, which poses a major limitation to the use of tree-like structure employed by phylogenetics. More generally, though, the knowledge of history of a given characteristic allows researchers to test hypotheses about its possible adaptive function. Do societies share a given characteristic, such as marital system, due to their common history or in response to ecological factors?

The final chapter of this section explores human cumulative culture through the lens of niche construction theory. It argues that cultural environment is an essential part of the human niche, which leads to various adaptations, including social learning mechanisms. These mechanisms, particularly imitation and teaching, are extremely common across human cultures. The diffusion of cultural content is further influenced by the effective population size, which involves the level of connectedness. Simulation studies show that a medium level of connectedness is most efficient for spreading cultural novelties, which has important implications for both the study of human societies and the evolution of culture in general.

Overall, this section clearly demonstrates how ideas about the evolution of culture have progressed from a marginalised field, previously overlooked by mainstream social sciences, into a flourishing and established area of scientific inquiry. Perhaps, in the near future, we will reach a point where one could rephrase Dobzhansky's (1973) famous aphorism for biology to 'nothing in culture makes sense except in the light of evolution'.

References

Dawkins, R. (1976). *Selfish gene*. Oxford University Press.

Dobzhansky, T. (1973). Nothing in biology makes sense except in the light of evolution. *The American Biology Teacher, 35*(3), 125–129.

Goodall, J. (1964). Tool-using and aimed throwing in a community of free-living chimpanzees. *Nature, 201*, 1264–1266.

Fisher, J., & Hinde, R. A. (1949). The opening of milk bottles by birds. *British Birds, 42*, 347–357.

Kawai, M. (1965). Newly acquired pre-cultural behavior of the natural troop of Japanese monkeys on Koshima Islet. *Primates, 6*, 1–30.

Sperber, D. (1985). Anthropology and Psychology: Towards an Epidemiology of Representations. *Man, 20*(1), 73–89.

Sperber, D. (1996). *Explaining culture: A naturalistic approach*. Blackwell.

Chapter 9
Culture: Uniquely Human?

Jitka Lindová

Abstract The subject of culture in (non-human) animals is important not only because it reveals interesting facts about the behaviour and cognition of various animal species but also because it hints at areas and aspects of human culture that are unique. This contribution mentions a variety of examples of animal behaviour which are often interpreted as cultural phenomena. Such phenomena span from learning of food preferences and routs, through vocal traditions and gesturing, social customs and hygiene habits, to tool use. The decisive factor in telling apart cultural characteristics from related phenomena is social learning, which should take place during the transmission of the behaviour in question. This is also why in this contribution, the author pays attention to laboratory studies of social learning and transmission of socially learned behavior among group members. Finally, we also discuss specific human characteristics that allow us to be the only animal to create a flexible adaptive environment by cummulating cultural traits.

Keywords Animal culture · Animal traditions · Social learning · Cummulative culture · Tool use

9.1 Introduction

Regardless of whether we think of it in terms of sophisticated tools and technologies, a wide array of social customs which define particular communities, or a complex system of symbols, culture is something that sets *Homo sapiens* apart from other animal species. Traditionally, culture has been considered an exclusively

J. Lindová (✉)
Department of Psychology and Life Sciences, Faculty of Humanities, Charles University, Prague, Czech Republic
e-mail: Jitka.Lindova@fhs.cuni.cz

J. Toman et al. (eds.), *Culture Evolving*, Biosemiotics 10,
https://doi.org/10.1007/978-3-032-16124-6_9

human achievement. Not long ago, scientists ascribed tool use and social customs only to humans, considering it a hallmark of human culture. Nowadays, it is clear that people are not the only animals to use tools: monkeys use stones to crack nuts, birds modify pieces of material to get at food they cannot reach directly, even dolphins use tools, and chimpanzees use a whole array of them. People are also not the only animal whose groups differ in terms of social customs passed down as traditions. Grooming or greeting customs have been described in apes and primates, not to mention the vocal traditions of numerous species, including whales or ravens.

Even so, no one doubts that human culture is a unique phenomenon: it has a more complex structure and dynamics, and its extent transcends anything found in other species. Even chimpanzees, the most culturally advanced non-human animals, 'sit naked in the jungle cracking nuts' (Laland et al., 2011). In some ways, human culture is extremely unlike anything seen in other animals. But it is hard to pinpoint what makes human culture principally different from the culture of other animals or to figure out what abilities (preferably intellectual ones) predestined humans—rather than some other animals—to create culture. What one can, however, state with certainty is that these human 'cultural' abilities evolved from a simpler state resembling what we can find in other apes (Laland et al., 2011), which is why the study of animal culture, or its precursors, is so important for better understanding of human culture.

The aim of this chapter is to review the available literature on animal culture, illustrating what culture in animals can look like and how it can be understood. We begin by exploring early efforts to characterize culture in animals, with a historical overview divided into three periods: initial observations, ethnographic studies, and the development of systematic research on cultural transmission. Next, we discuss the broad range of behaviors that may fall under the umbrella of animal culture, highlighting key principles that aid in understanding and describing it, such as social learning as a cognitive foundation, innovation as a source of cultural expression, transmission biases, and the role of conformity in the spread and maintenance of culture. Finally, we examine specific cultural traits, focusing on particular cultural areas that tend to be associated with certain animal taxa. In conclusion, we will attempt to draw insights on whether culture in animals evolves or can evolve, or, alternatively, why it might remain in a relatively simple form as described.

9.2 Definition and Key Characteristics of Culture

Our understanding of socially transmitted behaviour in animals is fast increasing, but we are still far from a consensus across scientific fields as to whether it deserves to be called 'culture'. That is because different disciplines vary substantially in how strictly or loosely they define culture and to what extent they require that culture should have all the hallmarks of human culture (cf. Chaps. 5, 7, 10, 11, 12, 13, 16, 17).

At the same time, there is a growing convergence within the biological sciences on what are considered the key aspects of culture. The zoologists Laland and Hoppitt (2003) define culture as group-typical behaviour patterns, which are shared within (but not so much between) groups, are socially learned, and socially, as opposed to genetically, transmitted. In the following, we therefore focus on these two aspects of animal cultural behaviour: group-level sharing and social learning.

Early attempts to conceptualise animal culture can be traced to McGrew and Tutin (1978), who revived an old remark by the cultural anthropologist Alfred Kroeber (1928) on what chimpanzee behaviour would have to look like to be considered cultural. They noted that it should be based on innovation, that is, on modifications of existing behavioural patterns or on inventions produced by individuals. It should also spread within a group, its application should be consistent, it should have the potential to spread to other groups, and it should persist in time and be handed down from one generation to the next. McGrew (1998) then added independence of such behaviour on the ecology of the environment in order to draw a distinction between cultural behaviour and behavioural differences between groups determined by ecological factors (such as the food available at a particular location). The criterion of being independent of local ecology had subsequently proven to be of key importance for ethnographical studies of chimpanzee cultures described below.

Nevertheless, cultural behaviour does not always need to meet all of the proposed criteria. For instance, not every cultural behaviour becomes a tradition handed across generations (as seen in fashion trends in human society; Ramsey, 2013). Cultural behaviour even does not need to spread into further communities or be independent of local ecological conditions (Schuppli & van Schaik, 2019; see also below). Therefore, while McGrew's list does identify some prominent characteristics of culture, we will not require that all these criteria be met in order to consider a behaviour cultural.

The study of animal cultures can help answer various questions about human culture. For instance, it can lead to generally valid findings about the individuals and circumstances that lead to innovations and about the role of such innovations in cultural evolution. It can help us identify the function of various types of social learning in cultural transmission and the impact of the dynamics of the spread of novelties in a population on the form of cultural behaviour or its stability.

Moreover, scientists often try to define what, in principle, sets apart human culture from culture in animals, what was the decisive step that animals did not take while humans did, the point that caused the 'ratcheting effect' in human culture. One such key moment clearly came when culture itself became subject to evolution. Closely related is the issue of accumulation of culture, that is, further development of adopted cultural behaviour and its enrichment by novel elements (Heyes, 2021). In the following, we therefore try to describe the limits of cultural behaviour and social cognition of animals and pinpoint the cognitive processes which are likely to be necessary preconditions of the more complex, accumulating, and evolving human culture.

In some cases, the study of cultural behaviour of animals even formulates theories about its ultimate function, which may be linked to the early phylogenesis of culture in humans. We will therefore investigate the innovations, forms of transmission, and cognitive processes that enable the emergence and persistence of culture, and the adaptive functions of culture in animals.

9.2.1 Early Evidence of 'Culture' in Animals

The beginnings of the study of animal culture in the twentieth century are related to sensational observations of unusual animal behaviour reminiscent of human cultural behaviour. Two such cases were documented in great detail, primarily with regard to the spread of a new behaviour. The first case involved great tits and how they learned to open milk bottles and consume the cream layer that settled on milk surface. It was first described in one London borough in 1921. By 1947, it spread into 30 locations inhabited by great tits all over Great Britain (Fisher & Hinde, 1949; Hinde & Fisher, 1951). Another famous observation came from Japan, where in 1953 a female called Imo (meaning 'potato' in Japanese), who lived in a group of Japanese macaques on the Koshima island, started to wash sweet potatoes provided by the locals in a creek. By washing them, she removed the annoying sand that was getting in her teeth as she ate them. The sweet potatoes were covered in sand because people simply dumped them for the animals on the beach. Soon, Imo's kin and other individuals close to her also started washing their potatoes. Eventually, the behaviour spread beyond this circle and was further improved when washing was moved to the sea, which probably gave the sweet potatoes a pleasantly salty taste. Within 8 years, the behaviour spread to almost entire local population (Hirata et al., 2008; Kawai, 1965; Kawamura, 1954).

The current significance of these classical cases is limited because the studies that described them say little about how the behaviour was transmitted. It seems that a simpler form of social learning played a role but some scientists expressed reservations as to whether learning how to open a bottle or wash a sweet potato can be viewed as a manifestation of animal culture given that the change in behaviour was motivated by human activity (introduction of a new food source). That inspired a novel behaviour which, however, did not significantly transcend the genetically determined behavioural patterns of the species in question. The novel behaviour moreover tended to spread in the population that experienced these ecological (food-related) conditions, which is completely different from particular communities maintaining their customs independently of the prevailing ecological conditions. On the other hand, both cases undoubtedly capture a nongenetic, social learning-based transmission of a particular behaviour within a population. In that, they meet the definition of cultural behaviour proposed by Laland and Hoppitt (2003).

9.2.2 *The Ethnographic Approach*

When Jane Goodall in 1960 observed a chimpanzee break off a twig and strip its leaves to use it to collect termites, that is, the production and use of a tool (Goodall, 1964), she opened a new chapter of research. Over time, many behaviours significantly similar to human ones were described in this ape, our closest relative, and subsequently also in other apes. Another conceptually important step was taken when McGrew and Tutin (1978) described a type of behaviour that has proven particularly informative for the study of animal culture, as it was not directly linked to subsistence or tool use. They identified a social custom among chimpanzees where two males hold their hands during grooming in a specific way. In contrast to the food-related cases in tits and macaques discussed above, there is no connection with the ecology of environment inhabited by a particular group and it is a behaviour that differs between groups but is stable within a group.

McGrew and Tutin called this behaviour cultural, which was at the time a daring step. They further framed this decision by setting certain standards for cultural behaviour in animals using the 'ethnographic' approach: a 'cultural behaviour' takes place only in some communities of the species and between-group differences in relation to the behaviour have no genetic basis and are not linked to the ecology of the environment. This approach was later applied to field observations of social customs and other socially transmitted behaviour in chimpanzees by researchers who produced an important joint 'ethnographic' publication describing 39 cultural traits in common chimpanzees from seven locations (Whiten et al., 1999). The contributors brought together strong arguments to show that the behaviours observed are independent of both ecology and genetics.

9.2.3 *Investigating the Mechanisms of Transmission*

Since the 1990s, scholars started to argue that the ethnographic approach is inadequate. For instance, the groups with different behaviours may be geographically isolated and consist of close relatives, which is why it may be hard to exclude the influence of ecology and genetics even where it seems unlikely. These researchers also emphasised that what matters is the manner of transmission and claimed that only imitation and teaching enable an accumulation of cultural traits that is typical for humans (Galef Jr., 1992; Tomasello, 1994; Tomasello et al., 1993). In chimpanzees, as well as in other animals, cultural traits are not uncommon but seem to be conserved in a simple form with little tendency to accumulate improvements or changes (on this, more below). On the other hand, Schuppli and van Schaik (2019) pointed out that the 'ethnographic' criteria for classifying a behaviour as cultural are excessively strict and may lead to a dismissal of most cultural traits in animals. They force us to omit all cultural adaptations to local ecological conditions and all universally present cultural traits. Other researchers, too, believe that ecology often creates conditions for the emergence of specific cultural behaviours, and that the role of such behaviors should not be dismissed in animals (Koops et al., 2014). Scholars therefore shifted their focus to social transmission and especially its complex forms, such as imitation.

What followed was a wave of laboratory experiments investigating whether chimpanzees and other species are cognitively equipped for the transmission of culture, that is, whether they use advanced forms of social learning (especially imitation) to transmit learned behaviours in a group. The initial results did not support the hypothesis of true imitation (e.g., Tomasello et al., 1987) but after year 2000, methodologically improved studies confirmed imitation in some animal species (e.g., Whiten et al., 2005). Moreover, some scholars started to argue that imitation is not a necessary precondition for cultural transmission and other factors may play an important role in this process. We shall return to the issue of imitation and teaching below.

Our current understanding of putative animal culture thus relies on two main kinds of sources: ethnographic records and a systematic study of social transmission. The former collect evidence on behaviours that could be viewed as cultural, while the latter helps explain the spread and maintenance of cultural behaviour. Research on transmission mechanisms is usually conducted in animals kept in captivity; however, some species—especially cetaceans, but also some birds and mammals—are extremely difficult to keep in captivity in numbers large enough to allow observation of group-level phenomena. In such cases, ethnographic data may be the sole source of information about their cultural behaviour, and transmission mechanisms can be inferred only from systematic observations of the spread of novel cultural elements (see below).

Recently, a systematic study of social transmission has moved also to field conditions, leading to highly ecologically valid results. Researchers use for instance trained individuals who demonstrate a learned behaviour to the population, and then they observe how the behaviour spreads (Aplin et al., 2015).

9.2.4 The Variety of Cultural Behaviour in Animals

Researchers have been discussing culture in the animal kingdom in connection with many areas of behaviour, including vocal traditions, socially transmitted preference for certain foods, migration routes, partners, food source locations, nesting or hunting locations, but also various social customs, gestures, and ultimately also technologies, including tool use. Species vary in the richness of cultural behaviours, which range from a single specific domain (such as possibly the song in songbirds) to numerous domains. In chimpanzees, we find traditional ways of using tools, traditional calls and gestures, and customs connected with mutual grooming or courting (Whiten, 2011). A similar range of social and food-related behavioural traits has been described in orangutans (van Schaik et al., 2003), and traditional hygienic practices, manners of locomotion, social customs, gestures, and other behaviours have been observed also in gorillas (Robbins et al., 2016). Researchers have described traditional behaviours in various domains in several species of New World monkeys: tool use but also social customs and social games have been described in capuchin monkeys (Moura & Lee, 2004; Perry et al., 2003), while Geoffroy's spider monkeys apart from their social traditions exhibit also particular cultural ways of drinking, locomotion, and food preferences (Santorelli et al., 2011). Vocal traditions and cultural behaviour related to hunting techniques have been reliably documented in cetaceans (Allen et al., 2013).

So far, traditions have been reliably documented in two or more areas of behaviour in just a handful of species, but it is likely that the list will grow. Based on existing findings regarding social learning in various species, observations of their behaviour in the field, and given the general rules of probability, Laland and Hoppitt (2003) assume that hundreds of vertebrate species may have culture. Finally, some studies have even explored cultural behaviour in various invertebrates and found that, for instance, female fruit flies can share a traditional preference for egg-laying substrates (Battesti et al., 2012).

9.2.5 Culture Transmission and Social Learning

It is generally accepted that only socially learned behaviour can be considered a cultural trait. Only socially learned behaviour—as opposed to behaviour learned individually—can be transmitted in a population and persist beyond the lifetime of an individual. Much more controversial is the issue of what types of social learning can lead to the formation of cultural traditions because different forms of social learning vary in transmission accuracy and cognitive requirements, that is, in two respects which are highly relevant from the perspective of culture.

Simpler forms of social learning include stimulus enhancement and local enhancement—in both cases, the demonstrator merely attracts the observer's attention to a particular object or location. The observer subsequently seeks out the stimulus (such as a food source) or the location and engages in a behaviour that is part of its innate repertoire. Observation of a particular behaviour in another individual can, moreover, increase the likelihood of identical behaviour in the observer if the observer concludes that the behaviour in that situation makes sense (facilitation of response) or is linked to a reward (observational learning of a relation between a reaction and reinforcement; Rendell & Whitehead, 2001). But such social learning is still based on associative processes that do not require individuals to go beyond previously acquired or innate motoric patterns and manipulations. Individuals simply engage in those behaviours in new situations. Some researchers believe that many putative cultural traits in animals rely primarily on individual learning where only the frequency of a particular behaviour is socially learned; according to this view, we are dealing with so-called socially mediated serial re-innovations within the zone of latent solutions (Bandini & Tennie, 2017).

Boyd and Richerson (1988), Galef Jr. (1992), Tomasello (1994), and many others ascribe a crucial role in cultural transmission to imitation, which is significantly more cognitively demanding. It involves a precise copying of the sequence and execution of the demonstrator's behavioural actions and an understanding that the demonstrator engages in this behavioural sequence to achieve a particular goal. Imitation enables a highly accurate transmission of behaviour and can thus lead to the maintenance of cultural traditions (Tomasello, 1999). In contrast, emulation, a related type of learning, is not so clearly associated with transmission accuracy. When emulating, the individual likewise understands the demonstrator's intention and can later achieve the same goal, but the technique by which it is achieved may differ from what was demonstrated. The emulator does not focus on faithfully copying the sequence of movements and often spontaneously chooses a technique that is common for that animal. The abovementioned researchers have argued that, unlike other forms of social learning, imitation is effective in learning complex sequences of behavioural actions, such as food acquisition consisting of several steps where each step on its own does not lead to the desired goal (cf. also 'process-opaque' tasks; Acerbi et al., 2011; Derex et al., 2012). Such complex techniques are typically the result of accumulation of advantageous modifications of existing behavioural traits (Dean et al., 2014); in other words, they form the basis of a cumulative culture that is typical of humans. At a certain point, the accumulated modifications reach a level of complexity that is so high that an individual cannot in its lifetime learn the technique non-socially, merely by individual learning (Tennie et al., 2020; Tomasello, 1994, 1999; Tomasello et al., 1993).

Teaching is another type of social learning that is important for cultural transmission. Caro and Hauser (1992) defined teaching as adjusting own behaviour in the presence of another individual, whereby this adjustment carries some costs to the animal (at a minimum some time) and its outcome is that another individual acquires a skill or understanding which it otherwise would acquire only later, not as well, or not at all. We find teaching in this sense in many life forms, including for instance ants, where more experienced individuals teach other ants particular routes by going through them at a slower pace with a 'student' while maintaining contact with their antennae to make sure the 'student' does not get lost (Franks & Richardson, 2006). It is, however, quite clear that teaching in ants and other animal species (meerkats, cats, leiothrices) is based on simpler cognitive and behavioural mechanisms than teaching in humans is (Galef Jr., 2009). In humans, it involves numerous complex cognitive processes, including an assessment of the student's knowledge and corresponding optimisation of the approach (Flynn, 2010). This human form—and perhaps even some simpler forms of teaching—can lead to a highly accurate transmission and transfer of complex techniques.

In the following, we move from thoughts about the forms of social learning that can function as mechanisms of cultural transmission to findings about the sources of cultural behaviours and then to the dynamics of spread, direction of transmission, and stabilisation of animal culture.

9.2.6 The Source of Culture: Innovation

The source of cultural behaviour are innovations, that is, new or significantly modified learned behaviours that did not previously appear in the population. In some cases, a behaviour can be transmitted by social learning, be independent of ecological conditions, and typical of a particular social group, and yet its source is not a behavioural innovation but rather an accidental change or error in the copying or presentation of a particular trait. In such cases, we speak about a cultural drift (cf. Chaps. 10, 11, 13).

Innovations are the product of individual learning of a particular individual. Innovators, like the abovementioned Japanese macaque Imo, can become a model for other members of a group who then learn the behaviour socially. Some individuals have a notable ability to produce innovations (Morrand-Ferron et al., 2011) and one might expect them to have more advanced cognitive abilities. Innovative solutions to complex problems appear more frequently in great apes, corvids, and parrots, that is, in species that have outstanding cognitive abilities, especially the ability to derive causal relations and create abstract rules (Auersperg et al., 2012). But inventiveness is also linked to certain more general cognitive abilities such as focus, the ability to inhibit behaviour, and to the motivation system, especially patience, neophilia, and play. In the case of simpler innovations, such as trying new foods, becoming an innovator might simply depend on the individual's level of activity, sensitivity, and the like (Tebbich et al., 2016).

Aside from this, it is known that animals can be to some extent driven to innovation by the circumstances, for instance by hunger. In other words, some circumstances are more likely to produce innovations (the 'necessity leads to innovations' hypothesis; Laland & Reader, 1999). In theory, Imo may have become an innovator simply because she was the hungriest member of her troop (Laland & Hoppitt, 2003). Curiously, in an interspecies context, it does not seem that the ability to produce complex innovative solutions is associated with the species maintaining cultural traditions. For instance, rooks in laboratory conditions excel at innovative solutions to puzzles that require the use of tools but in the wild, they do not use tools at all. In other words, rook populations do not maintain techniques of tool use in the form of cultural traditions. Scientists assume that what decides whether an innovation will become the foundation of a new cultural behaviour is its adaptive value (such as a new way of acquiring food) from the perspective of the population. For rooks, sophisticated tool use is apparently not important in their natural environment, because they acquire most of their food quite easily using their claws and beak (Kacelnik, 2009).

In line with the hypothesis that innovations are driven by circumstances is the observation that 'innovators' are usually the younger animals (Biro et al., 2003) and animals lower in the social hierarchy (Reader & Laland, 2001), that is, individuals who have a generally worse access to food sources. Nevertheless, in some species, the opposite is the case (Boogert et al., 2006). The spread of innovations from lower-raking or younger individuals is nevertheless not quite straightforward, because juveniles (at least in chimpanzees) tend to be imitated only by their peers or younger individuals (Biro et al., 2003), and lower-ranking individuals are less likely to be imitated by other members of their troop (Coussi-Korbel & Fragaszy, 1995). But sometimes, for instance if an innovation is advantageous in terms of biological fitness and proliferates horizontally (see below), the new behaviour can quickly spread even from a younger submissive innovator throughout the population and permanently influence how it interacts with its environment.

In the cumulative human culture, innovations often take the form of improvements on existing techniques (for instance of food preparation). In animals, though, it seems that modifications of traditional techniques are not so frequent (cf. Chap. 13). We shall see below that animals are highly conservative: if, for instance, chimpanzees adopt a certain way of reaching a goal, they are unwilling to change it even if a new technique would bring significant advantages (Hrubesch et al., 2009). The likelihood of innovation is also lowered by neophobia or by a strong preference for what one has over what one might acquire (Brosnan & Hopper, 2014). As a result, innovations in the animal kingdom seldom spread. We shall analyse this tendency to (non)spread of new behaviours in certain social situations in the following two sections.

9.2.7 The Direction and Bias in Culture Transmission

Cultural traits in animals can be transmitted vertically from parents to the offspring (e.g., along the maternal line), obliquely from older to younger (unrelated) members of the group, or horizontally, that is, within peer groups (for a more detailed overview of transmission types, see Chaps. 10, 11, 13). These forms differ in the resulting stability of the trait, which is greatest in vertical transmission, but also in transmission speed, which is highest in horizontal transmission (Acerbi & Parisi, 2006) (cf. Chaps. 10, 15).

Vertical transmission enables a sympatric coexistence of various cultural traditions, which can be practiced in individual kinship lines. If transmission takes place along the maternal line (e.g., in dolphins), the sharing of skills between mothers and their offspring cannot be explained solely by genetics because that would predict also an influence of the fathers. In this case, one can therefore distinguish innate behaviour from cultural traditions. Horizontal transmission, on the other hand, can lead to 'fashion waves', where relatively arbitrary variants spread quickly but their duration is limited—in extreme cases to just one season. This cultural form has been observed mainly in the spread of vocalisations, for instance in whales or birds.

The dynamics of spread is also influenced by transmission bias, that is, by preferences regarding when, what, and from whom one can learn. For instance, dominant individuals or individuals excelling in resource acquisition can be preferred as models. Selective copying of successful behaviours and successful individuals (who learn also individually) is more likely to lead to an accumulation of cultural innovations (Ehn & Laland, 2012). Evidence of such selective copying in animals is rather sparse, but for instance Kendal et al. (2015) have shown that chimpanzees tend to imitate both higher-ranking individuals (who need not be especially skilled at solving a given task) and individuals who excel in solving the task. In other words, chimpanzees show preferences for models similar to what is known in humans (cf. Chaps. 10, 11).

9.2.8 Stabilisation of Culture: Conformity

Animals, like humans, often simply follow the behaviour of the majority. This tendency to conformity is undoubtedly one of the strongest transmission biases in animals (cf. Chaps. 10, 11) and it applies both to learning new behaviours and to modification of acquired behaviours in the direction of a variant that is dominant in the population. Conformity in this sense has been demonstrated in chimpanzees (Watson et al., 2018), great tits (Aplin et al., 2015), and even insects (Whiten et al., 2021).

The advantage of conformity is that it accelerates the transmission of cultural forms of behaviour adopted by the majority of the group, which is significant for instance in the case of inexperienced youngsters or incomers from other groups. It thus contributes to the maintenance of traditional behaviours of the community, supports the sharing of skills and customs, and ultimately helps create the group's identity (cf. Chap. 15). Behaviours that function as identification signs of the community are copied most rapidly (Whitehead & Rendell, 2014) and in species such as chimpanzees, collective sharing of customs and group identity probably constitute a crucial evolutionary advantage (McGrew, 2009).

On the other hand, conformity can hamper the spread of innovations, and it may be in part responsible for conserving the chimpanzee culture on the level of simple activities (Hrubesch et al., 2009; Marshall-Pescini & Whiten, 2008b). Moreover, it can lead to the spread of 'fashion trends' and even to a proliferation of behaviours that compromise biological fitness and put the entire group at a disadvantage. Various 'fashion trends' have been described, for instance, in cetacean vocalisations (Rendell & Whitehead, 2001), while the spread of maladaptive traditions has been described in fish (copying of a more complicated route to a food source; Laland & Williams, 1998) (cf. Chap. 10).

Nevertheless, people, too, have a high preference for conformity, and the limited spread of innovations in animal species can have numerous causes ranging from differences in social learning and cognition to differences in population social structure. Having described some important characteristics of cultural behaviour and examples of various forms of transmission that determine the spread and maintenance of culture, the second part of the text focuses on selected animal taxa and behaviours that could be considered cultural.

9.3 Examples of Animal Culture

9.3.1 Learned Preferences of Location, Food, and Learning of Routes

If culture is understood to include even transmission via simple forms of social learning, leading to the establishment of behaviourally stable traditions within groups, it can be found in many species. Based solely on the criteria of social transmission and behavioural stability, even shared preferences for certain foods, partners, or locations may be considered cultural behaviour. An example of this is the social learning of food preferences in rats (cf. Chap. 10), where young individuals typically prefer foods eaten by older individuals (Galef Jr. & Clark, 1971), and such preferences vary widely across populations and locations (Fragaszy & Perry, 2003). This case is best understood as 'socially biased individual leaning', specifically stimulus enhancement, shaped by the local availability of food.

In many similar cases and in many groups of animals, ranging from insects to mammals, social learning takes the form of the presence of an experienced adult increasing the likelihood of a younger, inexperienced individual acquiring certain knowledge. For instance, in local enhancement, the attention of the inexperienced animal is attracted to the environment where the experienced individual operates. This represents another instance of socially biased individual learning, as the naïve individual still learns individually how to recognise and respond to such an environment (Fragaszy & Visalberghi, 2001; Galef Jr., 1995; Heyes, 1993).

The learning of preferences and routes is inherently specific in that it does not concern the behaviour itself: it merely affects the environment where, and with respect to whom or what, it is executed. Preferences or routes can aquire the status of a cultural trait only when they are learned socially, as individual learning alone cannot give rise to population-level stability and transmission. Even then, however, socially learned preferences and localities that merely reflect what other members of the group choose should not be considered a cultural trait unless there is some inherent advantage in the choice, such as a 'discovery' of a new food-rich location by a curious individual; otherwise, their spread is more plausibly interpreted as a case of cultural drift (see above).

Scholars have raised further objections as to why social learning of food preferences in rats and similar phenomena should not be considered cultural behaviour. For instance, Galef Jr. (2003) points out that one can speak of a 'tradition' *sensu proprio* only if the behaviour is actively performed by the demonstrators and transmitted further. That probably does not hold for food preferences and the like. In addition, preferences for particular locations in the species' habitat violate the condition of independence of local ecology. Last but not least, many of these preferences are characterised by limited spread among the rest of the population because if the preference (for food type or location) is shared by many individuals, the food source is fast depleted. Similar considerations apply to the 'cultural' behaviour of milk-bottle opening in great tits and sweet-potato washing in Japanese macaques (see above).

In some cases, however, especially when learning migration routes, the process is quite complicated. For instance, orcas make use of the seasonal availability of salmon in rather complex ways and the strategies of movement in a given location are specific for individual orca pods despite their sympatric cohabitation with other groups. Here, the transmitted behaviour amounts to a highly effective way of using the available locations, and what is shared are patterns of behaviour across a relatively large area (Nichol & Shackleton, 1996). Admirable are also the periodic migration routes of ungulates, such as bighorn sheep or elk. If a population in a given location dies out, the ability to migrate is lost even after the species' subsequent reintroduction and the new population starts to move across larger areas only gradually (Jesmer et al., 2018). In such cases, reported findings are based on field observations of group movements; while social transmission of route knowledge is therefore difficult to verify directly, it is highly likely.

In other species, we do have some evidence regarding the form of social transfer of the knowledge of tracks and migration routes. For instance, rock ants learn routes by 'running in a tandem' with a more experienced individual and, as noted above, this behaviour can be considered a simple form of teaching (Caro & Hauser, 1992). Various traditions in mammals and birds have a similar character. For instance, in beluga whales, southern right whales, and humpback whales, youngsters learn from their mothers, which results in strong maternal migration traditions (O'Corry-Crowe et al., 1997). Similar migration traditions probably also exist in some birds, such as the whooping crane (Mueller et al., 2013), but in other avian species social learning does not seem to play such a prominent role in choosing migration routes (cf. Chap. 13).

The potential benefits of social learning of routes can be considerable, as attested by a study on the effectiveness of pigeons' flights home. Sasaki and Biro (2017) showed that if pigeons fly home in pairs and the pairs are gradually alternated, the efficiency of routes continuously improves. Pigeons refine their route based on own innovation or because they adopted various shortcuts from partners they have flown with before; subsequent partners can then copy these improved route. What thus spreads in the population is an increasingly effective variant (or variants). In experimental conditions where pigeons are repeatedly challenged with the same task (to fly home from point A) and have a clear motivation to learn a behaviour that minimises travel distance, this process resembles an accumulation of collective knowledge and a form of cultural evolution.

9.3.2 Bird Dialects

Possibly the largest amount of evidence of socially transmitted behaviour has been gathered in relation to acoustic communication. Nevertheless, not all vocalisation sharing amounts to cultural behaviour. Researchers therefore try to differentiate between cultural vocal traditions, emerging through active copying of specific sounds within particular groups, and various spontaneously emerging shifts in the vocal repertoire of populations, which are the result of small mistakes in vocalisation learning and form vocal dialects. While vocal dialects are studied mainly in songbirds or bats, vocal mimesis, traditions, and cultural behaviour are being considered in relation to corvid, parrot, and cetacean vocalisations.

In songbirds, the mechanism of copying songs has been described as including imitation, copying, or vocal mimesis (Galef Jr., 1988). It is clearly an instance of social learning that results in a highly accurate copy (see, e.g., Slater & Ince, 1979), although it may be less cognitively demanding than motoric imitation because the individual can directly compare own sound with the overheard model. Close copying of the local song in combination with occasional modifications or errors means that the geographical distance between two individuals of a songbird species correlates with the difference in their song. Populations that inhabit different locations therefore often have different song dialects (Baker & Cunningham, 1985).

Similar to learning preferences for particular foods or locations, a newly learned song does not expand the behavioural repertoire of the species; the bird does not learn anything additional that would change how it, or its population, interacts with the environment. Social learning merely assists the completion of a particular form of innate behaviour (song) and, in this sense, song dialects should be understood as cultural drift rather than cultural behaviour.

At the same time, some secondary functions of the similarity of song in birds from a given location could make us view song varieties as cultural traditions. For instance, it has been shown that females prefer males who can convince them about their local origin by a typically local song (Riebel et al., 2015). If a local song serves as a marker of group membership or local identity, then the significance of bird dialects may be similar to the importance of vocal culture in corvids or parrots, and ultimately even comparable to the significance of culture in great apes, including humans (see below).

The social dimension of vocal similarity is particularly pronounced in corvids and parrots. These birds are highly skilled imitators of sounds of almost any kind, although they tend to prefer living models and own species (Pepperberg, 1992; Pepperberg et al., 2000). The ability of pet parrots to learn human vocal communication is undoubtedly linked to the fact that in the absence of conspecifics or better-suited models, they consider humans their 'own kind'. Parrots' copying of human words markedly declines if we keep several individuals together and do not suppress their natural behaviour (Giret et al., 2010).

Researchers have documented various ways in which individual repertoire in parrots and corvids can be influenced by the social group. In hooded crows, for instance, it has been observed that an individual newly introduced to a social group gradually adjusts its vocalisation by copying the calls of other group members and, as this happens, the newcomer starts to be better tolerated by the group (Brown, 1985). Similar observations have been made also in cockatoos (Scarl & Bradbury, 2009) and in budgerigars, where both females and males adopt vocalisations (especially contact calls) of other members of the group (Hile et al., 2000; Hile & Striedter, 2000).

On one hand, vocal novelties in these birds usually result from various accidental processes (such as changes in group composition, playing with sounds, or copying of environmental sounds) and are thus unlikely to have a particular function or carry any adaptive advantage. Therefore, they qualify as potential products of a cultural drift. On the other hand, in corvids and parrots the shared repertoire as a whole very likely serves to identify members of particular social groups (Farabaugh & Dooling, 1996). In other words, individual calls, which on their own are merely the result of cultural drift, jointly take on a derived adaptive function and the tendency to vocal copying in general is supported by selection. Various examples moreover indicate that what facilitates unification of the vocal repertoire of a group is conformity—and that is similar to the adoption of cultural customs in chimpanzees or humans.

In all of the abovementioned cases, the spread of a trait (call) was limited to a particular location inhabited by a particular group. On the other hand, a study of Enggist-Dueblin and Pfister (2002) on corvids showed that the transmission of calls is more pronounced within the sexes (which form an open and geographically undefined group) than within couples: one can thus speak of cultural traditions. In terms of stability, the sex-typical calls are not very long-lasting and it seems that—like in cetaceans (on whom more below)—they are somewhat prone to 'fashion waves'.

9.3.3 *Traditions in Cetaceans*

The most striking example of repertoire sharing in groups are found in whale vocalisations, which are often interpreted as cultural phenomena, in part perhaps thanks to the strange charisma of whales and their songs. Some cetaceans, for instance humpback whales, are unique in that they fully and uniformly share songs within populations spread across thousands of miles—and these shared songs change from one season to the next. What is specific is also the nature of these changes: in the northern hemisphere, the changes take the form of a gradual evolution, while in the southern hemisphere, researchers have observed 'revolutionary' replacements of one song for another. These changes spread eastwards through the Pacific from one group to the next rather like fashion waves. The western Pacific, near the Australian shores, is the home of larger populations which are probably a significant source of cultural novelties for the smaller populations in the east, in the central part of the ocean (Zandberg et al., 2021). Quite astonishing in comparison to birds is the speed at which new songs spread over vast distances. Nevertheless, nothing suggests that these vocal dialects are of a qualitatively different origin than the dialects of songbirds. The speed of horizontal spread could be simply due to the physics of sound propagation in deep sea or the consequence of the vast lengths of humpback whale migration routes. In humpback whales, like in many songbirds, singing is largely restricted to males, which implies a function in sexual selection, but the specific role of vocal copying in these cetaceans remains unclear (Herman, 2017).

The situation is different in cetaceans whose vocal traditions are transmitted vertically in sympatrically living groups. For instance, in orcas, who have a resident lifestyle, vocal dialects are linked to matrilineal groups although the members of these lines are in vocal contact also with other lines (Ford, 1991; Strager, 1995). Sperm whales, too, maintain their vocal dialects along the maternal line, although their groups live sympatrically and travel across entire oceans (Cantor et al., 2015; Watkins & Schevill, 1977). It seems quite clear that in this case, the sharing of vocalisation establishes a group identity and enables identification of clan members and their mutual communication. It is thus logical to view it as a cultural trait (Cantor et al., 2015).

Cultural behaviour of cetaceans is highly varied. Aside from social transmission of vocalisations and culturally determined preferences for certain foods, locations, and routes, cetaceans have cultural traits in practically all of the main areas known in other animals. When investigated in greater detail, the richness of cetacean culture may turn out to be comparable to culture in chimpanzees or orangutans. For instance, in many cetacean species we see traditions pertaining to hunting strategies. In sperm whales, it had been observed that they started to use, quite quickly and in great numbers, a new hunting technique based on lobtailing, where the whale beats the water surface with its tail around a school of fish, who are thus driven into a small space. In a part of sperm whale population, lobtailing had replaced an earlier technique, where a similar effect was achieved by a ring of bubbles (Weinrich et al., 1992). Emergence of this new technique was probably linked to changes in the composition of prey species but the nature of its spread, which has been described starting with the original occurrence in just one or two individuals and is very well documented by observation over three decades, points to social transmission.

Dolphins even seem to use tools. For instance, common bottlenose dolphins hunt for fish that live on the ocean floor using sea sponges. The sponge functions as a tool in the sense that the dolphin plucks the largest sponge it can find, places it on its nose, and swims with it close to the surface: the large shadow cast by the sponge then scares the fish near the ocean floor. According to Krützen et al. (2005), only one matrilineal clan engages in this behaviour but it lives sympatrically with other clans, forming an open social network. Other evidence points to traditional social behaviour in cetaceans (e.g., greetings rituals), socially transmitted standards of spatial behaviour (how they move, the amount of space between individuals in a group, distance transversed in a day, etc.), but these observations are so far more of an anecdotal nature (Rendell & Whitehead, 2001).

A very interesting category of cultural habits of cetaceans is interspecies collaboration during hunting. In Laguna, Brazil, common bottlenose dolphins were documented to collaborate with fishermen: the dolphins drive fish towards the fishermen's nets, let them know when they are finished, and then they feast on leftovers. A social analysis of this population had shown that the dolphins who engage in this behaviour have stronger social ties among each other than with other members of the population, who just hamper the efforts of both of the hunting parties (Daura-Jorge et al., 2012). It is possible that this social custom, too, is transmitted matrilineally because it has been practiced by part of the population for over three generations and one hundred years (Pryor & Lindbergh, 1990). Unsurprisingly, a corresponding technique on the part of fishermen is transmitted vertically by social learning from older generations (Daura-Jorge et al., 2012). Another case of cooperative hunting with fishermen has been described for instance in the Irrawaddy dolphin in the Ayeyarwady River in Myanmar (Tun, 2014).

Our knowledge of the variety and complexity of cetacean traditions is still incomplete and limited by the exclusively ethnographic nature of findings, which says little about the mechanisms of social transmission of dolphin or whale traditions. In contrast, when it comes to great apes (and especially chimpanzees), many cultural customs have been known for decades. Recent research focuses on laboratory study of complex social learning and the dynamics of culture transfer and addresses the limiting factors which prevent chimpanzees from developing a culture similar to the human one.

9.3.4 The Material Culture of Chimpanzees

Individual chimpanzee populations in their specific ways modify and use branches and sticks to bore and investigate holes, hunt for termites, acquire honey and resin, but also to engage in courtship (including loud manipulation with vegetation to attract the attention of a potential mate), invite to play, comb hair, drive off insects, or for grooming. They use leaves to wipe off food or to form mats and padding, while chewed leaves are used to collect liquids, and stones to break nuts. Chimpanzees have specific ways of engaging in courtship, play, and other social customs including dances, holding each other during grooming (by palms, fists, or holding onto the partner's forearm), hygienic habits that do not involve tool use, and population-specific gestures and calls (Pollick & de Waal, 2007; Whiten et al., 1999).

A wave of laboratory research into social learning in chimpanzees was initiated by the finding that despite this wide repertoire of cultural behaviours, chimpanzees find imitation very difficult. Tomasello et al. (1987) have shown chimpanzees how to use a stick to spear a food item. The chimpanzees then tried to get the food themselves. Some of their attempts looked like faithful copies of the demonstrated behaviour, but these were just very simple actions. When the demonstrator used a stick to get food from a less accessible place, the chimpanzees did not copy the behaviour fully. It seems they understood that one should use the stick as a tool but could not copy a more complex sequence of movements. Based on some further studies, Tomasello concluded that chimpanzees in such cases do not imitate but merely emulate, that is, they just learn what the tool is good for (Tomasello, 1996). In this, chimpanzees are thus fundamentally unlike humans, who are very good at imitation. This could be a significant limitation to transmission of cultural traditions because emulation does not necessarily lead to sufficiently faithful copying (Tomasello, 1999).

Learning by emulation rather than by imitation is not necessarily a less intelligent or less advantageous solution. It is based on understanding the temporal, spatial, or causal relation between the tool and solution of the problem (Tomasello & Call, 1997). Sometimes, the use of emulation rather than imitation has clear advantages, for instance when the demonstrator is showing a somewhat ineffective technique. Unlike children, who faithfully imitate even when some actions are clearly superfluous (overimitation; Lyons et al., 2007; Lyons et al., 2011), chimpanzees do not have such a strong tendency to imitation. Still, when it is unclear whether an action is needed to achieve the goal, they do resort to imitation. This has been demonstrated in a study where chimpanzees were twice shown an almost identical apparatus that was hiding a reward. The box enabled several operations with various movable parts and apertures on different sides but the upper and lower part were separated by a divider, so that operations in the upper part had no effect on what happened in the lower part where the reward was hidden. The sole difference between the two apparatuses was that the first was not transparent, so one could not see the divider, while the second was transparent. In the first case, the chimpanzees imitated actions of the demonstrator including those involving the upper part of the apparatus, while in the second version of the task, they skipped manipulations involving the upper part. Unlike children, who in both versions of the task faithfully copied all of the demonstrated actions, chimpanzees were efficiently switching between imitation and emulation (Horner & Whiten, 2005) (Fig. 9.1).

It is unclear to what extent emulation leads to a less faithful transmission than imitation. Theoretical models show that for simpler tasks, especially such where the relation between the process and its outcome is clear, emulation and imitation can lead to similarly stable and accurate results (Caldwell et al., 2012; Wasielewski, 2014).

Experimental studies confirm that, in some situations, chimpanzees use imitation, and the resulting transmission in groups can be faithful enough to preserve cultural traditions. Whiten et al. (2005) taught two chimpanzee females, each from a different group, two different ways of acquiring food from an apparatus. One female used a stick to prise off a divider, while the other used a stick to get the reward out through a hole. In each group, the solution demonstrated by the relevant female spread quickly and almost all members of the group used it. In other words, the chimpanzees were imitating, not emulating the solution. To make sure that transmission is accurate even beyond the immediate observers of the original demonstrator, the researchers then formed a transmission chain: a demonstrator showed the solution always to one individual, who became the demonstrator for the following individual, etc. The behaviour remained essentially unchanged over the subsequent six transmissions and further studies confirmed that such transmission functions even between populations and that this applies also to a technique requiring a sequence of two steps (Whiten et al., 2007).

A common objection to the hypothesis of cultural transmission in chimpanzees was that chimpanzees cannot imitate anything they could not figure out on their own, individually. Hopper et al. (2007) presented a large group of chimpanzees with the same stick and box apparatus as Whiten et al. (2005) did, and among the 56 individuals tested only one found an effective way of accessing the reward. The authors thus concluded that behaviours which spread among chimpanzees by

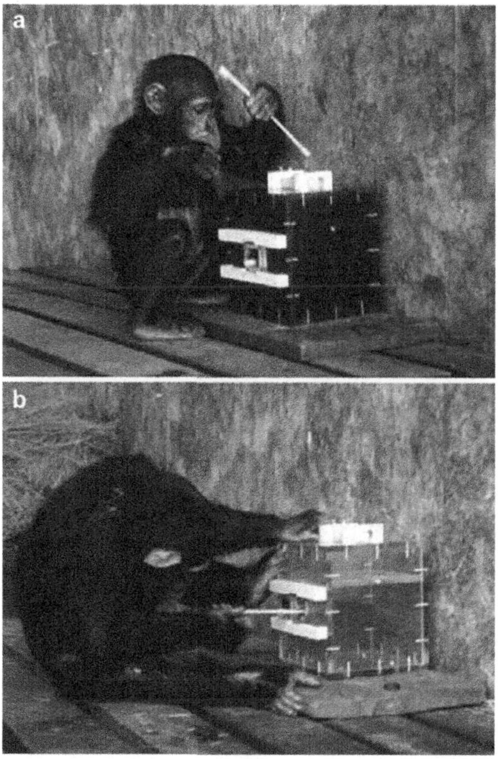

Fig. 9.1 Experiments on imitation versus emulation with chimpanzees: a) the subject inserts a tool into a top irrelevant hole of an opaque apparatus; b) the subject inserts a tool into the relevant hole in the front of the clear apparatus to retrieve the food reward (Horner & Whiten, 2005)

imitation tend to be so simple that over time, an innovator can find the solution by individual learning while most individuals learn this behaviour socially, by imitation.

Finally, some research has also challenged the assumption that chimpanzees (unlike people) pay more attention to the tool than to the hand handling it (as assumed by Tomasello, 1996). Hopper et al. (2008) used a design that employed 'ghosts', that is, invisible actors manipulating tools. It turned out that, except for simple actions such as moving an object left or right, chimpanzees (just like human children) were unable to imitate the outcome based solely on movement of the tool, without a living demonstrator. Field observations, too, indicate that chimpanzees pay attention to movements. Young chimpanzees learn how to use stones to crack nuts by observing their mothers and other older individuals. Crucially, when learning, they often spontaneously imitate the movements of limbs even without the appropriate tool or a nut to be cracked (Marshall-Pescini & Whiten, 2008a).

Although far less studied, another significant difference between humans and chimpanzees may be in the behaviour of the demonstrator (teacher), that is, in the tendency to teach. Dean et al. (2012) compared whether and how pairs of chimpanzees, capuchins, and children help each other learn how to open the door of an apparatus by a sequence of three steps. Children were notably better, and the difference was not only in their tendency to imitate. What was even more important was that child demonstrators supported, motivated (e.g., by rewards), and actively tried to teach the other child, for instance by using verbal instructions.

9.4 Is Animal Culture Subject to Evolution?

Recent research seems to confirm that chimpanzees conservatively adhere to the first solution learned. Hrubesch et al. (2009) first taught chimpanzees how to get food out of a maze using a stick, but then some individuals figured out that they could rattle the entire maze and access the food reward even faster. Researchers then prevented chimpanzees from using the original technique and observed whether they would learn the more effective rattling technique from the individuals who invented it. The chimpanzees, however, persisted in their efforts to apply the technique that no longer worked. Further studies confirmed that animals, and chimpanzees in particular, stick to the first-learned solution even when it is unsuitable, especially if they had invented it or it was demonstrated by a dominant individual (Hopper et al., 2011; Hrubesch et al., 2009; Marshall-Pescini & Whiten, 2008b). Interestingly, Lehner et al. (2011) found a greater flexibility in orangutans: they used a setup where the orangutans were getting syrup from a tube in different ways and gradually had to modify earlier learned techniques which in the subsequent tasks no longer worked.

Archaeological evidence likewise suggests that some chimpanzee traditions, such as the use of stones to crack nuts, are ancient (4300 years). Importantly, it seems they did not significantly change over time, they show no signs of accumulation, and at least sometimes, some individual could discover the technique in question on its own (Mercader et al., 2007).

The likelihood of spread of newly modified behaviours may also be limited by the dominant form of transmission. In chimpanzees, the primary model for skills such as nut cracking is usually the mother. Moreover, the skill seems to be learned within a sensitive period: individuals who do not learn how to crack nuts by the age of five just turn to gathering the leftovers of those who did learn it (Biro et al., 2003). Abilities such as learning from other models (older juveniles or adults), imitation of cultural customs even after reaching 5 years of age, or selective imitation of successful demonstrators are quite routinely observed in domesticated chimpanzees in laboratory conditions. They significantly increase the likelihood of spread of new behaviours (Watson et al., 2018) but they are not typical of chimpanzees in the wild (Hayashi & Matsuzawa, 2003; Hirata & Celli, 2003; Sousa et al., 2003).

On the other hand, naturally occurring modifications of existing cultural customs documented in capuchin monkeys suggest that animals sometimes can overcome their conservative tendencies. Perry et al. (2003) have described modifications of the so-called play with a toy, where two monkeys take turns trying to retrieve an object from each other's mouths. These variations involve subtle differences in technique and social coordination—such as whether the object is taken by hand or mouth, and whether the individuals engage in eye contact or tactile interaction. It is uncertain whether such modifications, which form through the inventiveness of juveniles, can be transmitted to the following generations as a tradition.

Importantly, it remains unknown whether eventual changes in animal culture tend to accumulate into more complex procedures and customs or, alternatively, animals socially learn only tasks whose complexity does not exceed what they could learn individually, non-socially. Some reports do point to an accumulation of cultural novelties but they are rare and pertain to relatively specific areas of behaviour. For instance, Sasaki and Biro (2017) have demonstrated an accumulation of knowledge in route learning by pigeons but that may be a specific area of social learning where accumulation can take place more easily.

A controversial finding is the accumulation of skills pertaining to the production of tools in New Caledonian crows. Hunt and Gray (2004) are convinced that these birds originally used broad twigs. Later, this behaviour was innovated, and the crows started to use slim twigs with an increasing amount of branching. But it seems that in crows the ability to use and even to modify such tools is largely innate (Kenward et al., 2005, 2006). Evidence of social learning of tool use in this species in the wild is effectively absent and the role of individual invention seems to be substantive (Holzhaider et al., 2010).

The most convincing evidence pointing to accumulation of cultural behaviour, albeit quite a rudimentary one, after all comes from the chimpanzees. It is the case of termite hunting technique observed in the Goualougo, where individuals not only make holes in termite mounds (by piercing them with a stick using an arm and a foot) but also crush the end of the stick to make a sort of brush, which they sink into the hole to catch more termites. Given that the first part of this behaviour has been observed elsewhere but creation of the brush-like end only in this locality, this behaviour probably emerged by modification in the sense of improvement of a pre-existing behaviour (Sanz et al., 2009).

9.5 Crossing the Threshold: Why Cumulative Culture Emerged Only in Humans

Laboratory studies and field observations of animal culture are largely motivated by a desire to understand why culture has developed to such an extent in humans but not in other animals. Unlike chimpanzees, humans have at some point in the past crossed a threshold after which culture started to accumulate. Since then, culture has been evolving and biological evolution of humans continued in coevolution with culture (Richerson & Boyd, 2005). Our current culture forms a specific niche for humans, a niche that we have actively constructed ourselves (Laland et al., 1996) (see Chap. 13). Human minds have adapted (and are shaped since birth by the cultural and social environment) to learning during ontogeny what is needed for a successful life in culture, as well as to supporting offspring and other group members in achieving the same.

The study of animal culture has so far only partially answered questions about the key predispositions for a highly developed culture such as we find in humans. It sheds some light on why, although animals are capable of socially transmitting behavioural traditions, cultural behaviour in animals does not have a tendency to accumulate, and why even in chimpanzees social learning appears limited to a level of complexity at which sufficiently curious and capable individuals could, at least occassionally, from time to time, discover the solution on their own. The ability to accumulate modifications is a key, and perhaps unique, characteristic of human culture (Tomasello, 1994; Boyd & Richerson, 1996).

It is starting to become apparent that these limitations in animals may be cognitive only in part. In chimpanzees, for instance, researchers were for a long time unable to prove the ability to imitate in laboratory conditions. It seemed, therefore, that imitation is a basic cognitive ability that is unique to humans and sets humans apart from animals. Nowadays, we know that, unlike children, chimpanzees flexibly switch between imitation- and emulation-based solutions depending on which is more effective in a given situation. The primary reason why chimpanzees do not use imitation in their natural environment is therefore probably not their cognitive inability to copy observed movements, but rather their different cognitive approach to problem-solving.

It should be noted that imitation or teaching are not necessary for the maintenance of some traditions typical of animals. In an unchanging environment or when, for instance, learning routes or preferences, it seems that stimulus or local enhancement, eventually combined with conditioning, function equally well (Matthews et al., 2010; Pradhan et al., 2012; van der Post & Hogeweg, 2008). Moreover, it turns out that imitation alone is insufficient for ensuring the stability of cultural behaviour in a population over time, as summarised by Claidière and Sperber (2010).

For culture to evolve towards greater complexity, other mechanisms and conditions must be in place. Scientists increasingly believe that social norms, prosocial behaviour, theory of mind, language, egalitarian tendencies, and the like may be co-responsible for the development of culture in humans and its stagnation in other species (Dean et al., 2014; Heyes, 2021; Tomasello, 2009; Whiten, 2017) (cf. Chap. 7).

Deacon (2012) argued that language is a complex intertwined system of symbols and syntactic rules that facilitate an efficient use and interpretation of these symbols. He proposed that humans are a 'symbolic species' that lives in an artificial niche formed by language. It is thus language that drives the adaptive advancement of various cognitive functions such as joint attention, empathy, a tendency to infer abstract rules, and an expectation of bidirectional transitive relations when learning associations. Furthermore, humans exhibit an extraordinary ability for vocal mimicry, articulatory and vocal control, and a natural inclination to perceive things as symbols. However, Deacon did not specifically address the question of why humans, unlike other species, were able to develop such a symbolic system.

Long-term language projects with apes had shown that, under certain conditions, they can learn language to a remarkable level. Although Deely (2002) argued that animals respond to signs without understanding them as representations (because they lack ‘semiotic consciousness’), evidence from interspecies communication studies suggests that some animals may engage with symbols more flexibly and meaningfully than his account implies (e.g., Pepperberg, 2010; Savage-Rumbaugh, 1996). Such studies challenge the idea that symbolic understanding is an exclusively human trait. Research shows that species such as the common chimpanzees, bonobos, and grey parrots can flexibly and creatively use lexigrams, vocal labels, or gestures not only to communicate about specific things, but also to apply them in new situations, transfer their meaning, and occasionally even inquire about new symbols or create them. Their significant limitation, however, is the absence of a tendency to name things in their surroundings and to share with others their opinion unless it is to ask for something. In a similar vein, a recent study by Roubalová et al. (2024) found that a key qualitative difference between the vocabularies of children and grey parrots kept as pets is that children, unlike parrots, include a greater proportion of object labels among their first words. It remains an open question whether this difference is due to an adaptation of human cognition in the symbolic niche that predisposes us to learning vocal labels with symbolic meaning, or whether some other preadaptation may have triggered this difference, such as, for example, the human tendency to share one’s views and goals with others. Further insight into this issue might come from Tomasello’s (2020) research synthesis which suggests that children, unlike the young of other animals, tend to adjust their emotional state and behaviour to the emotional state and behaviour of the adults in their group. Tomasello highlights this as a highly adaptive trait with respect to the ‘ultra-cooperative ontogenetic niche’ in which children grow up (cf. Chap. 13). Tomasello emphasises that cooperation, rather than the symbolic cultural system, acts as one of the key selection pressures in the evolution of human mind. Shared attention and cooperative communication are important dispositions which enable children to attune themselves to others.

In chimpanzees, we also miss evidence of teaching. It is possible that lack of support or pressure to learn from other members of the community is another factor that prevents imitation and accumulation of modifications of social customs from becoming the norm in chimpanzee society. In contrast, human social environment, fine-tuned to supporting and supervision of enculturation of other group members, creates a degree of psychological pressure, which in turn motivates naïve members of the group. Perhaps that is why people are less satisfied than other animals with existing solutions that may lead to the desired result but are suboptimal (Laland & Hoppitt, 2003). Animals tend to hold on to the existing solution (Hrubesch et al., 2009). After all, even in humans the ability to flexibly modify solutions is in evolutionary terms relatively new: it dates to about 1.6 million years ago, when Acheulean industry incorporated bilaterally worked tools (Goren-Inbar, 2011). Moreover, a tendency toward conservatism strongly persists in some recent human societies and individual behaviours (Gruber et al., 2011; Hanus et al., 2011) (cf. Chaps. 10, 15).

Finally, findings pertaining to culture in animals indicate what may have been the primary initial function of culture in humans, the function thanks to which people had crossed the threshold towards a complex and cumulative human culture. The likely ultimate function of culture in the communally living ape chimpanzee, just as in the communally living human ape (and probably also in other species of primates, birds, and cetaceans), is social identity. The particular cultural customs of chimpanzees probably function as cultural identifiers which jointly indicate who their bearers are and where they come from, just like people are identified by their dialect, dress, and the like (McGrew, 2009) (cf. Chap. 5). Whiten (2011) points out that chimpanzee culture is not limited to particular behavioural manifestations: each community is characterised by certain general cognitive representations that predetermine it to particular cultural solutions. It can be the focus on solving problems using tools (as in west African chimpanzees) or without tools (as in east African chimpanzees), eventually the tendency to use a particular tool (sticks versus leaves; Lycett et al., 2010). It is thus possible that the cultural identity of chimpanzees exists also on a conceptual level. In this crucial respect, the culture of chimpanzees—and perhaps also other animals—need not be all that different from human culture after all.

9.6 Conclusions

The study of animal culture suggests that human culture has evolutionary roots and offers insights into what early human culture may have looked like. Across many non-human species, individual groups maintain distinct traditions—whether in the form of tool use, social customs, migration routes, preferences, or vocalisations.

At the same time, it seems that although animals are capable of socially transmitting behaviour that may be described as cultural, this behaviour usually does not accumulate over generations, and what is socially learned does not appear to be substantially more complex than what animals can also acquire through individual learning. The ability to accumulate modifications seems to be a unique characteristic of human culture.

So far, research has only partly answered questions concerning the key predispositions for a highly developed culture such as that observed in humans. Many influential theories suggest that these predispositions do not necessarily need to be cognitive, and instead view the cognitive primacy of humans as a consequence of the pressures of a specific life niche—one in which culture plays a major role. For the onset of cultural evolution in humans, it was probably not only some complex forms of social learning and teaching that mattered, but also other dispositions: sensitivity to social norms, prosocial behaviour, theory of mind, language, and others. Most of these traits likely exist in humans due to the highly cooperative, social, and communal way of life of our species.

References

Acerbi, A., & Parisi, D. (2006). Cultural transmission between and within generations. *Journal of Artificial Societies and Social Simulation, 9*(1), 1-16.

Acerbi, A., Tennie, C., & Nunn, C. L. (2011). Modeling imitation and emulation in constrained search spaces. *Learning & Behavior, 39*(2), 104–114.

Allen, J., Weinrich, M., Hoppitt, W., & Rendell, L. (2013). Network-based diffusion analysis reveals cultural transmission of lobtail feeding in humpback whales. *Science, 340*(6131), 485–488.

Aplin, L. M., Farine, D. R., Morand-Ferron, J., Cockburn, A., Thornton, A., & Sheldon, B. C. (2015). Experimentally induced innovations lead to persistent culture via conformity in wild birds. *Nature, 518*(7540), 538–541.

Auersperg, A. M., Szabo, B., von Bayern, A. M., & Kacelnik, A. (2012). Spontaneous innovation in tool manufacture and use in a Goffin's cockatoo. *Current Biology, 22*(21), R903–R904.

Baker, M. C., & Cunningham, M. A. (1985). The biology of bird-song dialects. *Behavioral and Brain Sciences, 8*(1), 85–100.

Bandini, E., & Tennie, C. (2017). Spontaneous reoccurrence of 'scooping', a wild tool-use behaviour, in naïve chimpanzees. *PeerJ, 5*, e3814.

Battesti, M., Moreno, C., Joly, D., & Mery, F. (2012). Spread of social information and dynamics of social transmission within drosophila group. *Current Biology, 22*(4), 309–313.

Biro, D., Inoue-Nakamura, N., Tonooka, R., Yamakoshi, G., Sousa, C., & Matsuzawa, T. (2003). Cultural innovation and transmission of tool use in wild chimpanzees: Evidence from field experiments. *Animal Cognition, 6*(4), 213–223.

Boogert, N. J., Reader, S. M., & Laland, K. N. (2006). The relation between social rank, neophobia and individual learning in starlings. *Animal Behavior, 72*(6), 1229–1239.

Boyd, R., & Richerson, P. J. (1988). *Culture and the evolutionary process*. University of Chicago Press.

Boyd, R., & Richerson, P. J. (1996). Why culture is common, but cultural evolution is rare. In W. G. Runciman, J. M. Smith, & R. I. M. Dunbar (Eds.), *Evolution of social behaviour patterns in primates and man* (pp. 77–93). Oxford University Press.

Brosnan, S. F., & Hopper, L. M. (2014). Psychological limits on animal innovation. *Animal Behaviour, 92*, 325–332.

Brown, E. D. (1985). The role of song and vocal imitation among common crows (*Corvus brachyrhynchos*). *Zeitschrift für Tierpsychologie, 68*(2), 115–136.

Caldwell, C. A., Schillinger, K., Evans, C. L., & Hopper, L. M. (2012). End state copying by humans (*Homo sapiens*): Implications for a comparative perspective on cumulative culture. *Journal of Comparative Psychology, 126*(2), 161–169.

Cantor, M., Shoemaker, L. G., Cabral, R. B., Flores, C. O., Varga, M., & Whitehead, H. (2015). Multilevel animal societies can emerge from cultural transmission. *Nature Communications, 6*(1), 8091.

Caro, T. M., & Hauser, M. D. (1992). Is there teaching in nonhuman animals? *Quarterly Review of Biology, 67*(2), 151–174.

Claidière, N., & Sperber, D. (2010). Imitation explains the propagation, not the stability of animal culture. *Proceedings of the Royal Society B: Biological Sciences, 277*(1681), 651–659.

Coussi-Korbel, S., & Fragaszy, D. M. (1995). On the relation between social dynamics and social learning. *Animal Behaviour, 50*(6), 1441–1453.

Daura-Jorge, F. G., Cantor, M., Ingram, S. N., Lusseau, D., & Simões-Lopes, P. C. (2012). The structure of a bottlenose dolphin society is coupled to a unique foraging cooperation with artisanal fishermen. *Biology Letters, 8*(5), 702–705.

Deacon, T. W. (2012). Beyond the symbolic species. In T. Schilhab, F. Stjernfelt, & T. Deacon (Eds.), *The symbolic species evolved* (Vol. 6). Springer.

Dean, L. G., Kendal, R. L., Schapiro, S. J., Thierry, B., & Laland, K. N. (2012). Identification of the social and cognitive processes underlying human cumulative culture. *Science, 335*(6072), 1114–1118.
Dean, L. G., Vale, G. L., Laland, K. N., Flynn, E., & Kendal, R. L. (2014). Human cumulative culture: A comparative perspective. *Biological Reviews, 89*(2), 284–301.
Deely, J. (2002). *What distinguishes human understanding?* St. Augustine's Press.
Derex, M., Godelle, B., & Raymond, M. (2012). Social learners require process information to outperform individual learners. *Evolution, 67*(3), 688–697.
Ehn, M., & Laland, K. (2012). Adaptive strategies for cumulative cultural learning. *Journal of Theoretical Biology, 301*, 103–111.
Enggist-Dueblin, P., & Pfister, U. (2002). Cultural transmission of vocalizations in ravens, *Corvus corax*. *Animal Behaviour, 64*(6), 831–841.
Farabaugh, S., & Dooling, R. J. (1996). Acoustic communication in parrots: Laboratory and field studies of budgerigars, *Melopsittacus undulatus*. In D. E. Kroodsma & E. H. Miller (Eds.), *Ecology and evolution of acoustic communication in birds* (pp. 97–118). Cornell University Press.
Fisher, J., & Hinde, R. A. (1949). The opening of milk bottles by birds. *British Birds, 42*(11), 347–357.
Flynn, E. (2010). Underpinning collaborative learning. In B. Sokol, U. Muller, J. Carpendale, A. Young, & G. Iarocci (Eds.), *Self and social regulation: Social interaction and the development of social understanding and executive functions* (pp. 312–336). Oxford University Press.
Ford, J. K. (1991). Vocal traditions among resident killer whales (*Orcinus orca*) in coastal waters of British Columbia. *Canadian Journal of Zoology, 69*(6), 1454–1483.
Fragaszy, D. M., & Perry, S. (2003). Towards a biology of traditions. In D. M. Fragaszy & S. Perry (Eds.), *The biology of traditions: Models and evidence* (pp. 1–32). Cambridge University Press.
Fragaszy, D. M., & Visalberghi, E. (2001). Recognizing a swan: Socially-biased learning. *Psychologia, 44*, 82–98.
Franks, N. R., & Richardson, T. (2006). Teaching in tandem-running ants. *Nature, 439*(7073), 153–153.
Galef, B. G., Jr. (1988). Imitation in animals: History, definition, and interpretation of data from the psychological laboratory. In T. R. Zentall & B. G. Galef Jr. (Eds.), *Social learning: Psychological and biological perspectives* (pp. 3–28). Psychological Press.
Galef, B. G., Jr. (1992). The question of animal culture. *Human Nature, 3*(2), 157–178.
Galef, B. G., Jr. (1995). Why behaviour patterns that animals learn socially are locally adaptive. *Animal Behaviour, 49*(5), 1325–1334.
Galef, B. G., Jr. (2003). "Traditional" foraging behaviors of brown and black rats (*Rattus norvegicus* and *Rattus rattus*). In D. M. Fragaszy & S. Perry (Eds.), *The biology of traditions: Models and evidence* (pp. 159–186). Cambridge University Press.
Galef, B. G., Jr. (2009). Culture in animals? In K. N. Laland & B. G. Galef Jr. (Eds.), *The question of animal culture* (pp. 222–246). Harvard University Press.
Galef, B. G., Jr., & Clark, M. M. (1971). Social factors in the poison avoidance and feeding behavior of wild and domesticated rat pups. *Journal of Comparative and Physiological Psychology, 75*(3), 341–357.
Giret, N., Péron, F., Lindová, J., Tichotová, L., Nagle, L., Kreutzer, M., Tymr, F., & Bovet, D. (2010). Referential learning of French and Czech labels in African grey parrots (*Psittacus erithacus*): Different methods yield contrasting results. *Behavioural Processes, 85*(2), 90–98.
Goodall, J. (1964). Tool-using and aimed throwing in a community of free-living chimpanzees. *Nature, 201*(4926), 1264–1266.
Goren-Inbar, N. (2011). Culture and cognition in the Acheulian industry: A case study from Gesher Benot Ya'aqov. *Philosophical Transactions of the Royal Society B, 366*(1567), 1038–1049.
Gruber, T., Muller, M. N., Reynolds, V., Wrangham, R., & Zuberbühler, K. (2011). Community-specific evaluation of tool affordances in wild chimpanzees. *Scientific Reports, 1*(1), 128.

Hanus, D., Mendes, N., Tennie, C., & Call, J. (2011). Comparing the performances of apes (*Gorilla gorilla*, *Pan troglodytes*, *Pongo pygmaeus*) and human children (*Homo sapiens*) in the floating peanut task. *PLoS One, 6*(6), e19555.

Hayashi, M., & Matsuzawa, T. (2003). Cognitive development in object manipulation by infant chimpanzees. *Animal Cognition, 6*(4), 225–233.

Herman, L. M. (2017). The multiple functions of male song within the humpback whale (*Megaptera novaeangliae*) mating system: Review, evaluation, and synthesis. *Biological Reviews, 92*(3), 1795–1818.

Heyes, C. M. (1993). Imitation, culture, and cognition. *Animal Behaviour, 46*(5), 999–1010.

Heyes, C. M. (2021). Imitation and culture: What gives? *Mind & Language, 38*(1), 42–63.

Hile, A. G., Plummer, T. K., & Striedter, G. F. (2000). Male vocal imitation produces call convergence during pair bonding in budgerigars, *Melopsittacus undulatus*. *Animal Behaviour, 59*(6), 1209–1218.

Hile, A. G., & Striedter, G. F. (2000). Call convergence within groups of female budgerigars (*Melopsittacus undulatus*). *Ethology, 106*(12), 1105–1114.

Hinde, R. A., & Fisher, J. (1951). Further observations on the opening of milk bottles by birds. *British Birds, 44*, 393–396.

Hirata, S., & Celli, M. L. (2003). Role of mothers in the acquisition of tool-use behaviours by captive infant chimpanzees. *Animal Cognition, 6*(4), 235–244.

Hirata, S., Watanabe, K., & Masao, K. (2008). "Sweet-potato washing" revisited. In T. Matsuzawa (Ed.), *Primate origins of human cognition and behavior* (pp. 487–508). Springer.

Holzhaider, J. C., Hunt, G. R., & Gray, R. D. (2010). Social learning in New Caledonian crows. *Learning & Behavior, 38*, 206–219.

Hopper, L. M., Lambeth, S. P., Schapiro, S. J., & Whiten, A. (2008). Observational learning in chimpanzees and children studied through 'ghost' conditions. *Proceedings of the Royal Society B, 275*(1636), 835–840.

Hopper, L. M., Schapiro, S. J., Lambeth, S. P., & Brosnan, S. F. (2011). Chimpanzees' socially maintained food preferences indicate both conservatism and conformity. *Animal Behaviour, 81*(6), 1195–1202.

Hopper, L. M., Spiteri, A., Lambeth, S. P., Schapiro, S. J., Horner, V., & Whiten, A. (2007). Experimental studies of traditions and underlying transmission processes in chimpanzees. *Animal Behaviour, 73*(6), 1021–1032.

Horner, V. K., & Whiten, A. (2005). Causal knowledge and imitation/emulation switching in chimpanzees (*pan troglodytes*) and children. *Animal Cognition, 8*(3), 164–181.

Hrubesch, C., Preuschoft, S., & van Schaik, C. (2009). Skill mastery inhibits adoption of observed alternative solutions among chimpanzees (*Pan troglodytes*). *Animal Cognition, 12*(2), 209–216.

Hunt, G. R., & Gray, R. D. (2004). The crafting of hook tools by wild New Caledonian crows. *Proceedings of the Royal Society B, 271*(Suppl_3), S88–S90.

Jesmer, B. R., Merkle, J. A., Goheen, J. R., Aikens, E. O., Beck, J. L., Courtemanch, A. B., Hurley, M. A., McWhirter, D. E., Miyasaki, H. M., Monteith, K. L., & Kauffman, M. J. (2018). Is ungulate migration culturally transmitted? Evidence of social learning from translocated animals. *Science, 361*(6406), 1023–1025.

Kacelnik, A. (2009). Tools for thought or thoughts for tools? *Proceedings of the National Academy of Sciences of the USA, 106*(25), 10071–10072.

Kawai, M. (1965). Newly acquired pre-cultural behavior of the natural troop of Japanese monkeys on Koshima islet. *Primates, 6*(1), 1–30.

Kawamura, S. (1954). On a new type of feeding habit which developed in a group of wild Japanese monkeys (in Japanese). *Seibutsu-shinka, 2*, 11–13.

Kendal, R., Hopper, L. M., Whiten, A., Brosnan, S. F., Lambeth, S. P., Schapiro, S. J., & Hoppitt, W. (2015). Chimpanzees copy dominant and knowledgeable individuals: Implications for cultural diversity. *Evolution and Human Behavior, 36*(1), 65–72.

Kenward, B., Rutz, C., Weir, A. A. S., & Kacelnik, A. (2006). Development of tool use in New Caledonian crows: Inherited action patterns and social influences. *Animal Behaviour, 72*(6), 1329–1343.

Kenward, B., Weir, A. A. S., Rutz, C., & Kacelnik, A. (2005). Tool manufacture by naive juvenile crows. *Nature, 433*(7022), 121–121.

Koops, K., Visalberghi, E., & van Schaik, C. P. (2014). The ecology of primate material culture. *Biology Letters, 10*(11), 20140508.

Kroeber, A. L. (1928). Sub-human culture beginnings. *The Quarterly Review of Biology, 3*(3), 325–342.

Krützen, M., Mann, J., Heithaus, M. R., Connor, R. C., Bejder, L., & Sherwin, W. B. (2005). Cultural transmission of tool use in bottlenose dolphins. *Proceedings of the National Academy of Sciences of the USA, 102*(25), 8939–8943.

Laland, K. N., & Hoppitt, W. (2003). Do animals have culture? *Evolutionary Anthropology, 12*(3), 150–159.

Laland, K. N., Odling-Smee, F. J., & Feldman, M. W. (1996). The evolutionary consequences of niche construction: A theoretical investigation using two-locus theory. *Journal of Evolutionary Biology, 9*(3), 293–316.

Laland, K. N., & Reader, S. M. (1999). Foraging innovation in the guppy. *Animal Behaviour, 57*(2), 331–340.

Laland, K. N., & Williams, K. (1998). Social transmission of maladaptive information in the guppy. *Behavioral Ecology, 9*(5), 493–499.

Laland, K. N., Atton, N., & Webster, M. M. (2011). From fish to fashion: experimental and theoretical insights into the evolution of culture. *Philosophical Transactions of the Royal Society B: Biological Sciences, 366*(1567), 958–968.

Lehner, S. R., Burkart, J. M., & van Schaik, C. P. (2011). Can captive orangutans (*Pongo pygmaeus abelli*) be coaxed into cumulative build-up of techniques? *Journal of Comparative Psychology, 125*(4), 446–455.

Lycett, S. J., Collard, M., & McGrew, W. C. (2010). Are behavioral differences among wild chimpanzee communities genetic or cultural? An assessment using tool-use data and phylogenetic methods. *American Journal of Physical Anthropology, 142*(3), 461–467.

Lyons, D. E., Damrosch, D. H., Lin, J. K., Macris, D. M., & Keil, F. C. (2011). The scope and limits of overimitation in the transmission of artefact culture. *Philosophical Transactions of the Royal Society B, 366*(1567), 1158–1167.

Lyons, D. E., Young, A. G., & Keil, F. C. (2007). The hidden structure of overimitation. *Proceedings of the National Academy of Sciences of the USA, 104*(50), 19751–19756.

Marshall-Pescini, S., & Whiten, A. (2008a). Social learning of nut-cracking behavior in east African sanctuary-living chimpanzees (*pan troglodytes schweinfurthii*). *Journal of Comparative Psychology, 122*(2), 186–194.

Marshall-Pescini, S., & Whiten, A. (2008b). Chimpanzees (*Pan troglodytes*) and the question of cumulative culture: An experimental approach. *Animal Cognition, 11*(3), 449–456.

Matthews, L. J., Paukner, A., & Suomi, S. J. (2010). Can traditions emerge from the interaction of stimulus enhancement and reinforcement learning? An experimental model. *American Anthropologist, 112*(2), 257–269.

McGrew, W. C. (1998). Culture in non-human primates? *Annual Review of Anthropology, 27*(1), 301–328.

McGrew, W. C. (2009). Ten dispatches from the chimpanzee culture wars, plus postscript (revisiting the battlefronts). In K. N. Laland & B. G. Galef Jr. (Eds.), *The question of animal culture* (pp. 41–69). Harvard University Press.

McGrew, W. C., & Tutin, C. E. (1978). Evidence for a social custom in wild chimpanzees? *Man, 13*(2), 234–251.

Mercader, J., Barton, H., Gillespie, J., Harris, J., Kuhn, S., Tyler, R., & Boesch, C. (2007). 4300-year-old chimpanzee sites and the origins of percussive stone technology. *Proceedings of the National Academy of Sciences of the USA, 104*(9), 3043–3048.

Morand-Ferron, J., Cole, E. F., Rawles, J. E., & Quinn, J. L. (2011). Who are the innovators? A field experiment with 2 passerine species. *Behavioral Ecology, 22*(6), 1241–1248.
Moura, A. C. A., & Lee, P. C. (2004). Capuchin stone tool use in Caatinga dry forest. *Science, 306*(5703), 1909–1909.
Mueller, T., O'Hara, R. B., Converse, S. J., Urbanek, R. P., & Fagan, W. F. (2013). Social learning of migratory performance. *Science, 341*(6149), 999–1002.
Nichol, L. M., & Shackleton, D. M. (1996). Seasonal movements and foraging behaviour of northern resident killer whales (*Orcinus orca*) in relation to the inshore distribution of salmon (*Oncorhynchus* spp.) in British Columbia. *Canadian Journal of Zoology, 74*(6), 983–991.
O'Corry-Crowe, G. M., Suydam, R. S., Rosenberg, A., Frost, K. J., & Dizon, A. E. (1997). Phylogeography, population structure and dispersal patterns of the beluga whale *Delphinapterus leucas* in the western Nearctic revealed by mitochondrial DNA. *Molecular Ecology, 6*(10), 955–970.
Pepperberg, I. M. (1992). A review of the effects of social interaction on vocal learning in African grey parrots (*Psittacus erithacus*). *Netherlands Journal of Zoology, 43*(1-2), 104–124.
Pepperberg, I. M. (2010). Vocal learning in Grey parrots: A brief review of perception, production, and cross-species comparisons. *Brain and Language, 115*(1), 81–91.
Pepperberg, I. M., Sandefer, R. M., Noel, D. A., & Ellsworth, C. P. (2000). Vocal learning in the grey parrot (*Psittacus erithacus*): Effects of species identity and number of trainers. *Journal of Comparative Psychology, 114*(4), 371–380.
Perry, S., Baker, M., Fedigan, L., Gros-Louis, J., Jack, K., MacKinnon, K. C., Manson, J. H., Panger, M., Pyle, K., & Rose, L. (2003). Social conventions in wild white-faced capuchin monkeys: Evidence for traditions in a Neotropical primate. *Current Anthropology, 44*(2), 241–268.
Pollick, A. S., & de Waal, F. B. M. (2007). Ape gestures and language evolution. *Proceedings of the National Academy of Sciences of the USA, 104*(19), 8184–8189.
Pradhan, G. R., Tennie, C., & van Schaik, C. P. (2012). Social organization and the evolution of cumulative technology in apes and hominins. *Journal of Human Evolution, 63*(1), 180–190.
Pryor, K., & Lindbergh, J. (1990). A dolphin-human fishing cooperative in Brazil. *Marine Mammal Science, 6*(1), 77–82.
Ramsey, G. (2013). Culture in humans and other animals. *Biology & Philosophy, 28*(3), 457–479.
Reader, S. M., & Laland, K. N. (2001). Primate innovation: Sex, age and social rank differences. *International Journal of Primatology, 22*(5), 787–805.
Rendell, L., & Whitehead, H. (2001). Culture in whales and dolphins. *Behavioral and Brain Sciences, 24*(2), 309–324.
Richerson, P. J., & Boyd, R. (2005). *Not by genes alone: How culture transformed human evolution*. University of Chicago Press.
Riebel, K., Lachlan, R. F., & Slater, P. J. (2015). Learning and cultural transmission in chaffinch song. *Advances in the Study of Behavior, 47*, 181–227.
Roubalová, T., Jarůšková, L., Chládková, K., & Lindová, J. (2024). Comparing the productive vocabularies of Grey parrots (*Psittacus erithacus*) and young children. *Animal Cognition, 27*(1), 45.
Robbins, M. M., Ando, C., Fawcett, K. A., Grueter, C. C., Hedwig, D., Iwata, Y., Lodwick, J. L., Masi, S., Salmi, R., Stoinski, T. S., Todd, A., Vercellio, V. & Yamagiwa, J. (2016). Behavioral variation in gorillas: evidence of potential cultural traits. *PLoS One, 11*(9), e0160483.
Santorelli, C. J., Schaffner, C. M., Campbell, C. J., Notman, H., Pavelka, M. S., Weghorst, J. A., & Aureli, F. (2011). Traditions in spider monkeys are biased towards the social domain. *PLoS One, 6*(2), e16863.
Sanz, C., Call, J., & Morgan, D. (2009). Design complexity in termite-fishing tools of chimpanzees (*Pan troglodytes*). *Biology Letters, 5*(3), 293–296.
Sasaki, T., & Biro, D. (2017). Cumulative culture can emerge from collective intelligence in animal groups. *Nature Communications, 8*(1), 15049.
Savage-Rumbaugh, S. (1996). *Kanzi: The ape at the brink of the human mind*. Turner Publishing Company.

Scarl, J. C., & Bradbury, J. W. (2009). Rapid vocal convergence in an Australian cockatoo, the galah *Eolophus roseicapillus*. *Animal Behaviour, 77*(5), 1019–1026.

Schuppli, C., & van Schaik, C. P. (2019). Animal cultures: How we've only seen the tip of the iceberg. *Evolutionary Human Sciences, 1*, e2.

Slater, P. J. B., & Ince, S. A. (1979). Cultural evolution in chaffinch song. *Behaviour, 71*(1-2), 146–166.

Sousa, C., Okamoto, S., & Matsuzawa, T. (2003). Behavioural development in a matching-to-sample task and token use by an infant chimpanzee reared by his mother. *Animal Cognition, 6*(4), 259–267.

Strager, H. (1995). Pod-specific call repertoires and compound calls of killer whales, *Orcinus orca* Linnaeus, 1758, in the waters of northern Norway. *Canadian Journal of Zoology, 73*(6), 1037–1047.

Tebbich, S., Griffin, A. S., Peschl, M. F., & Sterelny, K. (2016). From mechanisms to function: An integrated framework of animal innovation. *Philosophical Transactions of the Royal Society, B: Biological Sciences, 371*, 20150195.

Tennie, C., Hopper, L., & van Schaik, C. P. (2020). On the origin of cumulative culture: Consideration of the role of copying in culture-dependent traits and a reappraisal of the zone of latent solutions hypothesis. In L. M. Hopper & S. R. Ross (Eds.), *Chimpanzees in context* (pp. 428–453). University of Chicago Press.

Tomasello, M. (1994). The question of chimpanzee culture. In R. W. Wrangham, W. C. McGrew, F. B. M. de Waal, & P. G. Heltne (Eds.), *Chimpanzee cultures* (pp. 301–377). Harvard University Press.

Tomasello, M. (1996). Do apes ape? In C. M. Heyes & B. G. Galef Jr. (Eds.), *Social learning in animals: The roots of culture* (pp. 319–346). Academic Press.

Tomasello, M. (1999). *The cultural origins of human cognition*. Harvard University Press.

Tomasello, M. (2009). The question of animal culture, plus postscript (chimpanzee culture, 2009). In K. N. Laland & B. G. Galef (Eds.), *The question of animal culture* (pp. 198–221). Harvard University Press.

Tomasello, M. (2020). The adaptive origins of uniquely human sociality. *Philosophical Transactions of the Royal Society of London B, 375*, 20190493.

Tomasello, M., & Call, J. (1997). *Primate cognition*. Oxford University Press.

Tomasello, M., Davis-Dasilva, M., Camak, L., & Bard, K. (1987). Observational learning of tool-use by young chimpanzees. *Human Evolution, 2*(2), 175–183.

Tomasello, M., Kruger, A. C., & Ratner, H. H. (1993). Cultural learning. *Behavioral and Brain Sciences, 16*(3), 495–552.

Tun, T. (2014). Castnet fishing with the help of Irrawaddy dolphin, *Orcaella brevirostris*, in Myanmar. *Proceedings of the Design Symposium on Conservation of Ecosystem (The 13th SEASTAR2000 Workshop)*, 2, 39–45.

van der Post, D., & Hogeweg, P. (2008). Diet traditions and cumulative cultural processes as side-effects of grouping. *Animal Behaviour, 75*(1), 133–144.

van Schaik, C. P., Ancrenaz, M., Borgen, G., Galdikas, B., Knott, C. D., Singleton, I., Suzuki, A., Suci Utami, S. & Merrill, M. (2003). Orangutan cultures and the evolution of material culture. *Science, 299*(5603), 102–105.

Wasielewski, H. (2014). Imitation is necessary for cumulative cultural evolution in an unfamiliar, opaque task. *Human Nature, 25*(1), 161–179.

Watkins, W. A., & Schevill, W. E. (1977). Sperm whale codas. *The Journal of the Acoustical Society of America, 62*(6), 1485–1490.

Watson, S. K., Lambeth, S. P., Schapiro, S. J., & Whiten, A. (2018). Chimpanzees prioritise social information over pre-existing behaviours in a group context but not in dyads. *Animal Cognition, 21*(3), 407–418.

Weinrich, M. T., Schilling, M. R., & Belt, C. R. (1992). Evidence for acquisition of a novel feeding behaviour: Lobtail feeding in humpback whales, *Megaptera novaeangliae*. *Animal Behaviour, 44*(6), 1059–1072.

Whitehead, H., & Rendell, L. (2014). *The cultural lives of whales and dolphins*. University of Chicago Press.

Whiten, A. (2011). The scope of culture in chimpanzees, humans and ancestral apes. *Philosophical Transactions of the Royal Society of London B, 366*(1567), 997–1007.

Whiten, A. (2017). Social learning and culture in child and chimpanzee. *Annual Review of Psychology, 68*(1), 129–154.

Whiten, A., Goodall, J., McGrew, W. C., Nishida, T., Reynolds, V., Sugiyama, Y., Tutin, C. E. G., Wrangham, R. W., & Boesch, C. (1999). Cultures in chimpanzees. *Nature, 399*(6737), 682–685.

Whiten, A., Harrison, R. A., McGuigan, N., Vale, G. L., & Watson, S. K. (2021). Collective knowledge and the dynamics of culture in chimpanzees. *Philosophical Transactions of the Royal Society B, 377*(1843), 20200321.

Whiten, A., Horner, V., & de Waal, F. B. M. (2005). Conformity to cultural norms of tool use in chimpanzees. *Nature, 437*(7059), 737–740.

Whiten, A., Spiteri, A., Horner, V., Bonnie, K. E., Lambeth, S. P., Schapiro, S. J., & de Waal, F. B. M. (2007). Transmission of multiple traditions within and between chimpanzee groups. *Current Biology, 17*(12), 1038–1043.

Zandberg, L., Lachlan, R. F., Lamoni, L., & Garland, E. C. (2021). Global cultural evolutionary model of humpback whale song. *Philosophical Transactions of the Royal Society B, 376*(1836), 20200242.

Chapter 10
Memetics: Rise, Fall, and Reconsideration

Jaroslav Flegr and Jan Toman

Abstract Emerging from a gene-centred analogy of cultural evolution, memetics rose to prominence between the 1980s and early 2000s before entering a period of relative decline. It conceptualises culture as a dynamic system composed of memes and their complexes, which are mutually cooperating and competing units expressed in human behaviour and propagated by imitation. Memes compete for a faster transmission and wider representation within a population's 'meme pool' either by benefiting their hosts or by spreading 'selfishly', even at the host's expense—this mirrors the dynamics observed in evolutionary parasitology. This parallel demonstrates both the strengths and the limitations of memetics, which inherited the foundational assumptions of the 'selfish gene' model. The waning influence of memetics can be attributed to the criticism that has been levied against its reductionism, biologicism, limited theoretical novelty, its assumption of particulate inheritance, and its neglect of broader cultural and social contexts. Nevertheless, informed by population genetics and evolutionary-parasitological models, memetics remains effective in explaining various specific cultural phenomena, especially the propagation of 'selfish' deleterious or destabilising cultural traits, which spread despite harming their hosts. At the same time, approaches from the philosophy of technology, including postphenomenology, emphasise that cultural evolution does not take place in a neutral medium: there exist various sociotechnical systems which actively shape which memes proliferate and at times amplify the spread of harmful variants. An integrative perspective that would enrich memetic principles by postphenomenological and even biosemiotic insights may thus provide the foundation of a renewed, 'neo-memetic' framework capable of addressing the dynamics of cultural evolution in an age of global digital interconnectivity.

J. Flegr · J. Toman (✉)
Department of Philosophy and History of Sciences, Faculty of Science, Charles University, Prague, Czech Republic
e-mail: jaroslav.flegr@natur.cuni.cz; tomanj@natur.cuni.cz

J. Toman et al. (eds.), *Culture Evolving*, Biosemiotics 10,
https://doi.org/10.1007/978-3-032-16124-6_10

Keywords Memetics · Cultural evolution · Biosemiotics · Philosophy of technology · Cultural transmission

10.1 What Is Memetics and what Are Memes

When studying cultural evolution, we can employ conceptual frameworks that focus either on competition and the selection of more viable variants or on the independent emergence and preferential retention of certain cultural variants grounded in common features of human cognition. Some of these frameworks assume that the elements of cultural evolution can be decomposed into discrete, independently analysable units, while other models resist such reduction (see Chaps. 11, 12, 13). Among the former type, the approach that most strongly asserts the selectionist and particulate nature of cultural processes is memetics. Over recent decades, memetics has followed a striking trajectory from modest beginnings, through a phase of intense enthusiasm, to a near-total abandonment by the mainstream scientific community. This chapter introduces the central tenets of memetics, explores the factors that contributed to its decline, and presents selected examples of cultural phenomena that memetics is uniquely well suited to explain. It concludes by asking whether the time has already come to revisit its core principles within a revised, more integrative theoretical framework that reflects the shortcomings of its earlier formulations.

As summarised by for instance Kundt (2015), the first efforts to apply an evolutionary framework to cultural phenomena date to the nineteenth century. Some of these attempts even predate the publication of Darwin's book *On the Origin of Species* (1859), while other concepts adopted Darwinian ideas but employed them selectively, often misinterpreting or omitting certain key aspects (cf. Chaps. 3, 4, 12, 18). Several such early theories had already highlighted the role of imitation in cultural transmission (Jahoda, 2002). Notable in this context is the work of Gabriel Tarde (1843–1904), a French criminologist, social psychologist, and sociologist, who proposed a theory of imitation and cultural diffusion that anticipated several themes later formalised within memetics. The idea that cultural phenomena are fundamentally particulate also has early formulations in the literature. As summarised by Rádl (2015: 284–285) and Laurent (1999), the earliest proponent of this view was probably the German naturalist Richard Semon (1859–1918), who in the early twentieth century coined the term 'mneme', deriving the name from Mneme, the Greek muse of memory. In Semon's theory, a mneme is a kind of memory trace left by experience that can be reactivated under various conditions. Semon also addressed the preservation and transmission of mnemes—he was, after all, a leading proponent of Lamarckian theories of organic memory (cf. Chap. 16)—but competition among the individual mnemes was not central to his inquiry (Rádl, 2015).

A renewed interest in the evolutionary character of culture emerged in the 1950s, laying the groundwork for contemporary memetics (Chvaja, 2020). Key figures in this resurgence include Leslie A. White (see Spuhler, 1959) and Edward O. Wilson (1975, 1978). Richard Dawkins, the founding figure of memetics, also acknowledges the influence of F. Ted Cloak, Luigi L. Cavalli-Sforza, and Jason M. Cullen. Memetics in its current form emerged in the mid-1970s and came to be recognised as a distinct field about a decade later. Interestingly, memetics originated more as an illustrative analogy than as a formal theory. In his influential work *The Selfish Gene* (1976), Richard Dawkins introduced memetics as a thought experiment designed to clarify the mechanisms of biological evolution by comparing them to a parallel cultural process. He argued that, like biological evolution, cultural evolution is driven by entities that replicate, inherit traits, and undergo variation. These variations affect the relative success of the entities, which in turn compete for a limited resource: in biological systems, for nutrients and energy, and in cultural systems, for space in the brains (or perhaps more aptly, the minds) of human hosts.

In biological evolution, the entities subject to selection were postulated to be genes, or rather their individual variants, alleles. By analogy, the units proposed to compete in cultural evolution were called 'memes', although probably without any direct reference to Richard Semon's earlier mneme concept. In coining the term, Richard Dawkins deliberately used the phonetic similarity between the words meme and gene, deriving meme from the Greek μῖμος (mimos), meaning 'to imitate'. The defining characteristic of memes is that they replicate through imitation, which fundamentally distinguishes them from genes, whose replication involves copying their physical carriers, DNA molecules. Elaborating on Dawkins (1976), a meme may thus be defined as *any structured unit of information capable of spreading from one human mind to another via some form of imitation.* This may include teachings, words, melodies, mathematical formulae, anecdotes, or even entire scientific theories.

Dawkins's book attracted widespread attention not only among natural scientists but also within the humanities, where scholars started to develop his initially modest idea into a more elaborate model of cultural evolution. The expansion of this field has been documented in detail by, for example, Vada (2015) and Chvaja (2020). The emergence of memetics as a formalised domain arguably started with Douglas R. Hofstadter's article 'Virus-like Sentences and Self-Replicating Structures' (1983) in the *Scientific American*; the text was later included a collection of articles called *Metamagical Themas: Questing for the Essence of Mind and Pattern*' (Hofstadter, 1985). During this formative period, further contributions were made by authors such as Lumsden and Wilson (1981), Wilson (1975, 1978), and Cavalli-Sforza and Feldman (1981).

However, the main boom of memetics took place in the second half of the 1990s. This era saw the publication of the online *Journal of Memetics—Evolutionary Models of Information Transmission* (1997–2005) and the appearance of three foundational books: Brodie's *Virus of the Mind* (1996), Lynch's *Thought Contagion* (1996), and Blackmore's *The Meme Machine* (1999). At the same time, thinkers such as Francis Heylighen, Klaas Chielens, Kate Distin, Robert Aunger, and Daniel

Dennett made significant theoretical contributions. A more recent overview of these developments can be found in Tim Tyler's *Memetics: Memes and the Science of Cultural Evolution* (2011). Meanwhile, the foundational ideas of memetics continue to inform, explicitly or implicitly, the work of a wide range of authors, including Daniel Dennett (1998), Pascal Boyer, Dan Sperber, Susan Blackmore, Aharon Kantorovich, and others, regardless of whether they identify with the memetic paradigm or distance themselves from it.

After the turn of the millennium, the field of memetics reached a turning point in its trajectory. A significant milestone in its departure from mainstream scientific discourse was the cessation of its dedicated journal in 2005. Although the journal never succeeded in establishing a large or cohesive research community, several of its final contributions are in retrospect strikingly pessimistic (see, for example, Edmonds, 2005). In hindsight, it is evident that at that time, a number of fundamental problems present since the field's inception had converged, having been exacerbated by field-specific tensions and personal animosities.

Among the most consequential issues was the excessive preoccupation with the ontology of memes, that is, debates concerning their nature, classification, and whether they ought to be regarded as products of cultural evolution or as structures rooted in the brain or mind. Unfortunately, this focus often came at the expense of empirical testing of hypotheses. From the outset, memetics was also encumbered by overreliance on analogies between genes (particularly 'selfish genes') and memes (Chvaja, 2020; Fomin, 2019), although genes and memes are not fully analogous. Unlike genes, memes could never function as quasi-independent, discrete replicators endowed with agency. This aligns with a prevailing view among semioticians and biosemioticians, who believe that memes are at best a reduced form of signs, entities that can replicate but lack key semiotic properties, such as interpretability (within a given paradigm, Kull, 2000) or autonomy (Deacon, 1999). Memes are often assumed to be transmitted independently of context and to operate in a context--free manner. However, further in this chapter, we shall argue that such assumptions are deeply reductive. Like genes, memes have always functioned in a dynamic relationship with the wider 'ecology' of other memes (or genes) within a population and, by extension, in interaction with the minds that interpret them.

A further complication is that, unlike genetic evolution, memetic evolution may follow a Lamarckian model, where acquired traits can influence the replication of cultural units. This contrast underscores a deeper problem: memes, originally introduced by Dawkins as a teaching device to illustrate the gene-centric view of evolution, were subsequently developed into a theoretical model by other scholars. But the selfish gene framework itself, which reduces all evolutionary dynamics to a competition among genetic variants for reproductive success, remains highly reductive even by the standards of biological science. In the humanities and social sciences, such reductionism met with even more scepticism. Finally, the field also suffered from lack of conceptual coherence. Even the most basic terms and assumptions were interpreted differently by different scholars, and that led not to an integrated interdisciplinary programme of memetics but rather to a proliferation of multiple incompatible versions of memetics.

A further major difficulty for memetics has been the testability of its hypotheses. This limitation becomes particularly apparent when contrasted with the related theory of dual inheritance (or the gene–culture coevolution theory) (Boyd & Richerson, 1985; cf. Chaps. 7, 10, 16). While both theories examine cultural dynamics, the dual inheritance theory focuses specifically on whether, and how, cultural variants contribute to the spread of certain alleles—and vice versa.[1] It addresses only those situations where culture is adaptive from a biological standpoint; this enables a formulation of testable predictions that can be tested by comparison with a clearly defined null hypothesis. By contrast, traditional memetics has typically focused on memes that either do not affect the biological fitness of their carriers or even spread despite reducing it. While these cases are undoubtedly intriguing, hypotheses concerning the spread of such memes are often difficult to differentiate from null expectations, particularly when working with cultural datasets derived from offline sources. Data sources more appropriate for investigating such hypotheses, such as social networks with well-mapped pathways of information transmission, became widely available only since the late 2000s, with the global spread of social media platforms. In this sense, one might argue that memetics was simply ahead of its time.

Memeticists moreover failed to dedicate sufficient efforts to integrating their approach with other, often more established, frameworks within cultural studies (Fomin, 2019). This oversight further deepened the scepticism already prevalent among scholars in humanities and social sciences, especially towards the biologisation of culture and, more broadly, towards diachronic accounts of cultural phenomena—a scepticism that is historically grounded (see Chaps. 3, 4, 5, 7). Another point of contention was related to the replicator model of cultural evolution itself, which often treats culture as a passive conduit rather than an active mediator. This model tends to obscure the complex economic, social, and political dynamics through which cultural elements are generated and maintained (Feenberg, 1999). As a result, the gap between the natural sciences and social sciences widened because while such reductionism may serve as a productive starting point in the natural sciences, in the social sciences it is frequently viewed as an unacceptable oversimplification.

Distrust of memetics was further exacerbated by its association with certain prominent figures who were not only scientists but also public proponents of scientism and atheism. These figures often criticised or even advocated the elimination of various religious traditions and practices (Blackmore, 1999; Dawkins, 2006). The field's decline may also be linked to its overly ambitious scope and perceived overreach. Some observers thus perceived memetics as a cultural 'theory of everything', a putative new science and paradigm intentionally detached from disciplinary tradition (see, for example, Brodie, 1996; Herrmann-Pillath, 2021). At the same time, the conceptual rifts over the understanding of culture deepened. Scholars in the natural sciences tend to understand culture broadly, that is, as encompassing the transmission of all non-genetic information (cf. 'animal cultures', cf. Chap. 9; 'a

[1] All phenomena described by dual inheritance theory also fall within the scope of memetics; this applies specifically to cases where genetic and memetic selection align.

general principle of meaning attribution', cf. Chap. 17; or 'epigenetic heredity of behaviour', cf. Chap. 16). In contrast, scholars in the humanities and social sciences increasingly favour a more exclusive view of 'cultures' as bounded social formations defined by shared values, identities, and institutions (cf. Chap. 5; Herrmann--Pillath, 2021). Socioeconomic factors may also have contributed to the marginalisation of memetics. Frequently cited influences include the rise of individualist ideologies in the Anglo-American world in the 1980s and 1990s, their retreat in the second decade of the twenty-first century, and the popularising tone adopted by many early memetic publications (Chvaja, 2020).

Despite these challenges, it should not be forgotten that memetics has provided a coherent framework for explaining the spread of cultural traits that either have no impact on biological fitness or actively harm their hosts. Such phenomena, particularly those involving a rapid dissemination of harmful cultural variants, remain challenging for other models of cultural evolution to accommodate. Nevertheless, their existence is increasingly visible in today's globally interconnected information environment, where they take various forms ranging from fake news and conspiracy theories to disinformation campaigns. For this reason, the following section will revisit the core principles of memetics. Drawing especially on insights from evolutionary parasitology and epidemiology, but also on ideas from semiotics and other adjacent fields (see, e.g., Fomin, 2019), we suggest that memetics may yet be reconstituted as a productive scientific framework capable of addressing some of the most urgent cultural phenomena of today.

10.2 Differences Between the Evolution of Memes and Genes

Cultural and biological evolution differ in several fundamental aspects. In biological evolution, a separation of functions has occurred between the replicator and the interactor. The replicator is the DNA molecule, whose two strands are copied in the process of replication to form two new molecules, and thus two new copies of a given gene. Individual gene variants, alleles, compete via their respective interactors, which are produced during individual development based on the genetic information that the alleles carry (Hull, 1980). The interactors are the bodies of organisms. A replicator that helps to create a higher-quality interactor—that is, an individual that is better at exploiting environmental resources under specific conditions and thus produces more offspring—will be represented in the next generation by a greater number of copies than a replicator that produces a less viable or less fertile interactor.

In memes, by contrast, the roles of replicator and interactor are typically embodied in the same entity, at least in all species other than modern humans. This entity is anything that can be imitated (see Chap. 9) by other organisms, usually conspecifics. A meme may therefore take the form of a melody, gesture, custom, or a particular hunting technique. In this way, advantageous behavioural modifications acquired during an individual's lifetime can be transmitted to others. With the emergence of language, and especially with the development of external memory media, there appeared in modern humans a secondary differentiation: in some cases, the meme itself (understood as the information it carries) came to function as a replicator, while its manifestation (in the form of behaviour, artefact, or belief) became the interactor because it is this outward expression that engages with the environment and shapes the meme's likelihood of being transmitted.

The evolution of memes can proceed via both the Darwinian and the Lamarckian mechanism or, rather, the distinction between the two becomes in this context somewhat irrelevant. For instance, if a particular melody is altered during imitation and the modified version turns out to be catchier or easier to remember, it may spread more effectively and eventually displace the original version of the meme. In biological evolution, by contrast, Lamarckian inheritance does not occur: if an individual increases muscle mass through training, this acquired trait is not transmitted to offspring and children of the trained individual are no stronger than those of an untrained one. Biological evolution is thus governed almost exclusively by Darwinian mechanisms (see, e.g., Flegr, 2002; cf. Chaps. 7, 16, 18). A gene may mutate during replication and if this altered variant leads, in the course of development, to a modified interactor, the resulting phenotype may affect the replicator's success in future generations. If the modified interactor reproduces more successfully than its unmodified counterpart, the new version of the gene will, in the population, gradually replace the original.

The Lamarckian dimension enables cultural evolution to proceed more rapidly than biological evolution. Another factor contributing to this faster pace is that while beneficial gene variants are typically transmitted only vertically, from parent to offspring, meme variants can also spread horizontally, among both related and unrelated individuals, and even across species. For example, in the 1920s, blue tits in the English village of Swaythling learned to open milk bottles and consume the cream from the top. This innovation spread throughout England within 30 years and was gradually adopted by several other bird species (Hinde & Fischer, 1952; Lefebvre, 1995; cf. Chap. 9). If this foraging behaviour had been genetically encoded, the spread of the corresponding allele would have taken tens or even hundreds of generations, and transfer to the gene pools of other species would have been most unlikely, relying solely on rare hybridisation events. Moreover, memes and their integrated groups actively shape their carriers' capacities to both transmit and receive cultural content. This is reminiscent of Ihde's (1990) demonstration of the means of technological mediation (from writing systems to telephones), which showed that each new medium not only enables transmission but also transforms the nature of what can be transmitted. Cultural evolution is thus not only faster than genetic evolution, but each new meme or memetic constellation also alters the landscape of cultural transmission itself.[2]

[2] Unlike the straightforward inheritance of acquired characters, this aspect of cultural evolution does have a parallel in biological evolution. It resembles what is known as the evolution of evolvability, that is, the process by which the relationship between genotype and phenotype is shaped

Another factor that accelerates cultural evolution is the preferential creation of advantageous meme variants. Like most new genes, many new memes arise from errors in copying, in this case, though, we are more likely to speak of errors in imitation. But the creation of memes often also involves insight. Numerous observations in birds and mammals suggest that individuals often solve new problems not via trial-and-error—as would be the case in natural selection—but by deliberately generating new behavioural strategies. For example, in a well-known ethological experiment, a chimpanzee faced with a cluster of bananas hanging from the ceiling first inspected the situation, then moved a crate beneath the fruit, and finally used a stick to knock it down (Pribram, 1969). The creation of new memes through insight is especially characteristic of species with highly developed cognitive capacities, such as humans.

Not only are advantageous memes preferentially generated—they are also preferentially transmitted, although neutral or even maladaptive variants can naturally spread as well (see below). Behaviours which benefit their carriers are more likely to be imitated than those that do not. In some animal species, specialised mechanisms seem to have evolved to facilitate such selective imitation. One such mechanism may involve a preferential imitation of individuals with high status in the social hierarchy of the group (prestige bias; see Henrich & Gil-White, 2001). Statistically, social status often correlates with the efficacy of behavioural strategies but under certain circumstances, these mechanisms can temporarily or even permanently fail, leading to an inefficient transmission of adaptive behaviours. An example of such a breakdown (observed in Japanese macaques; Kawamura, 1963) was discussed by Jitka Lindová in her chapter (cf. Chap. 9).

10.3 Evolution of Cultural Evolution

Cultural evolution takes place in many animal species, but its most elaborate form is found in humans, where the distinction between the interactor and the replicator became more pronounced, making cultural evolution even more closely analogous to biological evolution. As noted above, while the natural carrier of genes is the organism's DNA, the natural carrier of memes is the behaviour of members of a given species. This behaviour is controlled by the nervous system; individuals learn new behaviour patterns through imitation and other forms of social learning (cf. Chaps. 9, 13) and store the programs for their execution in memory.

Moreover, humans—and to a very limited extent also members of some other species—have acquired the ability to store information created during cultural evolution on external carriers. In the course of cultural evolution, humans have developed the capacity to record behaviours on external media, initially through linguistic

over evolutionary time to reflect various environmental selective pressures and to generate non-random, adaptively biased phenotypic variation.

expression and the production of artefacts, and later thanks to technologies such as video recordings, from which others can acquire memes directly. Speech enables us, among other things, to describe behavioural patterns symbolically and transmit them verbally to others. As a result, behaviour need not always be acquired through imitation; it can instead be conveyed through spoken language and, in modern humans, also through written texts or audio recordings.

Memes now exist in both active and inactive forms. Active memes—variously referred to as internal memes, i-memes, or memotypes—reside in human minds (are 'implemented in the meatware') and can directly influence the behaviour of their carriers (or hosts, to use the memetic terminology). Inactive memes, also known as external memes, e-memes, or mediotypes, are stored on external carriers from which they may be reproduced (e.g., by printing). Crucially, though, once stored, they cannot express themselves and therefore also cannot directly affect the hosts' behaviour (Heylighen & Chielens, 2009; McNamara, 2011). A special category of external memes are non-verbal artefacts, which are the outcomes of culturally transmitted behaviours. These may be simple enough to imitate through insight (a sharpened stick) or complex enough to require explicit instructions or a production process (a microchip). Importantly, the design, selection, and dissemination of external carriers are shaped by social and political dynamics. As Feenberg (1999) argues, the transmission of information is never entirely neutral: it is always influenced by pre-existing memetic structures and values.

Inactive memes, such as written texts or audio recordings, can be re-activated (converted by reading or listening into an active form) and then translated into behaviour. But they can also be copied without such interpretation: for example, a person unfamiliar with a language may reproduce a text without understanding it or a photocopier may replicate a text fully mechanically. As a result, cultural evolution now can in many cases approximate the replicator–interactor distinction familiar from biological evolution. Analogically to evolutionary drives such as molecular or meiotic drive, new variants can also emerge through directed changes shaped by the specific memetic background of the individual or the culture (this phenomenon is known as content bias; cf. Chap. 11). If a new variant is in some way more effective, it may spread and ultimately replace the original. Competition between memes can thus take place at multiple levels: between the behavioural patterns for expression and imitation and between the externally stored memes for the most frequent copying of their carriers.

Taken together, these observations suggest that species with complex cultures, especially humans, have created a complex and multilayered system of different meme types. This parallels what Hoffmeyer (2007) describes as semiotic scaffolding or with what evolutionary biologists call the evolution of evolvability (cf. Chap. 15). According to Hoffmeyer, this process leads to an evolving semiosphere, while in the view of biologists such as Watson and Szathmáry (2016), it enables open-ended evolution.

It is also important to recognise that one meme can carry multiple meanings—this is a critical observation that so far has not received much attention among memeticists. An arrow, for example, may be an artefact to imitate (as a weapon), a warning sign (a deterrent), or a navigational guide (indicating direction). The interpretation of a meme depends on cultural presuppositions, that is, the memes and meme complexes previously encountered, and on the broader context. From this perspective, some of the strongest biosemiotic critiques lose their edge. Kull (2000) suggests that cultural evolution should be viewed not as a replication of fixed content but as an ongoing conversation where each generation reinterprets the signs of the previous one in its own context. This emphasis on interpretation and context brings memetics and biosemiotics closer together. In this view, memetics approaches, or perhaps even crosses, the boundary into humanities. As Fomin (2019) proposes, memes might be best understood as degenerate semiotic signs[3] and memetics, when applied to human culture, as a form of evolutionary or diachronic semiotics. Attempts to define and categorise memes precisely remain contested and have often been the target of criticism (Chvaja, 2020) but the complexity of the system does not invalidate the memetic approach. In fact, Herrmann-Pillath (2021) had proposed an integration of biosemiotic insights into replicator-based models via self-replicating neurophysiological configurations ('neuromemes'). The proposed framework locates selection primarily in the neural dynamics, where sign-interpretations over time stabilise and thereby link memetic replication to biosemiotic notions of meaning.

As noted earlier, in non-human species, the symbolic encoding of memes in external carriers is extremely rudimentary. For instance, rats can transmit a meme concerning the toxicity of certain food, such as poisoned grain, through deposition of faeces on the food (Steiniger, 1950; cf. Chap. 9). They seem to have an innate tendency to both leave faeces on food that made them ill (or on food where they observed faeces previously) and to avoid such marked food. Young or naïve individuals can thus acquire survival-relevant information without direct demonstration. This form of transmission can persist long after the last witnesses of the poisoning event have died. On the other hand, spoken or written contents offer far more robust mechanisms for meme preservation. When the toxic food disappears from the environment, rats can no longer pass on the relevant information, and the meme may vanish. In contrast, human speech—and especially writing—enables a preservation of memes across generations even in the long-term absence of the referent. Still, the main difference between the rudimentary rat 'protolanguage' and human language lies in the symbolic vocabulary. Rats rely on pre-existing environmental objects (such as faeces or food), while humans create abstract symbols which are not tied to currently observable objects. Human language enables reference not only to individual things but also to classes of objects and to past and future entities. The ability to abstract and generalise makes human language a vastly more powerful medium of memetic transmission than any observed in non-human species.

[3] That is, sign-equivalents, deliberately reduced to their capacity for replication within the mechanistic paradigm of biology.

10.4 The Types of Memes and Mechanisms of Their Spread

Preferential spread of certain memes, that is, their evolutionary success, may be driven by a variety of causes and proceed through diverse mechanisms. One such cause is direct competition based on the meme's own substantive content: for example, a more entertaining anecdote is more likely to be retold than a dull one. This is called content bias (cf. Chap. 11). Another is indirect competition, in which memes influence the biological success of their carriers and thus 'hitchhike' with successful genes: for instance, a meme promoting family formation may spread more effectively than one encouraging lifelong celibacy. Memes can also compete based on their infectiousness as active elements, so that a display of a status symbol such as an expensive smartphone may be more readily imitated by peers than wearing one's grandfather's watch. Sorting in terms of stability plays a role as well: cigarettes that contain addictive substances are more likely to spread than non-addictive alternatives due to enhanced behavioural retention. Inactive memes are subject to memetic drive at the level of their carriers: for example, an entertaining cookbook is more likely to be published and circulated than a dry, technical one. Additionally, memetic evolution can be also influenced by a repeated but independent emergence of similar alterations (analogous to mutation pressure in genetics): for instance, the idea that 'a higher power rewards good and punishes evil' often mutates into a belief that 'if something bad happens, someone must have done something wrong'. Many of these mechanisms are acknowledged in other frameworks of cultural evolution, such as the attractionist models (cf. Chap. 11) and dual inheritance theory (cf. Chaps. 7, 10, 16). Memetics, however, has greater ambitions—its aim is to provide a unified explanation for all of the abovementioned cultural dynamics.

10.4.1 Direct Competition of Memes in Terms of the Attractiveness of Their Content

In cultural evolution, a population's meme pool becomes dominated by memes that manage to outperform their competitors in some respect. Many memes succeed because they appeal to members of a given population, who are therefore more inclined to imitate them. What is considered attractive or appealing within a population is shaped by both biological predispositions and cultural background. Biological evolution has predisposed us to favour stimuli that enhance our biological fitness by improving improve our health, fertility, or attractiveness to potential mates. Cultural evolution, on the other hand, has shaped the complex symbolic environments into which certain memes may fit more successfully than others. For instance, memes such as anecdotes, songs, or fashion trends which thrive in Saudi Arabia may not be equally successful in New York—and vice versa.

Just as biological evolution is shaped by evolutionary constraints, cultural evolution is influenced by cultural barriers, by limits and filters imposed by the sociocultural environment. These barriers determine which memes are more likely to emerge and spread in a particular population. This aspect of cultural evolution is highlighted by the concept of sociomorphic modelling (cf. Chap. 14). A good example is provided by the very idea of evolution or, to be precise, the idea of natural selection. Various authors have speculated why the concept, despite its apparent simplicity and the existence of ancient Greek and Roman proto-evolutionary notions (cf. Chaps. 1, 2), receded from the cultural and intellectual discourse and re-emerged only in the nineteenth century, at which point it was formulated three times and all these instances took place in Britain: it appeared in the works of Matthew, Darwin, and Wallace. A likely explanation is that the cultural background of Christian Europe, shaped since the late Hellenistic era and dominating through the early modern period, posed strong barriers to this kind of thinking. By contrast, early capitalist Britain provided a set of cultural constraints more conducive to the idea that nature could be modelled based on the logic of competition (Toman & Flegr, 2017).

10.4.2 Competition of Memes in Terms of Their Ability to Ensure the Reproduction of Their Carriers

In humans, many memes are transmitted vertically (solely or preferentially from parents to the offspring), and this leads to the emergence of a distinct and important category of 'biotrophic' memes. These are memes which directly or indirectly enhance the biological fitness of their carriers by improving health, increasing fertility, or promoting reproductive success. Their spread is facilitated by their capacity to hitchhike alongside the genes they help propagate.

Many biotrophic memes improve both biological fitness and the quality of life, for instance, those that promote balanced nutrition or the avoidance of environmental hazards. Other memes, however, can have more complex and ambiguous effects. Some memes (and genes) may reduce the quality of life while enhancing reproductive output, that is, the number of their copies to be passed down vertically. For example, certain forms of antisocial behaviour or even culturally modulated manifestations of specific mental disorders may increase the number of offspring by facilitating unstable or serial partnerships. An individual with a bipolar disorder or one prone to dysfunctional relationships may form and dissolve multiple unions, which potentially results in more offspring and more opportunities to pass on both genetic and memetic content, than an individual in a stable, monogamous relationship. This dynamic is particularly relevant in populations where the cultural norm for a woman is to have only one or two children.

The evolutionary trajectories of vertically transmitted memes often become closely linked to the fate of the genes carried by the same individuals. The spread of a particular allele can thus support the spread of a certain meme, and vice versa. In some species of social cetaceans (cf. Chap. 9), researchers have documented remarkably low mitochondrial genetic polymorphism. Subsequent analyses then showed that in these species, only females participate in the upbringing of the young, that is, in transmitting memes from parent to offspring. The most plausible explanation for the observed homogenisation of mitochondrial DNA is therefore a simultaneous spread of a particular mitochondrial lineage with a culturally advantageous meme (Whitehead, 1998, 1999).

Biotrophic memes are also highlighted in dual inheritance theory, which acknowledges that cultural variants beneficial to biological fitness need not be confined to vertical transmission (Boyd & Richerson, 1985). Under certain conditions, such memes may also be transmitted horizontally and spread by enhancing the fitness of entire social groups rather than just individuals (cf. Chaps. 7, 10, 13, 16).

10.4.3 Competition of Memes in Terms of Their Ability to Infect as Many Carriers as Possible

In contrast to biotrophic memes, which spread by enhancing the biological fitness of their carriers, necrotrophic memes propagate within a population's meme pool despite decreasing the reproductive success of their carriers. For example, a meme may increase an individual's chance of achieving a higher social status, thus enhancing their visibility and the likelihood that others will imitate them, while reducing their reproductive output. As noted above, it is often the behaviours of high-status individuals that are most widely imitated (Henrich & Gil-White, 2001), which is why such individuals may pass on few genes but a large number of memes.

In certain cultural environments, there can spread even memes that promote self-destructive behaviours, such as suicide or suicidal terrorism. For example, Mealey (2000) found that anorexia may develop as a result of imitating high-status women in the context of modern Western society. Some memes are so highly infectious (i.e., prone to being imitated) that they proliferate even when they reduce both the quality of life and biological fitness of their carriers to a point where individuals actively attempt to resist or discard them. Examples include memes associated with alcohol consumption or smoking. For instance, adolescents often imitate adult role models even when the behaviour is initially unpleasant or when they intuitively recognise that it may complicate their lives in the long term. Moreover, if the behaviour involves an addictive substance (discussed below), individuals may find it nearly impossible to abandon it even if they wish to.

Necrotrophic memes function as memetic analogues of ultra-selfish genetic elements, that is, genes or DNA segments that spread at the expense of their host's fitness. For example, the t-haplotype in mice includes tightly linked genes which ensure their own transmission by destroying sperm cells that do not carry the haplotype. A male carrying the t-haplotype may produce only half as much viable sperm, but nearly all his offspring will inherit the selfish element (Ardlie, 1998). Memetic counterparts to such genetic elements may attract disproportionate attention due to their striking, often dramatic nature; this is the case of reports about school shootings, suicide attacks, or viral patterns of drug use. Nevertheless, their importance may in fact be increasing in today's densely interconnected informational landscape.

The success of certain memes can also be attributed to their inherent mental processability, the relative ease with which particular memes are processed and copied by the human mind. This could be described as a form of cultural attraction (cf. Chap. 11). Simpler tunes, tools, or decorative patterns tend to spread more rapidly than more complex ones and, in current societies, the same principle applies to scientific theories, ideological frameworks, and explanatory models. As Kantorovich (2014) notes, simpler—albeit sometimes less accurate—narratives often outcompete more complex or nuanced alternatives. This preference for cognitive simplicity may also help explain some broader patterns in the evolution and competition of languages, such as their tendency to simplification or optimisation over time (Atkinson et al., 2018; Saldana, 2018). These dynamics are especially visible in the global success of certain languages, such as English and Spanish, whose expansion (along with expansion of their associated cultural spheres) can be attributed in part to their relatively high learnability, particularly in simplified but functionally adequate forms. Ultimately, this may extend even to morphological structures, that is, the systematic correspondences between parts of words and parts of meaning. Kirby et al. (2008) experimentally demonstrated that, when transmitted between individuals, artificial languages evolve towards greater transmissibility by becoming more structured and more learnable.

Language proficiency, or lack thereof, determines access to large segments of the global meme pool and plays therefore a central role in cultural identity and self-definition. The spread of a language sometimes occurs independently of genetic ancestry or kinship. For example, linguistic assimilation may go hand in hand with the expansion of one cultural sphere and a corresponding contraction of others. This dynamic is common in historical contexts of conquest: because children typically acquire language from their mothers and women are often spared in conflicts, linguistic and cultural assimilation of the conquerors into the local population can occur within just a few generations (cf. Chap. 5).

10.4.4 Competition of Memes in Terms of Their Stability

All the mechanisms discussed so far represent forms of selection, be it at the level of replicators or interactors. But there is also another evolutionary mechanism at work, one arguably more fundamental and universal: stability-based sorting (Toman & Flegr, 2017). This process involves competition between entities based on their persistence, that is, based on how long they endure before disappearing or transforming. Unlike selection, stability-based sorting does not presuppose a lineage between entities, it does not require inheritance or a parent–offspring relationship. Entities can emerge independently, much like individual snowflakes forming in a cloud (see Chap. 15).

Stability-based sorting operates at many levels of cultural evolution. In any cultural environment, we are more likely to encounter memes that are stable than those that are unstable and therefore either quickly vanish or easily mutate. For instance, a meme transmitted orally is generally more prone to transformation or loss than one conveyed in written form. As a result, in societies without writing, cultural evolution tends to proceed more rapidly than in literate ones, although the process is also influenced by numerous demographic factors (Derex & Mesoudi, 2020). The stability of certain memes is further reinforced by features of their biological or cultural context: as noted earlier, memes related to drug use may be stabilised by the presence of addictive substances, making it difficult for individuals to abandon them. In cultural systems, one meme may also stabilise another, so that for example a meme for a certain religious belief may be stabilised by an auxiliary meme prescribing the punishment or killing of apostates, thereby discouraging the abandonment of belief.

10.4.5 Competition of Memes in the Frequency of Copying their Abiotic Carriers—Memetic Drive

A distinct mechanism that drives the spread of memes in human populations is memetic drive. Unlike many other organisms, humans do not rely solely on non-linguistic forms of social learning, such as imitation: many memes are transmitted through spoken or written language. The abiotic carriers of these memes—such as linguistic content, scripts, or printed media—then often shape the direction of memetic evolution. For instance, if a word is difficult to pronounce or is infrequently used, it is more likely to be gradually replaced even if the new variant is less accurate or less useful to the meme's original function. Such changes may lead to a loss of meaningful content. For example, the species name of the American duck *Cairina moschata* is etymologically linked to its musky odour. In several European languages, the bird is called the 'musk duck', but in English there appeared in the eighteenth century an alternative name, 'muscovy duck'. Although it may have originally been understood as a variation of the 'musk duck', in modern English it connotes 'Moscow duck' although the species has no connection to Russia (Stahl, 2008). The meme 'muscovy duck', largely devoid of accurate referential content, nonetheless spread across English-speaking populations due to its exotic connotation, and was eventually reinforced by legitimising myths (such as a historically unsupported claim that they were imported by the Muscovy Company, an English trading enterprise active in the sixteenth century). Over time, this name became the standard: as of May 2025, a Google search returns over 3,500,000 instances of 'muscovy duck' but only 127,000 instances of 'musk duck'.

In such cases, it is the form of a meme, not its content, that drives its replication. A new variant of a message, be it oral or written, may spread not because it conveys useful information, but because it is perceived as aesthetically more pleasing or stylistically compelling. In interaction with other memes, it may even influence the prevailing notions of what is considered beautiful or interesting. The 'drive' towards aesthetic forms is thus not external to the meme but partly co-constituted by it (Ihde, 1990). This is particularly evident in songs, poems, or oral epics (see Chap. 11), which may be passed down for dozens or hundreds of generations. In such cases, the substantive content may have become outdated or even incompatible with current cultural norms, but the form ensures its continued transmission.

10.4.6 Competition of Memes in the Frequency of Their Independent Repeated Emergence—Mutation Pressure

The success of certain memes need not result from frequent imitation. It can also be due to mutation pressure, that is, their tendency to arise repeatedly and independently, either through a modification of existing memes or entirely *de novo*. These recurrent emergences are shaped both by the structure and function of the human nervous system and by the cultural environment, including the influence of existing memes and meme complexes. It thus constitutes a form of cultural attraction (cf. Chap. 11).

For example, the appealing meme that everyone should be equal in terms of opportunities can easily emerge as a mutation of a meme according to which everyone should be equal in terms of rights. In contrast, a mutation in the opposite direction (i.e., differentiation between rights and opportunities) requires more cognitive effort and is therefore less likely to arise spontaneously. This asymmetry in mutation frequency means that the meme advocating equality of opportunity can persist and spread within a population's meme pool even in the face of repeated negative experiences of societies which tried to implement this principle.

A striking example of how mutation pressure can shape memetic composition is the phenomenon known as the Mandela Effect, a form of collectively shared false memory (Prasad & Bainbridge, 2022). This term was coined by the American author and self-described paranormal researcher Fiona Broome in the late 2000s, who observed that many people in the Anglophone world incorrectly believed that Nelson Mandela, the South African political activist and former president, had died in prison in the 1980s. In reality, Mandela lived until 2013, dying at the age of 95. Even so, individuals reported detailed and vivid false memories concerning his supposed death decades earlier. This phenomenon illustrates how, under similar psychological and cultural conditions, brains tend to mutate memes in regular and predictable ways, often in line with culturally familiar narrative templates or expectations, such as the trope of a political dissident dying in prison. Such systematic distortions support the idea that mutation pressure—and not just selective imitation—plays a critical role in shaping the memetic landscape.

10.5 The Direction of Memetic Evolution

Memes spread in a population from carrier to carrier much like parasites spread among biological hosts. Many of the principles formulated by evolutionary parasitologists can thus be applied to the dynamics of cultural evolution (see, e.g., Ewald, 1994).

A key parameter that characterises memetic evolution and determines which memes will ultimately dominate a population's meme pool is the basic reproduction number (R_0). This figure indicates the average number of individuals not yet exposed to the meme who will be infected by a single meme carrier in the course of their lifetime. It assumes a naïve population in which no individuals are yet infected or resistant. The effective reproduction number (R_t) extends this model to real populations where some individuals are already infected or resistant. If R_t exceeds one, the meme spreads exponentially; if it is less than one, the meme will gradually fade out.

In the case of memes that reduce the lifespan of their carriers, selection favours variants that optimise their harmful effects, taking into account the average lifespan of the population, the frequency of infected and resistant individuals, and the mode of transmission. In populations with a low life expectancy (due to endemic disease, malnutrition, warfare, and the like) more harmful memes can spread successfully. When life expectancy is already low, the harm a meme causes to its host has minimal impact on R_0, which means that its chance to spread remains largely unaffected as long as it transmits before the host dies. In contrast, in populations where individuals live longer, memes that preserve their hosts are favoured. This logic implies that one effective way of counteracting the spread of harmful memes (such as those associated with terrorism or drug addiction) is to increase life expectancy by reducing violence and improving access to healthcare, sanitation, and basic nutrition.

Although this need not be apparent from watching contemporary news coverage, there has been a long-term global decline in both organised and interpersonal violence, particularly over the last millennium (Pinker, 2011). Just a few thousand years ago, an adult (especially a male) was more likely to die from violence than from any other cause. Since then, the proportion of violent deaths has steadily declined, albeit with fluctuations. The primary driver of this trend is the extension of human life expectancy that has been enabled by better access to food and energy, advances in medical care, and improvements in hygiene. But this dynamic operates through a positive feedback loop. A temporary surge in violence may lead to a reduction of the average lifespan, which in turn creates favourable conditions for the spread of more aggressive or 'violent' memes—and these, in turn, may further reduce life expectancy, thus reinforce the cycle and enable even more harmful memes to proliferate (Fig. 10.1).

The success of harmful (necrotrophic) memes also depends on the likelihood of transmission to uninfected individuals, which is in turn influenced by the level of 'immunity' present within the population. In the aftermath of events such as war or the fall of a totalitarian regime, the collective memory of consequences of violent ideologies may lead to a widespread resistance, which lowers the probability of

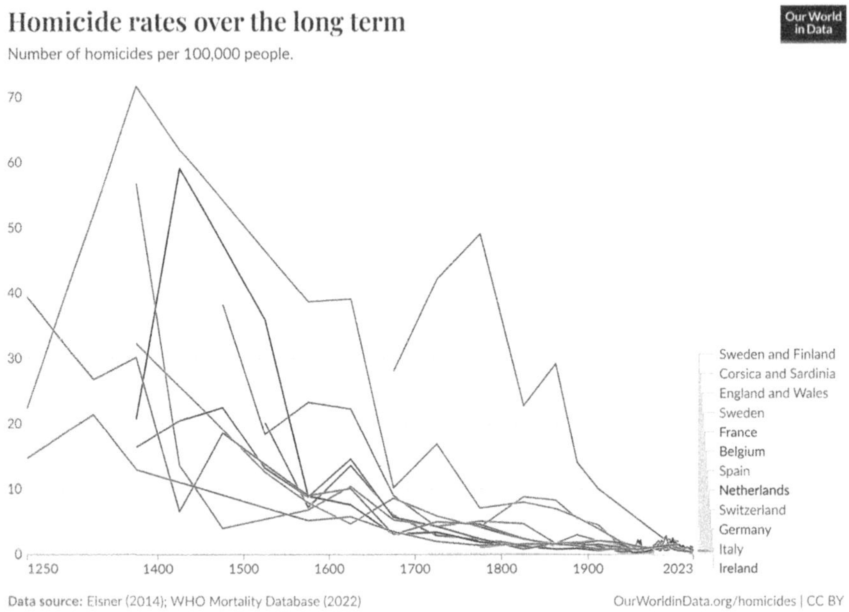

Fig. 10.1 Homicide rates in Western European regions, 1250–2023, showing a long-term decline in violence. Data from Eisner (2014), visualised by Our World in Data (https://ourworldindata.org/homicides)

spread of such memes. In such contexts, nonviolent or relatively beneficial memes tend to dominate. Over time, however, the proportion of resistant individuals declines, largely due to natural ageing and death of those whose firsthand experience conveyed the strongest 'immunity'. As this historical memory fades, the population becomes increasingly vulnerable to a re-emergence of harmful memes. This dynamic helps explain the cyclical recurrence of ideological violence throughout history.[4]

Memetic immunity also diminishes in a growing population, where newly born individuals dilute the proportion of resistant carriers. In such settings, the conditions become more favourable for the spread of harmful memes (such as those promoting violent behaviours or ideologies) than in stable or shrinking populations. To prevent the spread of harmful memes, educational efforts and critical thinking can function as a form of memetic vaccination by building resistance within the population. If the proportion of resistant individuals remains above a certain threshold, the effective

[4]The same dynamic was effectively documented by Fanta et al. (2019) in their study of settlement patterns in flood zones along Czech rivers. For approximately one generation following each major flood, new settlements were established in safer, elevated locations. But in the next generation, the awareness of flood risk declined, and new settlements gradually shifted closer to the riverbanks. Remarkably, this pattern recurred consistently over the entire study period which encompassed nine centuries.

reproductive number (R_t) stays below one, thereby preventing a large-scale memetic outbreak. But if this proportion drops below the threshold, the meme may begin to spread rapidly.

The mode of transmission also plays a significant role. Memes which are transmitted mostly vertically (from parents to offspring) are more likely to be biotrophic, that is, to enhance rather than diminish biological fitness. In contrast, memes transmitted horizontally, especially through imitating unrelated individuals, can harm their carriers without significantly impairing their ability to spread. Memes that spread through direct personal contact tend to be less harmful than those disseminated electronically across great distances between strangers. This is because the negative consequences of locally spread harmful memes often directly affect the disseminator and their kin, creating a selection pressure against such memes. Conversely, long-distance transmission can decouple the meme's harmful effects from the fitness of the disseminator, thus enabling necrotrophic memes to proliferate (cf. dual inheritance theory; Chaps. 7, 10, 16). As mobility, globalisation, and especially long-distance information flow accelerate, the likelihood of necrotrophic memes spreading increases.

The risk of spreading more harmful memes also rises with the increasing potential for superinfection, that is, with the growing number of distinct memes to which an average individual is exposed. The virulence of memes, that is, their capacity to cause harm, tends to negatively correlate with the probable lifespan of the infected individual. A necrotrophic meme harms its carrier and kin relatively less when it shortens the life of an individual who would soon die from other causes rather than one who would otherwise live long and continue to spread the meme and his genes. As a result, strongly necrotrophic memes, such as those promoting violent struggle against ideological opponents, are unlikely to spread widely in stable societies with a high life expectancy. Nevertheless, when individuals are exposed to numerous competing memes, some of which may drastically shorten host lifespan (for instance, by promoting spectacular suicidal terrorist acts), the selection pressure shifts. In such contexts, the fitness of a meme no longer depends primarily on its basic reproduction number (R_0) but rather on how many hosts it can infect per unit of time. This dynamic favours rapidly spreading, low-latency memes, which are often more harmful than the slower, more stable variants. Such conditions help explain why violent ideological movements (e.g., totalitarian regimes such as Nazism) are typically preceded by periods of societal breakdown and a simultaneous emergence of multiple contradictory extremist ideologies.

A similar effect can be produced by high memetic mutability, that is, the tendency of memes to vary quickly over time. In the case of superinfection, different memes enter the individual from the outside; in high-mutability environments, variants are generated internally, within individuals or tightly connected subpopulations. Under such conditions, even memes with a low R_0 but high transmission velocity (often corresponding to more harmful variants) can become dominant. For instance, orally transmitted ideologies are potentially more dangerous than ideologies rooted in written, canonised texts, which helps explain why institutionalised religions such as Christianity, Islam, or Judaism (all anchored in fixed scriptures) tend to be less harmful to their adherents than fringe sects or cults. Events such as the Jonestown mass suicide (1978) or the Waco siege (1993) illustrate the dangers associated with rapidly mutating, non-institutionalised belief systems.

Many of the risk factors facilitating the spread of increasingly necrotrophic memes discussed above—superinfection, mutability, and horizontal transmission—are intensified in the context of the current digital world, particularly in globalised, hyperconnected internet societies. Social media platforms such as YouTube, Facebook, X (Twitter), Instagram, and TikTok enable an unprecedentedly rapid, cheap, and targeted dissemination of information. This leads to a greater informational connectivity within populations but also to an intensive horizontal information flow among unrelated individuals across long distances. As of 2025, an estimated 5.2 billion of the world's 8.1 billion people use social networks. For many users, access to these platforms has increased their communicative reach by up to eight orders of magnitude and this trend is further amplified by the growing efficacy of machine translation, which enhances cross-cultural memetic flow.

This increase in informational connectivity among unrelated individuals creates conditions for harmful memes to spread across vast social and geographical distances. Moreover, digital technologies often produce a false sense of direct access to truth—what Don Ihde (1990) calls macro-perception—fostering the illusion of consensus. This perceived immediacy masks the interpretive filters and algorithmic infrastructures (micro-perceptions) that shape user experience. As a result, individuals with shared memotypes (completes set of memes carried by an individual, analogous to genotypes) tend to form selective clusters, which often leads to the formation of echo chambers: socially insulated environments where shared memes are continually reinforced while opposing views are excluded. Such conditions can lower community-level resistance to harmful memes. Once the threshold for outbreak is exceeded within one segment of society, the meme may cascade into others, creating conditions for a widespread contagion by dangerous memes.

As emphasised by the critical theory of technology and postphenomenology, digital platforms do not just transmit information: they mediate and co-construct user experience. Through their recommendation algorithms, interface design, monetisation models, and interaction architecture, these platforms actively configure behaviour. They do not simply host memes, they shape them. The result is the creation of conditions uniquely favourable to virality, including the spread of necrotrophic memes which exploit psychological vulnerabilities of users through mechanisms embedded in the technology's design (Feenberg, 2017) (Fig. 10.2).

Informational connectivity also increases the risk of superinfection by competing memes. Moreover, the process is subject to positive feedback dynamics. As noted earlier, environments saturated with necrotrophic memes (i.e., large volumes of harmful information) favour the proliferation of memes that can infect many hosts as rapidly as possible regardless of their potential harmfulness. This results in the normalisation of misinformation, manipulation, and outright falsehoods (Bak-Coleman et al., 2021). Additionally, digital environments such as social networks and online platforms introduce a greater degree of memetic instability compared to

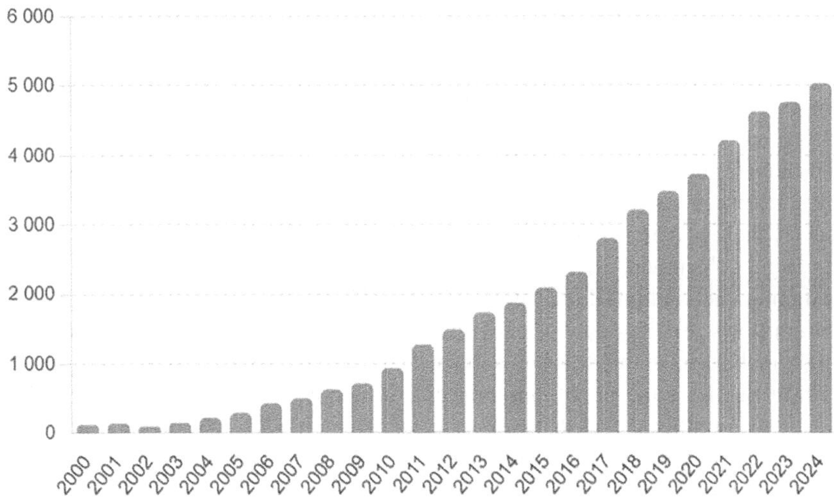

Fig. 10.2 Userbase of social media over time (in millions). Redrawn from Meltwater & We Are Social (2024)

earlier mass media ecosystems dominated by print and broadcast. These conditions probably contribute to a wide range of pathological societal phenomena, including radicalisation, institutional destabilisation, impaired collective problem-solving, and the spread of violent ideologies, fake news, and conspiracy theories. This canalisation of collective behaviour poses risks to democracy, scientific and technological progress, and the capacity to respond effectively to emerging global crises. The potentially harmful dynamics of digital environments where memes can spread widely regardless of their accuracy became acutely visible during the COVID-19 pandemic. The negative effects extended beyond harmless eccentricities, such as flat-Earth theory, to memes that significantly reduced the viability of their carriers, for instance by advocating a strict rejection of vaccination.

Given the relative novelty of these technologies—rooted in the development of ARPANET in 1969 but only widely adopted in the 1990s, with social networks gaining prominence in the early twenty-first century—it is challenging to predict their long-term impact on memetic dynamics and collective behaviour. Nevertheless, history offers precedents: prior increases in informational connectivity, such as those triggered by population growth, urbanisation, or the invention of the printing press, catalysed major transformations in social structure. These often gave rise to new mechanisms for stabilising collective behaviour, such as the emergence of organised religions or novel forms of governance. It is conceivable that the globally connected information society of the twenty-first century will evolve similarly profound systemic adaptations.

Efforts to combat pathological online dynamics often rely on interventions such as fact-checking, content moderation, banning or shadow banning,[5] and algorithmic filtering based on predefined value hierarchies. Unfortunately, such measures may produce unintended side effects and sometimes even amplify the very phenomena they aim to suppress; in short, they can backfire by leading to increased polarisation and radicalisation. As Bak-Coleman, Alfano, et al. (2021) have observed: 'We lack the scientific framework we would need to answer even the most basic questions that technology companies and their regulators face.' In this situation, it may be that the largely abandoned field of memetics could offer valuable conceptual tools for designing interventions that do not rely on rigid regulation or elitist control of information flows. Analogous to successful strategies in epidemiology, such strategies might include increased lifespan and societal stability, which tend to select against harmful memes, educational 'immunisation' through cultivation of critical thinking, prioritisation of local over distant interactions to limit the amplification of memetic contagion, prevention of the formation of echo chambers, where necrotrophic memes can incubate, and a stabilisation of the meme pool through the development of curated repositories such as online archives and encyclopaedias.

One promising approach in this direction is crowdsourced content moderation, exemplified by the 'Community Notes' feature introduced on the social media platform X (formerly Twitter). This mechanism harnesses the memetic dynamics for corrective purposes. Verified users can append to posts contextual notes whose visibility is then determined by an algorithm which rewards consensus across initially disagreeing voters. This system promotes critical engagement, disrupts echo chambers, and introduces negative feedback loops which amplify the circulation of high-quality, consensually validated information. Notably, such interventions do not necessarily reduce the spread of harmful memes directly (Chuai et al., 2023). Instead, they neutralise their influence through contextualisation, with the corrective notes themselves becoming successful counter-memes in a process analogous to the immunological memory of biological systems.

10.6 Rivalry Among Memetic Complexes

Neither genes nor memes exist in isolation. They tend to form larger structures known as memetic complexes or memplexes, systems of mutually reinforcing memes that increase each other's stability and transmissibility. These complexes typically emerge through stability-based sorting (cf. Chap. 15 for a more detailed discussion of this process). In most cases, it is not individual memes that compete within the cultural environment but entire memplexes of varying size and

[5] Such interventions may include preventing users from contributing for a fixed period of time, permanently banning them, or rendering their accounts or contributions effectively invisible to others.

complexity. In this process, efficiently transmitted or structurally stable memplexes tend to persist, while other ones vanish due to lack of adherents.

Some memetic complexes are mutually exclusive: for instance, a religion which commands 'Thou shalt believe in one god!' is inherently incompatible with polytheistic frameworks, but even such systems can exchange individual memes. This applies not only to religious traditions but also to secular ideologies. There are memplexes which thrive because they are for some reason resistant to extinction, while others succeed because they frequently re-emerge. Conversely, some memplexes decline rapidly due to internal fragmentation, because destabilising memes either evolve from within or enter from external sources. In this respect, conservative ideologies may be more stable than liberal ones due to their inherently lower tendency to change and typically lower internal variability. Other memplexes disappear entirely due to the extinction of the populations which carry them: these can disappear due to a demographic collapse (e.g., low reproduction rates), ecological catastrophe, or through a violent elimination.

Within a given memetic complex, individual memes have different functional roles. This is particularly evident in the case of religions, which are good examples of large and long-lived memplexes[6] (cf. Chap. 7). Many religious systems have endured for hundreds of generations and spread globally. Their durability is, at least in part, due to their design as interconnected meme systems that promote both internal stability and external transmissibility. Memplexes may include both biotrophic and necrotrophic memes, but long-term stability is generally associated with a biotrophic strategy, that is, with enhancing rather than reducing the fitness of hosts (cf. Chaps. 7, 16).

Many religious memes promote biological fitness, especially reproductive success. Some contribute to individual viability; this is the case of, for example, dietary laws prohibiting harmful or unhygienic foods or taboos against intragroup aggression, which protect both the individual and their kin who carry copies of the same memes and genes. More frequently, however, religious memes target fertility and promote higher reproduction rates by prohibitions against contraception, abortion, or masturbation. In some cases, they even impose practical barriers to these behaviours; for instance, circumcision has been interpreted as having such a function (Cox, 1995). Religions are typically transmitted vertically from parent to child (Cavalli-Sforza et al., 1982). For the long-term survival of such systems, it is advantageous that adherents produce numerous offspring. Additional memes help make the religion attractive and emotionally resonant: this is exemplified by, for instance, architectural design, visual art, music, and mutual assistance among believers (cf. Chaps. 7, 8). Moreover, it seems that human beings have an evolved tendency to seek hidden meaning or intentional agency in the world around them (Dawkins,

---

[6] Traditional story cycles (Oring, 2014), political ideologies, and even artistic styles can also serve as illustrative examples of memplexes. But large organised religions have a specific advantage: a long historical continuity along with incorporation of a wide array of interconnected memes, many of which directly influence the biological fitness of their hosts. As a result, their evolutionary dynamics is often especially informative and revealing.

2006). From an evolutionary standpoint, individuals who assumed that a snapping twig signalled a predator (rather than being just a random occurrence) were more likely to survive. Religions often satisfy this cognitive tendency by offering narratives that provide meaning, purpose, and comfort, particularly in times of suffering or existential uncertainty.

Some memes within a memplex function to increase its stability. This pertains most notably to canonised sacred texts which are transmitted across generations in fixed form and often preserved in ritual or archaic languages. This contributes to cultural inertia and makes such memes less susceptible to mutation. The wave of religious violence in Early Modern Europe was probably linked to the translation of sacred texts into vernacular languages. This development increased memetic volatility and created an environment more favourable to the spread of aggressive variants (cf. Chap. 18).

Other memes are dedicated to ensuring that adherents remain within the system. This is why many religions impose strict penalties for apostasy and promote complex rituals which reinforce group cohesion and facilitate the detection of nonconformity (cf. Chap. 10). Some religious or ideological memplexes also contain doctrines which are resistant to rational scrutiny, such as the Christian Trinity or the Marxist principle of negation of a negation. These can help identify and marginalise critically minded individuals who could otherwise challenge the coherence or transmissibility of the memplex.[7]

An essential component of religious memetic complexes is the presence of memes which facilitate an effective acquisition of new followers. As discussed earlier, religion is typically transmitted vertically, from parents to offspring. Many of its associated memes, rituals, and rules help ensure that transmission begins at the youngest age possible, during a developmental stage when humans are evolutionarily predisposed to easily, uncritically, and durably adopt memes from parental figures (cf. Chap. 7). On the other hand, successful religions also incorporate memes that promote a horizontal spread of the memetic complex to unrelated individuals. Such memes often compel believers to disseminate their faith beyond the confines of familial transmission, whereby such propagation may occur through personal example, persuasive discourse, or, in extreme cases, through coercion or violence (Fig. 10.3).

Religious and ideological memetic complexes may also spread via mechanisms analogous to group selection. This occurs when a given memplex increases intragroup cohesion, patriotism, discipline, or the willingness to sacrifice for fellow believers or deities. Societies, communities, or nations shaped by such complexes may expand at the expense of others due to enhanced collective efficacy. In biological evolution, the spread of genes that reduce individual fitness in favour of group

[7] This conclusion exemplifies the kind of statement that critics of memetics often dismiss as excessively scientistic. The appeal to logical consistency is, after all, a hallmark of modern rationalist thought, not a universal cognitive norm. Nevertheless, the perceived conflict between the two camps is ultimately illusory. Long-term success of any tradition is most likely shaped by the selection of individuals whose cognitive dispositions align with memes that promote particular modes of thought. Science is no exception: it merely selects in the opposite direction, favouring modern, 'domesticated' forms of logical reasoning.

Fig. 10.3 Organised religion, exemplified here by the Catholic Mass, serves as a good example of a memetic complex, i.e., a web of interconnected memes. Pexels (https://www.pexels.com/cs-cz/foto/cirkev-katolik-knez--katolicky-kostel-7219831/)

fitness is possible only under narrowly defined conditions (Flegr, 2023). Genes that promote altruism are typically outcompeted in the population by selfish variants, which leads to a decline in group-level altruism over time. In contrast, memes that promote conditional altruism face fewer evolutionary constraints. Because they are not limited to vertical transmission, such memes can spread horizontally to genetically unrelated individuals, thereby expanding their influence beyond kinship boundaries. Group selection may thus play a far more significant role in cultural evolution than in biological evolution; it can facilitate the success of memetic strategies that would be untenable in a strictly gene-centred framework.

10.7 Conclusions

Memetics experienced a rapid rise to prominence followed by an even swifter decline. This decline was driven by a combination of factors: an excessive focus on theoretically unproductive questions, difficulties in empirical validation, and social resistance not only from segments of the scientific community but even more so from scholars in humanities. As a theory that conceptualises memes as the basic units of culture, memetics was particularly in its more radical interpretations sometimes presented as a comprehensive theory of cultural evolution. Such claims, however, are easy to challenge. Culture encompasses not only replicable informational units with memetic properties but also artefacts and phenomena that do not behave like memes at all. Some cultural forms emerge from historical contingencies, political interventions, or power dynamics, while other forms are shaped by biological or environmental constraints. For example, the vocal range of musical traditions may vary according to the physiological characteristics of different populations, such as typical body mass, which influence the acoustic properties of the voice. Similarly, cultural artefacts will differ significantly across environments for purely ecological reasons: in humid regions, organic materials rapidly succumb to moulds or termites, while in arid or semi-arid areas, such materials may persist for centuries. This pattern may extend even to the phonetic and acoustic features of languages (Maddieson, 2018; Maddieson & Coupé, 2015). Cultural evolution thus most certainly does not fully coincide with the evolution of memes.

In its universalist ambition, memetics parallels other theories that seek to explain a wide range of complex phenomena through a single unifying lens, for example, the gene-centric model of biological evolution and the psychoanalytic framework of Freudian theory. Like these, memetics had sparked considerable controversy and provoked intense criticism. But as with those frameworks, it would be a mistake to dismiss it entirely. The insufficiency—or rather inadequacy—of gene-centrism to explain macroevolutionary patterns, symbioses, or the emergence of novel traits does not invalidate its explanatory power in accounting for the dynamics of selfish genetic elements or the evolution of adaptations that limit their spread. Similarly, the empirical shortcomings of Freudian psychology do not negate the centrality of sexuality in human behaviour. Perhaps we should adopt a similar stance toward memetics: while it is evident that memetics fails as a universal theory of culture, its core insights remain valuable, especially when it comes to better understanding the conditions under which cultural variants spread and persist. Many aspects of cultural evolution, especially the spread of harmful or maladaptive information in globally connected information society, are difficult—if not impossible—to grasp with other conceptual tools.

Crucially, the rise of social media, which coincided almost exactly with the decline of memetics in academic circles, has opened new empirical and theoretical frontiers. Digital platforms offer unprecedented opportunities for real-time tracking of large-scale horizontal information flows, thus enabling empirical testing of hypotheses which were previously inaccessible. Still, as Feenberg (2017) argues in his critical theory of technology, social media are not neutral platforms. They are sociotechnical systems shaped by design decisions, institutional priorities, and normative assumptions. These systems actively structure the environment in which memes and their complexes emerge, mutate, and spread (cf. the evolution of evolvability, Chap. 15).

Incorporation of insights from the philosophy of technology, attractionist models of cultural evolution (cf. Chap. 11), and biosemiotics into memetics seems to be the way towards a more nuanced 'neo-memetic' framework. Such approach would not only explain why certain memes succeed but also offer tools for understanding how platform design, cultural contexts, and community dynamics influence memetic processes. In doing so, neo-memetics could provide a powerful theoretical and practical perspective that may help policymakers, designers, and communities navigate and shape the memetic ecologies of digital society in more resilient, equitable, and reflexive ways.

References

Ardlie, K. (1998). Putting the brake on drive: Meiotic drive of t haplotypes in natural populations of mice. *Trends in Genetics, 14*(5), 189–193.

Atkinson, M., Smith, K., & Kirby, S. (2018). Adult learning and language simplification. *Cognitive Science, 42*(8), 2818–2854.

Bak-Coleman, J., Alfano, M., Barfuss, W., Bergstrom, C., Centeno, M., Couzin, I., Donges, J., Galesic, M., Gersick, A., Jacquet, J., Kao, A., Moran, R., Romanczuk, P., Rubenstein, D., Tombak, K., Van Bavel, J., & Weber, E. (2021). Stewardship of global collective behavior. *Proceedings of the National Academy of Sciences of the United States of America, 118*(27).

Blackmore, S. (1999). *The meme machine*. Oxford University Press.

Boyd, R., & Richerson, P. J. (1985). *Culture and the evolutionary process*. University of Chicago Press.

Brodie, R. (1996). *Virus of the mind: The new science of the meme*. Integral Press.

Cavalli-Sforza, L., & Feldman, M. (1981). *Cultural transmission and evolution: A quantitative approach*. Princeton University Press.

Cavalli-Sforza, L., Feldman, M., Chen, K., & Dornbusch, S. (1982). Theory and observation in cultural transmission. *Science, 218*(4567), 19–27.

Chuai, Y., Tian, H., Pröllochs, N., & Lenzini, G. (2023). *The roll-out of community notes did not reduce engagement with misinformation on Twitter*. arXiv preprint arXiv:2307.07960.

Chvaja, R. (2020). Why did memetics fail? Comparative case study. *Perspectives on Science, 28*(4), 542–570.

Cox, G. (1995). De Virginibus Puerisque: The function of the human foreskin considered from an evolutionary perspective. *Medical Hypotheses, 45*, 617–621.

Darwin, C. (1859). *On the origin of species by means of natural selection or the preservation of favoured races in the struggle for life*. John Murray.

Dawkins, R. (1976). *Selfish gene*. Oxford University Press.

Dawkins, R. (2006). *The god delusion*. Bantam Press.

Deacon, T. (1999). *Memes as signs*. The Semiotic Review of Books, 10(3), 1–3.

Dennett, D. C. (1998). *Memes: Myths, misunderstandings and misgivings*. Vol. 2017 Draft for Chapel Hill, October 1998. Retrieved 24 January, 2022, from https://ase.tufts.edu/cogstud/dennett/papers/MEMEMYTH.FIN.htm

Derex, M., & Mesoudi, A. (2020). Cumulative cultural evolution within evolving population structures. *Trends in Cognitive Sciences, 24*, 654–667.

Edmonds, B. (2005). The revealed poverty of the gene-meme analogy—why memetics per se has failed to produce substantive results. *A Memetics Compendium*, 483–485.

Eisner, M. (2014). From swords to words: Does macro-level change in self-control predict long-term variation in levels of homicide? *Crime and justice, 43*(1), 65–134.

Ewald, P. W. (1994). *Evolution of infectious disease*. Oxford University Press.

Fanta, V., Šálek, M., & Sklenicka, P. (2019). How long do floods throughout the millennium remain in the collective memory? *Nature Communications, 10*, 1105.

Feenberg, A. (1999). *Questioning technology*. Routledge.

Feenberg, A. (2017). *Technosystem: The social life of reason*. Harvard University Press.

Flegr, J. (2002). Was Lysenko (partly) right? Michurinist biology in the view of modern plant physiology and genetics. *Rivista di Biologia-Biology Forum, 95*, 259–271.

Flegr, J. (2023). Evolution by non-individual selection pressures. In T. K. Shackelford (Ed.), *Encyclopedia of sexual psychology and behavior*. Springer.
Fomin, I. (2019). Memes, genes, and signs: Semiotics in the conceptual interface of evolutionary biology and memetics. *Semiotica, 230*, 327–340.
Henrich, J., & Gil-White, F. J. (2001). The evolution of prestige: Freely conferred deference as a mechanism for enhancing the benefits of cultural transmission. *Evolution and Human Behavior, 22*(3), 165–196.
Herrmann-Pillath, C. (2021). Biosemiotic foundations of a Darwinian approach to cultural evolution. *Cultural Science Journal, 13*(1), 10–2478.
Heylighen, F., & Chielens, K. (2009). Evolution of culture, memetics. In R. Myers (Ed.), *Encyclopedia of complexity and systems science* (pp. 3205–3220). Springer.
Hinde, R. A., & Fischer, J. (1952). Further observations on opening of milk bottles by birds. *British Birds, 44*, 306–311.
Hoffmeyer, J. (2007). Semiotic scaffolding of living systems. In M. Barbieri (Ed.), *Introduction to biosemiotics: The new biological synthesis* (pp. 149–166). Springer.
Hofstadter, D. (1983). Virus-like sentences and self-replicating structures. *Scientific American, 248*, 14–19.
Hofstadter, D. (1985). *Metamagical themas: Questing for the essence of mind and pattern*. Basic Books.
Hull, D. L. (1980). Individuality and selection. *Annual Review of Ecology and Systematics, 11*, 311–332.
Ihde, D. (1990). *Technology and the lifeworld: From garden to earth*. Indiana University Press.
Jahoda, G. (2002). The ghosts in the meme machine. *History of the Human Sciences, 15*(2), 55–68.
Kantorovich, A. (2014). An evolutionary view of science: Imitation and memetics. *Social Science Information, 53*(3), 363–373.
Kawamura, S. (1963). The process of sub-cultural propagation among Japanese macaques. In C. H. Southwick (Ed.), *Primate social behaviour* (pp. 82–90). Van Nostrand.
Kirby, S., Cornish, H., & Smith, K. (2008). Cumulative cultural evolution in the laboratory: An experimental approach to the origins of structure in human language. *Proceedings of the National Academy of Sciences of the United States of America, 105(31), 10681–10686.*
Kull, K. (2000). Copy versus translate, meme versus sign: Development of biological textuality. *European Journal for Semiotic Studies, 12*(1), 101–120.
Kundt, R. (2015). *Contemporary evolutionary theories of culture and the study of religion*. Bloomsbury Publishing.
Laurent, J. (1999). A note on the origin of memes/mneme. *Journal of Memetics, 3*(1), 14–19.
Lefebvre, L. (1995). The opening of milk bottles by birds—Evidence for accelerating learning rates, but against the wave-of-advance model of cultural transmission. *Behavioural Processes, 34*, 43–53.
Lumsden, C., & Wilson, E. (1981). *Genes, mind and culture: The coevolutionary process*. Harvard University Press.
Lynch, A. (1996). *Thought contagion: How belief spreads through society*. Basic Books.
Maddieson, I. (2018). Language adapts to environment: Sonority and temperature. *Frontiers in Communication, 3*, 28.
Maddieson, I., & Coupé, C. (2015). Human spoken language diversity and the acoustic adaptation hypothesis. *The Journal of the Acoustical Society of America, 138*(3), 1838–1838.
McNamara, A. (2011). Can we measure memes? *Frontiers in Evolutionary Neuroscience, 3*(1), 9334.
Meltwater, & We Are Social. (2024). *Global Digital Report 2024*. Retrieved from https://www.meltwater.com/en/blog/digital-2024. Accessed 14 June 2024
Mealey, L. (2000). Anorexia: A "losing" strategy? *Human Nature, 11*, 105–116.
Oring, E. (2014). Memetics and folkloristics: The applications. *Western Folklore, 73*(4), 455–492.
Pinker, S. (2011). *The Bbetter angels of our nature*. Viking.

Prasad, D., & Bainbridge, W. A. (2022). The visual Mandela effect as evidence for shared and specific false memories across people. *Psychological Science, 33*(12), 1971–1988.
Pribram, K. (Ed.). (1969). *On the biology of learning*. Harcourt, Brace & World.
Rádl, E. (2015). *The history of biological theories*. BiblioLife.
Saldana, C. (2018). *Simplifying linguistic complexity: Culture and cognition in language evolution*. PhD The University of Edinburgh.
Spuhler, J. (1959). *The evolution of man's capacity for culture*. Wayne State University Press.
Stahl, P. (2008). Animal domestication in South America. In H. Silverman & W. Isbell (Eds.), *The handbook of south American archaeology* (pp. 121–130). Springer.
Steiniger, F. (1950). Beiträge zur Soziologie und sonstigen Biologie der Wanderratte. *Zeitschrift für Tierpsychologie, 7*, 356–379.
Toman, J., & Flegr, J. (2017). Stability-based sorting: The forgotten process behind (not only) biological evolution. *Journal of Theoretical Biology, 435*, 29–41.
Tyler, T. (2011). *Memes and the science of cultural evolution*. Mersenne publishing.
Vada, Ø. (2015). What happened to Memetics? *Emergence: Complexity & Organization, 17*(2), 1–5.
Watson, R., & Szathmary, E. (2016). How can evolution learn? *Trends in Ecology & Evolution, 31*(2), 147–157.
Whitehead, H. (1998). Cultural selection and genetic diversity in matrilineal whales. *Science, 282*, 1708–1711.
Whitehead, H. (1999). Culture and genetic evolution in whales. *Science, 284*, 2055–2055.
Wilson, E. (1975). *Sociobiology: The new synthesis*. Harvard University Press.
Wilson, E. (1978). *On human nature*. Harvard University Press.

Chapter 11
Cultural Attraction as One of the Pillars of Cultural Evolution

Pavlína Hillerová and Petr Tureček

Abstract During the second half of the twentieth century, there emerged two prominent branches of cultural evolutionism. The selectionist framework emphasised the importance of variation and selection of relatively stable cultural variants, while the school of thought known as cultural attractionism argued for the significance of transformation of one variant into another. Cultural attractor was usually described as the variant that is unwilling to change into other variants, while changes of other variants towards it are likely. Over time, the two directions of thought ceased arguing about the primacy of their explanatory frameworks and started to productively influence one another. The attractionists ultimately concluded that selection is a special case of attraction and described a superposition of several variants with relatively stable frequencies within a single population as the attractor of the entire dynamic system. This shift drew attention to the formal description of processes of cultural transmission and transformation. It is often difficult to determine whether what is dominant is systematic transformation or rather the selection of stable variants but in some cases (often in experimental settings), it is possible to estimate the relative contribution of these phenomena. Studies have shown that differences in the transformation probabilities (which are usually not 'guided' or even recognised by cultural agents) can explain some cultural phenomena, such as the emergence of structured grammar, very well.

P. Hillerová
Department of Philosophy and History of Sciences, Faculty of Science, Charles University, Prague, Czech Republic
e-mail: pavlina.hillerova@natur.cuni.cz

P. Tureček (✉)
Department of Philosophy and History of Sciences, Faculty of Science, Charles University, Prague, Czech Republic

Center for Theoretical Study, Charles University and Czech Academy of Sciences, Prague, Czech Republic
e-mail: petr.turecek@natur.cuni.cz

J. Toman et al. (eds.), *Culture Evolving*, Biosemiotics 10,
https://doi.org/10.1007/978-3-032-16124-6_11

Keywords Cultural attraction · Attractor · Transformative transmission · Transmission chain experiment · Evolutionary causal matrices

11.1 Introduction

What are scientists referring to when they discuss culture? The French philosopher and cognitive scientist Dan Sperber is one of those who pondered upon this question and this line of thought determined a large part of his career. During this fruitful period, he explored the possibility of cultural evolution independently of any parallels with biological mutation and selection. His view of the evolution of culture, which came to be known as attractionism, inspired a new area of research to which this chapter is dedicated.

In the following, we explore the theory of cultural attraction as a central framework within the field of cultural evolution. We examine its historical development, key theoretical components, and its relationship with selectionist approaches. We highlight how systematic transformation shapes culture and discuss the evidence that supports attractionist claims.

The chapter primarily serves as a review and synthesis of existing literature on cultural attraction theory, with emphasis on foundational work and subsequent theoretical refinements. Several examples (Box, Figs. 11.1, 11.7 and 11.8) and

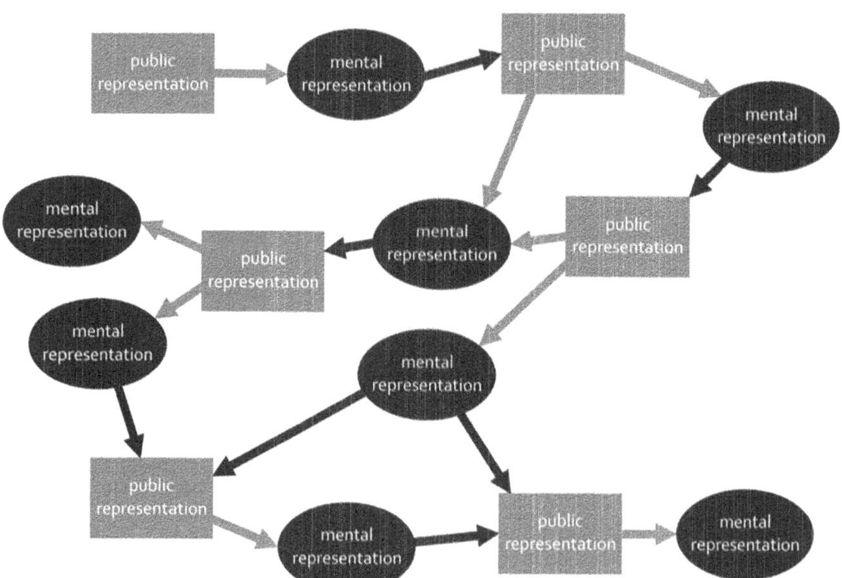

Fig. 11.1 A chain of mental and public representations, which according to Sperber constitute culture. Author: Petr Tureček

theoretical insights (the possibility of absence of variance in evolution through attraction, the extension of Causal Evolutionary Matrices to reflect frequency dependence) are, nevertheless, original.

The chapter opens with an outline of the origins of cultural attraction theory, with focus on Sperber's critique of anthropological and selectionist approaches. We introduce certain key concepts, such as epidemiology of representations, mental and public representations, or the role of transmission chains (and transformations that take place during transmission) in cultural evolution. Following this, we examine some experimental evidence illustrating the effects of psychological and ecological factors on an uneven probability of cultural transformation. We explain that differences in transformation probabilities render some cultural variants fleeting and others stable. In the concluding sections, we introduce the Causal Evolutionary Matrices, which were developed by Sperber's students, and discuss a reconciliation of attractionist and selectionist perspectives, showing how in current research these models became increasingly integrated.

11.2 The Beginning of an Alternative

Sperber's understanding of culture draws primarily on epidemiology and ontological naturalism. His work in this area was driven mainly by two reasons: the shortcomings of the then prevalent anthropological approaches and the inaccuracies of cultural concepts which are based on parallels with biology, especially genetics (Sperber, 1985, 1996; cf. Chaps. 10, 18).

In his critique of the anthropological approach to culture (cf. Chaps. 5, 7), Sperber focuses on three issues:

- Interpretive generalisation, that is, progressive generalisation from a specific phenomenon within a particular culture to all phenomena of that type across all cultures. Sperber argues that this seldom explains much and this approach is usually appropriate only for local phenomena.
- Structural analysis, which compares features shared by different phenomena. For instance, Claude Lévi-Strauss (1973) argued that myths are generated from other myths not only through imitation but also through systematic changes in their properties, such as swapping the protagonist's gender, changing the setting to a nearby mountain for a vertical direction or, conversely, following the protagonist into the world for a horizontal direction (Goldstein, 1987). Sperber does not find this approach inherently problematic but points out that systemic similarities and differences can be found between almost any two objects; he later illustrates this by comparing the Little Red Riding Hood with Hamlet, that is, two characters most unlikely to share a common origin or be derived from one another (Sperber, 1996).

- Functional explanations. Sperber argues that it is far too common to view a cultural phenomenon as sufficiently explained if the explanation points to its benefits to the society. While often true, it usually does not account for the phenomenon's origin. Sperber is highly critical of biological approaches to culture, finding the dual inheritance theory (DIT) and memetics (cf. Chaps. 7, 10, 16) as problematic as anthropological structuralism and functionalism. He claims that none of these perspectives accounts for the imperfection of humans as transmission devices. Although both DIT and memetics do employ a kind of cultural drift or occasional mutation, Sperber (1985, 1996) believes transformations occur very frequently, almost always (cf. Chap. 15).

11.3 Cultural Epidemiology

Sperber contrasts his 'epidemiology of representations', which encompasses mental and public representations, with both the memetic and the anthropological approach. Mental representations, that is, configurations of neurons and their activation patterns in the brain (triggered for instance by fairy tales such as that about the Little Red Riding Hood), lead to public representations, such as air vibrations that can be interpreted by other hearing beings as sequences of words—and, vice versa, words lead to specific brain activity in the listeners' hippocampus and ultimately to the formation of individually accessible memory traces. A mental representation never produces another mental representation (because telepathy is at least not yet part of our daily lives) and public representation does not automatically lead to its copy (except in rare cases when someone blindly copies a random document without examining its content). In short, what takes place in the process of cultural transmission is always some transformation (cf. Chap. 15). Instead of the fairy tale about Little Red Riding Hood, we might imagine anything—a melody, a way of holding the spade, or polyandry, a practice often found among mountain populations practicing a form of agriculture (Goldstein, 1987). The altitude itself obviously does not directly affect family organisation; along with other related factors, especially low soil fertility, it must first influence the minds of persons who live under these conditions (and lead to certain mental representations). Such influenced minds then increase (in public representation) the prevalence of a particular pattern of cohabitation that would not necessarily make sense under different conditions. Sperber and his students who adopted a more 'moderate' version of his teachings are not trying to turn anthropologists into epidemiologists but see it as necessary that this perspective be introduced into the study of culture.

Sperber's contemporaries Cavalli-Sforza and Feldman (1981) applied epidemiological models to model the transmission of culture. They highlighted the similarity of the modes of spread: for reproduction to take place, an 'infected' individual (in culture, e.g., an individual with some knowledge) must encounter a susceptible individual (capable of receiving and processing this information) (cf. Chap. 10). Aside from the basic model, they also adopted some epidemiological terminology, such as the notions of vertical and horizontal transmission (Cavalli-Sforza & Feldman, 1981). Sperber then adopted and slightly modified this idea when he pointed out that mental representations undergo, in the course of transmission, significantly more changes than envisaged by epidemiological models of infectious diseases. Still, epidemiological models seem suitable also because representations, like pathogens, can spread at varying speeds and persist for different lengths of time (cf. Chap. 15). Sperber also acknowledged the need for an individual approach to particular cultural phenomena, as is the case with diseases. For instance, even a layperson intuitively understands the difference between malaria (spread by an insect vector), measles (airborne disease that spreads directly through cough or sneeze droplets), and colon cancer (idiosyncratic occurrence in each patient, potentially mediated by heritable genetic predispositions but with no horizontal transmission). In a similar way, a funeral ritual should be approached differently than urban legends or the classification of colours (Sperber, 1996). There are differences in the speed and manner of spreading between different phenomena but also in the various factors that can influence their transmission and transformation (cf. Chap. 10).

11.4 Representations

As noted above, public representations are observable objects (a sculpture, a written sheet, an audio cassette, etc.) or phenomena (sound waves in the air, dance movements, light signals). Mental representations, on the other hand, are configurations of neurons and action potentials on their membranes—that is, thoughts, opinions, memories. In later works, Sperber and his followers (Claidière et al., 2014; Miton, 2022; Scott-Phillips et al., 2018) do not necessarily employ such a strictly materialistic conception (cf. Chap. 18) but the notion of a 'reciprocal causality' between behavioural manifestations and their interpretations remains clearly present. The form outlined above is idealised: in reality, public representations often acquire their intended meaning only in conjunction with other representations. If we hear the sound of a drum, we sometimes cannot be sure whether it is a mode of communication or merely a form of amusement and, similarly, when we see someone stacking bricks, we often cannot tell whether the result will be a wall, a house, modern art, or a store (Sperber, 1996). Some mental representations can be caused by several public representations (e.g., forming an opinion after hearing the arguments of both sides), while in other instances, one public representation creates many mental representations (for instance, mental representations in each member of a crowd addressed by a speaker; see Fig. 11.1). Some public representations even create no mental representations (such as when a public speech is not listened to).

Many representations, both public and mental, are found only in a single individual and for a very short time, for example an unspoken thought or a brief grimace. Because the mental representations of others are inaccessible to us, when we speak about the 'spread of an idea', we usually refer to the extensions and bifurcations of a transmission chain in which we can describe and categorise only the public representations. Public representations are rarely exact copies of one another: the possibility of direct mechanical copying or simple 'sharing' of an existing information pattern is a relatively recent invention. It is moreover often impossible to clearly distinguish related public representations from similar clusters using necessary and sufficient conditions. Instead, the public representations that belong to a victorious spread of a single 'viral meme' are linked by a network of bilateral resemblances, where each two public representations in the same chain share more than one feature. Such pairs belong to the realisation of a single phenomenon in our minds and models thanks to a 'family resemblance', a Wittgensteinian *Familienähnlichkeit* (Wittgenstein, 1953). An example of such network of public representations which are causally linked but resemble each other only vaguely can be found in BOX.

The semiotic approach is a useful way of looking at the reciprocal causation between mental and public representations (Herrmann-Pillath, 2021). The process of constructing a mental representation evokes the triadic nature of a sign, where interpretation inherently implies transformation and reconstruction, rather than direct copying. Sperber's stress on the 'material' aspect of mental representations can then be treated as the flipside of Deely's (2009) approach to signs: where Sperber insists that what he means by a 'mental representation' must be convertible to neuronal wiring, hormonal balance, or excitation patterns, Deely would argue that humans 'operate' at the highest level of a semiotic hierarchy, that is, that humans understand that signs are not just stimuli or memory traces but relational constructs. Deely's claim that the only objective reality we perceive is the one constructed in our minds aligns with the premise central to the whole field of cultural evolution, namely that people are primarily cultural animals, organisms that live in a world that is 'made of' representations.

Box: Make America Great Britain Again

As an example of public representations that can be interpreted as parts of a causal chain, let us consider a collection of jokes playing with the idea that the election of Donald Trump lands the United States in a situation so precarious that it would have been better had it never separated from the British Empire.

It is nearly impossible to pinpoint the first expression of this idea. What is certain, though, is that on 17 November 2011, there appeared a satirical Facebook profile titled 'Queen Elizabeth for President 2012'. After the 2012 elections, the page was renamed 'Queen Elizabeth for President 2016', and the page started to react to actual campaign activities.[1] It featured slogans such as 'I'm with Her Majesty' (a parody of 'I'm with Her', used by Hillary Clinton's supporters) and 'Make America Great Britain again!' (in response to Trump's 'Make America Great again!'). This revised slogan of the Republican candidate, in this wording, appeared in fact already somewhat earlier on the crowdfunding site Indiegogo,[2] where one could also buy a cap with this slogan. According to internet archives, the most famous graphical representation of the idea 'Make America Great Britain again' (see Fig. 11.2) appeared on the page of the Indian musician Suresh Singaratnam,[3] but it gained popularity only several months later, after being shared on popular sites such as 9GAG and Reddit. Around the same time, other manufacturers and creators, for instance on Etsy, started offering caps, T-shirts, and other items with the same slogan as Indiegogo. Dozens of images featuring the queen's likeness and the slogan spread across the internet. Once the idea of 'Make America Great Britain again!' was pulled from the digital realm, it started to be understood as a form of authorial humorous creation. The slogan inspired a trailer for the new season of 'The Royals' broadcasted by the American entertainment television network E!. The trailer attracted attention on 19 October 2016 and just 10 days later, shortly before American presidential elections, Andy Borowitz published a satirical fictional news piece in *The New Yorker* with the headline 'Queen Offers to Restore British Rule Over United States',[4] in which Queen Elizabeth II encouraged citizens of the United

---

[1] https://www.facebook.com/Queen-Elizabeth-for-President-2016-293577707331440/

[2] https://www.indiegogo.com/projects/make-america-great-britain-again-diversity#/

[3] https://www.facebook.com/sureshsingaratnam/photos/make-america-great-britain-again-elizabeth2016/10153675189583649/

[4] http://www.newyorker.com/humor/borowitz-report/queen-offers-to-restore-british-rule-over-united-states

States to write her name on the ballots on the election day. Concurrently, there was created a successful YouTube video MAKE AMERICA BRITISH AGAIN.[5] Shortly after the elections, a letter by comedian John Cleese began to virally spread on the internet; in it, Cleese on behalf of the Great Britain announces to the United States that Her Majesty Queen Elizabeth II with immediate effect revokes the independence of the United States because its inhabitants proved themselves incapable of electing a competent president.[6]

In reality, John Cleese wrote the abovementioned letter already in 2008, shortly after the election of Barack Obama, and we can thus place this public representation at the beginning of the transmission chain discussed in this section. The original Facebook page Queen Elizabeth for President 2012 was probably inspired by Cleese's letter.

Fig. 11.2 One of the public representations of the meme 'Make America Great Britain again!'

[5] https://www.youtube.com/watch?v=Vy9zxAywndk

[6] https://starrgazr.wordpress.com/2008/02/15/john-cleeses-letter-to-america/

11.5 Attractors

The theory of cultural attractors, closely intertwined with Sperber's cultural epidemiology, is based on the premise that humans are not perfect transmission devices. During the transmission of (the public representations of) cultural variants, what takes place is both interpretation and reconstruction, and systematic transformation can take place during either or both of these steps. An attractor is a state which transformations more frequently lead towards than away from (Sperber, 1996).

Imagine the cultural space as a two-dimensional chessboard, and a specific representation as a square on this chessboard; then also accept the relatively uncontroversial assumption that transmission does not produce an identical copy. There are eight possible directions in which the 'cultural offspring' of any particular representation might move. For simplicity, let us call any two neighbouring public representations separated by a single step of conversion 'cultural parent' and 'cultural offspring'. Cultural attractionists often used the metaphor of the 'telephone game', where we can clearly define what is meant by two consecutive cultural generations: in this instance, it is two consecutive instances of whispering (Morin, 2016b). If there is an attractor in the cultural space, the probabilities of moving to different neighbouring squares are not equal (see Fig. 11.3) or, rather, one should say that these unequal probabilities are what creates a cultural attractor.

An attractor does not provide a causal explanation of a phenomenon, nor is it a specific cultural variant. The existence of an attractor indicates that, in a given space of representations, there is a higher likelihood of a shift in a certain direction, which leads to a higher occurrence of variants around a certain point (Buskell, 2017; cf. Chap. 15). But this point does not have to be a specific cultural variant; it is rather the equivalent of summary statistics known from geometry or demography. It is like the centre of gravity, which is influenced by the density and shape of an object, on its own explains nothing but allows us to understand many aspects of movement, or like the average life expectancy, which does not kill anyone but helps us to compare longevity in different parts of the world. In a similar way, an attractor is not causally active but can help us describe phenomena that happen 'around it'. The theory of cultural attractors does not describe the evolution of a system in a predetermined direction (as in 'everything converges towards the attractor'). Instead, it captures the statistical evolution of a population of cultural units that tend to systematically cluster despite the stochasticity of the transmission process (Billiard & Alvergne, 2018).

Depending on whether we follow the original strict definition (Sperber, 1996) or opt for a more recent and looser interpretation (Claidière et al., 2014), an attractor is either a hypothetical point towards which variants transform more frequently (attractor *sensu stricto*) or a frequency equilibrium of various variants created by transformative transmission (*sensu lato*, see below).

An intriguing example illustrating the concept of a cultural attractor as a stable representation of a larger number of variants (thus *sensu lato*) is a study that tracked the frequency of different knots. Participants were asked to tie a double knot they saw in a recording. There were four possible types of knots: two reef knots (tying

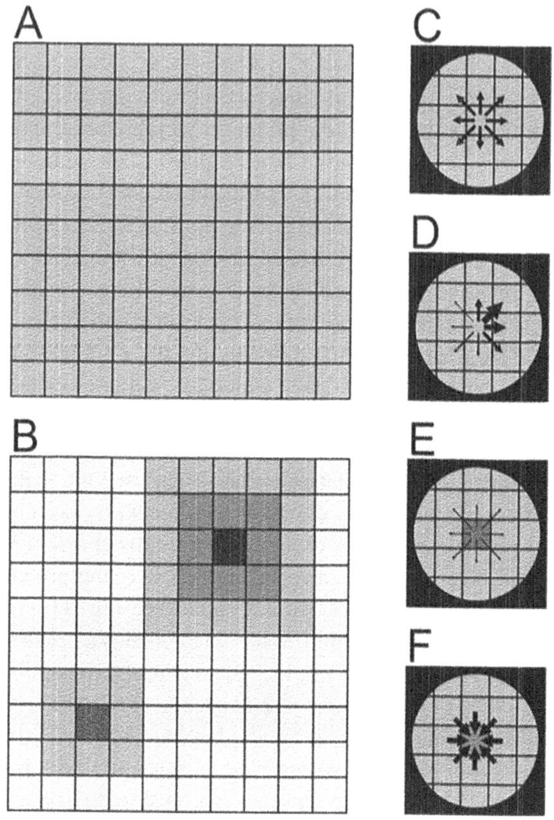

Fig. 11.3 Cultural attractors in a hypothetical (two-dimensional) cultural space. The shade of each square corresponds to the frequency of representation of a given cultural variant in the population (A, B). Although initially all cultural variants are equally represented (A), some will over time significantly prevail (B) because different transformations (C) are not equally probable (D). We can imagine a cultural attractor as a variant that is reluctant to transform into another (E), while transformation of other variants into it is probable (F). Author: Petr Tureček, inspired by Sperber (1996)

right over left, then left over right, or vice versa) and two granny knots (both knots are tied the same way, either twice right over left or twice left over right). Participants who were asked to tie a granny knot succeeded more often and most of their mistakes consisted in a full reversion of the knot (left over right instead of right over left, or vice versa). Those asked to tie a reef knot did not achieve such precise imitation—not just by reversing the knot. In over 30% of cases, instead of a reef knot, participants tied a granny knot (only 8% of participants managed to tie a reef knot instead of a granny knot) (Scanlon et al., 2019). In this case, an attractor emerged due to a tendency to make one type of mistake rather than another one: here, it was tying two identical knots rather than two different ones. The resulting frequency was also influenced by the fact that people tend to start knotting by crossing the left string over the right one. A model based on patterns of swapping knots after demonstration in arbitrarily long transmission chains was able to predict the basic frequency of double knots observed when scientists asked another group of participants to tie any double knot without demonstration. The individual 'micro-mechanisms' of imprecise transmission thus helped explain unequal frequencies of phenomena at the population level even though double knots are unlikely to be the subject of any significant selection pressure (cf. Chap. 10).

Another illustrative example could be Morin's (2013) study on the transformation of portrait paintings during the sixteenth century, when portraits with averted gaze were gradually replaced by portraits depicting persons looking directly at the painter. This was probably due to a preference for faces with a direct gaze (which is observable even in 3-day-old babies; Valenza et al., 1996), their easier recognisability, higher attractiveness, and easier identification. In this case, the attractor applied to a type of portrait and not to any specific painting. Nor is this an explanation of the phenomenon; it merely indicates a certain trend (Scott-Phillips et al., 2018). Nevertheless, it is not certain whether the change in the depicted individuals' gaze was truly just an attraction, that is, whether every apprentice exposed to the older type of portraits was more likely to shift towards paintings with a more direct gaze. Alternatively, one could claim that selection mechanisms played a role for example in the masters' choosing which examples from their collections they would show their apprentices. What is certain, however, is that a vast majority of cultural change occurred through generational replacement, not because 'old dogs learned new tricks'. Titian painted both averted and direct gazes with the same frequency in his twenties as he did in his sixties but younger painters showed a far greater tendency to paint their subjects captivatingly gazing right at the viewer.

11.6 Transmission Chains

Sequences of apprentices and students are crucial for studying cultural evolution, which is why they were 'boiled down to the essence' in an experiment (not unlike the aforementioned the 'telephone game') already at the very beginnings of development of cultural evolution as a field of study. Nearly one hundred years ago, Bartlett (1932) presented several students with a story about native Americans. The students were asked to read it (repeatedly) and write down what they remembered, thus creating the first generation of the transmission chain. The output then served as the template for subsequent participants, who read the creations of the first generation, wrote what they remembered, thus creating the second generation of the chain … and so on, for a certain number of generations. Bartlett found that details of the original story changed and some even disappeared. In this case, both the paper with the original story and each newly written variant of this story were public representations. The mental representations would be the versions of the story as they existed in the brains of the individual participants.

This design came to be known as the method of (linear) transmission chains and it is used to this day, albeit with modifications (Anglada-Tort et al., 2023; Uddenberg et al., 2023). Unless it is the subject of study, handwriting is nowadays not used. As the stimulus, researchers use stories, videos, drawings, or other forms, the mode of information transfer to the next generation is adapted accordingly, and researchers can use different numbers of generations. In fact, a significant part of research on cultural transmission and evolution is based on this experimental design, not necessarily in connection with the attractionist approach (Derex et al., 2024).

11.7 Transformations

Aside from public and mental representations, the attractionist framework tends to focus on changes that take place during transmission. While 'selectionists', that is, cultural evolutionists coming from the tradition of Boyd and Richerson (see, e.g., Boyd & Richerson, 1988; Richerson & Boyd, 2005), emphasise the importance of imitation as the equivalent of biological replication and random deviations from exact copying as the equivalent of genetic mutations (cf. Chaps. 10, 12), attractionists replace such two-step models with one-step transformative transmission (Sperber, 1996). They emphasise that transmission and change cannot be differentiated because, in culture, nothing 'replicates' without context, just like DNA polymerase proceeds along the nucleotides while attaching complementary bases to both strands of the double helix. All cultural elements are thus subjects to interpretation and their exact replication is thus unlikely (Kleisner & Tureček, 2017). Each step of the cultural chain from one public representation to another consists of information preservation and reconstruction. Inaccuracies, which take place during each of these phases, form one of the pillars of the attractionist approach. They may be somewhat random, as in mishearing or misspeaking, but a significant and more interesting part of inaccuracies is due to systematic distortion driven by psychological and/or ecological factors which act during transmission (Claidière et al., 2014; Griffiths et al., 2008). These influences cause disparities in the likelihood of changes of cultural variants during transmission, thus creating cultural attractors.

11.7.1 Psychological Factors

In Sperber's foundational work (1996), psychological factors encompass innate predispositions and tendencies present across human populations, such as the psychological underpinnings of intuitive biology (e.g., the ability to distinguish living things from non-living one), intuitive physics (such as awareness of the fact that one solid object cannot pass through another solid object), and intuitive psychology (for instance, some ability to discern which behaviour is intentional and which is not). In later works, psychological factors refer to all internal mechanisms, not only the innate ones (i.e., also skills, preferences, beliefs, knowledge, etc.), and it is assumed that newly acquired variants are more likely to be transformed so as to be consistent with already available information (Scott-Phillips et al., 2018). The likelihood of transformation is further influenced by several content biases, which have their equivalents in culturally selective pressures (cf. Chaps. 10, 13).

11.7.2 The Negativity Bias

Negativity bias has a significant influence on the transmission and transformation of representations. Most generally, it means that people pay more attention to information with a negative emotional load than to positive or neutral information (Fessler et al., 2014), so that everything from vomiting to unpleasant social interaction is more likely to capture our attention. From an evolutionary standpoint, this bias is explained by the notion that knowledge of such information, or rather of the process that led to the negative event, could protect us from similar situations in the future (Baumeister et al., 2001; Rozin & Royzman, 2001). Moreover, people not only prefer negative information, they even have a strong tendency to transform ambiguous formulations in a 'negative' direction. For example, the sentence 'The man took the older lady's bag' can mean either that he stole it or that he helped her carry it somewhere, but an experiment based on the design of transmission chains had shown that after several cultural generations, there was a significantly higher number of transformations to the version according to which the man stole the bag rather than helped the woman (Bebbington et al., 2017). The initial ambivalence was maintained only rarely.

Statements with emotional undertones (even positive ones, compared to emotionally neutral statements) thus form a significant group of general attractors. The increased likelihood of transformation towards the emotionally negative part of the spectrum can thus help explain the widespread practice of bloodletting across cultures. Although medically not very beneficial (and often quite dangerous), the combination of distasteful content and naive biology that appeals to treatment by 'removing bad blood' is cognitively attractive. Stories in which individuals were restored to health after accidentally cutting themselves were often transformed so that the cut was made intentionally (Miton et al., 2015).

Disgusting, warning, and horrifying plots are also found in most urban legends (Heath et al., 2001), that is, in modern tales dealing with current problems, fears, and dangers, often presented as absolutely true (often introduced by phrases such as 'A friend of a friend said…'). Such stories are usually either fictional or based on reality but heavily modified. Moreover, they are typically adapted to local conditions, so that for example a story where a syringe allegedly infected with HIV or another virus is intentionally stuck into a seat will be set on the subway in a city with a subway system but in a tram or bus in cities with no underground transport.

Urban legends have been the subject of many experimental studies (Eriksson & Coultas, 2014), and it has been found that aside from the abovementioned transformation of ambiguous information, the transmission of stories also involves an intensification of already negative information: for example, from 'a boy was attacked', through 'a boy was stabbed', all the way to 'a boy was murdered'. At the same time, modal particles which indicate uncertainty ('allegedly', 'probably', 'presumably', etc.) tend to disappear within the first two generations of the transmission chain (Stubbersfield et al., 2015). On the other hand, the loss of modal particles may be a consequence of the design, because participants in experiments may tend focus on the most important information and pay less attention to the fact that it is a story (Eriksson & Coultas, 2014). Experiments based on the transmission chains of urban legends and stories in general are—if we overlook the abovementioned problem—usually very successful and offer the possibility of studying various specific phenomena within cultural transmission. This design quite faithfully replicates the actual transmission of folklore, which tends to be oral or textual. Transmission also usually takes place either in one-on-one or one-to-several settings (Brednich, 2002); in other contexts, this experimental setting often does not feel authentic, which can significantly influence participants' behaviour and study results.

11.7.3 Systematicity and Lower Cognitive Demand

Cultural transmission leads to shifts towards greater systematicity. It is assumed that this is due to our memory being imperfect—and this imperfection is often viewed as an important source of likely changes (Morin, 2016b). By creating a structured system, we can remember more information and transmit it more faithfully.

Such increase in systematicity can be illustrated by a study on individuals who did not know the sign language and were asked to express the meaning of various terms using gestures (again within transmission chains). The results of their efforts were then used as the template for the next generation. Within just five generations, there emerged a simple 'language' where gestures for a prison, hair salon, and church always started with reference to a 'building' (for instance by joining hands to form a roof shape) followed by specification of the context, which was in turn common for, for example, a prison and a prisoner (holding imaginary bars) or for church and praying (clasped hands) (Motamedi et al., 2019). This study was based on classic experiments by Kirby et al. (2008), which addressed the increasing regularity of communication codes, convincingly demonstrating that an innate 'universal grammar' (defended for example by Chomsky, 1965) is not necessary for language evolution. The first participant in the experiment was given a set of images (typically combinations of three simple shapes, three colours, and three directions of movement) with random names (see Fig. 11.4), so that for instance a red square moving straight was assigned a completely different name than a blue square moving straight. The participant was asked to remember the names and then try to correctly name the objects. The answers obtained then, regardless of their correctness, served as the learning set for the next generation in the chain. Within ten generations of the chain, there emerged a simple system that used composition (e.g., the name of a spinning blue triangle differed from the name of a spinning red triangle by only one syllable), and participants make significantly fewer naming errors than their cultural ancestors who had to remember random clusters of letters (Kirby et al., 2008). In this case, the attractor was not individual words but a structured system that resembled simple grammar. The presence of a structure thus led to a significant increase in memorability and a decrease in the likelihood of transmission errors.

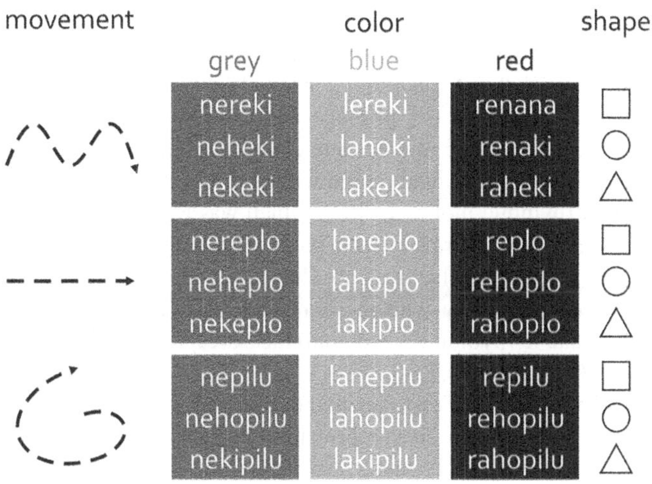

Fig. 11.4 An example of a labelling system: the result of one transmission chain over ten generations. In this case, participants were not allowed to assign the same name to two pictures the same. Moreover, they were only presented with one-half of the pictures for memorisation and had to infer the names of the remaining ones. The result is a structured system where the 'prefix' indicates colour, the 'root' of the word indicates shape, and the 'suffix' indicates movement. Redrawn from Kirby et al. (2008)

A similar trend was demonstrated for language also outside laboratory conditions, when researchers analysed various types of English texts from newspapers to books and political speeches published between 1800 to 2000. Over the centuries, concreteness increased in two ways: by a higher representation of words which are inherently more concrete due to their syntactic function (hence fewer articles but more nouns, adjectives, and verbs) and by increasing concreteness within the individual parts of speech (Hills & Adelman, 2015). This can likewise be explained by a tendency to reduce cognitive demands, increase learnability, and discard less important, even redundant, elements.

An analysis of 116 different writing systems found that vertical and horizontal lines are more frequently represented in current scripts than in their predecessors (Morin, 2018). Current writing systems moreover exhibit a high degree of vertical symmetry. Such features are more 'pleasing' to our evolutionarily honed cognitive apparatus: vertical and horizontal lines are more easily recognised by humans because they are abundantly represented in nature (Girshick et al., 2011). Moreover, vertically symmetrical objects are recognised more quickly and a preference for them appears from 1 year of age (Bornstein et al., 1981). Current writings systems may have thus emerged by transformations of characters which were more difficult for people in the past to recognise and remember, but the detected effects were not large, which seems to be due to motor limitations; a tired (Parush et al., 1998) or hurried (van Drempt et al., 2011) writer produces a less legible text, which is a factor acting in the opposite direction.

The emergence of a structured system can also be observed in music, that is, outside the context of language. For instance, researchers exposed participants to a drumming sequence that was randomly generated (in the first generation) or generated by a previous participant (in subsequent generations) and asked them to reproduce it. Within just eight generations, there emerged a sequence that resembled real music, was easily memorable, and reproducible even by non-musicians. All created sequences had a rhythmic motif (a 'riff'), which was repeated, gradually evolved, and slightly differed in each transmission chain (Ravignani et al., 2016). Something resembling music can thus arise from absolute chaos simply through transmission. These results are also in line with the hypothesis of universal aspects of music, which include a stable rhythm, division into measures, and similar lengths of tones (Savage et al., 2015).

The aforementioned list of experimental examples is by no means exhaustive but it convincingly shows that an uneven probability of change in the culturally transmitted material leads to the formation of systematic patterns whose established equivalents we see all around us. An understanding of the psychological processes involved in memorising and transmitting culture can help us predict how a given substrate of public representations will evolve and what the attractor state (the end state) will look like.

11.7.4 Ecological Factors

Another source of transformative cultural transmission are the external environmental influences, collectively known as ecological factors. A typical environmental influence is the climate (cf. Chaps. 10, 13). The spread of any sound is influenced by air temperature and humidity (Morton, 1975): for example, hot air distorts high-frequency sounds and more sophisticated 'noise' structures of the sound waves more than cooler air does, because higher temperatures lead to more microturbulence, and thus more random movement of air molecules between the speaker and the listener. Similar effects are caused by wind and rain sounds. This aligns with the observation that languages with more vowels (and thus a higher 'sonority' or sound penetrance) dominate in areas with higher temperatures (and perhaps humidity). Indigenous Australian and Niger-Congo languages thus have a significantly higher representation of vowels than Indo-European or Ural-Altaic languages (Maddieson, 2018). Other factors reflected in languages are the sounds made by other animals (e.g., birdsong) and vegetation density (Maddieson & Coupé, 2015).

Another significant factor is the availability of various raw materials. If, for example, the ingredients for preparing a dish are not available in a given location, people modify recipes and look for alternative ingredients—and such modified recipes gradually converge (Sperber & Claidière, 2008). Conversely, an abundance of some raw material can be associated with the invention of new tools for processing it, as in a special fork for mangoes (Petroski, 1994).

Local animals and plants undoubtedly also influence story variants. Various versions of the fairy tale of Little Red Riding Hood are widespread across most of

Eurasia and parts of Africa, but most variants in East Asia work not with a wolf but with either a tiger (Tehrani, 2013) or a 'panther' (Wilhelm, 1921). The antagonist is thus probably adapted to the salient potential threats in the area, and such variant is easier to remember. This is supported by the results of Bartlett's (1932) abovementioned original study, where transformations took place from objects less well-known to the participants to well-known objects (for example, from 'canoe' to a 'boat').

Environmental factors can also be viewed as cultural variants already present in the society where the transmission occurs, whereby the science dealing with how organisms transform their environment rather than passively adapting to it is called 'niche construction' (for more on this issue, see Chaps. 10, 13, 16, 17). A typical example of such factor is writing, which allows us to process and store vast amounts of information—and that subsequently enables a greater cultural complexity in areas with higher literacy rates. To remember a significantly higher amount of information, the availability of paper and a pencil is thus sufficient even for individuals who are not skilled writers (Houts et al., 2001).

11.8 The Reconciliation of Selectionists and Attractionists

A broad conception of cultural attraction was proposed by Claidière et al. (2014) who analysed the extent of the analogy between biological and cultural evolution. Proponents of memetics believe that both processes are fundamentally equivalent, and they usually require a clear identification of discrete replicators (see Chap. 10).[7] Other selectionists have loosened the analogy (Boyd & Richerson, 1988): they still emphasise selection, without which biological evolution could not occur, but they do not insist on discrete replicators. Claidière et al. (2014) then proposed that the science of replicators should be viewed as a special case of the selectionist model and they introduced four similarly nested explanatory frameworks (Fig. 11.5).

In accordance with the theory of cultural attractors, they emphasise that culture is a populational phenomenon which evolves even without selection, and the explanatory framework thus ought to be appropriately expanded. Selection can be characterised as a disproportion in the sum of all probabilities of transmission with transformation between two different cultural variants. If, in computer simulations of cultural transmission, we allow for a truly arbitrary assignment of the ratio of probabilities of transformations in cultural space (in this example, one-dimensional), we can easily find two squares whose ratio of transformations to neighbouring states

[7] This, however, does not necessarily apply to the first pitch of memetics as presented by Richard Dawkins in *The Selfish Gene* (cf. Chap. 10), where the vagueness of boundaries of different cultural variants is clearly articulated.

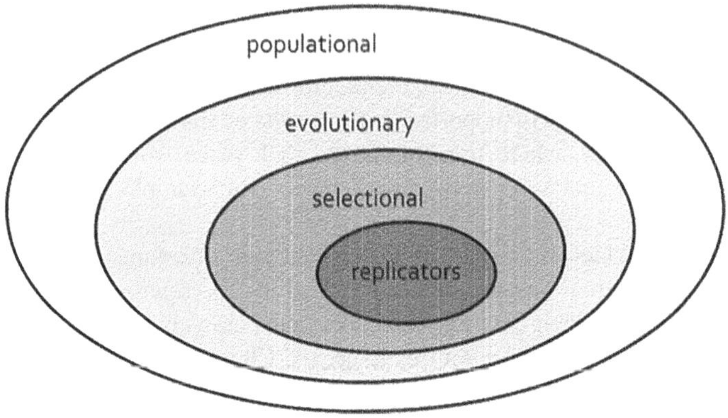

Fig. 11.5 Four nested explanatory frameworks according to Claidière et al. (2014). Redrawn

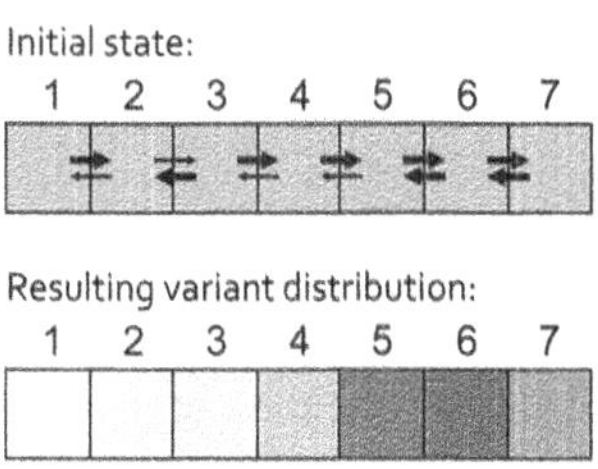

Fig. 11.6 Selection in combination with attraction. Although square 2 matches the basic definition of a cultural attractor (transformation into it is more likely than transformation from it), the representation of variants in the attractor around square 6 over time significantly prevails. This is due to a strong transformational flow from the left and a more significant contribution of square 6 to the frequency of neighbouring squares. Author: Petr Tureček

is the same. On the other hand, if we compare the sums of these transformations, we find that one is significantly different than the other—and that indicates an imbalance in the probability of transmission between these two variants in cultural space (Fig. 11.6).

Phenomena such as those mentioned in this chapter, but also other similar cases, can be mathematically described (and modelled) using causal evolutionary matrices (CEM; Claidière et al., 2014). A CEM contains information about how one entity (a variant of a trait, representation, etc.) at time t contributes to the frequency of all other entities of the system at time $t + 1$ (Table 11.1).

An illustrative example of application of CEM is the modelling of how pronunciation of the term 'data' evolved in the English language. English borrowed this

Table 11.1 A causal evolutionary matrix. I_{XY} expresses the average impact of type X element on the frequency of type Y elements over unit time. There is a potentially infinite number of elements between the rows and columns for elements B and Z. Author: Petr Tureček, based on Claidière et al. (2014).

	A(t + 1)	B(t + 1)	…	Z(t + 1)
A(t)	I_{AA}	I_{AB}	…	I_{AZ}
B(t)	I_{BA}	I_{BB}	…	I_{BZ}
…	…	…	…	…
Z(t)	I_{ZA}	I_{ZB}	…	I_{ZZ}

Table 11.2 Two different pronunciations of the word 'data'. Pronunciation [ˈdætə] contributes to the frequency of pronunciation of [ˈdeɪtə] 40 times more than the pronunciation [ˈdeɪtə] contributes to pronunciation [ˈdætə]. The contribution of both pronunciations to their own frequency is the same. Author: Petr Tureček, based on Claidière et al. (2014)

	[ˈdætə] (t + 1)	[ˈdeɪtə] (t + 1)
[ˈdætə] (t)	5	4
[ˈdeɪtə] (t)	0.1	5

word from Latin along with its original pronunciation [ˈdætə] but over time, the variant [ˈdeɪtə], which is compatible with common English pronunciation, became prevalent. According to the model presented by Claidière et al. (2014), both variants contribute to their own frequency at time $t + 1$ equally, but [ˈdætə] contributes to the frequency of [ˈdeɪtə] significantly more than [ˈdeɪtə] to [ˈdætə] (4 vs 0.1). In a simplified way, we can imagine the situation so that when one person pronounces [ˈdætə], it leads to five other people—for instance among the audience of a lecture—pronouncing [ˈdætə] and four other people, for example note-takers who focused on writing and then tried to impress their friends at home, pronouncing [ˈdeɪtə]. Because the original Latin pronunciation [ˈdætə] is more frequently used by experts who use this term in casual conversation, the original row sum is biased in favour of [ˈdætə] in a ratio of 9 to 5.1. Nevertheless, starting with an initial frequency of [ˈdeɪtə] at 0.001, after 40 steps, the ratio of usage of pronunciations stabilises at 14% of [ˈdætə] and 86% of [ˈdeɪtə] (Table 11.2).

Selection can also be modelled as a special case of attraction, where each variant of a trait at time t gives rise only to the same variants of the trait at time $t + 1$ (Table 11.3).

Table 11.3 CEM selection. Because each element increases only its own frequency, allele B (the highest number on the diagonal) will over time prevail regardless of the original representation of alleles in the population. If mutations between alleles were allowed, we would not see any zeros on the off-diagonal. Author: Petr Tureček, based on Claidière et al. (2014).

	Allele A (t + 1)	Allele B (t + 1)
Allele A (t)	4	0
Allele B (t)	0	5

Table 11.4 CEM of the spread of a story in a population. A person who knows the story tells it on average to six listeners. Each of these listeners hears it on average five times before telling the story themselves (one narration leads to the creation of 0.2 mental representation). Author: Petr Tureček, based on Claidière et al. (2014).

	Mental rep. (t + 1)	Public rep. (t + 1)
Mental rep. (t)	0	6
Public rep. (t)	0.2	0

Another special case of the CEM is the Sperber's classical causal chain between mental and public representations, because public representations lead only to the creation of mental representations and vice versa. The relationship between mental and public representations can be illustrated by the example of spread of an orally transmitted story (Table 11.4).

The fact that a system can be modelled by CEM allows us to include it in an evolutionary explanatory framework. Such system can be assigned an attractor, that is, a final state which the system enters and from which it does not depart unless the parameters in CEM change. For more complex phenomena, such as frequency-dependent selection (Flegr, 2009), CEM cells need to include functions with additional variables. Some such systems do not reach a completely stable attractor but a periodically oscillating one, and they can be even used to model biological and ecological phenomena (such as systems of relative population states of predators and prey). A model of conformity bias (cf. Chap. 13) is shown in Table 11.5 and a

Table 11.5 CEM of conformity bias. When one of the elements slightly predominates (even by sheer chance), its frequency will continually increase due to the negative impact on the frequency of the second element. Author: Petr Tureček, based on Claidière et al. (2014).

	Behaviour A (t + 1)	Behaviour B (t + 1)
Behaviour A (t)	1	−0.1
Behaviour B (t)	−0.1	1

Table 11.6 CEM of the spread of the Little Red Riding Hood story with a mistake and a reconstruction, Variant A corresponds to a story where both the grandmother and Little Red Riding Hood are saved, while in variant B, only one of them emerges from the wolf's belly. Listeners tend to the correct variant B (public representation B contributes more to the mental representation A than to itself). The attractor of the system is a state where nearly 19% of retellings (Pub. A or Pub B.) contain a logical error (Pub. B) although only 3% of narrators (Ment. A or Ment. B) believe it should be that way (Ment. B). Author: Petr Tureček, based on Claidière et al. (2014).

	Ment. A (t + 1)	Ment. B (t + 1)	Pub. A (t + 1)	Pub. B (t + 1)
Ment. A (t)	0	0	5	1
Ment. B (t)	0	0	1	5
Pub. A (t)	0.6	0	0	0
Pub. B (t)	0.5	0.1	0	0

reconstruction of evolution of the story of Little Red Riding Hood and its mathematical expression using CEM is depicted in Table 11.6.

A minor problem with this approach is that if CEM were to accurately describe evolutionary processes, they would always need to include a vast number of rows and columns. For instance, Table 11.4 describes well the fairytale's transmission but not its evolution. Table 11.6 allows for capturing the evolution but public and mental representations can take only two predetermined values, which resembles reality only remotely. The table describes only a 'competition' between two alternatives.

Such CEM can be used to demonstrate certain common elements of biological and cultural evolution but in practice, their application is rather limited. In their current form, they mostly resemble the mathematical models (and, moreover, models after linearisation around an equilibrium point) used, for example, in ecology to study the dynamics of population processes. But evolution significantly depends on stochastic processes as well. Mutations or recombination and separation of alleles are a prime source of random terms in biological inheritance but, since the emergence of biosemiotics, thinking in terms of 'attention' and 'interpretation' had become applicable also to discussions about the genetic code and its transcription. It is becoming apparent that biological sciences could benefit from the tools and narratives developed within the framework of cultural evolution. For decades, we have tried to 'biologise culture'—now is perhaps the time to 'culturalise biology' (Kleisner & Tureček, 2017).

11.9 Interaction Between Attraction and Selection

Interaction between the processes of selection and attraction, which cannot be simply decided based on the observed result, is illustrated by the abovementioned example of a gradual shift towards direct gaze in Renaissance portraits. Similarly, cognitive attraction and selection probably operate in the same direction in the cultural evolution of toys, which—although intended for children—are designed by adult designers and purchased by adult customers. A classic example is the teddy bear: a comparison between the first models made around 1900 and the contemporary ones shows a remarkable change in proportions (Figs. 11.7 and 11.8). Over the past century, this popular toy had transformed and acquired markedly neotenous features (i.e., shapes and proportions similar to children or non-human infants) which we consider cute (Miton, 2022). Most importantly, this was achieved by increasing the size of teddy bear's forehead relative to the rest of the head (Borredon et al., 2025).

The popular character Mickey Mouse evolved in a similar direction. The original version from the 1920s, derived from Oswald the Lucky Rabbit (also from Walt Disney's workshop), had an elongated face and long ears (cf. Chap. 15). Mickey was overall rounder than Oswald and over time became even cuter, with bigger eyes and larger ears (Gould, 2008). But still, the original Mickey looks almost frightening compared to its current version. Its success was undoubtedly also helped by the unmistakable silhouette of its head (a large circle with two smaller circles as ears) and the historical context of the Great Depression, when the working class projected

Fig. 11.7 The original plush teddy bear from 1903. National Museum of American History (https://americanhistory.si.edu/collections/nmah_491375)

Fig. 11.8 One of current versions of a teddy bear. Photograph by Marina Shatskikh from Pexels (https://www.pexels.com/photo/brown-bear-on-fence-1212118/)

its deepest desires onto this cheerfully squeaky mouse (sailing on a steamboat, lots of free time, the joyful experience of non-alienated work).

It is not clear whether every young designer spontaneously designs teddy bears which are just a bit cuter than the teddy bears of their predecessor (as the attractionist model would assume) or whether the development is steered by seasoned marketers who present designers with successful models to emulate (the cutest ones from the previous 'generation', as a purely selectionist model would assume). The case of Mickey Mouse might be taken to support the attraction aspect because, within a single series of cartoons with the animated mouse, we find no variability in his appearance. The possibility of evolution despite absence of variability is one of the strongest arguments in favour of the attractionist approach to cultural evolution, and it is this aspect that most clearly distinguishes it from the selectionist approach. Selection requires populational variance to alter the populational mean. This was at the core of early criticism of Darwinism formulated even before mutation-enhanced Mendelism was established as the source of both stability and novelty in biological inheritance (Bulmer, 2004).

Based on the accumulation of similarly convincing evidence, selectionists started to emphasise that their model, too, can account for a gradual transformation independent of selection: in their case, it takes the form of 'guided variation' (Acerbi & Mesoudi, 2015). In guided variation, deviations from the transmitted cultural variants do not occur randomly (in this, it resembles models of psychological or environmental attraction), but selectionists originally envisaged it as an active process of improvement of culturally transmitted artefacts, sociocultural regulations, and ideas (Morin, 2016a). Nevertheless, other scholars (for instance, Taleb, 2011) pointed out that invention does not create any more targeted changes than random variation would, because (for example in the development of computers) it often happens that an innovation created with a certain intention eventually prevails in a completely different way than originally intended.

Sperberian attractionists initially emphasised the phenomenon of distortion or inaccurate copy of any stimulus inserted into a transmission chain (or its 'wild' equivalent). It is thus no coincidence that typical precursors to transmission experiments are the telephone game or the example of a broken copier that leaves black smudges on copies (Morin, 2016b). They transmit inaccurately to the extent that the copies and copies of copies modify the content so much that the defect ceases to manifest itself (and one ends up copying entirely black pages). Nowadays, though, the selectionists and attractionists have put aside their differences because in the formal models, regardless of whether they include CEM and call the equilibrium between potentially competing variants 'attractor' or not, the differences between their related systems largely disappeared.

11.10 Conclusions

As research covered in this chapter shows, it is often difficult to tell whether the dominant factor influencing a particular outcome of cultural evolution and change is a systematic transformation or rather selection between available variants. But, especially in controlled experiments, one can assess the relative impact of attraction—and it is certainly not null. Large variations in transformation probabilities, which typically occur without any deliberate action of the cultural participants, make cultural evolution stand out against the background of biological evolution at whose core is precise replication.

Nevertheless, even if we dismiss the stress on the relative end-state proportion of variants acting as an attractor as a post-hoc update that had distorted the strong original idea of attractionism, this approach had highlighted the need of a clear formalisation of many hazily defined intuitions—and that has benefited the field of cultural evolution as a whole.

References

Acerbi, A., & Mesoudi, A. (2015). If we are all cultural Darwinians what's the fuss about? Clarifying recent disagreements in the field of cultural evolution. *Biology & Philosophy, 30*(4), 481–503.

Anglada-Tort, M., Harrison, P. M. C., Lee, H., & Jacoby, N. (2023). Large-scale iterated singing experiments reveal oral transmission mechanisms underlying music evolution. *Current Biology, 33*(8), 1472–1486.e12.

Bartlett, F. C. (1932). *Remembering: A study in experimental and social psychology*. Cambridge University Press.

Baumeister, R. F., Bratslavsky, E., Finkenauer, C., & Vohs, K. D. (2001). Bad is stronger than good. *Review of General Psychology, 5*(4), 323–370.

Bebbington, K., MacLeod, C., Ellison, T. M., & Fay, N. (2017). The sky is falling: Evidence of a negativity bias in the social transmission of information. *Evolution and Human Behavior, 38*(1), 92–101.

Billiard, S., & Alvergne, A. (2018). Stochasticity in cultural evolution: A revolution yet to happen. *History and Philosophy of the Life Sciences, 40*(1), 9.

Bornstein, M. H., Ferdinandsen, K., & Gross, C. G. (1981). Perception of symmetry in infancy. *Developmental Psychology, 17*(1), 82–86.

Borredon, Q., Bulamac, Z., Crozat, C., Dayre, E., Fuchs, E., Hallo, M., et al. (2025). Did teddy bears culturally evolve to be cuter? A preregistered replication. *Journal of Cognition and Culture, 25*(1–2), 114–127.

Boyd, R., & Richerson, P. J. (1988). *Culture and the evolutionary process*. University of Chicago Press.

Brednich, R. W. (2002). Where they originated. Some contemporary legends and their literary origins. *Folklore: Electronic Journal of Folklore*, (20–22), 7–16.

Bulmer, M. (2004). Did Jenkin's swamping argument invalidate Darwin's theory of natural selection? *The British Journal for the History of Science, 37*(3), 281–297.

Buskell, A. (2017). What are cultural attractors? *Biology and Philosophy, 32*(3), 377–394.

Cavalli-Sforza, L. L., & Feldman, M. W. (1981). *Cultural transmission and evolution: A quantitative approach*. Princeton University Press.

Chomsky, N. (1965). *Aspects of the theory of syntax*. MIT press.

Claidière, N., Scott-Phillips, T. C., & Sperber, D. (2014). How Darwinian is cultural evolution? *Philosophical Transactions of the Royal Society B: Biological Sciences, 369*(1642), 20130368.

Deely, J. (2009). *Purely objective reality*. Mouton de Gruyter.

Derex, M., Edmiston, P., Lupyan, G., & Mesoudi, A. (2024). Tradeoffs-, control conditions, and alternative designs in the experimental study of cultural evolution. *Proceedings of the National Academy of Sciences of the United States of America, 121*(48), e2322886121.

Eriksson, K., & Coultas, J. C. (2014). Corpses, maggots, poodles and rats: Emotional selection operating in three phases of cultural transmission of urban legends. *Journal of Cognition and Culture, 14*(1–2), 1–26.

Fessler, D. M., Pisor, A. C., & Navarrete, C. D. (2014). Negatively-biased credulity and the cultural evolution of beliefs. *PLoS One, 9*(4), e95167.

Flegr, J. (2009). *Evoluční biologie (2. vyd.)*. Academia.

Girshick, A. R., Landy, M. S., & Simoncelli, E. P. (2011). Cardinal rules: Visual orientation perception reflects knowledge of environmental statistics. *Nature Neuroscience, 14*(7), 926–932.

Goldstein, M. C. (1987). When brothers share a wife. *Natural History, 96*(3), 39–48.

Gould, S. J. (2008). A biological homage to Mickey mouse. *Ecotone, 4*(1), 333–340.

Griffiths, T. L., Kalish, M. L., & Lewandowsky, S. (2008). Theoretical and empirical evidence for the impact of inductive biases on cultural evolution. *Philosophical Transactions of the Royal Society B: Biological Sciences, 363*(1509), 3503–3514.

Heath, C., Bell, C., & Sternberg, E. (2001). Emotional selection in memes: The case of urban legends. *Journal of Personality and Social Psychology, 81*(6), 1028–1041.

Herrmann-Pillath, C. (2021). Biosemiotic foundations of a Darwinian approach to cultural evolution. *Cultural Science Journal, 13*(1), 10–2478.

Hills, T. T., & Adelman, J. S. (2015). Recent evolution of learnability in American English from 1800 to 2000. *Cognition, 143*, 87–92.

Houts, P. S., Witmer, J. T., Egeth, H. E., Loscalzo, M. J., & Zabora, J. R. (2001). Using pictographs to enhance recall of spoken medical instructions. *Patient Education and Counseling, 43*(3), 231–242.

Kirby, S., Cornish, H., & Smith, K. (2008). Cumulative cultural evolution in the laboratory: An experimental approach to the origins of structure in human language. *Proceedings of the National Academy of Sciences of the United States of America, 105*(31), 10681–10686.

Kleisner, K., & Tureček, P. (2017). Cultural and biological evolution: What is the difference? *Biosemiotics, 10*, 127–130.

Lévi-Strauss, C. (1973). Structuralism and ecology. *Social Science Information, 12*(1), 7–23.

Maddieson, I. (2018). Language adapts to environment: Sonority and temperature. *Frontiers in Communication, 3*, 28.

Maddieson, I., & Coupé, C. (2015). Human spoken language diversity and the acoustic adaptation hypothesis. *The Journal of the Acoustical Society of America, 138*(3), 1838–1838.

Miton, H. (2022). *Cultural attraction* [Preprint]. PsyArXiv. https://doi.org/10.31234/osf.io/qs2et

Miton, H., Claidière, N., & Mercier, H. (2015). Universal cognitive mechanisms explain the cultural success of bloodletting. *Evolution and Human Behavior, 36*(4), 303–312.

Morin, O. (2013). How portraits turned their eyes upon us: Visual preferences and demographic change in cultural evolution. *Evolution and Human Behavior, 34*(3), 222–229.
Morin, O. (2016a). Reasons to be fussy about cultural evolution. *Biology and Philosophy, 31*(3), 447–458.
Morin, O. (2016b). *How traditions live and die*. Oxford University Press.
Morin, O. (2018). Spontaneous emergence of legibility in writing systems: The case of orientation anisotropy. *Cognitive Science, 42*, 664–677.
Morton, E. S. (1975). Ecological sources of selection on avian sounds. *The American Naturalist, 109*(965), 17–34.
Motamedi, Y., Schouwstra, M., Smith, K., Culbertson, J., & Kirby, S. (2019). Evolving artificial sign languages in the lab: From improvised gesture to systematic sign. *Cognition, 192, 103964*.
Parush, S., Pindak, V., Hahn-Markowitz, J., & Mazor-Karsenty, T. (1998). Does fatigue influence children's handwriting performance? *Work, 11*(3), 307–313.
Petroski, H. (1994). *The evolution of useful things*. Vintage Books.
Ravignani, A., Delgado, T., & Kirby, S. (2016). Musical evolution in the lab exhibits rhythmic universals. *Nature Human Behaviour, 1*, 7.
Richerson, P. J., & Boyd, R. (2005). *Not by genes alone: How culture transformed human evolution*. University of Chicago Press.
Rozin, P., & Royzman, E. B. (2001). Negativity bias, negativity dominance, and contagion. *Personality and Social Psychology Review, 5*(4), 296–320.
Savage, P. E., Brown, S., Sakai, E., & Currie, T. E. (2015). Statistical universals reveal the structures and functions of human music. *Proceedings of the National Academy of Sciences of the United States of America, 112*(29), 8987–8992.
Scanlon, L. A., Lobb, A., Tehrani, J. J., & Kendal, J. R. (2019). Unknotting the interactive effects of learning processes on cultural evolutionary dynamics. *Evolutionary Human Sciences, 1*, e17.
Scott-Phillips, T. C., Blancke, S., & Heintz, C. (2018). Four misunderstandings about cultural attraction. *Evolutionary Anthropology, 27*(4), 162–173.
Sperber, D. (1985). Anthropology and psychology: Towards an epidemiology of representations. *Man, 20*(1), 73–89.
Sperber, D. (1996). *Explaining culture: A naturalistic approach*. Blackwell.
Sperber, D., & Claidière, N. (2008). Defining and explaining culture (comments on Richerson and Boyd, Not by genes alone). *Biology & Philosophy, 23*(2), 283–292.
Stubbersfield, J. M., Tehrani, J. J., & Flynn, E. G. (2015). Serial killers, spiders and cybersex: Social and survival information bias in the transmission of urban legends. *British Journal of Psychology, 106*(2), 288–307.
Taleb, N. N. (2011). *The black swan: The impact of the highly improbable*. Penguin Books.
Tehrani, J. J. (2013). The phylogeny of little red riding hood. *PLoS One, 8*(11), e78871.
Uddenberg, S., Thompson, B. D., Vlasceanu, M., Griffiths, T. L., & Todorov, A. (2023). Iterated learning reveals stereotypes of facial trustworthiness that propagate in the absence of evidence. *Cognition, 237*, 105452.
Valenza, E., Simion, F., Cassia, V. M., & Umiltà, C. (1996). Face preference at birth. *Journal of Experimental Psychology: Human Perception and Performance, 22*(4), 892.
van Drempt, N., McCluskey, A., & Lannin, N. A. (2011). A review of factors that influence adult handwriting performance. *Australian Occupational Therapy Journal, 58*(5), 321–328.
Wilhelm, R. (1921). *Chinese fairy book*. Frederick A. Stokes company.
Wittgenstein, L. (1953). *Philosophical investigations*. Blackwell.

Chapter 12
Cultures as Species: Tree Thinking and the Phylogenetic Approach in Linguistics, Archaeology, and Anthropology

Pavel Duda

Abstract Ever since the publication of *On the Origin of Species*, linguists, archaeologists, and anthropologists have explored the potential of the Darwinian framework (descent with modification) for explaining cultural diversity. Cultural phylogenetics, a rapidly developing field that applies methods from evolutionary biology to the study of culture, has made significant contributions to historical linguistics, archaeology, cultural anthropology, and behavioural ecology by providing insights into the processes that shape languages, material culture, traditions, and social norms. By reconstructing patterns of relatedness among cultural entities, phylogenetic methods enable researchers to infer underlying historical processes such as grammatical and lexical change, the evolution of artefact design, ancient human migrations, and the spread of cultural innovations. These methods also provide a means of accounting for common ancestry in cross-cultural comparisons. This chapter provides a historical and methodological review of cultural phylogenetics from the early comparative approaches to the emergence of computational phylogenetic methods, including tree-building techniques and comparative analyses. It highlights key studies in which these methods have been applied to diverse cultural domains and discusses their implications for understanding the dynamics of cultural change. In doing so, it underscores the broader significance of cultural phylogenetics as a bridge between the biological and social sciences, demonstrating how evolutionary principles can illuminate the mechanisms that drive cultural diversity and change.

P. Duda (✉)
Department of Zoology, Faculty of Science, University of South Bohemia, České Budějovice, Czech Republic
e-mail: dudapa01@prf.jcu.cz

J. Toman et al. (eds.), *Culture Evolving*, Biosemiotics 10,
https://doi.org/10.1007/978-3-032-16124-6_12

Keywords Cultural phylogenetics · Cultural evolution · Phylogenetic comparative methods · Tree thinking in anthropology · Cultural transmission

> The formation of different languages and of distinct species, and the proofs that both have been developed through a gradual process, are curiously parallel. (Charles Darwin, 1871: 89)

12.1 Curious Parallels

In his book *Darwin's Dangerous Idea*, the American philosopher Daniel Dennett likened Darwin's theory of evolution to a universal acid—a hypothetical liquid so corrosive that it eats through anything it touches. 'It eats through just about every traditional concept, and leaves in its wake a revolutionised worldview, with most of the old landmarks still recognisable, but transformed in fundamental ways' (Dennett, 1996: 63). This metaphor aptly captures the far-reaching impact of Darwinian theory, which continues to reshape not only biology but also fields as diverse as linguistics, archaeology, anthropology, and the social sciences. Ever since the publication of *On the Origin of Species* (Darwin, 1859), scholars have debated whether evolutionary processes extend beyond biology, and the discussions and research they inspired have fundamentally influenced our understanding of cultural diversity and change.

This chapter traces the historical development of cultural phylogenetics, starting with Darwin's own reflections on the parallels between biological and linguistic evolution. It then moves on to the early adoption of evolutionary frameworks in linguistics and anthropology and highlights the ideas of key figures such as August Schleicher and Edward Burnett Tylor. This is followed by a discussion of methodological advancements, particularly the incorporation of phylogenetic methods into the study of cultural evolution. Case studies from linguistics, archaeology, and anthropology illustrate how these methods have been used to reconstruct historical relationships between languages, artefacts, and social norms. In this context, this chapter also addresses debates about the role of horizontal transmission and the extent to which cultural evolution follows a tree-like structure. Finally, it concludes by reflecting on the broader implications of cultural phylogenetics, emphasising its potential to bridge the exact sciences and the humanities and generate new insights into the evolution of culture.

It is somewhat ironic that *On the Origin of Species* says little about how new species originate through the splitting of evolutionary lineages, a process known as cladogenesis. Darwin focused mainly on anagenesis, the gradual transformation and persistence of existing species as interbreeding populations of organisms. Even so, as evidenced by the fact that the only illustration in his book is a schematic 'tree of life' diagram (Darwin, 1859: 116–117, Fig. 12.1a), he clearly regarded cladogenesis as an important outcome of the evolutionary processes he described.

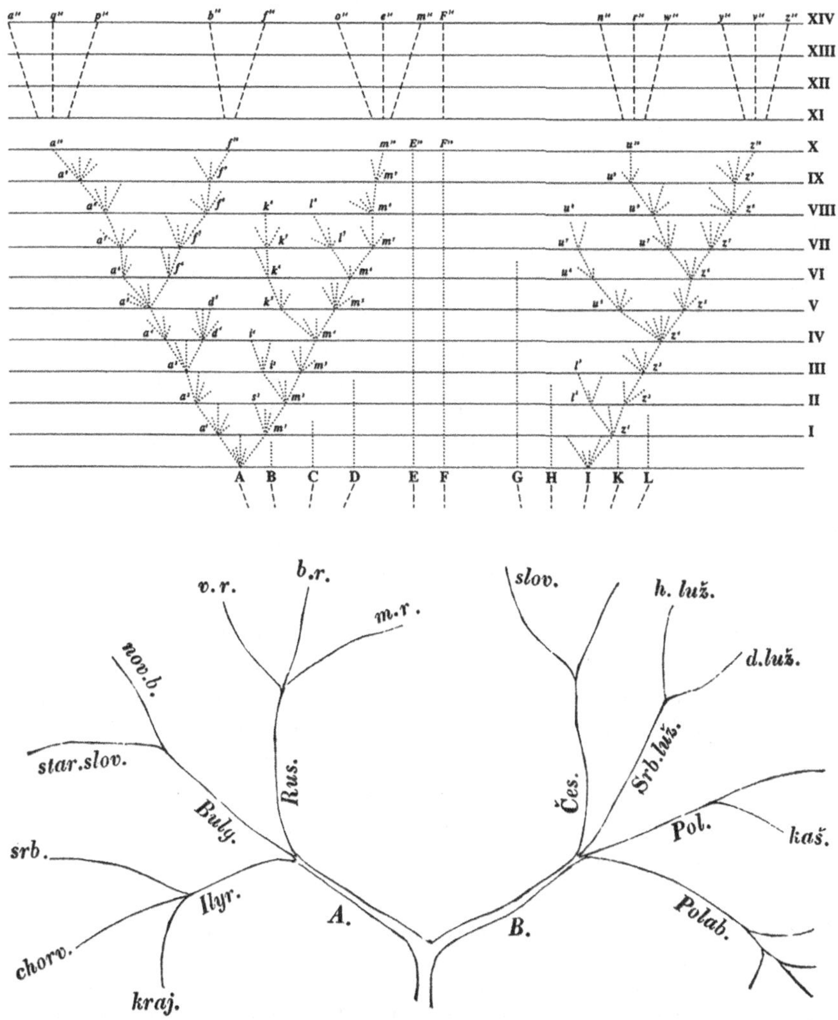

Fig. 12.1 (**a**) The 'tree of life' diagram by Darwin (1859: 116–117). (**b**) The tree of Slavic languages by Čelakovský (1853: 3)

Darwin also advocated a natural classification of organisms based on their phylogenetic relationships, comparing it to the classification of languages:

> If we possessed a perfect pedigree of mankind, a genealogical arrangement of the races of man would afford the best classification of the various languages now spoken throughout the world; and if all extinct languages, and all intermediate and slowly changing dialects, had to be included, such an arrangement would be the only possible one. (Darwin, 1859: 498)

This demonstrates not only Darwin's conviction that cultures can evolve in a manner analogous to biological populations and species, but also his realisation that biological and cultural systems may evolve in a correlated fashion.

The idea that languages originate from a common ancestor predates the theory of evolution by natural selection. In fact, the first family trees of cultural phenomena—such as languages or manuscripts—are about as old as the first phylogenies of biological species (see Chap. 2). Ernst Haeckel (see Chap. 4), who championed and popularised Darwin's work in Germany, sent a copy of *The Origin of Species* to the linguist August Schleicher, his colleague at the University of Jena. In response, Schleicher (1869) wrote an open letter in which he expressed his fascination with Darwin's ideas but noted that some of the principles outlined in the book had already been applied in the study of languages. Often regarded as a founding figure of historical comparative linguistics, Schleicher (1853a, 1853b) had previously illustrated the historical development of Indo-European languages using family trees. It is even possible that he adopted the idea from Czech linguist František Ladislav Čelakovský, with whom he had worked at the Charles University in Prague before moving to Jena in 1857 (List et al., 2014). Čelakovský's book, published posthumously in 1853, contains a family tree of Slavic languages (Čelakovský, 1853: 3, Fig. 12.1b).

Darwin came to know Schleicher later, after his open letter to Haeckel was published in English under the name *Darwinism Tested by the Science of Language* (Schleicher, 1869). Long before this, however, Darwin had been influenced by the British astronomer John Herschel, whose ideas he encountered during his formative years and whom he met in person in Cape Town in 1836, at the end of Darwin's voyage on HMS Beagle (Cannon, 1961). In a letter to the geologist Charles Lyell (who was also a source of inspiration for Darwin), Herschel argued that the time required for the emergence of differences such as exist between present-day languages must have been considerably longer than what is implied by biblical chronology, and that the world must therefore be much older than is generally assumed (cf. Chap. 2):

> [W]hen we see what amount of change 2000 years has been able to produce in the languages of Greece & Italy or 1000 in those of Germany France & Spain we naturally begin to ask how long a period must have lapsed since the Chinese, the Hebrew, the Delaware & the Malesass [Malagasy] had a point in common with the German & Italian & each other. – Time! Time! Time! we must not impugn the Scripture Chronology, but we must interpret it in accordance with whatever shall appear on fair enquiry to be the truth for there cannot be two truths. (Cannon, 1961: 308)

Darwin read this letter in 1837, a year after returning from the voyage of the *Beagle* (Burkhardt & Smith, 1986).

Herschel was not the only person to believe that the study of languages could provide insights into human prehistory. William Jones, a British judge who worked in Bengal between 1783 and 1794, noticed similarities between Sanskrit, Persian, Classical Greek, and Latin, from which he concluded that these languages were descended from a common ancestor (Jones, 1786). This common ancestor later became known as Proto-Indo-European. But even long before that, in the sixteenth century, the French scholar Joseph Justus Scaliger recognised that some traits were passed down between languages and understood their importance for language classification. Scaliger classified languages in Europe into four groups based on the word for God: *deus*, *god*, *bog* and *θεός* [*theos*]. In doing so, he successfully identified the Romance, Germanic, Slavic, and Greek branches of the Indo-European language family (Pedersen, 1931).

The study of stemmatics applies the idea of reconstructing past relationships from present-day patterns of variation to written manuscripts. Family trees of manuscripts—so-called *stemmatā cōdicum*—are constructed using a method called recension, pioneered by the German philologist Karl Lachmann. The guiding principle of recension is that a 'community of error implies community of origin' (Reynolds & Wilson, 2013), which is analogous to the principle of inferring molecular phylogeny from shared genetic mutations in a homologous sequence of nucleotides. It is then assumed that versions that share the same errors are derived from the same ancestral version, the hyparchetype. The earliest known published stemma is the family tree of the *Västgötalagen* (Westrogothic law), a Swedish provincial law compiled in the early thirteenth century and published by Carl Johan Schlyter (Collín & Schlyter, 1827).

The publication of Darwin's seminal work sparked an exchange of ideas about development in general and evolution in particular among thinkers and scientists active in both biology and the social sciences, which was facilitated by the fact that these two domains were not as clearly separated as they are today (cf. Chaps. 3, 5, and 7). After all, some prominent figures in the social sciences were also biologists. The most influential of these nineteenth-century polymaths was Herbert Spencer, a sociologist, anthropologist, and biologist who used Darwin's term 'natural selection' in his two-volume treatise *The Principles of Biology* (Spencer, 1864/1867), while Darwin adopted Spencer's term 'survival of the fittest' in later editions of *The Origin*. The eminent linguists August Schleicher and Jacob Grimm were also passionate amateur botanists (Chisholm, 1911; McElvenny, 2019). Another of these versatile researchers was Bashford Dean, an American zoologist and museum curator who studied the morphology and systematics of Devonian armoured fishes (placoderms), sharks, and chimaeras, but also the history of medieval and modern weapons and armour. To this day, he remains the only person to have been employed simultaneously at the American Museum of Natural History and the Metropolitan Museum of Art in New York (La Rocca, 2014).

Dean illustrated the historical relationships of weapons and armour using elaborate tree-like schemes that captured both anagenetic and cladogenetic processes, incorporated the time axis, and displayed the characters he used to reconstruct the phylogeny (Lyman & O'Brien, 2006, Fig. 12.2). From the late nineteenth century onwards, not only linguists but also ethnographers and archaeologists used family trees to depict mutual relationships among objects they studied. One of them was Augustus Henry Lane Fox, a British collector of artefacts who later adopted the name Pitt-Rivers; his museum in Oxford was the first to display archaeological and ethnographic items according to their presumed phylogenetic relationships (Lane Fox, 1875). Another notable figure was the British archaeologist Henry Balfour, the first curator of the Pitt Rivers Museum, who studied the evolution of various tools, weapons, and decorative art (Balfour, 1893).

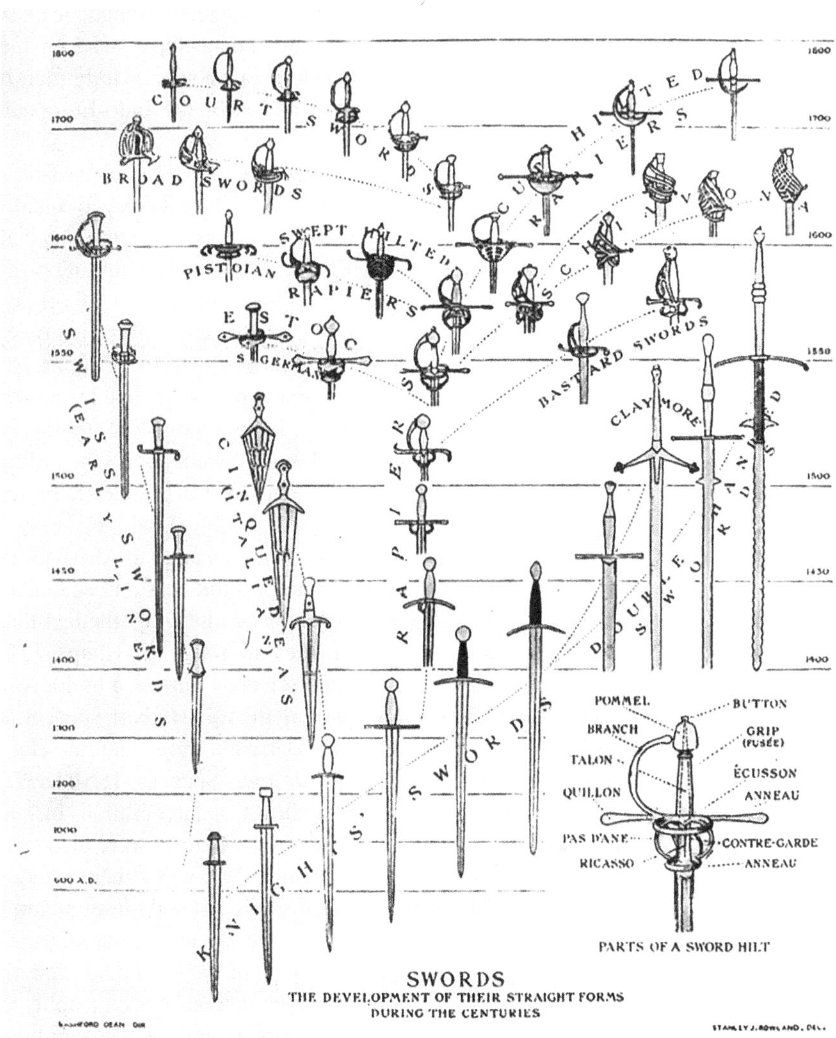

Fig. 12.2 The tree of medieval and modern swords. Bashford Dean compared the evolution of armour and weapons with biological evolution, see Lyman & O'Brien, 2006: 84. Wikimedia Commons (https://commons.wikimedia.org/wiki/File:Evolution_of_swords.jpg)

12.2 Evolutionism in Anthropology and Its Decline

Darwin's writings had a profound influence on the emerging fields of social and cultural anthropology. This applies especially to *The Descent of Man* (1871), which contains references to cultural evolution, including speculations about a competition between primitive societies through advancements in weaponry or 'the struggle for life' between grammatical structures or languages. Evolutionism was the first approach in anthropology that strove to explain human social and cultural diversity without relying on the concept of degeneration—that is, the idea that 'primitive' societies were examples of cultural regression. Evolutionist anthropologists, including Lewis Henry Morgan, Edward Burnett Tylor, and others, argued that 'primitive cultures' represent different stages of cultural evolution (Morgan, 1871; Tylor, 1871). While these anthropologists were inspired by the idea of evolution, they actually drew very little on its central concept of natural selection. Their views were based not on Darwin's theory but rather on Spencer's, which equated evolution with progress (Freeman, 1974). They believed that societies progress along a linear sequence of similar stages of increasing complexity—from savagery through barbarism to civilisation—ultimately arriving, each at its own pace, at the same final stage (Morgan, 1871). Unlike the models developed by linguists and archaeologists, which relied to varying degrees on a branching model of evolution, the schemes of social evolution were strictly unilineal and progressive, advancing, for instance, from promiscuity to monogamy, from matriarchy to patriarchy, or from polytheism to monotheism (cf. Chaps. 2, 3, 5, 7, and 14).

In devising these sequences of cultural evolution, evolutionist anthropologists largely disregarded cultural diffusion (borrowings) as a source of cultural change, as well as convergence or parallel evolution, reversals and their broader implications (cf. Chaps. 5 and 7). At a meeting of the Royal Anthropological Institute, Tylor presented his pioneering cross-cultural study of 350 societies, which demonstrated an association between socio-political complexity on the one hand and descent and post-marital residence rules on the other (Tylor, 1889). He found that less complex societies, organised into bands or tribes, are often matrilineal and matrilocal, while more complex societies, organised into chiefdoms or states, tend to be patrilineal and patrilocal. Tylor interpreted this as evidence of a functional link between civilisation and patriarchy, and as support for the hypothesis that the ancestral societies were mostly matriarchal. During the discussion that followed, Francis Galton, who was present in the audience, pointed out that a cross-cultural association between two traits does not necessarily indicate a functional association. Societies cannot be treated as independent when investigating these cross-cultural trends, because they are historically related to each other. In Tylor's own words:

> The difficulty raised by Mr. Galton [is] that some of the concurrences might result from transmission from a common source, so that a single character might be counted several times from its mere duplicates, is a difficulty ever present in such investigations. (Tylor, 1889: 272).

In other words, if a number of societies inherit a particular combination of traits from an ancestral society, all these co-occurrences could be pseudoreplications rather than evidence of adaptation. This inability to differentiate between 'historical' and 'functional' associations—between homologies and convergences—plagued cross-cultural research for decades afterward (Mace & Pagel, 1994; Naroll, 1965).

In the early twentieth century, unilineal and progressive views of cultural evolution faced fierce opposition from the emerging school of cultural relativism, which was represented by Franz Boas and his intellectual followers (Lewis, 2001). While Boas appreciated Darwin's focus on diversity and change—which resonated with his own views on the fluidity and uniqueness of cultures—he strongly rejected the 'armchair speculations' of Tylor (1871), Morgan (1871), and others regarding social evolution, arguing that culture develops independently of the biological characteristics of human populations. He maintained that culture, language, and race are mutually independent and not causally related determinants of human existence (Boas, 1940). This anti-evolutionist backlash in anthropology coincided with a broader retreat from Darwinism across various fields of science—including, to some extent, even biology—in the early decades of the twentieth century (see Chaps. 5 and 7).

While the unilineal models were rejected for their lack of empirical support, branching models of cultural evolution faced criticism for their inability to account for cultural diffusion, including cultural borrowings, acculturation, creolisation, and language shift (Moore, 1994). The frequency of the fusion of distinct cultural lineages (syngenesis) was seen as constituting the primary distinction between the tree-like biological evolution and reticulate cultural evolution, as illustrated by the famous scheme proposed by Alfred Kroeber (1923: 68, see Fig. 12.3).

The view that biological evolution is primarily characterised by branching, while cultural evolution is dominated by blending, became deeply ingrained across disciplines. This view is exemplified by biologist Stephen Jay Gould's statement:

> Biological evolution is a system of constant divergence without subsequent joining of branches. Lineages, once distinct, are separate forever. In human history, transmission across lineages is, perhaps, the major source of cultural change. (Gould, 1991: 65)

While phylogenetic trees became the leading paradigm for describing organismal evolution, their prominence in historical-comparative linguistics began to wane. Johannes Schmidt (1872) pointed out that the family tree model could not adequately capture the relationships between various Indo-European languages. He therefore proposed the wave theory, according to which lexical and grammatical changes spread across neighbouring speech communities like waves. This leads to the formation of linguistic areas (sprachbunds), groups of unrelated languages that share vocabulary and structural features. At the turn of the century, 'cultural diffusionism'—which emphasised the spread of cultural traits through interaction rather than descent with modification—emerged as the prevailing paradigm in anthropology.

Despite the growing emphasis on cultural diffusion, a parallel development of evolutionary biology and historical linguistics continued into the twentieth century.

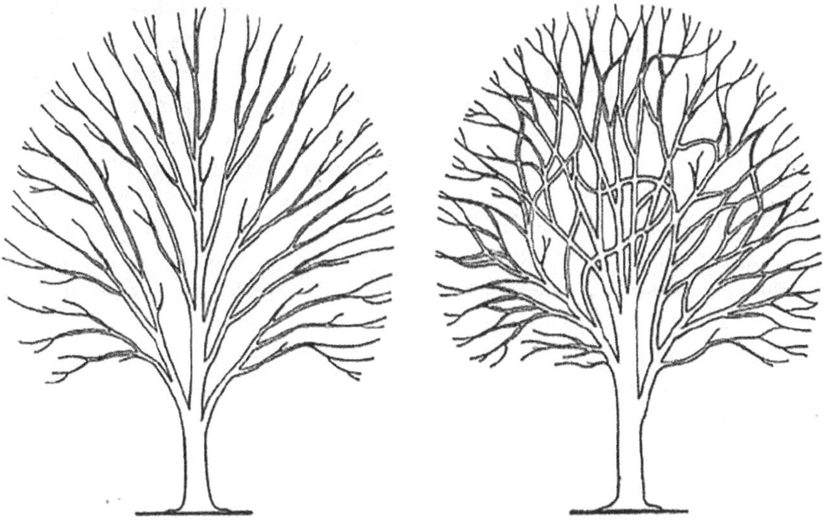

FIG. 18. THE TREE OF LIFE AND THE TREE OF THE KNOWLEDGE OF GOOD AND EVIL—THAT IS, OF HUMAN CULTURE

Fig. 12.3 A visual representation of the difference between the branching tree of life (left) and the reticulate tree of human culture (right), as illustrated by Kroeber (1923)

Building on the pioneering efforts of August Schleicher (1853a, 1853b), Karl Brugmann refined the comparative method by introducing a distinction between retentions and innovations: retentions are shared traits that were already present in the common ancestor, while innovations refer to shared characters that were not. Brugmann recognised that shared innovations provide much stronger evidence of linguistic relationships than shared retentions do. In essence, he outlined the key principles of phylogenetic systematics (cladistics) decades before their formalisation. His concepts are closely parallel to those introduced over 50 years later by the German zoologist Willi Hennig, who spoke of plesiomorphies (retentions) and apomorphies (innovations), or symplesiomorphies (shared retentions) and synapomorphies (shared innovations) (Atkinson & Gray, 2005).

12.3 Towards the Modern Science of Cultural Phylogenetics

The late 1950s and early 1960s saw major methodological breakthroughs in evolutionary biology, with the development of numerical taxonomy (phenetics) by Sokal and Sneath (1963), and the introduction of the molecular clock hypothesis by Zuckerkandl and Pauling (1962). But even earlier, in the 1950s, the linguist Morris Swadesh introduced analogous methods—namely, lexicostatistics and glottochronology—for reconstructing mutual relationships between languages (Atkinson & Gray, 2005; Swadesh, 1955). These methods estimated linguistic divergence by calculating pairwise distances based on the percentage of shared vocabulary and then inferring the divergence date under the assumption of a constant rate of lexical replacement.

Despite initial enthusiasm, Swadesh's methods faced strong criticism and gradually fell out of favour due to their methodological limitations and flawed assumptions (Blust, 2000; Campbell, 2004). The conversion of lexical data into a pairwise distance matrix led to information loss, which in turn hampered the method's ability to accurately reconstruct language relationships. Moreover, the clustering algorithm tended to group together languages that evolved at a similar rate rather than those that were actually related, introducing systemic errors into the topology of the inferred trees. Furthermore, the assumption that lexical changes occur at a constant rate across different languages, time periods, and semantic domains is unrealistic and leads to unreliable divergence estimates.

In the mid-twentieth century, interest in evolution was revitalised in anthropology with the emergence of the neoevolutionist school (Sahlins & Service, 1960), represented by Leslie White, Julian Steward, Marshall Sahlins, and Elman Service (see Chap. 7). Neoevolutionist anthropologists sought to build on the work of nineteenth-century evolutionists while rejecting the notion of 'primitive' and 'advanced' cultures. Instead, they emphasised multilineal evolution—that is, the idea that societies adapt to similar environments through pathways that may differ, be it in terms of technological advancements or changes in social organisation. These transformations can result either from independent innovations or from cultural borrowings. Later, cultural materialists such as Marvin Harris argued that cultural norms, including taboos, are best understood as adaptations to the local environment (Harris, 1974). Nevertheless, the work of these evolutionarily-minded anthropologists was never grounded in a formal, widely applicable theoretical framework. In the 1970s, Richard Dawkins reignited interest in cultural evolution by coining the term 'meme' as a cultural analogue to gene in the last chapter of *The Selfish Gene* (Dawkins, 1976). Yet while the 'meme of a meme'—that is, the idea of a meme as such—became deeply entrenched in the digital and internet culture (Shifman, 2013), the science of memetics has never developed into a sustained empirical research programme, and the term has slowly faded from the scientific literature (see Chap. 10).

A formal theory of cultural evolution was finally developed in the 1980s by Luigi Luca Cavalli-Sforza and Feldman (1981) and Boyd and Richerson (1985) (cf. Chaps. 7 and 11). Boyd and Richerson defined culture simply as 'information capable of affecting individuals' behaviour that they acquire from other members of their species through teaching, imitation, and other forms of social transmission' (Richerson & Boyd, 2008: 5), where the term 'social transmission' (or 'cultural transmission') refers to the non-genetic transfer of learned information from one individual to another. Tylor (1871: 1) famously defined culture as 'that complex whole which includes knowledge, belief, art, morals, law, custom, and any other capabilities and habits acquired by man as a member of society'. Boyd and

Richerson's new definition enabled a breakdown of 'that complex whole' into (culturally transmitted) units, thus finally opening the door to a formal analysis of the processes of cultural evolution.

Researchers described these processes using quantitative, mathematical models based on population genetics and epidemiology, but, crucially, they did not simply apply biological models to cultural phenomena. Instead, their models allowed for different modes of cultural transmission, including, for instance, vertical transmission (from parents to offspring), horizontal transmission (from peers), and oblique transmission (from older, unrelated individuals) (Cavalli-Sforza & Feldman, 1981). The processes they described included cultural analogues of mutation, drift, and extinction, but they also modelled processes that have no population genetic equivalents, such as transmission biases (e.g., prestige-biased social learning or conformity). Importantly, these models did not rely on particulate inheritance based on discrete units of cultural transmission like memes: they allowed for non-particulate, blending inheritance and associated modes of cultural transmission (e.g., one-to-many transmission). Cavalli-Sforza, Feldman, Boyd, and Richerson did for cultural evolution what Fisher, Haldane, and Wright did for Darwinism half a century earlier: they transformed it into a science capable of formulating testable predictions.

These models focused on the cultural equivalent of what biologists call microevolution, that is, on changes within populations. In the following decade, phylogenetic methods were also for the first time applied to the study of cultural macroevolution (changes at the level of populations and species) to reconstruct the evolution of languages, oral and written traditions, artefacts, and cultural norms. Just as in biology, this involves the construction of the most likely evolutionary tree given the data, while differentiating between homologies and homoplasies. Alongside phylogenetic inference (tree-building) methods, phylogenetic comparative methods were introduced to anthropology by Mace and Pagel (1994). These methods combine cultural data with an underlying phylogeny based on genetic or linguistic data and are used either to reconstruct the evolutionary history of various cultural traits or to test hypotheses about functional association (adaptation) while controlling for common ancestry; they thus provide a principled solution to Galton's problem. In biology, this issue is known as phylogenetic autocorrelation. In the 1980s, biologists developed methods to account for this problem using phylogenies (Felsenstein, 1985), but Mace and Pagel adapted these methods for use in anthropology, leading to a wave of ambitious studies of cultural macroevolution (Gray & Watts, 2017; Mace & Holden, 2005).

Meanwhile, the rapid development of population genetic and computational phylogenetic methods, and their application to the study of human prehistory, led to a renewed interest in reconstructing a global phylogeny of human populations. The anthropologist and geneticist Cavalli-Sforza joined forces with Joseph Greenberg and Merritt Ruhlen, two proponents of long-range linguistic comparison. After comparing phylogenies of human populations based on blood group antigens, protein polymorphisms, and other so-called 'classical markers' with trees based on linguistic classification, they concluded that genes and languages do indeed coevolve on a global scale (Cavalli-Sforza et al., 1988, 1992) (cf. Chap. 5). 'It was exciting to discover that we had confirmed a conjecture made by no less a pioneer than Charles Darwin…that if the tree of genetic evolution were known, it would enable scholars to predict that of linguistic evolution', remarked Cavalli-Sforza (1991) in an allusion to Darwin's famous quote in *The Origin of Species*. Eventually, though, this initial assessment proved to be overly optimistic: it turned out that genetic and linguistic trees are mismatched as often as they are aligned (Barbieri et al., 2022). Nevertheless, while the early studies on gene-language coevolution were later criticised for using low-quality genetic data and controversial linguistic classifications (Bolnick et al., 2004; Campbell, 2015), both formal and informal assessments of gene—language coevolution have since become commonplace in the population-genetic literature (Pakendorf, 2014).

The quality of genetic data rapidly improved as the field moved on from classical markers to DNA sequences, genome-wide data (single nucleotide polymorphisms and microsatellites), and eventually to whole genomes. As the data grew in both size and quality, analytical methods evolved accordingly. Distance-based methods were gradually abandoned in favour of character-based methods such as maximum parsimony, which avoid many of the methodological pitfalls associated with earlier techniques. Over time, parsimony was largely replaced by more sophisticated Bayesian and maximum likelihood methods of phylogenetic inference. Although originally developed to reconstruct phylogenies from morphological and molecular data, these methods were soon applied to linguistic and cultural datasets as well (Mace & Holden, 2005; Mace et al., 2005).

12.4 Language Phylogenies

Language has served as the 'model organism' of cultural phylogenetics since the field's inception. It is particularly well suited to phylogenetic analysis because it meets the criteria of a selectively neutral trait and is transmitted with high fidelity, primarily through vertical cultural transmission. Numerous parallels between biological and linguistic evolution are readily apparent (Pagel, 2009).

While phylogenies of species are based on either morphological or molecular homologous characters, language phylogenies are typically constructed based on homologous words. These words, inherited in a direct line of descent from an ancestral term in a protolanguage, are known as cognates. An example of a set of cognates in Indo-European languages is the word for 'brother': English *brother*, Dutch *broer*, German *Bruder*, Icelandic *bróðir*, Swedish *bror*, Lithuanian *brālis*, Czech *bratr*, Polish *brat*, Russian *брат* (*brat*), Welsh *brawd*, French *frère*, Italian *fratello*, Romanian *frate*, Latin *frāter*, Ancient Greek *φράτηρ* (*frátēr*), and Sanskrit *b^hrā̆tar*. All these words are homologous, having descended from the ancestral Proto-Indo-European term **b^hréh2tēr* (Fortson, 2010).

While a cognate is equivalent to what is called homology, determining which words are true cognates is analogous to sequence alignment in molecular phylogenetics. It can be difficult to distinguish homologies from homoplasies, that is, cognates from non-cognates, because cognates are not always obvious. For example, *brother* (English) and *broer* (Dutch) are clearly cognates, but the Armenian word *ełbair* is also a cognate although it is less obvious, just as the human clavicle is homologous not only to the chimpanzee clavicle but also to the furcula (the wishbone) of a chicken. Over a period of 7000–9000 years, extensive phonetic and semantic changes make identification of true cognates increasingly difficult; beyond a certain threshold, it becomes nearly impossible (Nichols, 1992). This process can lead to the emergence of 'false cognates', words that appear related but are not. For example, the word *akh* (brother) exists in both Arabic and Mongolian but this resemblance is purely coincidental and not evidence of a linguistic relationship between Semitic and Mongolic languages. In biological terms, such false cognates constitute a form of homoplasy.

In molecular evolution, there exist two types of homologies in DNA sequences: orthology, which arises from a speciation event, and paralogy, which results from a duplication event within a species. The linguistic equivalent of such paralogous sequences is a class of cognates called doublets—that is, pairs of words with different meanings that coexist in the same language, such as *skirt* and *shirt*. In some cases, a cognate may be lost or replaced by a non-cognate. For example, the Spanish word for *brother* is *hermano*, which (along with its counterparts *irmão* in Portuguese and *irmán* in Galician) does not derive from the Latin noun *frāter* but from the adjective *germanus*, which took over the meaning of brother in Proto-Ibero-Romance (Fortson, 2010).

Horizontal transmission (lexical borrowings) is widespread across languages. Modern English, for example, contains numerous loanwords from unrelated languages. The word *tomato* originates from Nahuatl (Aztec), while *ketchup* comes from Min Chinese (Hokkien). *Coffee* derives from Arabic (via Turkish), and *café* from French, while *sugar* traces its roots to Sanskrit, passing through Persian and Arabic before reaching English. Other borrowed words include *pal* from Romani, *jabber* from the Australian language Woiwurrung, and *taboo* and *tattoo* from Austronesian languages of the Pacific Islands (Haspelmath & Tadmor, 2009; Durkin, 2014).

The presence of numerous loanwords that do not fit the tree-like model of language evolution poses a challenge for phylogenetic inference. To address this, researchers focus on words from the core vocabulary, which tend to be resistant to borrowing and change only rarely. The core vocabulary includes personal and interrogative pronouns, time and place adverbs, descriptive and comparative adjectives, quantifiers, kinship terms, body parts, common animals, natural phenomena, basic actions and states, and other words that are broadly shared across environments and are largely independent of cultural or technological advancement. The best-known such word list was compiled by Morris Swadesh (1955), and its shortest version contains just 35 words: *I*, *you*, *it*, *who*, *what*, *one*, *two*, *fish*, *dog*, *louse*, *blood*, *bone*, *egg*, *horn*, *tail*, *ear*, *eye*, *nose*, *tooth*, *tongue*, *hand*, *know*, *die*, *give*, *sun*, *moon*, *water*, *salt*, *stone*, *wind*, *fire*, *year*, *full*, *new*, and *word* (Haspelmath & Tadmor, 2009).

Phylogenetic analysis of character matrices indicating the presence or absence of cognates (e.g., *brother* [1], *broer* [1], *hermano* [0]) in the basic vocabulary enables researchers to reconstruct language phylogenies that are minimally affected by horizontal transmission. While the English lexicon is a varied assemblage of words derived from numerous languages, including Celtic, Latin, Old Norse, and Norman French, but also Arabic, Persian, or Hindi, the core vocabulary remains relatively stable (Pagel et al., 2007) as evidenced by the fact that the 200-word version of Swadesh's list contains only 12 loanwords. A phylogenetic analysis of this basic vocabulary correctly classifies English as a West Germanic language, closely related to Frisian, a language spoken mostly in the coastal regions of the Netherlands and Germany (Gray & Atkinson, 2003).

By tracing changes in individual words along the branches of the inferred language tree, linguists can reconstruct parts of the protolanguage and thereby gain insights into the lives and cultures of its ancient speakers. For example, Proto-Indo-European included words such as *g*^{w}*ōus* ('cattle'), *éḱwos* ('horse'), *yugóm* ('yoke'), *k*^{w}*ek*^{w}*lo-* ('wheel'), and *weǵh-* ('wagon') (Fortson, 2010), which suggests that Proto-Indo-Europeans were mid-Holocene nomadic pastoralists who domesticated horses and invented wheeled transport. These linguistic reconstructions align with the 'Steppe hypothesis' proposed by Gimbutas (1956) and later developed by David Anthony (2007), which places the Proto-Indo-European homeland in the Pontic steppe north of the Black Sea.

Phylogenetic analyses of lexical data have been used to reconstruct language trees for Indo-European (Gray & Atkinson, 2003), Austronesian (Gray & Jordan, 2000), and Bantu languages (Holden, 2002), among others, to test competing hypotheses about their origins and expansions. Most such analyses are limited to individual language families because words evolve so rapidly that all cognates are thought to be replaced by non-cognates after a span of 7000–9000 years (Nichols, 1992). Moreover, when Pagel et al. (2007) traced lexical changes in the Indo-European language tree, they found that most words have about a 50% chance of being replaced by non-cognates every 2000–4000 years. The rate of cognate replacement varies greatly, by as much as one hundred times, depending on the word meaning. Some words have a much longer 'linguistic half-life', persisting for 10,000–20,000 years or more. This variation is largely determined by word frequency in everyday speech: the words used most frequently are replaced least often. The most conservative words include pronouns, special adverbs, and numerals (e.g., *I*, *you*, *here*, *how*, *not*, *there*, *what*, *two*, *five*), probably because they are crucial for conveying meaning and are subject to strong purifying selection. As a result, languages can retain mutual intelligibility long after their divergence. For example, high-frequency English words are more likely to be of Old English origin than other parts of the English vocabulary. Just as some homologous genes persist across speciation events, some homologous words survive long after language splits.

The existence of such 'ultraconserved' words raises the possibility of reconstructing deep language relationships beyond individual language families, potentially supporting the existence of macrofamilies such as Eurasiatic, Amerind, or Dené-Caucasian, as proposed by Greenberg, Ruhlen, and other proponents of mass lexical comparison (Ruhlen, 1987). Pagel et al. (2013) conducted a phylogenetic analysis of frequently used words across seven language families hypothesised to belong to the Eurasiatic macrofamily. Their results suggest that some cognates may have persisted for approximately 15,000 years, surviving since the end of the last Ice Age. However, these findings remain highly contentious (Heggarty, 2013), and the number of highly conserved words is likely too small to provide robust phylogenetic inferences.

Another approach to extending the temporal limits of linguistic reconstruction is to focus on the structural features of language, that is, not on vocabulary but on grammar, which is thought to evolve more slowly (Nichols, 1992). Grammatical structures have been used to reconstruct the phylogeny of languages of Melanesia, which include both numerous closely related Austronesian languages and some highly divergent 'non-Austronesian' (Papuan) isolates (Dunn et al., 2005). Similar methods have also been applied to European languages, including Indo-European and Uralic languages as well as the language isolate Basque (Longobardi et al., 2015).

Grammatical structures are more tightly integrated than lexical or phonological features, making them more resistant to evolutionary change. For the same reason, they are also thought to be more resistant to borrowing. However, their tight integration means that a change in one structure often triggers changes in others. Moreover, while grammatical features are theoretically less prone to borrowing, some actually readily diffuse between languages in contact. In addition, the abstract nature and limited number of possible character states make the identification of grammatical homologies challenging. Most importantly, they offer relatively little character information. For example, there are only six possible ways to order the subject, object, and verb across languages. As a result of all these factors, comparisons of lexical and grammatical evolution suggest that grammatical features change, on average, faster than the core vocabulary (Greenhill et al., 2017). They also exhibit higher rates of homoplasy and more bursts of contact-induced change, which indicates that they are, overall, no more conservative than lexical data.

Some phylogenetic analyses of lexical data bypass the identification of cognates altogether and instead treat words from the core vocabulary as molecular sequences of letters, thereby maximising the available character information. These automated approaches—essentially computational versions of Joseph Greenberg's mass lexical comparison—have provided support for some proposed macrofamilies, such as Eurasiatic (Jäger, 2015), but they often struggle with the accuracy of the proposed language phylogenies, especially of families that evolve at highly variable rates. As a result, these methods tend to produce tree topologies with numerous artefacts. In short, while computational phylogenetics has indeed revolutionised historical-comparative linguistics, the goal of reconstructing a 'perfect pedigree of mankind' remains elusive.

12.5 Cultural Phylogenies of Written, Oral, and Artefact Traditions

Some studies have extended the tradition of stemmatics by applying phylogenetic methods to manuscript analysis (Howe et al., 2001). The first major literary work to undergo such analysis was Chaucer's *Canterbury Tales* (Barbrook et al., 1998). Although Chaucer's original manuscript—written presumably in the late fourteenth century—has been lost, dozens of later versions survive. Cladistic and network analyses of 58 manuscript versions of 'The Wife of Bath's Prologue' from the fifteenth century aimed to identify the version closest to the lost original. The findings suggest that Chaucer's original manuscript was not a finalised or fair copy, but a working draft that contained his notes on passages to be deleted or added, as well as alternative drafts of certain sections. As a result, the descendant manuscripts varied significantly and formed five distinct clades.

Oral traditions such as fairy tales, legends, and myths can be analysed using similar approaches. Jamshid Tehrani (2013) applied phylogenetic and network methods to a reconstruction of the cultural history of the *Little Red Riding Hood*, a classic fairy tale in which a young girl visits her grandmother's house, only to be deceived and eaten by a wolf disguised as the old woman. The best-known version appears in *Grimms' Fairy Tales* (Grimm & Grimm, 1812), but an older, darker, and more moralistic version was written by Charles Perrault in the late seventeenth century. However, there exist many versions of the story across Eurasia and Africa; they are catalogued as ATU 333 in the Aarne-Thompson-Uther Index of folktales (Aarne & Thompson, 1961; Uther, 2004). These versions differ significantly, and many even omit the red hood and the girl's nickname. Some are difficult to distinguish from related tales, such as the *Wolf and the Kids* (ATU 123), a widely known folktale, or *The Tiger Grandmother*, an East Asian variant that blends elements of both stories. Tehrani (2013) analysed 58 versions of these folktales using a character matrix of 72 plot features ('motifs'). Parsimony, Bayesian analyses, and network analyses revealed that ATU 333 (*Little Red Riding Hood*) and ATU 123 (*The Wolf and the Kids*) share a common ancestor from which the *Little Red Riding Hood* diverged in the Early Middle Ages. African versions of *The Wolf and the Kids* trace their origins to Middle Eastern variants and their similarities to the *Little Red Riding Hood* are due to convergent evolution. An exception is the Ibo-language version of ATU 333 in Nigeria, which is closely related to European versions. *The Tiger Grandmother* may have emerged through hybridisation of East Asian versions of ATU 123 and European versions of ATU 333 in the seventeenth century.

Just as phylogenetic methods can be applied to fiction, they have also been used to study music, including folk music (Díaz-Báñez et al., 2004), classical music (Windram et al., 2014), and even modern electronic music (Youngblood et al., 2021). Still, most such analyses focus on artefact traditions, thus continuing in the footsteps of Pitt Rivers, Balfour, and Dean. Phylogenetic methods have been used to reconstruct the evolutionary histories of hand axes, arrowheads, carpets, baskets, musical instruments, fine art, houses, boats, and various other cultural artefacts

(Lycett, 2015). These studies explore the modes of cultural transmission, infer the historical development of artefacts, classify them into meaningful groups, and can even reconstruct population movements associated with the spread of particular material culture. One notable example is the phylogenetic analysis of canoe design in Polynesia, aimed at tracing the expansion routes of Austronesian-speaking people across the region (Rogers et al., 2009). This study found that functional traits—that is, those that would affect the canoe's performance—exhibited phylogenetic signal as strong as genetic data, while symbolic traits, which are not subject to the same selective pressures, showed higher levels of homoplasy, indicating a greater susceptibility to cultural convergence and borrowing.

12.6 Using Phylogenies to Study Cultural History and Adaptation

Cultural anthropologists and behavioural ecologists use phylogenies of human populations and languages to reconstruct the evolutionary history of cultural practices and to test hypotheses about cultural adaptations. In archaeology, phylogenetic comparative studies integrate cultural data with independently derived phylogenies (usually a time-calibrated language tree). These methods are often employed to infer the history of cultural traits by reconstructing ancestral states; in other words, they can help determine how deep in time particular traditions extend (Mace & Holden, 2005; Gray & Watts, 2017). Ultimately, such analyses help researchers trace the evolution of social organisation, subsistence strategies, religious beliefs, rituals, and oral traditions (cf. Chaps. 5 and 7).

One example of such a study is the work of Da Silva and Tehrani (2016), which traced the evolution of fairy tales along the phylogenetic tree of Indo-European languages. Unlike Tehrani's earlier study (2013), which focused on the cultural history of a single tale rather than the people telling it, this analysis sought to determine whether Indo-Europeans have been telling the same stories for thousands of years. The question of how deep the roots of European fairy tales extend has been debated ever since the nineteenth century, when Jakob and Wilhelm Grimm observed striking similarities between the 'German' fairy tales they had collected and the oral traditions of Slavic, Persian, Indian, and Arabic cultures (Grimm, 1884).

Da Silva and Tehrani (2016) examined the folktales listed in the Aarne–Thompson–Uther Index, mapping them onto a tree of Indo-European languages. Their study focused specifically on tales about magic that feature supernatural beings and enchanted objects, including well-known stories such as *Hansel and Gretel* (ATU 327A) and *Beauty and the Beast* (ATU 425C). Many of these tales exhibited a strong phylogenetic signal, with 14 tracing back to a common ancestor of the western branch of Indo-European languages (Balto-Slavic, Germanic, Celtic, and Romance) and four even tracing back to the common ancestor of all Indo-European languages. The story most likely present in the Proto-Indo-European language is *The Smith and the Devil* (ATU 330), a tale that follows a blacksmith who sells his soul to the devil in exchange for mastery of his craft but later uses his powers to outwit the devil and escape the bargain. Variants of this story appear across Indo-European languages, with some featuring Death or a genie instead of the devil. These findings provide insight into the technological knowledge of Proto-Indo-Europeans and further support the steppe hypothesis, which posits a Bronze Age origin of the Indo-European languages (cf. Chap. 5).

Phylogenetic comparative methods also make it possible to test for correlated evolution, that is, to examine whether changes in one cultural trait are associated with changes in another. These methods allow researchers to evaluate long-standing hypotheses about evolutionary sequences proposed by the nineteenth- and twentieth-century anthropologists. For example, Opie et al. (2014) used ethnographic data and the phylogeny of Bantu languages in sub-Saharan Africa to test George Peter Murdock's (1949) main sequence theory, which posits that any transformation of a social system begins with a shift in the rules of post-marital residence, which in turn influence other aspects of social organisation. Their phylogenetic reconstruction identified multiple instances of change in both residence and descent rules throughout Bantu history, including reversals, challenging the idea that social evolution follows a rigid, unidirectional path.

A similar study by Haynie and Bowern (2016) examined the evolution of colour terminology in the Pama-Nyungan languages in Australia, testing Berlin and Kay's (1969) sequence model of colour term development. According to Berlin and Kay, languages progress through a fixed, unidirectional sequence in which basic colour terms are gradually added but never lost. Contrary to expectations, Haynie and Bowern's phylogenetic analysis revealed that this is not the case: basic colour terms are, in fact, often lost, challenging the assumption that the expansion of colour lexicon follows an irreversible trajectory.

Methods such as those mentioned above enable a reconciliation of 'Darwinian' and 'Spencerian' approaches in anthropology, that is, those that emphasise branching and convergent evolution versus those that stress a linear and progressive evolution (cf. Chap. 11). Currie and Mace (2011) examined the evolution of political organisation in Austronesian societies of Maritime Southeast Asia and the Pacific, with the aim of testing Spencerian theories of social complexity within a broader Darwinian framework. The Austronesian expansion, which began in Taiwan over 5000 years ago, led to the rise of complex chiefdoms and small states as Austronesians spread across more than 7000 km, from the Philippines all the way to Polynesia (Bellwood, 2005). Chiefdoms emerged in places such as Samoa and Tonga, while states arose in Sumatra (Srivijaya), Java (Majapahit), Hawaii (Hawaiian Kingdom), and Madagascar (Merina Empire). Yet, some Austronesian societies, such as the Māori of New Zealand, experienced a decline in political complexity. The study's findings were remarkably consistent with the theories of Spencer and Tylor, showing that political organisation in the region evolved through a regular sequence of incremental steps, with increases in complexity being more common than decreases. Moreover, while higher levels of political complexity developed gradually, collapses in complexity occurred more abruptly (cf. Chap. 15).

While some anthropologists focus on reconstructing cultural history and inferring processes of cultural evolution, others are more interested in finding functional associations by testing hypotheses about cultural adaptations using ethnographic data. In their analyses, these researchers use trees (based on linguistic or genetic data, or sometimes on linguistic classification) as a 'phylogenetic control'. Controlling for phylogeny in a regression analysis helps to reveal consistent patterns of cultural evolution, such as associations between social organisation and local environmental conditions (Mace & Pagel, 1994; Mace, 2014; cf. Chaps. 7 and 13). As noted above, researchers conducting cross-cultural studies have been aware of Galton's problem since the nineteenth century. In the twentieth century, they tried to mitigate it by using carefully selected subsamples of ethnographically documented societies, especially from the *Standard Cross-Cultural Sample* (SCCS; Murdock & White, 1969), which contains 186 societies from across the world selected from Murdock's *Ethnographic Atlas* (Murdock, 1967) to represent relatively independent cultural clusters. However, reducing the sample to societies presumed to be only distantly related cannot fully eliminate phylogenetic autocorrelation, because the remaining societies are still interconnected to varying degrees.

Minocher et al. (2019) tested various socioecological hypotheses about marriage patterns across societies, specifically the prevalence of polygyny, using a tree of the SCCS societies as a phylogenetic control. To address the challenge of inferring a global phylogeny of human populations from lexical data, their study combined genetic and linguistic trees into a single consensus tree, a 'supertree' (Duda & Zrzavý, 2016) with its topology constrained by linguistic classification. This analysis revealed that cross-cultural variation in polygyny is largely explained by socioecological factors such as pathogen stress and intergroup violence, which exacerbate differences in male health, wealth, and status. Importantly, the tree itself accounted for roughly twice as much variance as all the socioecological factors combined, underscoring the importance of phylogenetic control, as failure to account for common ancestry can lead to both false positives and false negatives. Marriage patterns are apparently highly conservative across societies, reinforcing the need for phylogenetic methods in cross-cultural research. Ultimately, as Mace (2014: 447) put it, these methods allow us to 'put prehistory back into anthropology'—a development that would have pleased the discipline's founders.

12.7 How Tree-Like Is Cultural Evolution?

The dichotomy between branching, tree-like biological evolution and reticulate, bush-like cultural evolution is a false one. Phylogenetic methods provide a formal means for addressing the long-standing question of how tree-like cultural evolution truly is. Specifically, they allow researchers to evaluate the relative contributions of vertical versus horizontal transmission (cultural diffusion) to cultural variation. Collard et al. (2006) compared the 'treeness' of cultural datasets (including archaeological objects, traditions, and cultural practices) and biological datasets (derived from molecular, morphological, and behavioural data) using the consistency index and the retention index. These indexes measure the degree of agreement between a character matrix and the resulting phylogenetic tree. The consistency index quantifies the proportion of character transformations that conform to the tree topology. The retention index reflects how many of the shared derived traits (synapomorphies) occur along the internal branches of the tree, supporting its overall topology. A comparison of these measures across biological and cultural datasets did not indicate that cultural datasets are inherently less tree-like than biological ones. These findings challenge the assumption that biological evolution is primarily driven by branching, while cultural evolution is dominated by blending (cf. Chaps. 7 and 15).

The relative contribution of branching and blending in the evolution of technology and artefact design seems to vary depending on the type of trait, especially between functional and symbolic, or stylistic, traits (Rogers et al., 2009). These contributions can also shift over time, and it is possible to measure such changes by sampling different 'temporal strata'. Tehrani and Collard (2002) conducted a cladistic analysis of Turkmen textiles and showed that the frequency of horizontal transmission (cultural borrowings) increased following certain sociopolitical changes in Turkmen history, specifically after their defeat by Russia in 1881. As the Turkmens were forced to abandon their traditional nomadic pastoralism and settle, they became increasingly dependent on the local market. The design of textiles produced after 1881 exhibits a weaker tree-like signal, reflecting an increase in horizontal transmission as carpets and rugs became commercial commodities. Once textile design ceased to function primarily as an expression of clan or family identity and became market-driven, different clans started to copy external design elements, thus introducing conflicting signals into the dataset. Yet even so, vertical transmission remained the dominant process in the evolution of Turkmen textiles.

Still, phylogenetic trees are a poor representation of cultural history when cultural traits change rapidly or are highly susceptible to horizontal transmission. Rivero and O'Brien (2014) analysed the design of engraved stone plaques from megalithic funerary monuments in the south-western Iberian Peninsula dating to the Late Neolithic and Early Bronze Age. These plaques are decorated with geometric, zoomorphic, or anthropomorphic motifs, whereby each plaque features one or two holes, suggesting they were worn as necklaces. A widely accepted hypothesis posits that the plaques functioned as mnemonic devices recording genealogies and serving as heraldic markers of family or clan affiliation—but cladistic analysis did not support this interpretation. The plaque design exhibited a high degree of homoplasy and produced a poorly resolved tree, indicating that rather than adhering to strict lineage-based traditions, their creators had considerable creative freedom.

Another approach to assessing the relative importance of different types of cultural transmission in the evolution of cultural diversity is to examine their correlation with various proxies for vertical and horizontal transmission. The pioneering studies of Guglielmino et al. (1995) and Hewlett et al. (2002) analysed variations in 47 cultural traits across 277 societies in sub-Saharan Africa. Their findings indicated that various aspects of social organisation—such as descent, residence, and the rules of property inheritance—strongly correlate with language classification (used as a proxy for phylogeny), suggesting that these traits are transmitted mostly vertically. In contrast, other cultural practices, such as initiation rites or religious beliefs, were more strongly associated with geographic distance, suggesting a significant role of horizontal transmission or cultural diffusion.

Finally, it seems that phylogenetic analyses of cultural traits are not limited to human societies; this was demonstrated by Lycett et al. (2007), who analysed cultural behaviours in chimpanzee populations across western, central, and eastern Africa. Cladistic analyses of 65 behaviours—including tool use—across seven chimpanzee populations produced a cladogram that did not align with the phylogeny of these populations, suggesting that these behaviours are not genetically determined. Moreover, their distribution did not correlate with local environmental factors, indicating that those behaviours are transmitted mainly vertically (cf. Chap. 9). Notably, the CI and RI values for this dataset were comparable to those for human cultural datasets and to cross-species datasets based on molecular, morphological, and ethological data.

12.8 Conclusions

The strength and promise of cultural phylogenetics lie in the formal quantitative approach and its ability to integrate data and methods from disciplines that have been asking similar questions and facing comparable methodological challenges. The application of phylogenetic methods, originally developed within biology, to anthropology has allowed researchers to gain completely new and illuminating perspectives on a vast body of ethnographic and anthropological data accumulated over generations. The implementation of these methods on cultural data had led to valuable insights that would have remained beyond reach had we not ventured across the boundaries of the exact sciences and humanities.

Phylogenetic methods can be applied to any entity or trait, be it biological or cultural, as long as it exhibits variation, inheritance, selection, and descent with modification. These are powerful tools not only for testing long-standing hypotheses but also for generating new ones. The story of cultural phylogenetics is therefore not just a tale of curious parallels between biological and cultural evolution, but also a reflection of the intellectual trajectories of biology and the social sciences—of how these disciplines once split apart and have now found their way back to each other.

References

Aarne, A., & Thompson, S. (1961). *The types of the folktale: A classification and bibliography* (2nd rev. ed.). Suomalainen Tiedeakatemia. (FF Communications, No. 184).

Anthony, D. W. (2007). *The horse, the wheel, and language: How Bronze-Age riders from the Eurasian steppes shaped the modern world*. Princeton University Press.

Atkinson, Q. D., & Gray, R. D. (2005). Curious parallels and curious connections. Phylogenetic thinking in biology and historical linguistics. *Systematic Biology, 54*(4), 513–526.

Balfour, H. (1893). *The evolution of decorative art: An essay upon its origin and development as illustrated by the art of modern races of mankind*. Percival & Company.

Barbieri, C., Blasi, D. E., Arango-Isaza, E., Sotiropoulos, A. G., Hammarström, H., Wichmann, S., Greenhill, S. J., Gray, R. D., Forkel, R., Bickel, B., & Shimizu, K. K. (2022). A global analysis of matches and mismatches between human genetic and linguistic histories. *Proceedings of the National Academy of Sciences of the United States of America, 119*(47), e2122084119.

Barbrook, A. C., Howe, C. J., Blake, N., & Robinson, P. (1998). The phylogeny of The Canterbury Tales. *Nature, 394*(6696), 839–839.

Bellwood, P. (Ed.). (2005). *The Austronesians: Historical and comparative perspectives* (2nd ed.). ANU Press.

Berlin, B., & Kay, P. (1969). *Basic color terms: Their universality and evolution*. University of California Press.

Blust, R. (2000). Why lexicostatistics doesn't work: The 'universal constant' hypothesis and the Austronesian languages. In C. Renfrew, A. McMahon, & L. Trask (Eds.), *Time depth in historical linguistics* (pp. 311–332). The McDonald Institute for Archaeological Research.

Boas, F. (1940). *Race, language and culture*. Macmillan.

Bolnick, D. A. W., Shook, B. A. S., Campbell, L., & Goddard, I. (2004). Problematic use of Greenberg's linguistic classification of the Americas in studies of Native American genetic variation. *The American Journal of Human Genetics, 75*(3), 519–522.

Boyd, R., & Richerson, P. J. (1985). *Culture and the evolutionary process*. University of Chicago Press.

Burkhardt, F., & Smith, S. (1986). *The correspondence of Charles Darwin* (Vol. 2, pp. 1837–1843). Cambridge University Press.

Campbell, L. (2015). Do languages and genes correlate? *Language Dynamics and Change, 5*(2), 202–226.

Campbell, L. (2004). *Historical linguistics: An introduction* (2nd ed.). Edinburgh University Press.

Cannon, W. F. (1961). The impact of uniformitarianism: Two letters from John Herschel to Charles Lyell, 1836–1837. *Proceedings of the American Philosophical Society, 105*(3), 301–314.

Cavalli-Sforza, L. L. (1991). Genes, peoples and languages. *Scientific American, 265*(5), 104–110.

Cavalli-Sforza, L. L., & Feldman, M. W. (1981). *Cultural transmission and evolution: A quantitative approach (Monographs in population biology, 16)*. Princeton University Press.

Cavalli-Sforza, L. L., Minch, E., & Mountain, J. L. (1992). Coevolution of genes and languages revisited. *Proceedings of the National Academy of Sciences of the United States of America, 89*(12), 5620–5624.

Cavalli-Sforza, L. L., Piazza, A., Menozzi, P., & Mountain, J. (1988). Reconstruction of human evolution: Bringing together genetic, archaeological, and linguistic data. *Proceedings of the National Academy of Sciences of the United States of America, 85*(16), 6002–6006.

Čelakovský, F. L. (1853). *Čtení o srovnavací mluvnici slovanské [Lectures on comparative Slavic grammar]*. České museum; František Řivnáč.

Chisholm, H. (Ed.). (1911). *Encyclopædia Britannica* (Vol. 12, 11th ed., pp. 603–605). Cambridge University Press. Grimm, Jacob Ludwig Carl.

Collard, M., Shennan, S. J., & Tehrani, J. J. (2006). Branching, blending, and the evolution of cultural similarities and differences among human populations. *Evolution and Human Behavior, 27*(3), 169–184.

Collín, H. S. and Schlyter, C. J. (Eds.). (1827). *Corpus iuris Sueo-Gotorum antiqui: Samling af Sweriges gamla lagar, på Kongl. Maj:ts. nådigste befallning* (13 vols., pp. 1827–1877). Haeggström.
Currie, T. E., & Mace, R. (2011). Mode and tempo in the evolution of socio-political organization: Reconciling 'Darwinian' and 'Spencerian' evolutionary approaches in anthropology. *Philosophical Transactions of the Royal Society B-Biological Sciences, 366*(1567), 1108–1117.
da Silva, S. G., & Tehrani, J. J. (2016). Comparative phylogenetic analyses uncover the ancient roots of Indo-European folktales. *Royal Society Open Science, 3*(1), 150645.
Darwin, C. (1859). *On the origin of species by means of natural selection, or the preservation of favoured races in the struggle for life*. John Murray.
Darwin, C. (1871). *The descent of man*. John Murray.
Dawkins, R. (1976). *The selfish gene*. Oxford University Press.
Dennett, D. C. (1996). *Darwin's dangerous idea: Evolution and the meanings of life*. Simon & Schuster.
Díaz-Báñez, J. M., Farigu, G., Gómez, F., Rappaport, D., & Toussaint, G. T. (2004). El compás flamenco: A phylogenetic analysis. In *Bridges: Mathematical connections in art, music, and science* (pp. 61–70).
Duda, P., & Zrzavý, J. (2016). Human population history revealed by a supertree approach. *Scientific Reports, 6*(1), 29890.
Dunn, M., Terrill, A., Reesink, G., Foley, R. A., & Levinson, S. C. (2005). Structural phylogenetics and the reconstruction of ancient language history. *Science, 309*(5743), 2072–2075.
Durkin, P. (2014). *Borrowed words: A history of loanwords in English*. Oxford University Press.
Felsenstein, J. (1985). Phylogenies and the comparative method. *American Naturalist, 125*(1), 1–15.
Fortson, B. W. (2010). *Indo-European language and culture: An introduction* (2nd ed.). Wiley-Blackwell.
Freeman, D. (1974). The evolutionary theories of Charles Darwin and Herbert Spencer. *Current Anthropology, 15*(3), 211–237.
Gimbutas, M. (1956). *The prehistory of Eastern Europe. Part I: Mesolithic, Neolithic and Copper Age cultures in Russia and the Baltic area*. Peabody Museum.
Gould, S. J. (1991). *Bully for Brontosaurus: Reflections in Natural History. W. W. Norton & Company.*
Gray, R. D., & Atkinson, Q. D. (2003). Language-tree divergence times support the Anatolian theory of Indo-European origin. *Nature, 426*(6965), 435–439.
Gray, R. D., & Jordan, F. M. (2000). Language trees support the express-train sequence of Austronesian expansion. *Nature, 405*(6790), 1052–1055.
Gray, R. D., & Watts, J. (2017). Cultural macroevolution matters. *Proceedings of the National Academy of Sciences of the United States of America, 114*(30), 7846–7852.
Greenhill, S. J., Wu, C.-H., Hua, X., Dunn, M., Levinson, S. C., & Gray, R. D. (2017). Evolutionary dynamics of language systems. *Proceedings of the National Academy of Sciences of the United States of America, 114*(42), E8822–E8829.
Grimm, J., & Grimm, W. (1812). *Children's and Household Tales. Realschulbuchhandlung.*
Grimm, W. (1884). *Preface. In Children's and Household Tales (3rd ed.). George Bell.*
Guglielmino, C. R., Viganotti, C., Hewlett, B., & Cavallisforza, L. L. (1995). Cultural variation in Africa: Role of mechanisms of transmission and adaptation. *Proceedings of the National Academy of Sciences of the United States of America, 92*(16), 7585–7589.
Harris, M. (1974). *Cows, pigs, wars, and witches: The riddles of culture*. Random House.
Haspelmath, M., & Tadmor, U. (Eds.). (2009). *Loanwords in the world's languages: A comparative handbook*. De Gruyter Mouton.
Haynie, H. J., & Bowern, C. (2016). Phylogenetic approach to the evolution of color term systems. *Proceedings of the National Academy of Sciences of the United States of America, 113*, 13666–13671.

Heggarty, P. (2013). Ultraconserved words and Eurasiatic? The "faces in the fire" of language prehistory. *Proceedings of the National Academy of Sciences of the United States of America, 110*(35), 14126–14127.

Hewlett, B. S., de Silvestri, A., & Guglielmino, C. R. (2002). Semes and genes in Africa. *Current Anthropology, 43*(2), 313–321.

Holden, C. J. (2002). Bantu language trees reflect the spread of farming across sub-Saharan Africa: A maximum-parsimony analysis. *Proceedings of the Royal Society B-Biological Sciences, 269*(1493), 793–799.

Howe, C. J., Barbrook, A. C., Spencer, M., Robinson, P., Bordalejo, B., & Mooney, L. R. (2001). Manuscript evolution. *Trends in Genetics, 17*(3), 147–152.

Jäger, G. (2015). Support for linguistic macrofamilies from weighted sequence alignment. *Proceedings of the National Academy of Sciences of the United States of America, 112*(41), 12752–12757.

Jones, W. (1786). The Third Anniversary Discourse, delivered 2 February, 1786 by the President, at the Asiatick Society of Bengal: "On the Hindús". In L. Teignmouth (Ed.), *The Works of Sir William Jones: With the life of the author by Lord Teignmouth* (pp. 24–46). Cambridge University Press.

Kroeber, A. L. (1923). *Anthropology: Culture, patterns & processes* (1st ed.). Harcourt, Brace & World.

La Rocca, D. J. (2014). *A look at the life of Bashford Dean. Now at the Met, Metropolitan Museum of Art*. Archived from the original on 15 June 2014. Retrieved 15 June 2014.

Lane Fox, A. H. (1875). 'On the principles of classification adopted in the arrangement of his anthropological collection, now exhibited in the. *Bethnal Green Museum' Journal of Anthropological Institute, 4*(1875), 293–308.

Lewis, H. S. (2001). Boas, Darwin, science, and anthropology. *Current Anthropology, 42*(3), 381–406.

List, J. M., Nelson-Sathi, S., Geisler, H., & Martin, W. (2014). Networks of lexical borrowing and lateral gene transfer in language and genome evolution. *Bioessays, 36*(2), 141–150.

Longobardi, G., Ghirotto, S., Guardiano, C., Tassi, F., Benazzo, A., Ceolin, A., & Barbujani, G. (2015). Across language families: Genome diversity mirrors linguistic variation within Europe. *American Journal of Physical Anthropology, 157*(4), 630–640.

Lycett, S. J. (2015). Cultural evolutionary approaches to artifact variation over time and space: Basis, progress, and prospects. *Journal of Archaeological Science, 56*, 21–31.

Lycett, S. J., Collard, M., & McGrew, W. C. (2007). Phylogenetic analyses of behavior support existence of culture among wild chimpanzees. *Proceedings of the National Academy of Sciences of the United States of America, 104*(45), 17588–17592.

Lyman, R. L., & O'Brien, M. J. (2006). Sedation and cladistics: The difference between anagenetic and cladogenetic evolution. In C. P. Lipo, M. O'Brien, J. M. Collard, & S. Shennan (Eds.), *Mapping our ancestors* (pp. 65–88). Aldine.

Mace, R. (2014). Human behavioral ecology and its evil twin. *Behavioral Ecology, 25*(3), 443–449.

Mace, R., & Holden, C. J. (2005). A phylogenetic approach to cultural evolution. *Trends in Ecology and Evolution, 20*(3), 116–121.

Mace, R., Holden, C. J., & Shennan, S. (Eds.). (2005). *The evolution of cultural diversity: A phylogenetic approach. UCL Press.*

Mace, R., & Pagel, M. (1994). The comparative method in anthropology. *Current Anthropology, 35*(5), 549–564.

McElvenny, J. (2019). Materialism in the field of language: Schleicher, Marx, and the biology of language. *Theoretical Linguistics, 45*(3–4), 177–193.

Minocher, R., Duda, P., & Jaeggi, A. V. (2019). Explaining marriage patterns in a globally representative sample through socio-ecology and population history: A Bayesian phylogenetic analysis using a new supertree. *Evolution and Human Behavior, 40*(2), 176–187.

Moore, J. H. (1994). Putting anthropology back together again: The ethnogenetic critique of cladistic theory. *American Anthropologist, 96*, 925–948.

Morgan, L. (1871). *Ancient society*. Henry Holt.
Murdock, G. P. (1949). *Social structure*. Macmillan Company.
Murdock, G. P. (1967). *Ethnographic atlas*. University of Pittsburgh Press.
Murdock, G. P., & White, D. R. (1969). Standard cross-cultural sample. *Ethnology, 8*(4), 329–369.
Naroll, R. (1965). Galton's problem: The logic of cross-cultural analysis. *Social Research*, 428–451.
Nichols, J. (1992). *Linguistic diversity in space and time*. University of Chicago Press.
Opie, C., Shultz, S., Atkinson, Q. D., Currie, T., & Mace, R. (2014). Phylogenetic reconstruction of Bantu kinship challenges Main Sequence Theory of human social evolution. *Proceedings of the National Academy of Sciences of the United States of America, 111*(49), 17414–17419.
Pagel, M. (2009). Human language as a culturally transmitted replicator. *Nature Reviews Genetics, 10*(6), 405–415.
Pagel, M., Atkinson, Q. D., & Meade, A. (2007). Frequency of word-use predicts rates of lexical evolution throughout Indo-European history. *Nature, 449*(7163), U717.
Pagel, M., Atkinson, Q. D., Calude, A. S., & Meade, A. (2013). Ultraconserved words point to deep language ancestry across Eurasia. *Proceedings of the National Academy of Sciences of the United States of America, 110*(21), 8471–8476.
Pakendorf, B. (2014). Coevolution of languages and genes. *Current Opinion in Genetics & Development, 29*, 39–44.
Pedersen, H. (1931). *The discovery of language—Linguistic science in the nineteenth century*. Indiana University Press.
Reynolds, L. D., & Wilson, N. G. (2013). *Scribes and scholars: A guide to the transmission of Greek and Latin literature*. Oxford University Press.
Richerson, P. J., & Boyd, R. (2008). *Not by genes alone: How culture transformed human evolution*. University of Chicago Press.
Rivero, D. G., & O'Brien, M. J. (2014). P Phylogenetic analysis shows that Neolithic slate plaques from the southwestern Iberian Peninsula are not genealogical recording systems. *PLoS One, 9*(2), 18.
Rogers, D. S., Feldman, M. W., & Ehrlich, P. R. (2009). Inferring population histories using cultural data. *Proceedings of the Royal Society B-Biological Sciences, 276*(1674), 3835–3843.
Ruhlen, M. (1987). *A guide to the world's languages: Volume I, classification*. Stanford University Press.
Sahlins, M. D., & Service, E. R. (Eds.). (1960). *Evolution and culture*. University of Michigan Press.
Schleicher, A. (1853a). Die ersten Spaltungen des indogermanischen Urvolkes [The first splits of the Indo-European prehistoric people]. *Allgemeine Monatsschrift für Wissenschaft und Literatur, 3*.
Schleicher, A. (1853b). O jazyku litevském, zvláště ohledem na slovanský [On the Lithuanian language, and specifically on Slavic]. *Časopis Českého Musea, 27*, 320–324.
Schleicher, A. (1869). *Darwinism tested by the science of language*. John Camden Hotten.
Schmidt, J. (1872). *Die Verwantschaftsverhältnisse der indogermanischen Sprachen*. Hermann Böhlau.
Shifman, L. (2013). *Memes in digital culture*. MIT Press.
Sokal, R. R., & Sneath, P. H. A. (1963). *Principles of numerical taxonomy*. W. H. Freeman.
Spencer, H. (1864/1867). *Principles of biology* (Vol. 2). Williams & Norgate.
Swadesh, M. (1955). Towards greater accuracy in lexicostatistic dating. *International Journal of American Linguistics, 21*(2), 121–137.
Tehrani, J. J. (2013). The phylogeny of little red riding hood. *PLoS One, 8*(11), e78871.
Tehrani, J., & Collard, M. (2002). Investigating cultural evolution through biological phylogenetic analyses of Turkmen textiles. *Journal of Anthropological Archaeology, 21*(4), 443–463.
Tylor, E. B. (1871). *Primitive culture: Researches into the development of mythology, philosophy, religion, art, and custom* (Vol. 2). John Murray.

Tylor, E. B. (1889). On a method of investigating the development of institutions; applied to laws of marriage and descent. *The Journal of the Anthropological Institute of Great Britain and Ireland, 18*, 245–272.

Uther, H.-J. (2004). The types of international folktales: A classification and bibliography. In *Based on the system of Antti Aarne and Stith Thompson* (Vol. 3, FF Communications No. 284-286). Suomalainen Tiedeakatemia/Academia Scientiarum Fennica.

Windram, H. F., Charlston, T., & Howe, C. J. (2014). A phylogenetic analysis of Orlando Gibbons's Prelude in G. *Early Music, 42*(4), 515–528.

Youngblood, M., Baraghith, K., & Savage, P. E. (2021). Phylogenetic reconstruction of the cultural evolution of electronic music via dynamic community detection (1975–1999). *Evolution and Human Behavior, 42*(6), 573–582.

Zuckerkandl, E., & Pauling, L. (1962). Molecular disease, evolution, and genetic heterogeneity. In M. Kasha, & B. Pullman (Eds.), *Horizons in biochemistry* (pp. 189–225). Academic Press.

Chapter 13
Human Cumulative Culture: Bridging Cognition and Population Dynamics

Jan Havlíček

Abstract Humans create the cultural environments they inhabit, and this has profound and unprecedented consequences for their evolution. The main goal of this chapter is to review current evolutionary perspectives on the cognitive and demographic factors which support the spread of new cultural content. I start by introducing the concepts of cultural niche and cumulative culture, two features which are characteristic of humans but rare in other species. Then I focus on key social learning adaptations—especially flexible imitation and various forms of teaching—and show why this distinctive combination of cognitive mechanisms seems crucial for efficient cultural transmission. Finally, I highlight the role of demographic factors (such as the degree of social connectedness) in the diffusion of cultural content and demonstrate their implications for cultural complexity and diversity. Taken together, these insights from multiple disciplines contribute to a more nuanced understanding of what it means to be a cultural species.

Keywords Cumulative culture · Overimitation · Learning bias · Teaching · Conectedness

13.1 Introduction

The study of culture is the primary subject of various disciplines in the social sciences, including ethnology and social anthropology. Researchers in these fields are often sceptical of grand theories (or metatheories) due to their speculative nature (see Chaps. 5, 7, 8, and 12). This scepticism applies also to the theory of evolution, which was employed by late nineteenth century evolutionary anthropologists such

J. Havlíček (✉)
Department of Zoology, Faculty of Science, Charles University, Prague, Czech Republic
e-mail: jhavlicek@natur.cuni.cz

J. Toman et al. (eds.), *Culture Evolving*, Biosemiotics 10,
https://doi.org/10.1007/978-3-032-16124-6_13

as Edward B. Tylor or Lewis H. Morgan. These scholars were inspired by evolutionary ideas popular at that time, such as the concept of progressive evolution, which assumes that organisms inevitably evolve from simple to more complex forms. Tylor, for instance, analogously postulated that cultures develop from primitive to sophisticated ones in a ladder-like fashion, but such ideas had little empirical support, neither in biology nor in anthropology. Moreover, evolutionary theory was misused when the principles of selection were employed to justify the present social order and scientific racism (see Chap. 3). For all these reasons, evolutionary concepts in anthropology became after the Second World War simply *salonunfähig*.

Nevertheless, at least since the 1970s, a small number of scholars tried to model cultural processes and gene-culture evolution analogically to population geneticists and their work has laid the theoretical foundations to the field of cultural evolution. At that time, research in this area included few empirical studies and was based mainly on modelling. This began to change around the turn of the millennium, and during the last two decades, we have witnessed an enormous boom in studies on cultural evolution. They draw on various fields and methods, including traditional simulation and modelling, developmental and comparative psychology, experimental cognitive psychology, sociolinguistics, or field ethnography. The diversity of methodological approaches reflects the diversity of theoretical concepts employed in these works, which span from gene-culture coevolution, memetics, and cultural attractions all the way to cultural phylogenetics, to name just a few. On one hand, this led to an unprecedented influx of new findings and insights one could only have dreamed about at the beginning of the millennium but, on the other hand, it also resulted in various debates about terms and concepts, which made the field somewhat blurred, at least for outsiders.

The main aim of this chapter is to provide an overview of recent findings and theoretical concepts related to cultural evolution. First, I introduce the key concepts of niche construction and cumulative cultural evolution. Second, I describe the main psychological mechanisms involved in cultural transmission, including imitation, teaching, and social learning biases. Subsequently, I focus on the demographic factors that affect cultural transmission, such as the level of connectedness and human social structure and, finally, I provide a short overview of future directions for a more complex understanding of cultural evolution.

13.2 Environmental Inheritance

The theory of biological evolution is based on genetic forms of inheritance, recently complemented by various forms of epigenetic inheritance (Markoš & Švorcová, 2019) (see Chap. 16). But some scholars argue that, not only humans but in fact many organisms also inherit their environment that is the product of their activities (Laland et al., 2016), and such environmental changes effected by the organisms lead to changes in selective pressures. The active role of organisms in their evolution is highlighted by the niche construction theory. Niche construction can be reactive or active: reactive niche construction refers to the environmental changes caused by previous generations of conspecifics, while active niche construction includes processes such as migration, selection of the environment depending on the organisms' characteristics, and purposeful changes in the environment (such as the building of beaver dams). While these general characteristics apply to varying extent across different species, humans are characterised by extensive reliance on cultural products (Odling-Smee & Laland, 2011) (Fig. 13.1). One may argue that cultural environment is analogous to the ecology of a given species, so that while some species, such as badgers, build burrows, others, such as humans, build houses.

One key difference between species ecology and cultural environment might be in how quickly they usually change. Major environmental changes are usually slow, taking place over many generations, so that organisms can have enough time to evolve biological adaptations to cope with them. In contrast, as famously noted by Bowlby (1983), cultural changes occur at such breakneck speed that biological evolution may not be an adequate mechanism to keep up with them, which might leave humans in some respects maladapted to their current environment. Many scholars would agree with such a statement: for instance, due to the amount of time spent reading and looking at mobile screens, myopia is found in almost 90% of Chinese and Korean adolescents while it is rare in some non-industrial societies, for instance among the Inuit (Dolgin, 2015). Others would argue that the pace of biological

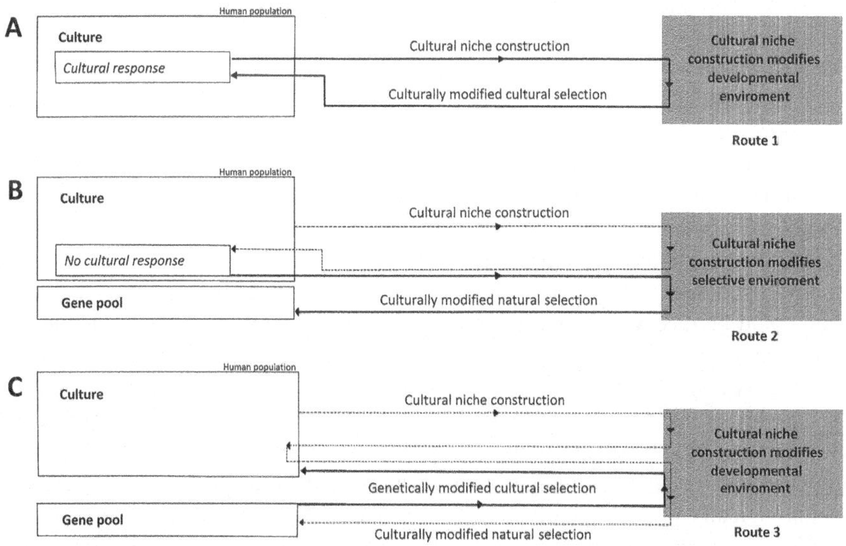

Fig. 13.1 Three routes to feedback from cultural niche construction: (**a**) cultural processes modify the environment but are responded to through further cultural niche construction; (**b**) a failure to respond at the cultural level triggers a genetic response to cultural niche construction; and (**c**) a changed genetic constitution within the population modifies the developmental environment, triggering further innovation and cultural learning. Redrawn from O'Brien and Laland (2012)

evolution is not as slow as one might expect, and people have already adapted to various cultural inventions (Laland et al., 2010). For instance, people living in societies where the diet includes large amounts of starch commonly have multiple copies of amylase, the enzyme that breaks down starches (Perry et al., 2007). Similarly, the persistence of active lactase enzyme in adulthood is widespread in populations which domesticated milking animals such as goats and cattle (Bersaglieri et al., 2004).

Another efficient way of dealing with cultural change is through cultural innovation, for instance by inventing glasses to address myopia, but such innovations create new changes in the human environment and call therefore for further 'cultural adaptations'. Although highly speculative, this mechanism might create a positive feedback loop that would lead to an explosion of culture-based environmental changes. What is clear, though, is that the extent of current human-made environmental changes is so vast and pervasive that some geologists and ecologists argue that we live in the Anthropocene era (Lewis & Maslin, 2015). Whether this is an idiosyncratic trait of a single culture or a more general characteristic of cultural evolution remains an open question.

13.3 Cumulative Cultural Evolution

We have now ample of evidence showing that animals belonging in various taxa use social learning to maintain particular behaviours across generations (for review, see Van Schaik, 2010). This allows for the emergence of cultural evolution, which is to some extent independent of its physical bearers and contributes to behavioural variation across populations of a given species. Examples of such behaviours include tool use in capuchin monkeys, singing in songbirds, or acoustic communication in whales (see Chap. 9). Many of these behaviours, such as tool use, were traditionally believed to be unique to humans; theorists should therefore reconsider in what respect, if any, human culture is special.

In most non-human species, cultural traditions are relatively simple, such that they could be discovered by a single individual, and they show little evidence of subsequent elaboration (but cf. Whiten, 2019), which clearly contrasts with the complexity and successive development often seen in human culture. To emphasise this quality of human culture, scholars started to speak about it being 'cumulative'. Boyd et al. (2011) have repeatedly argued that various technologies used for obtaining food, processing it, building shelters, or making clothing are so complex even in technologically relatively simple human societies, such as the Inuit one, that no single person could invent them by individual learning. Instead, individual technologies evolve across numerous generations of learners in small steps, where each step involves either invention or learning error; the next generation then builds upon the results of the previous generation and these effects accumulate over time. It should be noted, however, that various authors use the notion of a cumulative culture relatively loosely.

More recently, Mesoudi and Thornton (2018) tried to identify the core and extended criteria for cumulative cultural evolution. They claim that the core criteria involve a behavioural change followed by social learning that leads to improved performance, with the previous steps being reiterated to generate sequential improvement over time. The extended criteria involve a diversification of cultural lineages, recombination across lineages, cultural exaptation, and cultural niche construction. Clearly, such core criteria neatly fit technologies and other utilitarian cultural products whose performance can be easily measured. On the other hand, it may be misleading to apply them to many other cultural products, which result from sequential actions across generations of learners but their utility is either unclear and potentially absent; this may pertain to folk tales, clothing design, or perhaps some social rules. Along similar lines, it seems that in genetic evolution, where utility plays a key role, selection may be the dominant process, but in other cases, other evolutionary processes prevail. Equating selection and evolution, as Mesoudi and Thornton (2018) apparently did, would mean overlooking a number of important cultural phenomena (cf. Chaps. 11 and 15).

13.4 Social Learning

Cultural evolution relies on various forms of social learning and understanding these cognitive processes is thus fundamental for understanding cultural evolution. Much has been written about the different forms of social learning that makes human cultural evolution specific. They include most notably imitation and teaching, which thus deserve a closer attention.

13.4.1 Imitation

There is a relative consensus among scholars of cultural evolution that imitation is a key cognitive mechanism involved in the human cumulative culture. But while some argue that it is the central mechanism (Tennie et al., 2009), others more soberly list imitation alongside a host of other socio-cognitive capabilities (Legare, 2017). The term 'imitation' is sometimes used in a broad sense of copying, but cognitive psychologists apply it more narrowly to refer to a form of social learning where the learner observes the model's behaviour and, based on that, performs the behaviour. This distinguishes imitation from other forms of social learning, such as emulation, where learners aim to achieve the same goal as the models using their own means, including individual learning (Tomasello et al., 1993) (cf. Chap. 9). Some scholars emphasise that imitation is cognitively demanding because it must solve the 'correspondence problem', meaning the learner must convert the observed behaviour, which cannot be perceived through proprioception, into their own movement that is controlled by proprioception—and that is a process the learner cannot observe

(Heyes & Ray, 2000). Importantly, there are apparently large differences in the fidelity of imitation not only between individuals but also between different activities of a single individual and there is currently no consensus on what level of fidelity amounts to imitation (cf. Chap. 10).

If imitation functions primarily as an instrumental goal, one may expect that learners would selectively imitate behaviours which seem to lead to the given goal. It was indeed reported that 14-months-old infants can selectively imitate a model's behaviour (Gergely et al., 2002). In this particular experiment, the adult model switched a lightbox with their forehead, either with their hands occupied or with their hands held free but placed on the table. Children imitated the forehead switching in 69% of the free hands condition but only in 21% of the hands-occupied condition. Humans also have a tendency to imitate seemingly unnecessary actions of the model: this phenomenon came to be known as overimitation (or indiscriminate imitation), which emphases that the learner imitates all activities of the model regardless of their effect (Lyons et al., 2007) (cf. Chap. 9). This was demonstrated in a pioneering study where the model demonstrated a set of actions using either an opaque or transparent puzzle box (Horner & Whiten, 2005). In the transparent box condition, it was clear that some of the presented actions were not needed to reach the reward. Even so, preschool children tended to imitate all of the demonstrator's actions in both conditions. Chimpanzees, on the other hand, imitated all actions only in the opaque box condition, where inefficiency of some of the actions was not apparent, but skipped them in the transparent box condition.

Since this initial study, overimitation has been demonstrated in both children and adults under various conditions in several dozen studies (for a review, see Hoehl et al., 2019). A high frequency of overimitation has also been shown by several studies conducted in non-Western cultural settings, for instance in the South African !Kung San or in Australian Aboriginal children (Nielsen & Tomaselli, 2010). The frequency of indiscriminate imitation tends to rapidly increase from toddler age to preschoolers, and then steadily rise to adult levels, but this pattern can vary depending on cultural beliefs about children and parenting, as shown, for instance, in the Aka pygmies (Berl & Hewlett, 2015) (cf. Chap. 8).

It has been suggested that imitation may stem from two general motivations: (1) instrumental, which includes learning a new procedure to achieve a given goal, and (2) social, which involves affiliation with the model (Over & Carpenter, 2013). Some scholars emphasise that the social motivations include the normative aspect of cultural learning, where individuals imitate all actions of the model irrespective of their relevance to the goal because they expect that this is how it ought to be done. Several studies with preschoolers have shown that they tend to protest if the unnecessary action is skipped, and sometimes even claim that if such unnecessary action was omitted, the whole sequence was done wrong (Kenward et al., 2011; Keupp et al., 2013). This indicates that they hold beliefs about the normativity of the entire action sequence. Note that such reactions would not be expected if overimitation were performed solely to affiliate with the model. Researchers tend to agree that instrumental and social motivations play a role in most real-life settings, although their relative importance may vary, for instance, depending on the likeability or familiarity of the model. Moreover, in most children and perhaps also many adults it may be impossible to disentangle the instrumental and normative motivations. This applies to various cultural beliefs often referred to by the umbrella term 'magical thinking' (Subbotsky, 2004), which posit that if the normative procedure is not correctly followed, it might impact the outcome of the action (that is, the instrumental aspect) by the action of various magical powers or deities (cf. Chap. 7).

Several studies investigated whether overimitation is an 'error in matrix' or, rather, a behaviour that brings some benefits. Some authors emphasised that many aspects of human cultural activities are causally opaque (Richerson & Boyd, 2005), meaning that outsiders cannot recognise which parts of the procedure are relevant to the desired goal and which are not. Under such conditions, it can be beneficial to copy all actions of the model even though it may occasionally also lead to an imitation of irrelevant actions (see Chaps. 8 and 9).

13.4.2 Social Learning Biases

Humans are highly social beings, which leads to plentiful opportunities for social learning, but do humans sample new information randomly, depending on who they are interacting with, or is there a systematic underlying pattern? Decades of social psychology research show that people tend to learn from high-status individuals; this tendency is known as 'prestige bias' (cf. Chaps. 7, 9, and 10). The underlying assumption is that higher status is perceived as indicating that the model is using a successful strategy and the learner will benefit from learning from that individual. Other scholars nevertheless suggest that one should differentiate social status, prestige, and competence (Henrich & Gil-White, 2001), because these three terms refer to different forms of cues learners rely on. In the case of status, one implicitly assumes that a given behaviour led to the status. For instance, if a hunter has many rare and precious animal skins, observers will assume the hunter has superior hunting skills. In the case of prestige, one assumes that the model's reputation is based on a mastery of skills pertinent to the activity in question: to develop the example above, the learners will take into consideration that their social peers consider the model to be a good hunter. Finally, one can directly observe model's behaviour, which provides direct cues about the model's competence; in this instance, it would amount to going for a hunt together.

As noted above, many human activities are so complex that it is extremely difficult to estimate which of the model's activities led to success and which are just incidentally performed by the model but have no direct effect on the learned behaviour. The unclear link between cause and effect is also sometimes referred to using the term 'opacity'. To follow our hunting example, it is difficult to determine whether the hunting success of a given person is due to the tools, technique of tool use, tracking, hunting time, attire, tattoos, the amount of prayer, smoking, or many other factors. In such circumstances, the learner cannot be sure whether skipping one of the activities, such as smoking or having tattoos, would not cause the hunt to fail, which is why the safest thing is to imitate them all. Theoretical models show that *en bloc* imitation of prestigious individual can be, on average, an adaptive learning strategy (cf. Chap. 9). Occasionally, though, *en bloc* imitation of prestigious individuals may lead to the spread of maladaptive cultural contents (Richerson & Boyd, 2005) (cf. Chap. 10). Importantly, people also tend to imitate the views of high-status individuals even in domains not directly related to their expertise. For instance, the political views of medical doctors (a high-status job) are often considered more relevant than the views of individuals with lower-status jobs although political science has little to do with medical expertise (Richerson & Boyd, 2005).

Other forms of social learning bias depend on the frequency of occurrence of the learned content: the cases where more frequently occurring behaviour is imitated preferentially are known as instances of 'conformity bias', while preferential imitation of the less frequently occurring behaviour is known as 'anticonformity bias'. Existing research consistently shows that conformity bias can be found in different domains and cultural settings (for review, see Mesoudi, 2009). Conformity bias has a strong impact on the diffusion of new cultural elements: it has been shown that diffusion of novelties typically follows a flat S-shape, indicating a slow initial increase in the novel behaviour (Henrich, 2001), but after the behaviour is adopted by majority of the population, it reaches a flex point and, due to the conformity bias, quickly becomes prevalent.

Both the prestige and the conformity bias lead to a long-term unification within societies while maintaining variation across societies (Henrich & Boyd, 1998). It has been, for instance, shown that linguistic diversity tends to increase in small communities with a low level of connectedness; an extreme example of this is the linguistic diversity of Papua New Guinea, which has over 9 million inhabitants speaking over 850 languages (Kik et al., 2021).

13.4.3 Teaching

Human social learning often involves active participation of the model who facilitates learning to the student. This behaviour is commonly known as teaching but various scholars either explicitly or implicitly define teaching in different ways depending on the perspective they adopt. For instance, learning psychologists emphasise the cognitive skills which are necessary preconditions of teaching, such as a theory of mind or joint attention. This approach also views the intentionality of the teacher's activity as a key characteristic. Some social anthropologists reserve the term 'teaching' only for activities of passing knowledge in formal schooling institutions such as those known from the European or traditional Chinese school systems (Lancy & Grove, 2010). Evolutionarily oriented researchers, on the other hand, emphasise the functional aspect of teaching, i.e., that it is a behaviour which evolved to facilitate learning in others. It has been, for instance, argued that teaching involves features of cooperative behaviour with response-dependent pay-off and coordinated interaction of the teacher and the learner (Thornton & Raihani, 2008).

In the last two decades, there emerged several reports on teaching in species other than humans. Particularly instructive is the case of meerkats, which often forage on venomous scorpions. Adults provide to juveniles partly immobilised scorpions; the level of immobilisation depends on the age of the youngster, which serves as a reliable cue to their level of experience (Thornton & Mcauliffe, 2006). Such behaviour would clearly fit the functional definition of teaching but perhaps not the psychological definition, which emphasises the intentionality of such behaviour.

To address the various mechanisms involved in teaching, Kline (2014) developed a taxonomy of teaching types: (1) Teaching by social tolerance, which involves allowing the learner to observe the teacher's behaviour, whereby tolerance of learners should be higher than the regular level of social tolerance, and it should be context-dependent. (2) Teaching by opportunity provisioning, where the teacher provides the learner with an opportunity to observe or perform a given activity. (3) Teaching by stimulus or local enhancement, where the teacher directs the learner's attention to specific stimuli or places that can be subsequently explored by individual learning strategies. (4) Teaching by evaluative feedback, where the teacher provides positive or negative reinforcement (e.g., desirable food vs. physical punishment) to the learner in response to the learner's success or failure to perform a given task. (5) Direct active teaching, where the teacher manifests relevant information and the learner interprets it.

Kline's (2014) taxonomy is helpful, because it does not restrict teaching to humans and can therefore serve as the starting point of a comparative approach. This is important because it allows us to investigate not only the cognitive skills involved but also the social and environmental contexts that can lead to the emergence of different types of teaching. By doing so, we might be able to identify the adaptive problems faced by individuals in a given species. Other scholars have argued that the difference between direct active teaching and other forms of teaching is more substantial (Caldwell et al., 2018). Other forms of teaching are usually restricted to a single behavioural domain; for instance, the abovementioned example of learning how to deal with dangerous prey in meerkats is restricted to their foraging behaviour. The teaching process can be made easier or its complexity reduced by splitting a complex activity in smaller parts: this approach is sometimes referred to as providing scaffolding. It is expected that scaffolding behaviour has evolved to solve specific problems and its transfer to another domain might be therefore limited. Domain-specific scaffolding occurs in humans as well. Perhaps the most thoroughly investigated phenomenon of this kind is infant-directed speech, which involves a spontaneous accommodation of one's own vocal characteristics when speaking to an infant or toddler. Infant-directed speech is characterised by a high pitch, simplified vocabulary, repetitive sequences, and exaggerated intonation (Golinkoff et al., 2015), whereby all these components provide scaffolding for the child to facilitate learning the mother tongue but are not generalised to other domains.

Such examples contrast with direct active teaching, which exhibits high levels of generalisability and flexibility, so that similar psychological mechanisms can be employed regardless of whether the teaching concerns hunting skills, weaving, or playing a musical instrument. While other forms of teaching need not require a functional theory of mind, for active direct teaching it appears indispensable on the side of both the teacher and the learner. The learner should understand that the teacher wants to demonstrate the activity to them and the teacher should have some idea about the learner's ignorance of the subject being taught. It has further been proposed that active direct teaching involves 'natural pedagogy', that is, a suite of mechanisms including an ostensive communication of pedagogical intent, eye contact, and joint attention (Csibra & Gergely, 2011).

Social anthropologists who study various forms of acquisition of local knowledge are sceptical about the universal use of active direct teaching. This scepticism may partly arise from the different use of terminology, but it is clear that learning by instruction is prevalent in societies where formal schooling exists but may be limited or missing in other communities. Social anthropologists further emphasise that cultural beliefs regarding children's abilities to learn and about the appropriate forms of learning dramatically affect how local knowledge is acquired. For instance, Mayan children are expected to be attentive to the activities of adults but rarely receive instructions and reasoning pertaining to the activities they observe (Rogoff, 2003).

13.5 Directions of Transmission

Any type of social learning can take several directions, including vertical, horizontal, and transversal transmission. Each of these has specific consequences for the speed and form of cultural change. Vertical transmission signifies that social learning occurs between parents and their offspring and, because transmission from parents to children (but not vice versa) is analogous to genetic transmission, one might expect that cultural content transmitted vertically will support biological fitness (cf. Chaps. 7 and 10). This could be the case of, for instance, the pronatalist beliefs in some religions. One of the consequences of prevalence of vertical transmission is that it preserves cultural variation between biologically unrelated individuals, whereby the spread of particular variants depends on the reproduction of their bearers (Cavalli-Sforza et al., 1982). On the other hand, the evolutionary change of vertically transmitted content will be relatively slow because it takes a generation to spread further.

Both horizontal transmission, which refers to social learning between individuals belonging to a particular cohort, and transversal transmission, that is, learning between non-related individuals across generations, are much faster than vertical transmission. Because they are not associated with genetic transmission, they may allow for the diffusion of cultural content that could interfere with biological fitness (see Chap. 10). While horizontal transmission creates differences between cohorts, transversal transmission tends to homogenise cultural variation. What is currently not clear is whether the prevailing direction of transmission is domain-specific or whether it is societies that differ in the prevailing direction of transmission. Some earlier studies indicate that in hunter-gatherer societies, vertical transmission might dominate (Hewlett & Cavalli-Sforza, 1986) but recent detailed studies suggest that the overall picture might be more complex. First of all, the direction of learning not only varies across domains but also fluctuates during the individual's ontogeny: in early childhood, the vertical direction is dominant only to be overtaken by horizontal transmission from mid-childhood to adolescence. Hunter-gatherer children spend most of their time in age-mixed peer groups and a majority of social learning occurs there. Only later, in early adulthood, they tend to refine their skills and acquired knowledge from 'specialists', and this learning is characterised by transversal transmission (Reyes-García et al., 2016). As a result, at least some such skills may be an amalgamation of learning from different individuals.

13.6 The Effect of Demography

Cumulative cultural evolution depends not only on the cognitive processes involved in the transmission of cultural content but also on various demographic features of a given population. Population size seems to be a prime candidate and, indeed, earlier models which relied heavily on the analogy with genetic evolution argued that cultural drift may affect cultural diversity in small populations. This basically means that due to random errors in cultural transmission, small populations may suffer from cultural loss. More complex cultural traits, such as artefacts whose production involves numerous steps and the use of different technologies, are particularly prone to cultural loss. The most frequently reported example of a loss of cultural complexity is the case of Tasmania: archaeological records show a gradual decline in both the complexity and diversity of various artefacts, including bone tools, warm clothing, and some subsistence activities such as fishing after Tasmania separated from mainland Australia at latest 10,000 years before the present (Henrich, 2004).

Still, cultural diversity also depends on the level of innovation—which in turn depends on the population size. One should therefore bear in mind that it is not simply the population size that matters when it comes to cultural transmission but rather the effective population size, which takes into account the level of connectedness between individuals belonging to a given population. Connectedness, or lack thereof, may take different forms. Structural barriers include limitation of physical contact due to physical distance or geographical barriers, such as large rivers. For instance, it has been shown that differences in the camp structure of the Hadza and BaYaka hunter-gatherers affect contact as well as teaching opportunities. While in BaYaka camps, houses are close to each other, allowing children to interact and learn from non-kin, Hadza camps are much larger, and children's social interaction is largely limited to family members (Lew-Levy et al., 2020).

In contrast, behavioural barriers involve various aspects of the 'human factor', including language barriers, cultural norms, or rivalry (Derex & Mesoudi, 2020). Theoretical models that studied various levels of connectedness have shown that while a high level of connectedness leads to homogenisation of the population, low levels of connectedness hamper the spread of innovation. It is the medium level of connectedness that is most efficient for maintaining cultural diversity (Derex et al., 2018). Experimental studies used online game-building tools (arrowheads and fishing nets) of varying complexity in groups of different sizes. It was found that in larger groups, cultural variation was maintained, and the complex task did not suffer from information loss (Derex et al., 2013). In contrast, Fay et al. (2019) found that artefact performance (in this case, the distance flown by a paper plane) did not increase with the number of models participants were exposed to. The authors speculate that the lower performance under the multiple model condition was due to the participants' cognitive overload.

Several archaeological studies also found no relationship between population size and the diversity of tools (Collard et al., 2016). As noted above, what may matter more is the level of connectedness and not simply the population size, but estimating effective population size in historical populations is difficult and may thus lead to biased conclusions (Derex & Mesoudi, 2020) (cf. Chap. 5). On the other hand, other research found that population size is not related to the complexity of folk tales (Acerbi et al., 2017). It is therefore possible that if there is an association between effective population size and cultural complexity, it might be restricted to cultural products with primarily utilitarian functions, such as tools. The complexity of some non-material cultural products, such as folk tales or rituals, need not be directly related to their utility and their evolutionary trajectories may thus depend on factors other than the effective population size.

A better understanding of the role of demography in cultural evolution may also shed light on some more general aspects of human evolution. In particular, various hypotheses tried to explain the sudden rise in the diversity and complexity of various artefacts during the Late Stone Age in Africa. Earlier studies often attributed these changes to an increase in the cognitive capacity of modern humans but current modelling shows that the expansion of the tool kit during that time may have been rather due to a rise in the population size or its connectedness, eventually a combination of both of these components of the effective population size.

13.7 Social Structure

In contrast to apes, the social structure of most human societies is notoriously difficult to specify and is often characterised by vague terms such as diverse, complex, or fluid. It is generally accepted among scholars that the ancestral mode of subsistence in modern humans is hunting and gathering. It might, therefore, be informative to focus on the social dynamics of hunter-gatherer societies. Their current societies still exhibit a relatively high variation in family size, camp size, level of migration and other characteristics, but some typical features nevertheless emerge (Chapais, 2013). First, the social structure is typically multilevel and multilocal. Second, the elementary level consists of a family unit, which typically involves reproducing individuals, their juvenile offspring, and a varying number of other kin such as grandparents, aunts, etc. Third, these family units are part of a broader multifamily federation, including wider kin and non-kin. Finally, this structure shows a high level of temporal fluidity, so that either individuals or whole family units may reside in different camps depending on the season, various types of cultural festivities, or individual needs such as health issues, labour, etc.

What is important from the perspective of social transmission is that not all individuals are in an equal amount of mutual contact. Several recent studies have investigated social networks in various hunter-gatherer communities using electronic devices to measure close contact (Migliano et al., 2020; Salali et al., 2016) and they found that, for instance, in both the Agta people of the Philippines and the BaYaka people of Congo, around 30% of all interactions were with non-relatives, and most of these interactions were restricted to one to four friends. Importantly, interactions with friends started to form already in early childhood (around 3 years of age) and their frequency tended to increase until teenage years (Migliano et al., 2017).

13.8 Conclusions

In this chapter, I have emphasised that cultural products, both material and symbolic, are a key part of the human environment and in this respect, humans actively contribute to the evolution of their own species. I have further introduced the concept of cumulative culture and argued that it might be excessively narrow to equate it with cultural selection because other processes also play a role. The spread of cultural content is based on social learning processes, and it seems unlikely that a single learning mechanism is responsible for the evolution of human cumulative culture. Instead, it might be a unique combination of psychological propensities, including not only social cognition but also a high level of social tolerance and equality. Finally, cultural diffusion depends on various demographic factors among which the level of connectedness plays a key role.

To conclude, the case of human culture exemplifies the harmfulness of the traditional divide between life sciences and humanities (along with the rivalries and tribalism related to such a split). This was most eloquently formulated by Henry Plotkin (2010), who remarked that 'Mankind's natural place is in culture, and culture is a part of human biology because it is our biology that gives us the ability to enter into culture. For this reason, any contrast or opposition that is made between biology and culture, or between genes and culture or between evolution and culture, is an expression of a wholly wrong conception of the causal structure of the world'.

References

Acerbi, A., Kendal, J., & Tehrani, J. J. (2017). Cultural complexity and demography: The case of folktales. *Evolution and Human Behavior, 38*(4), 474–480.

Berl, R. E. W., & Hewlett, B. S. (2015). Cultural variation in the use of overimitation by the Aka and Ngandu of the Congo Basin. *PLoS ONE, 10*(3), e0120180.

Bersaglieri, T., Sabeti, P. C., Patterson, N., Vanderploeg, T., Schaffner, S. F., Drake, J. A., Rhodes, M., Reich, D. E., & Hirschhorn, J. N. (2004). Genetic signatures of strong recent positive selection at the lactase gene. *American Journal of Human Genetics, 74*(6), 1111–1120.

Bowlby, J. (1983). *Attachment: Second edition (Attachment and loss series)* (Vol. 1). Basic Books.

Boyd, R., Richerson, P. J., & Henrich, J. (2011). The cultural niche: Why social learning is essential for human adaptation. *Proceedings of the National Academy of Sciences of the United States of America, 108*(Suppl. 2), 10918–10925.

Caldwell, C. A., Renner, E., & Atkinson, M. (2018). Human teaching and cumulative cultural evolution. *Review of Philosophy and Psychology, 9*(4), 751–770.

Cavalli-Sforza, L. L., Feldman, M. W., Chen, K. H., & Dornbusch, S. M. (1982). Theory and observation in cultural transmission. *Science, 218*(4567), 19–27.

Chapais, B. (2013). Monogamy, strongly bonded groups, and the evolution of human social structure. *Evolutionary Anthropology, 22*(2), 52–65.

Collard, M., Vaesen, K., Cosgrove, R., & Roebroeks, W. (2016). The empirical case against the 'demographic turn' in palaeolithic archaeology. *Philosophical Transactions of the Royal Society B: Biological Sciences, 371*(1698), 20150242.

Csibra, G., & Gergely, G. (2011). Natural pedagogy as evolutionary adaptation. *Philosophical Transactions of the Royal Society B: Biological Sciences, 366*(1567), 1149–1157.

Derex, M., & Mesoudi, A. (2020). Cumulative cultural evolution within evolving population structures. *Trends in Cognitive Sciences, 24*(8), 654–667.

Derex, M., Boyd, R., Beugin, M. P., Godelle, B., Raymond, M., Mesoudi, A., & Morgan, T. J. H. (2013). Experimental evidence for the influence of group size on cultural complexity. *Nature, 503*(7476), 389–391.

Derex, M., Perreault, C., & Boyd, R. (2018). Divide and conquer: Intermediate levels of population fragmentation maximize cultural accumulation. *Philosophical Transactions of the Royal Society B: Biological Sciences, 373*(1743), 20170062.

Dolgin, E. (2015). The myopia boom. *Nature, 519*(7543), 276–278.

Fay, N., De Kleine, N., Walker, B., & Caldwell, C. A. (2019). Increasing population size can inhibit cumulative cultural evolution. *Proceedings of the National Academy of Sciences of the United States of America, 116*(14), 6726–6731.

Gergely, G., Harold, B., & Király, I. (2002). Rational imitation in preverbal infants. *Nature, 415*(6874), 755–756.

Golinkoff, R. M., Can, D. D., Soderstrom, M., & Hirsh-Pasek, K. (2015). (Baby) talk to me: The social context of infant-directed speech and its effects on early language acquisition. *Current Directions in Psychological Science, 24*(5), 339–344.

Henrich, J. (2001). Cultural transmission and the diffusion of innovations: Adoption dynamics indicate that biased cultural transmission is the predominate force in behavioral change. *American Anthropologist, 103*(4), 992–1013.

Henrich, J. (2004). Demography and cultural evolution: How adaptive cultural processes can produce maladaptive losses—The Tasmanian case. *American Antiquity, 69*(2), 197–214.

Henrich, J., & Boyd, R. (1998). The evolution of conformist transmission and the emergence of between-group differences. *Evolution and Human Behavior, 19*(4), 215–241.

Henrich, J., & Gil-White, F. J. (2001). The evolution of prestige: Freely conferred deference as a mechanism for enhancing the benefits of cultural transmission. *Evolution and Human Behavior, 22*(3), 165–196.

Hewlett, B. S., & Cavalli-Sforza, L. L. (1986). Cultural transmission among Aka pygmies. *American Anthropologist, 88*(4), 922–934.

Heyes, C. M., & Ray, E. D. (2000). What is the significance of imitation in animals? *Advances in the Study of Behavior, 29*, 215–245.

Hoehl, S., Keupp, S., Schleihauf, H., McGuigan, N., Buttelmann, D., & Whiten, A. (2019). 'Over-imitation': A review and appraisal of a decade of research. *Developmental Review, 51*, 90–108.

Horner, V., & Whiten, A. (2005). Causal knowledge and imitation/emulation switching in chimpanzees (*Pan troglodytes*) and children (*Homo sapiens*). *Animal Cognition, 8*(3), 164–181.

Kenward, B., Karlsson, M., & Persson, J. (2011). Over-imitation is better explained by norm learning than by distorted causal learning. *Proceedings of the Royal Society B: Biological Sciences, 278*(1709), 1239–1246.

Keupp, S., Behne, T., & Rakoczy, H. (2013). Why do children overimitate? Normativity is crucial. *Journal of Experimental Child Psychology, 116*(2), 392–406.

Kik, A., Adamec, M., Aikhenvald, A. Y., Bajzekova, J., Baro, N., Bowern, C., Colwell, R. K., Drozd, P., Duda, P., Ibalim, S., Jorge, L. R., Mogina, J., Ruli, B., Sam, K., Sarvasy, H., Saulei, S., Weiblen, G. D., Zrzavy, J., & Novotny, V. (2021). Language and ethnobiological skills decline precipitously in Papua New Guinea, the world's most linguistically diverse nation. *Proceedings of the National Academy of Sciences of the United States of America, 118*(22), e2100096118.

Kline, M. A. (2014). How to learn about teaching: An evolutionary framework for the study of teaching behavior in humans and other animals. *Behavioral and Brain Sciences, 38*, e31.

Laland, K. N., Odling-Smee, J., & Myles, S. (2010). How culture shaped the human genome: Bringing genetics and the human sciences together. *Nature Reviews Genetics, 11*(2), 137–148.

Laland, K., Matthews, B., & Feldman, M. W. (2016). An introduction to niche construction theory. *Evolutionary Ecology, 30*(2), 191–202.

Lancy, D. F., & Grove, M. A. (2010). The role of adults in children's learning. In D. F. Lancy, J. Boc, & S. Gaskins (Eds.), *The anthropology of learning in childhood* (pp. 145–180). AltaMira Press.

Legare, C. H. (2017). Cumulative cultural learning: Development and diversity. *Proceedings of the National Academy of Sciences of the United States of America, 114*(30), 7877–7883.

Lewis, S. L., & Maslin, M. A. (2015). Defining the anthropocene. *Nature, 519*(7542), 171–180.

Lew-Levy, S., Kissler, S. M., Boyette, A. H., Crittenden, A. N., Mabulla, I. A., & Hewlett, B. S. (2020). Who teaches children to forage? Exploring the primacy of child-to-child teaching among Hadza and BaYaka hunter-gatherers of Tanzania and Congo. *Evolution and Human Behavior, 41*(1), 12–22.

Lyons, D. E., Young, A. G., & Keil, F. C. (2007). The hidden structure of overimitation. *Proceedings of the National Academy of Sciences of the United States of America, 104*(50), 19751–19756.

Markoš, A., & Švorcová, J. (2019). *Epigenetic processes and the evolution of life*. Taylor and Francis.

Mesoudi, A. (2009). How cultural evolutionary theory can inform social psychology and vice versa. *Psychological Review, 116*(4), 929–952.

Mesoudi, A., & Thornton, A. (2018). What is cumulative cultural evolution? *Proceedings of the Royal Society B: Biological Sciences, 285*(1880), 20180712.

Migliano, A. B., Page, A. E., Gómez-Gardeñes, J., Salali, G. D., Viguier, S., Dyble, M., Thompson, J., Chaudhary, N., Smith, D., Strods, J., MacE, R., Thomas, M. G., Latora, V., & Vinicius, L. (2017). Characterization of hunter-gatherer networks and implications for cumulative culture. *Nature Human Behaviour, 1*(2), 0043.

Migliano, A. B., Battiston, F., Viguier, S., Page, A. E., Dyble, M., Schlaepfer, R., Smith, D., Astete, L., Ngales, M., Gomez-Gardenes, J., Latora, V., & Vinicius, L. (2020). Hunter-gatherer multilevel sociality accelerates cumulative cultural evolution. *Science Advances, 6(9)*, eaax5913.

Nielsen, M., & Tomaselli, K. (2010). Overimitation in Kalahari bushman children and the origins of human cultural cognition. *Psychological Science, 21*(5), 729–736.

O'Brien, M. J., & Laland, K. N. (2012). Genes, culture, and agriculture: An example of human niche construction. *Current Anthropology, 53*(4), 434–470.

Odling-Smee, J., & Laland, K. N. (2011). Ecological inheritance and cultural inheritance: What are they and how do they differ? *Biological Theory, 6*(3), 220–230.

Over, H., & Carpenter, M. (2013). The social side of imitation. *Child Development Perspectives, 7*(1), 6–11.
Perry, G. H., Dominy, N. J., Claw, K. G., Lee, A. S., Fiegler, H., Redon, R., Werner, J., Villanea, F. A., Mountain, J. L., Misra, R., Carter, N. P., Lee, C., & Stone, A. C. (2007). Diet and the evolution of human amylase gene copy number variation. *Nature Genetics, 39*(10), 1256–1260.
Plotkin, H. (2010). *Evolutionary worlds without end.* Oxford University Press.
Reyes-García, V., Gallois, S., & Demps, K. (2016). A multistage learning model for cultural transmission: Evidence from three indigenous societies. In H. Terashima & B. S. Hewlett (Eds.), *Social learning and innovation in contemporary hunter-gatherers* (pp. 47–60). Springer.
Richerson, P. J., & Boyd, R. (2005). *Not by genes alone.* University of Chicago Press.
Rogoff, B. (2003). *The cultural nature of human development.* Oxford University Press.
Salali, G. D., Chaudhary, N., Thompson, J., Grace, O. M., van der Burgt, X. M., Dyble, M., Page, A. E., Smith, D., Lewis, J., Mace, R., Vinicius, L., & Migliano, A. B. (2016). Knowledge-sharing networks in hunter-gatherers and the evolution of cumulative culture. *Current Biology, 26*(18), 2516–2521.
Subbotsky, E. (2004). Magical thinking in judgments of causation: Can anomalous phenomena affect ontological causal beliefs in children and adults? *British Journal of Developmental Psychology, 22*(1), 123–152.
Tennie, C., Call, J., & Tomasello, M. (2009). Ratcheting up the ratchet: On the evolution of cumulative culture. *Philosophical Transactions of the Royal Society B: Biological Sciences, 364*(1528), 2405–2415.
Thornton, A., & Mcauliffe, K. (2006). Teaching in wild meerkats. *Science, 313*(5784), 227–229.
Thornton, A., & Raihani, N. J. (2008). The evolution of teaching. *Animal Behaviour, 75*(6), 1823–1836.
Tomasello, M., Kruger, A. C., & Ratner, H. H. (1993). Cultural learning. *Behavioral and Brain Sciences, 16*(3), 495–552.
Van Schaik, C. P. (2010). Social learning and culture in animals. In P. Kappeler (Ed.), *Animal behaviour: Evolution and mechanisms* (pp. 623–653). Springer.
Whiten, A. (2019). Cultural evolution in animals. *Annual Review of Ecology, Evolution, and Systematics, 50*(1), 27–48.

Part IV
Biosemiotic Reflections upon the Relationships Between Nature and Culture

Jana Švorcová

For a long time, cultural and biological evolution were viewed as entirely distinct and separate realms, each with its own evolutionary dynamics, sources of variation, and forms of transmission (e.g., horizontal vs vertical). Cultural evolution, including specific cultural adaptations as one of its evolutionary outcomes, is typically faster than biological evolution. In cultural evolution, what is transmitted are practices, ideas, or technological novelties rather than genetic information. It also seems more plastic, even allowing for a return to a previous state (e.g., social norms). Moreover, environmental and cultural contexts play different roles in biological and cultural evolution. Culture is heavily influenced by agents who intentionally modify their surroundings and cultural development, while biological evolution relies mostly on random changes. There is no creativity in biological evolution, at least not in the traditional sense of the term, and only cultural evolution is cumulative.

Nevertheless, each of these claims can be countered or at least heavily questioned: The rate of evolution, including biological evolution, can vary greatly and range from gradual to punctualistic. This applies to cultural evolution, too, although it is undeniable that cultural evolution is generally faster than the biological one. As we are about to see in this section, not only genetic information is transmitted in the course of biological evolution; epigenetic and ecological inheritance play significant roles as well. The concept of phenotypic plasticity – the ability of an organism with the same genome to develop differently in response to different conditions – has also made a comeback, indicating that not only cultural evolution can be plastic. On top of that, although we typically do not attribute intentionality to organisms, their individual accumulated experiences over time must also be taken into account, because these experiences, too, also guide their future evolution. And ultimately, if we view organisms as agents, even biological evolution can be viewed as creative.

J. Švorcová
Department of Philosophy and History of Sciences, Faculty of Science, Charles University, Prague, Czech Republic

The views presented by the authors of this section align with the subjects outlined just above. Our position is informed by biosemiotics, which broadens our comprehension of semiotic processes beyond human cognition. The concept of semiosis implies that all living entities have an inherent capacity for meaning attribution, choices, and habitual way of being – and this has an impact on the transmission of cultural traits and on the influence of cognition on evolutionary dynamics. Biosemiotics thus disrupts the conventional boundaries between biological and cultural evolution, nature and nurture, and even the dichotomy between the body and mind. The leitmotif of all the following chapters is an emphasis on the intrinsic similarity of processes within culture and nature, and on the extent to which the processes of cultural and biological evolution are deeply and organically intertwined. The topics discussed in the following chapters – such as the processes underlying both biological and cultural evolution, the constant interaction of nature, science, and culture, a simultaneous evolution of genes, epigenetics, and culture, analogies between evolutionary biology and historiography, or the epigenetic stabilisation of cultural traits—present different approaches to this core theme.

In the first chapter, Petr Hampl shows how nature and culture, science and the human society constantly influence and co-create each other. The natural world of biological forms serves as the primary foundation for culture, while culture and society in turn influence our scientific understanding and perception of nature. This constant interplay reinforces the mutual relationships between culture and nature, as aptly demonstrated by sociomorphic modelling, a concept introduced by Ernst Topitsch in the 1950s in order to explore the complex relationships between science, culture, and society. According to sociomorphic modelling, human thought—including scientific and philosophical inquiry—is shaped by permanent conceptual forms rooted in our everyday experiences and by interactions with our sociological, technical, and biological environment. Scientific knowledge is thus deeply intertwined with cultural contexts, which shows that nature and culture symbiotically co-create each other.

These relationships are demonstrated on a specific model from entomology, on the example of ants, which have been used as a metaphor and model for human society throughout history. This historical overview demonstrates that scientific interpretations of nature are deeply influenced by the cultural and societal contexts of their times, forming a reciprocal relationship where culture and science continually shape and validate each other. This perspective challenges the Enlightened view of objective science by suggesting that anthropomorphism is always and inevitably an integral part of human thought. In order to understand science, one must therefore always acknowledge its cultural dimensions, which influence and validate scientific representations.

In the following chapter, Jan Toman explores the parallels between cultural and biological evolutions through the lens of stability-based sorting (SBS), a fundamental process inherent in both cultural and biological systems. SBS implies that the proportion of more stable elements tends to increase over time, while less stable elements vanish. This process shapes the trajectory of evolution across various domains, including organisms, cultural systems, and civilisations. In cultural evolution, SBS influences the persistence and spread of cultural elements; in particular, it helps explain why simpler and more contextually adaptable elements often outperform complex or less adaptable ones. This principle extends beyond individual elements to entire cultures, where stable but potentially detrimental practices can accumulate over time, and highlights the dynamic nature of cultural evolution compared to its biological counterpart.

Aside from SBS, the text explores the associated concepts of 'frozen plasticity' and 'frozen evolution', elucidating the punctuationalist nature of both biological and cultural evolutions. Still, cultural evolution exhibits a greater adaptability and flexibility than biological evolution due to its dynamic and Lamarckian nature which allows for an intentional creation of new elements and leads to fewer constraints compared to biological systems,

The third chapter, authored by Jana Švorcová and Karel Kleisner, focuses on the interconnectedness of cultural and biological evolution, suggesting that they mutually influence each other in both a bottom-up and top-down manner, without any one process being primary. This chapter partially advocates for the concept of biocultural evolution as proposed by Boyd and Richerson, which emphasises the close relationship between biological inheritance and socially anchored existence shaped by symbols and influenced by technologies. But the analysis goes deeper, beyond the influence of culture on the genetic background: it also incorporates some recent insights into the epigenetic inheritance of behaviour or the niche construction theory. Studies which demonstrate the epigenetic inheritance of traumatic experience with famine, warfare, or fear in general support the claim that culturally induced influences on our biological nature can be stored as biological traces for future generations. The authors also argue that epigenetic processes can be regarded as the missing link between nature and nurture. Moreover, through a historical exploration of the concept of organic memory – which was popular in the nineteenth century – this chapter also examines why the cultural and biological aspects of evolution have been considered sometimes in separation and at other times as one integrated whole.

In the fourth chapter, Anton Markoš investigates the distinction between nonliving natural processes, often described through the timeless models of physics and chemistry, and the realm of the living, defined by closures. Closures separate living entities from their surroundings, fostering unique dynamics, compositions, and symbolic abilities manifested in the processes of memory and experience. The author argues that within a network of communicating organisms, there exists a mutual understanding among organisms belonging to different developmental lineages and this understanding is based on fortuitous short-term evolutionary trends rather than strict laws. This perspective aligns with a theory which suggests that *culture is inherent to all life forms*, that culture influences the unique expressions of species, which are shaped by individual and evolutionary experiences, community indoctrination, and environmental context. All this leads to a dynamic evolution of appearances and cultural expressions, that is, to a situation that contrasts with passive evolutionary models like McShea's zero-force evolutionary law. Like historiography, evolutionary biology interprets the history of life through the lens of symbolic communication and community interactions. This highlights the intricate relationships between individuals and communities in biological contexts, reflects isomorphism with human cultures, and emphases their inseparability and mutual development. The author further explores several case studies which serve as an argument for moving beyond the reductionist approach of linear biology.

The last chapter, authored by Jan Horský, serves as the grand finale of the book because it analytically reflects the main topics of several of the previous chapters. According to the author, culture is in historiography often construed as a semiotic system that encompasses various elements such as symbols, values, and behaviours, which are transmitted through social learning rather than by genetic inheritance. Historical sciences extend this notion to include group-typical behaviour patterns, which leads to discussions about the diffusion of such patterns and to challenges when it comes to differentiating between social learning and environmental influences. The aim of historiography is primarily to understand the internal cultural contents and their evolutionary changes; to this purpose, historians use interpretative hermeneutics to connect comparative explanations with perspectives more inclined towards attractionism than selectionism, akin to Darwinian natural selection.

Historical sources, such as texts or artefacts, are interpreted within cultural contexts. Historical sciences organise these traces into developmental lines akin to evolutionary lineages in biology. But the nature of these lines is still debated and the question of whether they are interpretatively imposed or inferred from surviving traces remains open. An establishment of the ontological foundations for developmental lines in historical sciences is further complicated by nominalism, the challenge of addressing the relationship between historiography and past reality. Further, Horský discusses whether biological communities construct stories about themselves or whether biologists create narratives about the development of life. He explores analogies between the development of biological species and human cultures and considers the possibility of somatisation of cultural evolution, which would further blur the boundaries between nature and culture.

Chapter 14
Mirroring Nature: The Role of Sociomorphic Modelling in the Perception of Nature

Petr Hampl

Abstract The chapter introduces Ernst Topitsch's concept of sociomorphic modelling and advocates for the semiotic study of human culture and society while paralleling Bruno Latour's approach of examining individual actors and their relationships. Topitsch's work builds on Jakob von Uexküll's ideas, which explore how organisms create their own 'worlds' through interpreting their environment. Topitsch extends this to the human culture: he suggests that to comprehending cultural and societal developments, we need to understand the network of meanings and relationship. This approach is applied to the history of science, in particular on the example of myrmecology, which shows how scientific fields evolve within their cultural and social contexts.

Keywords Sociomorphic modelling · Actor-network theory · Social insects · Ernst Topitsch

14.1 Introduction

The aim of this chapter is to reflect on the nature and origin of culture, its variability, and on certain analogies between nature and culture as well as natural and cultural phenomena. Is culture truly separate from nature? Does it have its own specific nature? How is culture connected to our understanding of nature? Can we observe nature impartially, and can one speak *sine ira et studio* about observing nature?

In the following, culture is understood as a set of non-biologically transmitted information (see Chaps. 9 and 11), while nature is conceived of as independent of

P. Hampl (✉)
Department of Philosophy and History of Sciences, Faculty of Science, Charles University, Prague, Czech Republic
e-mail: p.hampl@email.cz

J. Toman et al. (eds.), *Culture Evolving*, Biosemiotics 10,
https://doi.org/10.1007/978-3-032-16124-6_14

culture and the human mind, but I also discuss nature in the form of knowledge of nature and natural phenomena, which are closely linked to culture. At the intersection between these conceptions of nature and culture, we can find the idea of nature as 'naturalness', traditionally understood as the inner essence of an organism which necessarily depends on interpretation or understanding—and thus also on specific shared knowledge and contemporary culture.

Institutionalised scientific knowledge is an important part of modern society's culture. While in prescientific or pre-Modern societies, cosmology and religion were important for culture, science occupies since the Age of Enlightenment a highly significant place in Western culture, partly filling the role earlier played by religion and mythical cosmological ideas. Modern culture uses science as a tool to extract facts about the world, which then (ideally) help improve society and culture itself. Scientific practice thus legitimises culture and modern societies: it endows them with meaning and enables their further development within a stable framework of rational engagement with the world. According to Bruno Latour (1993), scientific laboratories are the centre of this activity, since that is where nature is purified, objectified, and transformed into facts—and thus artefacts, according to the Latin *facere*—which culture can then absorb. These absorbed facts are subsequently 'metabolised' (integrated into culture) and used to create culture's own image of nature. Although culture, society, and science distance themselves from nature and present themselves as objective and impartial agents, they are dependent on it, incapable of existing without nature and the surrounding objective world.

Latour calls the facts which are produced by science but co-create cultural content 'hybrids': without them, modern humans could not understand their place in the world, but due to their hybrid nature these facts can also disrupt modernity. One of the clearest examples of this is the theory of evolution. The scientific worldview, which had significantly influenced contemporary culture and worldview, is largely the result of transfer of the social image of nineteenth-century English society onto nature.

In the following, I will argue that the relationship between nature and culture, science and the human society, is neither unidirectional nor one-sided—science does not stand alone and culture does not emerge from nothing. One of the main sources of culture is the natural, objectified world, and, conversely, culture and society shape scientific knowledge and our perception of nature. Culture and nature are thus in a mutually reinforcing, reciprocal relationship. We focus on sociomorphic modelling, which traces this mutual relationship between science and culture, and explore how sociomorphic modelling can be used to reflect on science and the creation of scientific knowledge.

14.2 From Nature to Culture and Back Again

In the history of European culture, the worlds of culture and nature have traditionally been understood as distinct, if not outright contradictory. People have set themselves apart from nature, claimed for themselves an exceptional status (cf. Chaps. 2, 4, and 9). We will, however, show that the human and natural worlds intertwine. Human activities, including culture and science, are always partly anthropomorphic. From the perspective of the Enlightenment tradition and scientific claims of objectivity, this anthropomorphism might be seen as negative but it is difficult, if not downright impossible, to entirely free human thinking from it. In fact, I will show that anthropomorphism seems to be an integral part of human thought. Science should thus acknowledge its presence and consciously work with it, rather than try to expunge it at all costs. In recent years, this intertwining of culture with science—or rather the cultural nature of science and our image of nature—has been noted by historians of science thanks to the work of Bruno Latour but also others.

Various scholars have explored the cultural nature of human thought and associated activities, such as science, and consequently also the subject of the cultural character of the image of nature, animals, and humans themselves. The history of science thus helps us uncover the history of culture and map its development. Especially relevant in this context is the work of Donna J. Haraway, a biologist and historian of science, who examines the cultural and social contexts of zoology. In the *Primate Visions* (Haraway, 1989), she traced the historical development of knowledge about primates. Taking the history of primatology as her example, she demonstrated that the twentieth-century interest in primates has also been driven by efforts to answer questions about human origins and the nature of human races. The ideals of (post)colonial Western society had influenced how scientists thought about primates and the fossilised remains of our ancestors. According to Haraway, evolution was presented as a heroic journey into new lands, colonisation of new continents, and progress from barbarism to a globalised capitalist society (see Chaps. 3 and 7). Viewed through the lens of contemporary societal views, primates and fossils helped create a narrative which culture subsequently accepted as factually valid.

Haraway addressed the study of culture and science also in other works on cultural history, such as *When Species Meet* (Haraway, 2007), which focuses on domesticated animals from the perspective of intertwining of their existence with humans and human culture. Domesticated animals reflect our society's demands, preferences, and desires. From household pets, whose various breeds correspond to the various emotional needs of contemporary humans, to laboratory and farm animals, which directly satisfy human needs as a food source or aid our thirst for knowledge. Animals are thus products of culture—but they also co-create it. Haraway speaks of a mutual contamination of culture by nature and *vice versa*. As a result, it is challenging, if not impossible, to separate the natural from the cultural: every time we embark on studying culture, we must consider its natural component.

Lorraine Daston examined the contemporary image of nature against the background of its historical cultural development and showed that science is part of culture, like any other cultural content, and it is therefore subject to the same laws of evolution. The image of nature represented by particular animals has a variable moral and cultural authority, becoming the measure and tool by which humans create their self-image and worldview. Culture and nature cannot be separated: they must be understood together as one scientific issue. Daston also addressed the issue of morality and ethics and their origin in nature. Morality, as a set of values and rules and a significant cultural phenomenon that co-shapes our behaviour, can be associated with various parts of nature which underpin it in different epochs (Daston & Stolleis, 2008; Daston & Vidal, 2013).

Boria Sax (2000, 2001, 2003) approaches the history of science in a similar way. He studied animal mythologies and animals in myths and legends and tried to show how culture and society co-create nature and the image of various animals presented by science. He also touched upon how the image of a particular animal can influence the creation of the human image and, *vice versa*, transform the culture from which it emerged. These thoughts align with concepts proposed by Ernst Topitsch, which are discussed below.

Other noteworthy authors include Harriet Ritvo, one of the first theorists to address the relationship between human culture and the image of various animals and plants (Ritvo, 1997, 2010), Charlotte Sleigh, who wrote a monograph on the cultural history of ants (Sleigh, 2007), Erica Fudge and her work on the cultural history of animals (Fudge, 2002, 2019), or John F.M. Clark, who explored the cultural history of social insects, especially bees (Clark, 1998, 2009). The Reaktion Books have dedicated to this topic an entire series on the cultural history of biology.

The work of the abovementioned scholars demonstrates that science, like other human activities, is a distinct sociocultural practice with its own rules, history, and social context. It cannot exist without a connection to the culture of its time. When studying nature, we must therefore consider the social and cultural contexts since they not only shape but also validate scientific representations of the world. Besides natural facts as such, culture, too, is thus essential for the creation of scientific knowledge.

That does not imply some boundless anthropomorphism or resignation on objective knowledge. From the perspective of Latour's concept of actor-network theory, which views the connection between culture and nature as a tangle of morphisms, we can see that this connection may be entirely natural and inherent to human thought and engagement with the world, and that without it, no intellectual or cultural activity would be possible. Questions about our humanity, cultural contents, and science involve interwoven morphisms: zoomorphism intertwines with anthropomorphism, technomorphisms, theomorphisms, and various psychomorphisms. According to Latour, if there is anything that defines humanity, it is this complex network of cultural, natural, and social relationships.

Still, culture is not an entity *per se*; it emerges and transforms along with the network of relationships of all human activities encompassed in it. It is thus fundamentally a symbiosis of various cultural contents, a unique organism that persists and evolves with each transformation. Latour's perspective on the relationship between culture and society and on the network of relationships among the actors in this network thus essentially subordinates the study of natural sciences, including humans, to cultural studies. This is particularly evident in life sciences. Natural sciences that deal with living nature are not just about animals or nature *per se*. They do not merely inquire into the essence of a group of animals or plants. Rather, through the close interconnection between the natural and the cultural they ultimately relate to humans and their place in the world. With the modern incorporation of humans into nature, this ambition of science has deepened. The image of humans is thus constructed together with the image of animals—or even with its help. Science thus helps culture establish itself and culture, in its turn, assists science in finding the right questions.

14.3 Sociomorphic Modelling

Recent interpretations of the role of animals and nature in the development of culture and the creation of cultural content outlined above are not the first attempts to explain the complex relationships between science, culture, and society. The intertwining of anthropomorphism with zoomorphism and the mutual dependence of science and culture have also been addressed in the 1950s by the philosopher and sociologist Ernst Topitsch (1919–2003) in his concept of sociomorphic modelling. Initially concerned with myths and their interpretation, Topitsch wanted to understand their formation and the rules they follow. Influenced by the neo-Kantian philosophy of Ernst Cassirer (cf. Chap. 18), he gradually realised that he was in fact looking for certain permanent features of human thought applicable not only to the study of mythology but also to philosophy and, of course, science, whose role in the modern society is analogical to myths and mythologies.

Topitsch called his method a 'genetic-historical analysis of metaphysical thought'. He sought the outlines of conceptual forms that have a similar presence and validity in culture, art, politics, as well as in myths, philosophy, and science. He was significantly influenced by Cassirer but also by the Czech Austrian legal theorist Hans Kelsen, who, inspired by Kant's rationalistic philosophy, tried to base law on the firm foundations of normative science. He introduced the concept of a 'pure law' or 'pure legal theory' and rejected the notion of a natural origin of legal norms because of their temporal and societal conditioning, erroneous derivation from presumed naturalness, and a confusion between facts and norms (the is—ought problem) (Kelsen, 1928).

Topitsch's concept of models in philosophy and science was also influenced by Jakob von Uexküll's theory of Umwelt (von Uexküll & Kriszat, 1934), which suggests that each organism has its own species-specific lived world. Each such world consists of elements and properties that enable the individual to interact and communicate with its environment, thus securing its survival. The Umwelt therefore forms the basis for experiencing and relating to the world in a way that is always specific to the particular organism. Topitsch interpreted this theory as implying that the relationship between humans and their Umwelt is the foundation of any further

Fig. 14.1 The surroundings and Umwelt of the honeybee as perceived by von Uexküll in his book *Streifzüge durch die Umwelten von Tieren und Menschen* (von Uexküll & Kriszat, 1934)

worldview formation—and therefore also a key element of philosophy, science, and the entire culture. The fact that humans do not control their Umwelt but must adapt, limit themselves, and deal with the world around them is a creative impulse for our thinking about the world and our existence within it. This applies to purely philosophical questions about death, free will, suffering, and the meaning of existence, but also to entirely practical matters handled by craft, technology, and science (Topitsch, 1958: 295–296) (Fig. 14.1).

This connection with Jakob von Uexküll's thinking makes Ernst Topitsch's work highly relevant to contemporary biosemiotics. Both von Uexküll and Topitsch emphasise the perception and interpretation of signals in specific environments. They also stress that organisms, including humans, do not perceive objective reality: they create a subjective interpretation of the world through their unique sensory abilities and experiences (von Uexküll, 1909, 1920; Topitsch, 1958: 8). Both thinkers conceive of organisms as active interpreters of their environment, as not just passive recipients of information but active agents engaged with their surroundings. This view of organisms as subjects that have their own meaning systems implies the need to analyse communication patterns and their semantic structures in different animal species, humans, and entire cultures. Both von Uexküll and Topitsch understand life as a continuous process of interpreting and responding to signals from the external environment. This dynamic interaction between organisms, the environment, and their Umwelts provides the basis for studying symbolic systems and communication forms across the living world—and that is the key theme of biosemiotic inquiries. Topitsch's ideas thus provide an inspiration for understanding life processes as acts of communication that can be analysed using semiotic methods.

Topitsch also refers to the philosopher Wilhelm Dilthey, specifically to the part of his work that deals with our knowledge, cognition, and experiences as inherently related to humans. A worldview as an image of the world that is the goal of all knowledge arises from our needs to orient ourselves in the world and to understand it as a challenge, but it is also shaped by the needs and limitations of our bodies and our intellect.

The foundation of the formation of our worldviews, science, and culture, lies according to Topitsch in established forms of thought that have their basis in the elementary events of our lives, our *Dasein*:

> Man must first understand what things mean to him and what he can expect from them, and therefore how to behave towards them. (…) The result is a certain evaluative elementary relationship to these things, which is the basis of the thought forms that man uses to interpret and understand the world. (…) Thus, things and processes of our daily life serve as models for our understanding of the world. (Topitsch, 1958: 3)

Through these relationships with the world arises our conception of the world, which reflects our being in the world and establishes our life through cognition, religion, or myths into a cosmic harmony with the surrounding world. That is also supposed to be one of the basic functions of our thinking from the evolutionary perspective, because only functional and practically focused thinking could succeed in evolution. Knowledge of the world thus reflects our being in the world, and humans dealing with their environment create a store of learned and accustomed thought forms, which they automatically use to understand the world around them. These thought forms subsequently offer themselves as models for analogically interpreting the world, our behaviour, our surroundings including natural phenomena—and ultimately the entire cosmos. We can then study these models for interpreting the world and thus better grasp the thought forms underlying the development of science, culture, and human society.

Topitsch distinguishes between three basic models differing by what they relate to in our lives: the biomorphic model, which works with the biological component of our life, the sociomorphic model, which stems from our social life and social norms, and the technomorphic model, which is based on the practical aspect of our life, such as work, craftsmanship, or technology. Although stemming from a seemingly abstract concept, the sociomorphic and biomorphic models play a crucial role in understanding cultural and scientific content. Biomorphic modelling relies on models of our individual development: growth, adolescence, transitional rituals, aging, and death. According to Topitsch, this is what was primarily reflected in ancient mythologies (those in which the world originated from an egg, from a pair of a man and a woman symbolised by the sun and the earth, the father in heaven, etc.).

In science, biomorphic sources are found less frequently but they can be nevertheless present, for instance in research into the origin of life (e.g., ignition by lightning and animation of the primordial matter as a kind of magical fertilisation of the earth). Topitsch also observes that, in Greek mythology, one can see a clear connection between biomorphic and sociomorphic cultural components. Gods are born, they have paternal and familial relationships, but these familial and other relationships also reflect the contemporary Greek society. Christianity, on the other hand, is predominantly biomorphic: this is apparent in its focus on the birth and death of Jesus and his divine nature linked to the Father.

As indicated above, sociomorphic models apply to religion, but more broadly, they reflect our social relationships, to the *Dasein* of other people, the common order of the entire known society, nation, humanity—or perhaps just a family. Topitsch interprets them first based on primarily religious motifs and then uses them to interpret scientific thought. In effect, he uses the model of a nuclear family with an all-knowing and all-powerful father who watches over the entire lineage to interpret monotheistic religions. Concepts such as sin, punishment, salvation, and guilt refer to the fundamental relationships in families and communities: sin is an offense against the father of the lineage or community, and condemnation is punishment for violating social rules. The world is thus a family, a community, kinship, or tribe of interconnected individuals, and the order of the world is the order of power and subordination, rights and obligations towards the world (i.e., to other members of the tribe), rewards and punishments, in short, the social system in which a person is integrated and finds one's place in the cosmos (cf. Chaps. 7 and 8).

According to Topitsch, the sociomorphic models apply also to other parts of human culture, including science and scientific knowledge. Society is one of the main models or patterns through which the world is interpreted—this is something one finds already in the ancient religions. In the realm of science, the very concept of natural law once again refers to the unbreakable laws or a system of rules that regulate a community or family, a system that is internalised as an integral part of one's lived world. The laws of nature are thus primarily a way in which our social mind tries to interpret natural phenomena as orderly, meaningful, reliable, but also as having immense power over our lives.

Aside from the aforementioned example of scientific theories about the origin of life, Topitsch also mentions investigations into the human nature as strongly sociomorphic. According to him, the relationship between society and scientific knowledge is, in these areas, twofold and reciprocal. Topitsch talks about double transmission (*Doppelübertragung*) or reverse transmission (*Rückübertragung*). Both these concepts indicate that during the process of scientific research, one begins to see social reality in nature—and subsequently, nature as a model is transposed back onto the social reality. On the one hand, we interpret nature through the prism of social models—and, on the other hand, we base social models on nature. In doing so, we create a circle of interpreting questions about human nature, the essence of humanity and society.

What does the concept of sociomorphic modelling ultimately do, how does it interpret the relationship between science, culture, and society? Topitsch uses a neo-Kantian idea to demonstrate the common basis of myth, philosophy, religion, and science, to find their underlying 'thought forms' (patterns). On this basis, he wants to develop a conception that could explain the formation of culture and the rules of its transformation, as well as the formation and development of science and philosophy. His modelling concept is limited to three sources: biomorphic, sociomorphic, and technomorphic. Compared to various current concepts of cultural and scientific development, it may seem outdated, but it does have its merits.

It illustrates very well the interconnection between the development of culture and science. Culture and society serve here as the framework for all further considerations—be it scientific or philosophical—but both are also recipients of scientific and philosophical thought, which they in turn incorporate into themselves. According to Topitsch, the sociocultural context of science is indispensable, and it also flows back and helps shape culture.

Science and the scientific view of the world (as well as the worldviews of philosophy or religion) thus always have a strong sociocultural component, and the image of the world created by science or philosophy always to some extent reflects the contemporary state of culture and society. This does not accuse science of anthropomorphism: rather, it is an attempt to delineate the space within which science operates and a way of explaining its specific development. This concept is thus helpful and beneficial for thinking about the history of science and the cultural context of scientific theories.[1]

14.4 Cultural Natural History

To better understand the application of sociomorphic modelling to the history of natural sciences, and to try and grasp the development of culture in different epochs and the interrelation between science, culture, and society, let us take a brief excursion into the history of entomological literature. Although elements of sociomorphic modelling can be discerned in various scientific disciplines which, at first glance, need not seem related to questions concerning humans, when examining the history of particular fields, one can sometimes clearly see how the cultural and societal content of a given period imprinted itself onto scientific knowledge, which then in turn legitimised the scientific theories and methods. Highly illustrative of this phenomenon are writings on social insects, which, despite their anatomical and physiological distance from humans, can resemble human societies in their social organisation. The study of social insects exemplifies very well how their interpretation and scientific (or pre-scientific) understanding intertwined with contemporary notions about humans—and, in turn, how insights about insects were subsequently applied to humans, helping to establish or legitimise cultural and societal ideas of the time. We can thus observe a reinforcing loop of culture legitimising itself from nature and nature from culture.

Social insects, particularly ants, can serve as a mirror reflecting culture and society. In antiquity, their colonies were interpreted as the prototype of ideal farmers, diligently working in the fields, gathering food for the winter. Although the Mediterranean harbours a large variety of ant species with different life strategies, ancient authors focused only on the ants that gather seeds. No other species besides harvester ants seemed to resemble the life of ancient humans to such an extent.

[1] An attempt to apply sociomorphic modelling to the history of biology was made by Hans M. Peters (1960) and, even before Topitsch, Adolf Portmann (1947) considered the history of natural sciences in the same way. In sociology and social sciences, sociomorphic modelling has been recently explored, for example, by Ossewaarde (2014).

Apart from labour, virtues such as honesty, morality, and combativeness were also viewed in antiquity as important. For instance, Hesiod (2004) in his *Works and Days*, provided agricultural instructions using ants as examples. Plato and Aristotle viewed them as virtuous and industrious, akin to humans. Seneca admired their martial spirit and according to Pliny and the 11th book of his *Historia animalium*, ants engaged in market exchanges, festivities, and gatherings to get mutually acquainted. Ant communities were viewed as so interconnected that, according to Pliny, they gathered for the funerals of their dead and Plutarch in his *De sollertia animalium* even mentions commercial exchanges between different colonies.

The society of social insects was essentially seen as familial units that work together and live in a single nest, much like ancient humans live in their communities. Occasionally, they deal with attacks from enemies, whom they fight with all the determination and virtue they apply to work during times of peace. Ancient authors claimed that ants have a good knowledge of astronomy, agriculture, and trade, that they live social lives and suffer from no internal conflicts. Ancient thinkers saw in ants, with their diversified societies capable of strategic efforts, a reflection of their own society. Ants were a perfect model of ancient humans and, conversely, served as a model of righteous and virtuous existence, thereby helping to legitimise the cultural and societal rules of their time.

With Christianity, there came a cultural and societal transformation and with it a different view of nature. Different phenomena and laws were selected from nature, and they became the measure of righteous life and legitimised the life of righteous Christians. Medieval scholars drew on ancient sources but adapted the image of ants so as to reflect medieval society and its Christian virtues. They shifted the emphasis from warfare, trade, and celebrations to piety, humility, and honesty. Ants thus became a living metaphor of Christian virtues, a model for every devout medieval person. Nature was perceived as a microcosm reflecting the entire Creation, and scholars looked in it for examples and rules for humans to follow with the aim of mirroring God's perfection. Animals, especially social ones, were viewed as symbols and metaphors for saints and Christian dogmas.

Christian thinkers saw nothing in nature that would contradict this schema: they selected from the multitude of natural phenomena those that confirmed and legitimised their culture and societal rules. They managed to completely overlook even some highly conspicuous phenomena such as ant wars, occupation of new territories, or slave-like behaviour and conflicts within the colony. The reality of ants as creatures was reduced and modelled after Christian virtues and medieval society, which was then retroactively legitimised as unchanging and created by God.

In the Modern Era, the situation changed once again. Scholars no longer focused on agricultural or religious aspects of social insects but on their architectural abilities, such as the building of 'cities' and complex street systems. Modern scholars were not attracted to the agricultural or rural reality of ant societies: they saw in the natural reality of social insects the facts that surrounded them, facts they experienced in their Modern culture. People of the Modern Era, scholars and readers of encyclopaedic works, lived in a world of Modern urban culture where entirely different economic and social relationships prevailed (Enenkel & Smith, 2010; Harrison, 1998).

Once again, their observations about ants reflected the cultural and societal realities of the times. From farmers and models of Christian life, ants became craftsmen with various urban occupations so that, at this time, we also find various descriptions of the complexity of their 'cities'. For instance, Aldrovandi (1638) elaborated on the beauty of their sewage system, urban festivals, and funeral rituals. He praised their cultural wealth and civilisational sophistication as if describing the advantages of modern urban life. Works of other authors, such as Simone Majoli (1592) or Muffet et al. (1634), depict ants in a similar vein, portraying ant professions, beautiful structures reminiscent of basilicas, or the intricate network of ant urban streets and alleys (Fig. 14.2).

From the eighteenth century onwards, social insects became part of the debate on the nature of animal soul, which emerged from disputes between the rationalists and empiricists about the essence of human soul. Ants, as well as other animals, appeared in these debates as exemplary organisms to which both sides referred and used them to support their arguments. The image of socially living animals thus not only reflected the evolution of contemporary culture and social debates but also served as a material for legitimisation and proof.

René Antoine Ferchault de Réaumur (1683–1757)—as well as other French authors such as Noël-Antoine Pluche (1688–1761), Pierre André Latreille

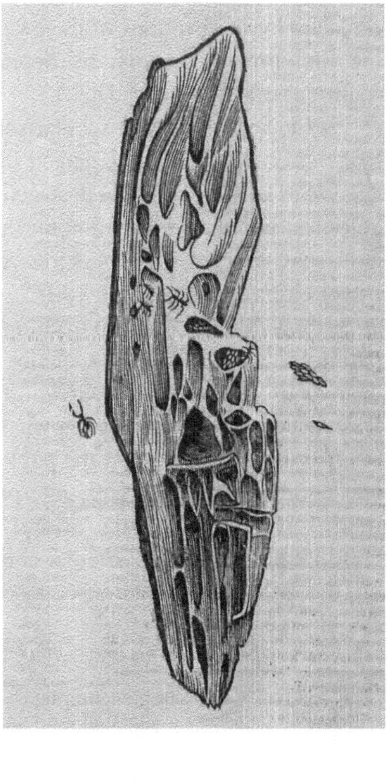

Fig. 14.2 An illustration of the 'cities' of unspecified ants from Aldrovandi's *De Animalibus Insectis* (Aldrovandi, 1638)

(1762–1833), Charles Bonnet (1720–1793)—viewed insects through an entirely different lens. They wanted to know whether, and if so to what extent, ants are capable of emotions, mutual love, care for their offspring, whether they show respect for each other or are merely mindless mechanistic machines. These scholars thought about what binds the perfectly functioning ant community together and how their ability to coexist could be transferred to human society (Réaumur, 1740, 1926).[2] Notably, these questions came to prominence among thinkers who lived during the complex social changes of the French Revolution. For instance, Pierre A. Latreille pondered upon the origin of a harmonious social order and saw in the ant community a reflection of human society with its conflicts, enslavement of poor workers, but also harmonious coexistence for the greater good of the whole (Latreille, 1802).

On the other hand, English authors, such as William Gould (1747), lived in entirely different social conditions in monarchist England. They were aware of different problems and their image of social insects at work differed correspondingly. Gould described the insects' loyalty to the queen, their respect for institutions of the monarchy, and natural love for the royal lineage, reflecting a hierarchical ant colony reminiscent of an aristocratic society, rather than the egalitarian and revolutionary spirit of late eighteenth-century France.

In the nineteenth century, debates about the nature of the human and animal soul continued but the rapid social changes of this tumultuous century of revolutions, wars, and newly formed states manifested themselves also in the natural sciences. The image of social insects in the works of entomologists mirrored these changes to a remarkable extent. Old descriptions of ant morality and exemplary diligence were replaced by descriptions of their warlike expeditions, predatory behaviour and, for the first time in history, by descriptions of enslavement and conflicts in their 'national states' (Huber, 1820).

As European societies secularised, ants lost their divine nature as well and, much like the Europeans who engaged in wars and revolutions, these insects, too, were described as embarking on destructive military expeditions, destroying surrounding colonies, and quarrelling among themselves within nests. For the very first time, ants were described as ingenious war strategists, the most valiant fighters of the insect empire (Forel, 1874, 1907).

Alongside this, their civilisational advancement was also highlighted, again in parallel to modern human societies (cf. Chap. 3). Aside from warfare, descriptions of their life highlighted their farming and husbandry skills. The farming of aphids was discovered; although known much earlier, this phenomenon captivated the interest of researchers and the public at large only now. The cultivation of fungi by leaf-cutter ants was correctly interpreted and some authors placed them at a 'higher rung of civilisation' than the 'primitive' peoples of overseas colonies (Lubbock, 1882).

Remarkably, some particular contemporary political currents also left their mark on the study of social insects. Some researchers who sympathised with the

[2] The text was written in French in late 1742 and early 1743 but remained in a manuscript until rediscovered much later and published, jointly with a translation, in 1926.

emerging socialism studied ants as model organisms to understand the biological basis of socialist communities. They examined ant colonies and asked themselves what constitutes the basis of socialism in the animal kingdom and how socialism could be implemented among humans. For the first time, systematic experiments were conducted where different nests were mixed, and scientists studies the ants' reactions to incomers from other ant colonies. Based on this, scholars inquired into the origins of nationalism, studied how nations assimilate, and what helps maintain social stability.

Descriptions of social insects also reflected the social problems emerging in the contemporary industrial society. Alcoholism, unemployment, social unrest, and economic disparities seemed like harbingers of general decline and even biological regression and degeneration of the whole society. Socially pathological phenomena haunted intellectuals of the time who feared that certain social classes or races would eventually threaten the rest of society and compromise its 'genetic' potential (cf. Chap. 3). All this was reflected also in contemporary descriptions of ant colonies: for the very first time, scholars spoke about 'social parasitism' in ants, about 'nonworking' ants and 'pests' living at the expense of the common good. This went hand in hand with thoughts about 'cleansing' the society, about this process in social insects, and how it could be applied to humans. Researchers thus associated eugenics or racial hygiene with the study of social insects and pointed out that an efficient society naturally rids itself of unsuitable individuals, cares for its 'body', and eradicates all seeds of 'disease'.

This eugenic view of social insects was expressed most prominently in Germany, where the debate was influenced by the social and political situation after the First World War. Rising nationalism was accompanied by racially hygienic considerations about the need to cleanse society from all harmful influences. In this situation, the perfect society of social insects was inspiring and seemed worthy of emulation. It was viewed as a model example of application of socially and racially hygienic measures. Karl Escherich, who focused on insect pests, became the icon of ant research in Germany (Escherich, 1906).

Escherich, who joined the Nazi Party already in 1921, interpreted social insects and their pests based on his idealised vision of the human society and its presumed pests. He was fascinated by the absolute subordination of individuals in eusocial communities to the queen, by their willingness to sacrifice own life for the good of the community. Eschenrich expressed admiration for the ant and termite queens who represent the ultimate purpose of life for millions of individuals in the community; their attitude thus resembles the fascist celebration of a totalitarian Leader without whom the nation could survive. Also in line with the Nazi ethos, he spoke about 'pests' who parasitise the society, sap its strength, or even threaten its survival. As examples, he used some frequently studied social parasites of the ants and bees, whereby by 'pests' he meant all groups that allegedly harm the majority society. In the 1930s and 1940s, when he taught at the University of Munich, it became clear that he meant also German citizens whom the Nazi regime identified as enemies (Escherich, 1934, 1935).

Aside from Escherich, August Forel, another significant figure of myrmecology, was in his eugenic views also inspired by insect communities. He never had anything to do with Nazi ideology, but his political views were strongly influenced by the study of ants—and his study of ants was strongly influenced by his socialist interpretation of human society. His ants were thus presented as socialists in a perfectly functioning socialist society. His position vastly differed from Escherich's but both depictions of ant societies demonstrate the same sociomorphic bias.

Aside from dozens of works on neurology, intelligence, behaviour, or the systematics of ants, Forel also wrote a popular work *Le Monde Social des Fourmis* (Forel, 1921), where he explained the essence of eusociality and presented thoughts on how one could, based on this concept, reform a human society. This sociomorphic model of ant and human society had a significant impact in the socialist and eugenic circles in his native Switzerland but in France, but it did not spread much into Germany because in the Weimar Republic the situation was different.

As a convinced socialist, Forel saw socialist elements even in nature—and subsequently confirmed them through his socialist approach. In his writings, ants exemplified harmonious coexistence and cooperation, mutual aid, and all-encompassing altruism. From their evolutionary success, he inferred that this way of life is favoured by evolution, and to succeed in evolution, every organism (including humans) must eventually reach this socialist phase.

Forel was convinced that altruism and mutual assistance should be introduced into and supported in human society, which would thus become more ant-like. Following the example of ant societies, Forel wanted to transform the entire human society and create a kind of human 'supercolony', a perfect socialist society living in the 'United States of the Earth', where all nations and ethnicities would coexist in peace and mutual respect, much like ant supercolonies.

He realised that it would not happen on its own, and social engineering in the form of highly specific and strict eugenic measures would be required. In his books, Forel thus asked 'What can we do to be more like ants and yet remain human?' (Forel, 1922: 31). Above all, we would need to streamline our society and reduce the influence of less 'valuable' individuals who disrupt the socialist vision. As tools for regulating the inferior members of society, Forel proposed surgical castration, sterilisation by X-rays, or removal of ovaries or uterus in women, and obliteration of sperm ducts in men. Ultimately, Forel's sociomorphic vision of human society based on the ant model would thus require measures as drastic as those in Escherich's vision of ant society as a totalitarian fascist state.

The last sociomorphic transformation of ant communities took place after Second World War, when the image of social insects once again clearly reflected the shifts in contemporary Western society. The main actor of this change was Edward O. Wilson, one of the most prominent entomologists of the time. In his sociobiological interpretation of insect communities, ants resemble a capitalist society or corporation, where the queen is the top manager, while workers compete for power and access to resources and form competing clans. Any semblance of altruism and common work is just the consequence of a calculated game for the most successful investment in their genes. Love and mutual care are conditioned by a specific type

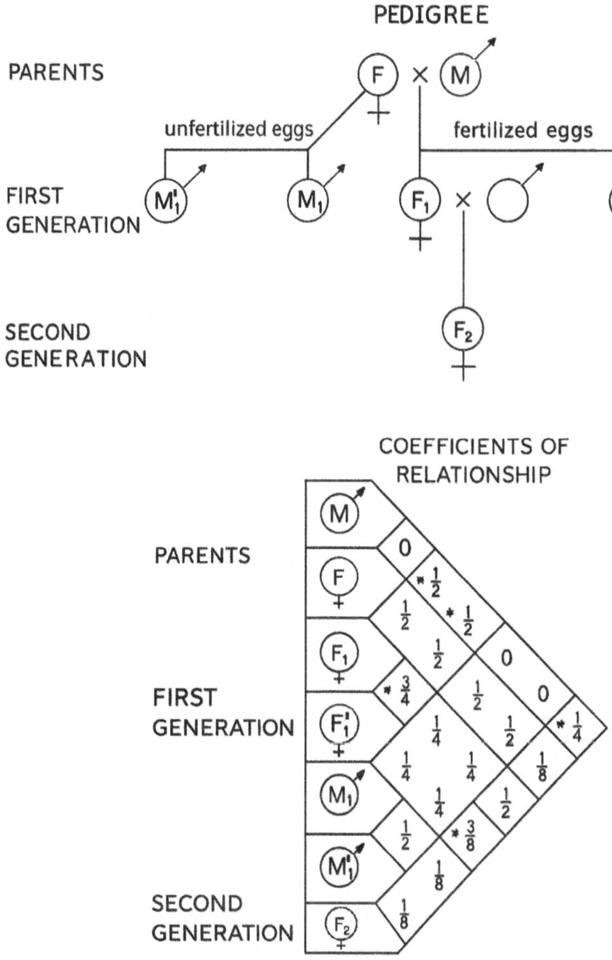

Fig. 14.3 The schema of William Hamilton's theory of kin selection as applied to eusocial communities of haplodiploid insects. Redrawn from Wilson (1971)

of relatedness, which essentially forces members of the anthill to altruism and work for the collective. The entire ant world is thus driven by a desire to acquire and improve the (genetic) capital.

Wilson's ants are not members of a socialist community who help each other out of selfless love and all-encompassing altruism, nor are they members of a totalitarian state who would sacrifice their personal interests for the benefit of the Leader and the whole. Instead, they are just vehicles of their genes, which are more easily passed to the next generation due to their specific haplodiploid inheritance. Altruism in eusocial communities is based on a genetic evolutionary 'calculation', depersonalised mathematics of gains and losses resembling a well-prepared business plan of a 'corporate colony' (Wilson, 1971) (Fig. 14.3).

This brief overview of the history of myrmecology shows some of the potential of using Topitsch's principle of sociomorphic modelling in the study of the history of natural sciences. Cultural and social phenomena interact with science, especially in fields touching upon the essential questions about humans, human nature, and the foundations of society. In such areas, we can discern a mutual exchange and mixing of cultural and scientific contents. Nature, in this case one group of social insects, has been interpreted from a clearly contemporary cultural perspective, and the resulting description was thus shaped by what the society saw in nature.

Social insects have also inspired biosemiotic studies, which suggest that insect communities should be investigated as complex interpretive structures. From this perspective, a community is based on the flow of information from individual workers, who interpret their surroundings in a particular way, to higher-order complex structures. Eventually, these structures appear as if they 'consciously' perceived and interpreted their environment and responded to it. While individual ants have limited cognitive abilities, the collective of an ant colony or termite mound functions as a higher-order cognitive system that creates a representational map of the world around them, relates to other organisms in complex ways, and reshapes their environment to meet their needs (Sharov & Tønnessen, 2021; Turner, 2016).

Scientific facts thus seem to have a significant cultural component, serving as a sort of *exemplum* from the vast realm of nature. This *exemplum* is then applied back onto culture, confirms its contemporary form, and legitimises those parts of culture that had been reflected in nature like a mirror. But that does not mean that the image, reflected in this cultural mirror, is necessarily distorted or unscientific. As shown above, it has scientific and factual relevance and can meet the requirements of scientific objectivity. It merely extracts from nature certain facts and truths which are relevant at a particular time to a particular cultural context.

This mutually beneficial relationship appears not only in clearly anthropomorphic descriptions of nature, such as fables, bestiaries, or popular scientific narratives about animal life. Anthropomorphisation of nature and zoomorphism of culture are also inherent in scientific works produced at specific times, just more cryptically. For instance, in Wilson's and Forel's descriptions of the ant life, there is nothing that should be considered factually incorrect or inconsistent with reality, but the two scientists' portrayal of social insects (and humans) nevertheless results in diametrically different images of communal life.

14.5 Conclusions

In natural sciences, all inquiry is ultimately motivated by the desire to understand who we are, the qualities of the world we live in, and our place in it. Scientific knowledge always has an anthropological component and relates to humans and their being-in-the-world, their *Dasein*. All sciences are therefore to some extent anthropomorphic. People are historical, cultural, and social beings—and all these aspects are reflected in science. Through science, its developments and changes, one can thus follow the process of the development and transformation of culture and society.

Our image of nature, and ultimately of ourselves, is filtered through contemporary cultural and social patterns. It resembles a Rorschach test, interpreted anew at each stage of cultural and social development. Contemporary projections of our social life, our ideas of good and evil, the essence of humanity and the world, are imposed onto nature and considered a true image of the world around us.

This image is subsequently incorporated back into culture and helps to shape it. The whole process resembles looking into a mirror, each time through a slightly different filter. The particular tint of the reflection is best seen in questions that are central and pressing for people of the given period, be it the issue of the nature of God and his relationship to humans, morality, or human nature. The hue of the reflection demonstrates the anthropomorphic nature of science (and, in Topitsch's conception, the philosophy or myth) as well as the zoomorphic nature of culture. Culture, like all human activity, is—in Latour's words—a tangle of 'morphisms', the result of a complex network of relationships among a variety of cultural and social actors.

Untangling this network is not simple and multiple paths often need to be followed simultaneously. Ernst Topitsch therefore advocated the study of contemporary myths, philosophies, sciences, and religions, just as Bruno Latour advised that one should follow the individual actors, the agents of action, and their relationships. Topitsch's work, especially his concept of sociomorphic modelling, thus calls for a semiotic study of human culture and society. It closely follows upon the ideas of Jakob von Uexküll, who studied the worlds and 'cultures' of biological entities in a similar manner. Topitsch's approach to the study of culture, philosophy, morality, or science necessarily involves the study of meaning networks, relationships between the individual actors, their interpretations of the surrounding world, and the creation of new 'cultures'.

In this way, Topitsch's work creates a platform for an interdisciplinary approach that connects insights from semiotics with analysis of human societies and cultures. Similar to von Uexküll's examination of how individual organisms interpret their environment and create their own 'worlds', Topitsch focuses on how humans interpret their culture and society and form new systems of meaning. His work is thus an important contribution not only to semiotics but also to sociology, anthropology, and philosophy.

The concept of sociomorphic modelling (and various concrete examples of its application in a brief excursion into the history of science) was introduced in order to demonstrate how culture is integrated into thinking about nature and how the image of nature in turn shapes culture. Culture is not separate from nature, it does not stand against it, nor is it an independent human invention opposed to an immutable and 'non-cultural' nature. Culture and nature are closely interconnected and symbiotically linked, but the examples above do not show a distortion of knowledge; rather, they illustrate an inherent aspect of how humans construct meaning. Recognition of sociomorphic modelling not only provides insight into the development of science but also invites us to critically reflect upon the narratives we create about nature and society. Understanding this dynamic is crucial, because it allows us to see science not merely as a neutral, objective endeavour but as a deeply embedded cultural practice through which humans construct meaning, interpret the world, and continuously reshape their understanding of reality.

References

Aldrovandi, U. (1638). *De animalibus insectis libri septem, cum singulorum iconibus ad vivum expressis*. Ferronius.

Clark, J. F. M. (1998). 'The complete biography of every animal': Ants, bees, and humanity in nineteenth-century England. *Studies in History and Philosophy of Science Part C: Studies in History and Philosophy of Biological and Biomedical Sciences, 29*(2), 249–267.

Clark, J. F. M. (2009). *Bugs and the victorians*. Yale University Press.

Daston, L., & Stolleis, M. (Eds.). (2008). *Natural law and laws of nature in early modern Europe: Jurisprudence, theology, moral and natural philosophy*. Ashgate.

Daston, L., & Vidal, F. (Eds.). (2013). *The moral authority of nature*. University of Chicago Press.

Enenkel, K. A. E., & Smith, M. S. (Eds.). (2010). *Early modern zoology. The construction of animals in science, literature and the visual arts* (Vol. 2). Brill.

Escherich, K. (1906). *Die Ameise. Schilderung ihrer Lebensweise*. F. Vieweg und Sohn.

Escherich, K. (1934). *Termitenwahn. Rede gehalten beim Antritt des Rektorats der Ludwig-Maximilians-Universität am 25. November 1933*. Albert Langen – Georg Müller.

Escherich, K. (1935). *Biologisches Gleichgewicht. Eine zweite Münchener Rektoratsrede über die Erziehung zum politischen Menschen*. Albert Langen – Georg Müller.

Forel, A. (1874). *Les fourmis de la Suisse. Systématique, notices anatomiques et physiologiques, architecture, distribution géographique, nouvelles expériences et observations de moeurs*. H. Georg.

Forel, A. (1907). *Die psychischen Fähigkeiten der Ameisen und einigen anderen Insekten; mit einem Anhang über die Eigentümlichkeiten des Geruchsinnes bei jenen Tieren*. Ernst Reinhardt.

Forel, A. (1921). *Le Monde Social des Fourmis du Globe comparé à celui de l'homme*. Libraire Kundig.

Forel, A. (1922). *Mensch und Ameise*. Rikola Verlag.

Fudge, E. (2002). *Animal*. Reaktion Books.

Fudge, E. (2019). *Brutal reasoning: Animals, rationality, and humanity in early modern England*. Cornell University Press.

Gould, W. (1747). *An account of English ants*. Printed for A. Millar.

Haraway, D. (1989). *Primate visions: Gender, race, and nature in the world*. Routledge.

Haraway, D. J. (2007). *When species meet*. University of Minnesota Press.

Harrison, P. (1998). The virtues of animals in seventeenth-century thought. *Journal of the History of Ideas, 59*(3), 463–484.

Hesiod. (2004). *Theogony, works and days, shield*. Johns Hopkins University Press.

Huber, P. (1820). *The natural history of ants*. Longman, Hurst, Rees, Orme, and Brown.

Kelsen, H. (1928). *Die philosophischen Grundlagen der Naturrechtslehre und der Rechtspositivismus*. R. Heise Verlag.

Latour, B. (1993). *We have never been modern*. Harvard University Press.

Latreille, P. A. (1802). *Histoire naturelle des fourmis*. L'Impremierie de Crapelet.

Lubbock, J. (1882). *Ants, bees and wasps*. D. Appleton & Company.

Majoli, S. (1592). *Dies caniculares hoc est. Colloquia tria et viginti physica, nova et penitus admiranda ac summa iucunditate concinnata*. Aloysy Zanetti.

Muffet, T., Wotton, E., Gesner, C., & Penn, T. (1634). *Insectorum sive minimorum animalium Theatrum*. Thomas Cotes.
Ossewaarde, M. (2014). The role of technomorphic and sociomorphic imagery in the long struggle for a humanistic sociology. In R. Bod, T. Weststeijn, & J. Maat (Eds.), *The making of the humanities, Vol. III: The modern humanities* (pp. 597–608). Amsterdam University Press.
Peters, H. M. (1960). Soziomorphe Modelle in der Biologie. *Ratio, 3*(1), 22–37.
Portmann, A. (1947). *Vom Bild der Natur. Ein Beitrag der Lebensforschung zu aktuellen Fragen*. Friedrich Reinhardt.
Réaumur, R. A. F. (1740). *Mémoires pour servir à l'histoire des insectes* (Vol. 5). L'Imprimerie Royale.
Réaumur, R. A. F. (1926). *The natural history of ants*. Knopf.
Ritvo, H. (1997). *The platypus and the mermaid, and other figments of the classifying imagination*. Harvard University Press.
Ritvo, H. (2010). *Noble cows and hybrid zebras. Essays on animals and history*. University of Virginia Press.
Sax, B. (2000). *Animals in the Third Reich: Pets, scapegoats, and the Holocaust*. Continuum.
Sax, B. (2001). *The mythical zoo. The encyclopedia of animals in world myth, legend and literature*. ABC-CLIO.
Sax, B. (2003). *Crow*. Reaktion Books.
Sharov, A., & Tønnessen, M. (2021). *Semiotic agency. Science beyond mechanism*. Springer.
Sleigh, C. (2007). *Six legs better: A cultural history of myrmecology*. Johns Hopkins University Press.
Topitsch, E. (1958). *Vom Ursprung und Ende der Metaphysik: Eine Studie zur Weltanschauungskritik*. Springer.
Turner, J. S. (2016). Semiotics of a superorganism. *Biosemiotics, 9*(1), 85–102.
von Uexküll, J. (1909). *Umwelt und Innenwelt der Tiere*. Springer.
von Uexküll, J. (1920). *Theoretische Biologie*. Verlag von Gebrüder Paetel.
von Uexküll, J., & Kriszat, G. (1934). *Streifzüge durch die Umwelten von Tieren und Menschen*. Springer-Verlag.
Wilson, E. O. (1971). *The insect societies*. Harvard University Press.

Chapter 15
Stability-Based Sorting in Cultural Evolution: Macroevolutionary Phenomena as a Model for Explaining Cultural Dynamics

Jan Toman

Abstract The evolution of culture can be analysed within various theoretical frameworks, including selectionism, attractionism, diffusionism, functionalism, and neutral evolution. Nevertheless, an often overlooked but fundamental factor in all historical systems, including culture, is stability-based sorting (SBS). SBS is based on the assumption that, over time, the more stable or frequently recurring elements accumulate, while the less stable elements gradually disappear. This makes even natural selection a specific case of SBS. In biological evolution SBS accounts for several well-documented but enigmatic phenomena. In cultural evolution, where the variation-generating processes are more diverse and embedded in complex and dynamic contexts, SBS exerts a broader but more subtle influence. As a result, its effects in cultural evolution are less pervasive than in biology but nonetheless shape a wide range of non-obvious cultural patterns, such as the accumulation of safeguards against short-term advantageous changes that would undermine long-term stability (e.g., musical styles), the persistence of cultural elements that may be neutral or even disadvantageous in the short term but exhibit long-term stability (e.g., agriculture, writing, conservatism), and the punctuated nature of cultural evolution, alongside of instances of evolutionary stasis, modularisation, machinification, and increasing complexity. We can also extend SBS to interpretive and sign-based processes, underscoring how cultural elements stabilise through ongoing meaning-making. This highlights the often 'invisible' yet essential role of SBS in both micro- and macro-level cultural dynamics.

Keywords Stability-based sorting · Macroevolution · Punctuated equilibria · Evolutionary trends · Frozen evolution theory

J. Toman (✉)
Department of Philosophy and History of Sciences, Faculty of Science, Charles University, Prague, Czech Republic
e-mail: tomanj@natur.cuni.cz

J. Toman et al. (eds.), *Culture Evolving*, Biosemiotics 10,
https://doi.org/10.1007/978-3-032-16124-6_15

15.1 Introduction

Cultural evolution can be broadly defined as a cumulative set of changes that take place in culture over time. While this broad definition depends on how culture itself is conceptualised (see Chaps. 5, 7, 9, 16 and 18), its strength lies in its inclusivity, by not being restricted to particular phenomena. It applies regardless of whether culture is viewed as an exclusively human construct or as a broader phylogenetic phenomenon and both when it is defined strictly in terms of material artefacts or includes also the nonmaterial aspects of behaviour. Furthermore, this definition is neutral with respect to the specific cognitive processes, customs, linguistic practices, religious traditions, artistic expressions, or, more broadly, the socially or otherwise transmitted information that may be considered part of 'culture.'

Within this framework, cultural evolution can be analysed through various, not necessarily mutually exclusive, theoretical lenses. These include selectionism and attractionism (see, e.g., Boyd & Richerson, 1985; Sperber, 1996; Chaps. 10 and 11), diffusionism and functionalism (see, e.g., Boas, 1940; Radcliffe-Brown, 1952; Chaps. 5, 7 and 12), but also cultural convergence and neutral evolution (see, e.g., Boyer, 1998; Bentley et al., 2004).

This chapter examines a fundamental process that operates in both biological and cultural evolution irrespective of the theoretical framework adopted. And although this process has so far received limited scholarly attention, it plays a significant role in shaping cultural evolution, particularly over long periods of time. This phenomenon is known as stability-based sorting (SBS; Toman & Flegr, 2017b).

15.2 Stability-Based Sorting

In all evolving systems, there emerges over time a consistent trend: the more stable elements tend to accumulate, while the less stable elements gradually diminish. More specifically, SBS can be defined in two ways: (1) *as the differential survival of entities based on their persistence before they either disappear or transform*, or (2) as *a process in which the less variable or recurrently emerging elements accumulate and ultimately dominate the system and surpass those which are more prone to change* (Toman & Flegr, 2017b).

SBS is a widely occurring phenomenon. It does not provide a comprehensive explanation for the emergence and evolution of adaptive traits, as natural selection does, nor does it directly account for the origin or dissemination of new cultural variants. Moreover, SBS is not a force in the same sense as, for example, gravitation. Rather, it functions as an ontological or thermodynamic principle—it is a fundamental property of evolving systems. Despite its conceptual simplicity, or perhaps precisely because of it, SBS has a profound influence on cultural evolution (Fig. 15.1).

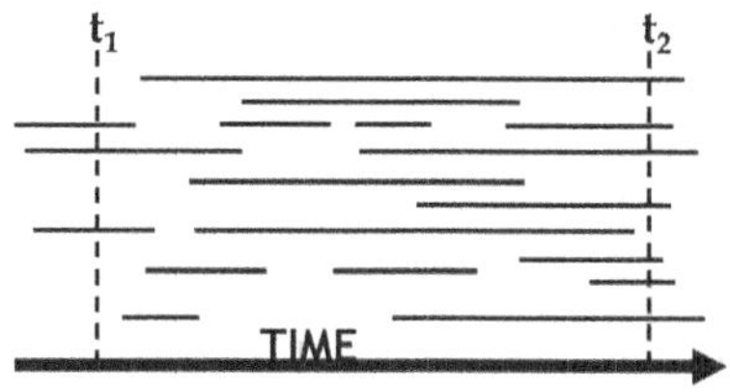

Fig. 15.1 Stability-based sorting. A process in which less changeable or constantly emerging elements accumulate in a system and dominate over elements which are more unstable and more prone to change. Author: Jan Toman

The simplicity and generality of SBS probably explain why many researchers across various fields have failed to fully appreciate its significance. As a principle, it applies universally to all open and closed, expanding, stagnating, and contracting systems that undergo evolution, have a history, or retain a memory of past states (Toman & Flegr, 2017b). Cultural evolution, with all its manifestations, undoubtedly is such a system (cf. Chaps. 16 and 17).

SBS operates regardless of whether the system consists of material entities (e.g., artefacts) or nonmaterial ones (e.g., phonemes or words; see Chap. 12), including memes (see Chap. 10). Furthermore, the origin of the sorted entities is not important. While scientists often conceptualise the emergence of one entity from another, for instance in reproduction or purposeful imitation (cf. Chaps. 9 and 10), many entities in our universe likely emerge independently, rather like snowflakes in a snowstorm or traditional decorative patterns (cf. Chap. 11).

The emergence and persistence of entities can be influenced by various external environmental factors, which in the case of cultural elements include human cognitive processes, social dynamics, and broader cultural contexts (cf. Chaps. 5, 7, 11 and 13). A particularly intriguing case of mutually independent yet conditioned emergence involves cultural elements that arise by accident or mistake, for instance due to some cognitive bias. It is important to note that SBS significantly differs from selection in that it requires neither the emergence of entities from one another through propagation or replication, nor the presence of heritable and variable characteristics.

A common objection to SBS is that it lacks empirical testability, serving instead as a tautological or metaphysical concept which ultimately merely notes that stable elements persist while changeable ones transform or disappear. This observation is logical (Grand, 2001, pp. 34–38; Pross, 2012; Shcherbakov, 2012; Pascal & Pross, 2014) but, as Shcherbakov (2013) argues, if SBS is a tautology, it is the 'Great Tautology', the most fundamental description of how novelty arises in the universe. In alignment with this interpretation, Toman and Flegr (2017b) have demonstrated that selection—the key driver of biological evolution and the only known process capable of producing and refining adaptive traits within realistic timescales—is, in fact, a special case of SBS.

The simplest and most widespread manifestation of SBS is sorting based on *static stability*, that is, the tendency of certain elements to persist over time due to their resistance to change. The likelihood that an element will either disappear or transform into something else varies significantly. It is also influenced by diverse factors, which range from basic environmental resilience (e.g., protection against weathering) and institutional importance (e.g., legal frameworks crucial to societal functioning) all the way to the cognitive appeal of specific cultural elements, such as enduring melodies, stories, or decorative motifs (in analogy to content bias; Boyd & Richerson, 1985).

Beyond static stability, SBS also encompasses dynamic stability, which governs systems where elements are linked through genealogical relationships, meaning that new elements arise from pre-existing ones. Examples range from biological organisms to cultural phenomena such as orally transmitted anecdotes, hunting techniques, and popular songs—all of which are propagated or created in relation to pre-existing elements or patterns. In such systems, sorting occurs based on the interplay between the likelihood of an element's disappearance and its capacity to replicate and generate derivative forms (cf. Chaps. 10 and 11). In essence, dynamic stability amounts to selection, since the most enduring and prolific variants become over time dominant (Fig. 15.2).

A prime example of entities sorted through dynamic stability includes organisms, their communities, species, and higher clades, all of which are subject to natural selection in biological evolution. On the hand, one could also consider various units of information transmitted among individuals in cultural evolution (see e.g., Boyd & Richerson, 1985; Sperber, 1996; Tyler, 2011), such as expansive religions, sensational stories that spread more rapidly than more 'pedestrian' narratives (cf. Chaps. 10 and 12), rituals and their components, hunting and production techniques, musical styles, anecdotes, languages, and the like. Although these elements may exhibit relatively low static persistence, they can proliferate extensively and spread horizontally in societies based on content bias (cf. Chap. 11). But even in systems governed by dynamic stability, static stability remains a key factor. For instance, a story's longevity may be enhanced not only by its sensational appeal and frequent

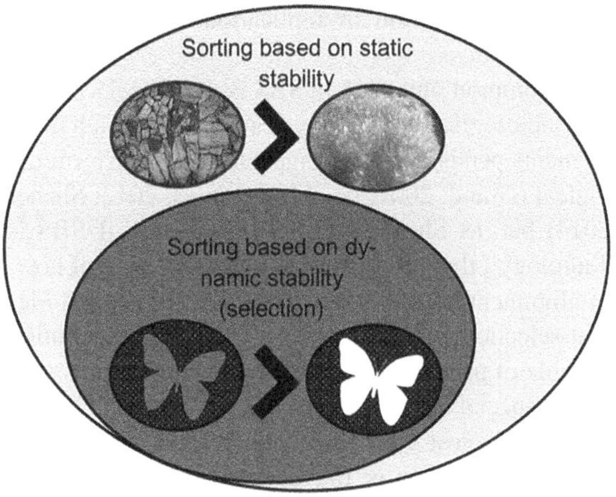

Fig. 15.2 Relationship between stability-based sorting and selection. Selection can be understood as a specific instance of stability-based sorting that occurs in systems whose elements are connected by genealogical relationships and exhibit heredity and variation. Author: Jan Toman

retelling but also by its documentation in a written form and preservation in archives (see Chap. 10).

In the natural sciences, the concept now referred to as SBS has been discussed by numerous scholars. Perhaps the best-known formulation comes from Richard Dawkins, who proposed that classic natural selection (based on a differential survival of offspring that exhibit heritable variation) was preceded by a more fundamental process of 'survival of the stable', that is, the 'selection of stable forms and a rejection of unstable ones' (Dawkins, 1976, p. 13). Neither we nor Dawkins view this as in any way mysterious; it is rather just an inevitable outcome of any evolving system. Several other researchers have explored similar ideas (see, e.g., Lotka, 1922a, b; Simon, 1962; Wimsatt, 1980; Van Valen, 1989; Michod, 2000; Grand, 2001; Maynard Smith & Szathmáry, 2010).

The prevailing consensus suggests that SBS is relatively insignificant compared to selection. It is typically viewed as a weaker variant of natural selection (see, e.g., Okasha, 2006, p. 214; Godfrey-Smith, 2009, pp. 40, 140). Among the few scholars who attribute a greater significance to SBS, the key areas of discussion include its relation to natural selection (Pascal & Pross, 2014; Pross, 2003, 2012), its role in the origin of life (Bourrat, 2014), its role as the foundation of biological fitness (Bouchard, 2008), and its involvement in the emergence of complex systems independent of natural selection. This is particularly relevant in studies of stable ecosystems, including planetary-scale systems within the framework of the Gaia theory (Bouchard, 2008; Doolittle, 2014, 2017; Lenton, 2004; Lenton et al., 2018; Nicholson et al., 2018). We have explored these topics in more detail elsewhere (Toman & Flegr, 2017b). In the field of cultural evolution, scholars have observed similar patterns, although most research in this area focuses on specific manifestations of SBS within distinct cultural systems. We will examine some of these below.

15.3 Sorting of Cultural Variants

15.3.1 Stabilising Dynamics of SBS in Cultural Transmission

The elements of cultural evolution exhibit several distinctive characteristics relevant to SBS. In some ways, they behave similarly to genes: they can be conceptualised as discrete elements that spread through human populations by imitation (and other forms of social learning) and compete for transmission to the greatest number of hosts (though this is a simplified model that relies on several assumptions; see Chaps. 9, 10 and 13). Nevertheless, compared to genes, these cultural elements exhibit a Lamarckian dynamic, that is, they accumulate acquired changes that can influence their success. They also evolve far more dynamically and not necessarily towards a clear selective advantage (see Chaps. 10, 11 and 18), because cultural change can be driven not only by random modifications (analogous to genetic mutations) or deliberate interventions by conscious agents (such as improving a cooking recipe). Rather, the emergence and developmental trajectory of cultural elements is also shaped by the functioning of the human brain, the biologically inherent structures of human society, and the broader cultural and historical context (Boyd & Richerson, 1985; Sperber, 1996; cf. Chaps. 11, 16 and 17).

This aligns with Hoffmeyer's (1996) claim that living systems actively interpret signs, meaning that cultural variants are not simply transmitted but continually reinterpreted. In this light, SBS emphasises that what is selectively sorted (and thus accumulated) over time are stable sign-objects. A complementary perspective, offered by Herrmann-Pillath (2021), suggests that selection may also occur at the neuronal level, where one can posit the formation of so-called neuromemes, self-replicating neurophysiological configurations. In this sense, the brain is not a passive conduit but an active participant in the process of shaping which cultural variants persist or fade away. By reinforcing individual preferences, learned associations, and meaning-making processes, these neural mechanisms can reinforce the patterns that emerge on the population level. Some changes are thus more likely to prevail than others, for instance a contraction of frequently used words, the association of certain sensory modalities with a shared valence (e.g., a low tone signifying 'down'), preferences for certain types of social leaders, or the emergence of recurring archetypes.

This has several notable consequences. Some dramatic transformations—such as the reversal of meaning in polar opposites ('awfully good') in language evolution, or the rapid spread of telecommunication and the internet in societal history—can occur more easily than other, seemingly more modest shifts, such as changes in the emotional valence of colours (cf. Chap. 18) or adoption of the metric system. Moreover, cultural elements are typically less clearly delineated than genes and alleles: although genes do interact and may even spread horizontally under certain conditions, such interactions are far more common and influential in cultural evolution. Cultural units, even those belonging to vastly different domains (such as preaching and modern media, or typeface design and political campaigning), often combine and influence both their own transmission and the transmission of other cultural units (see Sperber, 1996; cf. Chap. 10).

In cultural evolution, SBS serves as a conserving mechanism. Unlike selection or intentional cultural innovation, it does not by itself generate any beneficial elements. Rather, SBS gradually sorts and accumulates variants that exhibit a greater contextual stability: in short, SBS favours variants that persist under given conditions. In some cases, SBS may even hinder evolutionary changes driven primarily by selection (Toman & Flegr, 2017b). A prime example of this phenomenon is the notable persistence of traditional measurement systems, where the sheer number of users has prevented some countries from transitioning to the International System of Units despite its evident advantages.

In cultural elements linked to biological fitness, SBS plays a similarly influential role; in some instances, it can even lead to a fixation of variants that reduce individual fitness. This is evident in certain medically questionable rituals, hazardous narratives, or addictive behaviours—for instance, tobacco smoking, which remains highly persistent (difficult to quit) due to the addictiveness of nicotine. The role of SBS in cultural evolution may even surpass its significance in biological evolution, because the cultural elements which spread at the expense of biological fitness do so far more easily and faster in cultural systems than in biological ones (cf. Chap. 10).

Similar to its function in biological evolution, SBS can lead to an accumulation of cultural elements which have safeguards against certain short-term advantageous changes that would ultimately compromise their long-term stability. If a particular trait of a cultural element slightly inhibits its immediate spread but enhances its long-term persistence, elements exhibiting this trait will tend to accumulate. This can be seen in music styles that have safeguards against excessive complexity, because overly intricate compositions, while initially appealing, may become less favoured over time. Accordingly, simpler and more easily performed songs tend to exhibit greater stability even at the expense of more elaborate temporarily popular musical pieces (*sensu* Sperber, 1996); this trend is exemplified in the simplification of campfire songs (Tureček & Havlíček, 2017). The same principle applies also for instance to hunting techniques, which resist becoming overly complicated, to clothing, which avoids excessive impracticality, and to various cultural designs, which endure due to their resilience against opportunistic changes. Cultural elements that lack these safeguards ultimately disappear from the mainstream discourse, leaving behind elements we often consider 'timeless'. Nevertheless, cultural evolution remains a highly dynamic, interconnected, and nonlinear system, so that the accumulation of stable elements can sometimes paradoxically make transformations of some of them or entire systems more likely, because stable elements may later be repurposed or recombined in unexpected ways (cf. Chaps. 11 and 18). Even so, the general patterns outlined above still apply.

The historical predominance of elements with limited immediate success but high long-term stabilising potential may also be seen as a fixation of cultural safeguards whose legacy persists into the present. A prime example is the transition to agriculture, particularly the cultivation of grains. From the perspective of individuals, early agriculture had numerous disadvantages, such as increased susceptibility to the spread of parasites and diseases (Larsen, 2006). In the long term, however, it enabled the stable existence of large, structured, and settled societies, leading to the rise of cities and states. Some scholars hypothesise that early agricultural societies did not transition from farming to large populations but rather vice versa. Large gatherings, such as religious congregations, may have required a stable food source, which in turn made grain cultivation a functional necessity (Schmidt, 2012)—and once established, this shift became irreversible.

A similar trajectory can be observed in the spread of writing, which may have proliferated not thanks to some immediate practical benefits but because literate societies tend to exhibit greater stability in terms of ritual continuity, governance, and linguistic consistency (Ong, 1982; cf. Chap. 10).[1] This is analogous to the

[1] It is important to note, however, that the pace of cultural evolution is shaped by various demographic factors (Derex & Mesoudi, 2020).

spread and fixation of sexual reproduction among eukaryotes, which, despite certain short-term disadvantages, enhances long-term survivability of sexual species in changing environments (Toman & Flegr, 2017a). More broadly, social conservatism may serve as a parallel mechanism that acts as a safeguard against disruptive cultural shifts (see McAuliffe, 2016).

At an even higher level, entire cultures and 'civilisations' are subject to SBS (for the complexities of defining civilisations, cf. Chaps. 5 and 18). Over time, more stable civilisations accumulate, while less stable ones disappear, regardless of their temporary success. History provides numerous examples of cultures which over-exploited their resources and collapsed despite the initial prosperity (Diamond, 2005, but see also, e.g., Mulrooney et al., 2024). Some argue that the current (post) modern culture exhibits similar vulnerabilities: for continued growth and survival, it must either discover new resources, improve efficiency, or curtail consumption. Extreme examples include civilisations that acquired the ability to self-destruct through weapons of mass destruction or those whose dense populations enabled a rapid spread of extremely virulent diseases. Given that modern civilisation has only recently reached this stage, such self-destructive tendencies may provide a potential explanation of the Fermi's paradox, that is, the question of why we have not yet found any sign of existence of extraterrestrial civilisations (Webb, 2002).

Overall, SBS seems to have a profound impact on cultural evolution, potentially surpassing its role in biological evolution, where selection remains the dominant process and the explanatory role of SBS is limited to several narrowly defined mysterious phenomena (Toman & Flegr, 2017b). On top of that, many cultural systems (including agriculture, writing, and conservatism) have accumulated numerous additional adaptive and non-adaptive elements which obscure their original role as safeguards of stability. A good example is the abovementioned literacy, which, aside from playing a role in administration, also took on functions in education, entertainment, and many other areas of life. Still, this does not alter the fact that without its safeguarding and stability-enhancing role, writing may never have gained widespread adoption.

15.3.2 Specific Mechanisms of SBS in Cultural Evolution

We have already noted that while elements of cultural evolution are subject to selection, their evolutionary dynamics are far more fluid than those of specific alleles in biological evolution. The elements of cultural evolution exhibit a considerable diversity in variability, heritability, and substrate. Consequently, the mode and intensity of their transmission can likewise significantly vary (Boyd & Richerson, 1985; see Chaps. 9 and 13). Moreover, cultural elements are typically influenced by more intricate contextual factors, including the cultural background that shapes their meaning, transmission, and transformation (Sperber, 1996). For example, a face covering can hold entirely different meanings depending on the context, for instance, in a classic Western film, at a masquerade ball, in the Islamic culture, or during a pandemic.

In cultural evolution, static stability may also manifest as convergence towards an attractor point (see Sperber, 1996; cf. Chap. 11), which reduces the likelihood of transformation into an alternative form. But this attractor point is not always straightforward: given the dynamic nature of cultural evolution and its contextual dependencies, development towards an apparent attractor may be subject to nonlinear influences. As a result, it can prove difficult—if not impossible—to identify a single stable attractor point. A notable example is the tendency of modern societies to evolve towards democratic governance, which is broadly appealing to most citizens but remains relatively unstable and prone to internal disruptions (Tilly, 2007).

In recent years, scholars in biological evolution have increasingly focused on processes that shape the emergence of phenotypic variation (see, e.g., Toman & Flegr, 2018a, b). In cultural evolution, however, the intensity and diversity of analogical processes is probably even greater than in biological evolution. That is why a fitting analogy for cultural evolution may be found not in the strictly gene-centric neo-Darwinian model but rather in the broader perspectives of developmental and evolutionary biology (Evo-Devo), in the extended evolutionary synthesis (see Chap. 16), or in biosemiotics (see Chap. 17). These interdisciplinary approaches highlight multiple systems of heredity, multiple levels of selection, the context-dependence of allelic effects, developmental plasticity, organismal integration, the role of symbiotic relationships, the evolution of evolvability, its similarity with learning, and other macroevolutionary processes (Laland et al., 2015; Markoš & Švorcová, 2019; Watson & Szathmary, 2016). As emphasised by, for instance, Hoffmeyer (1996), from the perspective of the biosemiotic approach current organisms exhibit a level of complexity and competence far exceeding that of simple programmable machines, making the mechanistic metaphor deeply flawed. Organisms actively interpret their environments using internal memory, environmental signals, and a variety of interpretative codes and systems operating on both individual and evolutionary timescales. By examining how signs become stabilised or destabilised, we can observe clear parallels between these processes and the dynamics of stability-based sorting (SBS) in cultural evolution.

In biological evolution, identical traits emerge independently only under specific conditions, such as when identical mutations occur at the same genomic loci or when similar functional pathways are recruited. By contrast, in cultural evolution, identical elements emerge far more frequently and independently; examples include weapon designs, melodies, decorative motifs, narrative structures, and much more. As previously discussed, our shared biological and cultural substrates often facilitate the emergence and spread of similar variations. Compared to biological evolution, cultural evolution therefore features more frequently occurring but dynamically unstable elements, which emerge repeatedly due to common cognitive biases and social influences. A notable example is human phytophylia, a universal preference for floral patterns in decorative arts and clothing (Eibl-Eibesfeldt, 1989). Similarly, visual representations based on the *Kinderschema* (Lorenz, 1970)—such as cartoon characters, plush toys, and domesticated animal breeds with childlike features—illustrate how biological predispositions shape cultural patterns. In this sense, the human brain, in conjunction with social and cultural contexts, functions analogously to the external environment. Just as low temperatures and high humidity facilitate the independent formation of snowflakes, the neurological settings of the human brain predispose us to perceive specific facial features and proportions as aesthetically pleasing, which has led to a recurrent emergence of certain visual styles across cultures.

In comparison to the dynamically stable and repeatedly emerging elements, purely statically stable elements have been rare throughout most of human cultural evolution. A reversal of this trend occurred only with the expansion of cultural bearers (Ostdiek, 2011), especially after the invention of writing. Writing enabled a long-term recording of large numbers of cultural elements, including stories, production techniques, and ritual instructions, in a 'frozen' textual form (Ong, 1982). It also facilitated the codification of further elements, such as language rules and legal codes (see Chap. 10). On the other hand, this recorded cultural material is still just the tip of the iceberg. Cultural evolution remains inherently dynamic, even in literate societies, albeit operating at different levels. For instance, one can purchase a 'frozen' novel in a bookshop, but the literary genre itself—such as science fiction or Nordic noir—continues to evolve dynamically.

Unlike biological evolution, cultural elements can emerge deliberately and *de novo*, either through unrestricted recombination or via selective inspiration by preexisting elements. J. R. R. Tolkien (1984) famously compared this process to cooking a soup where the existing ingredients are transformed into something new and greater than the sum of its parts. A portion of newly created cultural elements, such as popular songs, novels, and films, thus emerges intentionally, by drawing on previously successful cultural forms (Blackmore, 1999, pp. 59–62). This principle extends to technological innovations, which exhibit an autocatalytic character, so that the more successful variants promote the development of even more beneficial successors. Technological revolutions thus act as drivers of scientific progress, profoundly influencing the trajectory of cultural evolution. For instance, modern automated surveillance and social assessment systems, which are expanding in China and other countries, may contribute to the emergence of highly stable social structures.[2] In biological evolution, such processes either do not occur at all or manifest themselves only to a limited extent over long timescales (Watson & Szathmary, 2016).

[2] This example also warns against a simplistic view of technology. Technologies are not just passive tools that undergo continuous refinement: they actively shape our abilities to transmit and receive cultural content while being in turn influenced by social, economic, and political dynamics. In doing so, they reshape the broader cultural landscape and influence its future trajectory (Feenberg, 1999).

Cultural evolution therefore seems to be far more volatile than biological evolution, exceeding it not only in its speed but also in the mechanisms and pathways through which change occurs. This volatility suggests that cultural elements are at a far greater risk of experiencing changes rapid enough to obscure their historical relationships. One possible exception in biology are viruses, whose high rates of mutation and horizontal gene transfer similarly obscure their deep evolutionary history (Claverie, 2020). Moreover, while biological systems always preserve the information required for self-replication, thus ensuring a core continuity of identity, cultural entities lack a directly analogous, continuously preserved 'core of identity' or 'primordial function' (Sharov, 2016). As noted above, certain cultural elements persist because they are entertaining, frequently emerging, necessary for survival, or required by society. Rather than conceptualising cultural evolution as a simple branching tree, it may thus be more accurate to view it as a densely interconnected network.

Box: Stability-Based Sorting Explains Patterns in Biological Evolution
In biological evolution, SBS contributes significantly to the accumulation of genetic, developmental, functional, and structural elements which are effectively resistant to change (Toman & Flegr, 2017b, 2018a, b). At the genetic level, this process is particularly evident in sexually reproducing organisms, where the effect of alleles on biological fitness depends on the presence of other alleles of the same gene but also on interactions with other genes within an individual's genome and the population's gene pool. This context-dependent impact of alleles on phenotypic expression and overall fitness leads to the long-term persistence of numerous genetic variants, a phenomenon driven by balancing selection. Under various conditions, this persistence can be advantageous (see Toman & Flegr, 2017a) but this process also constrains adaptive evolution by slowing down both the fixation and elimination of gene variants.

Because most traits are shaped by multiple genes (a phenomenon known as epistasis) and individual genes can in turn influence multiple traits (pleiotropy), genetic elements form a highly interconnected network that governs phenotypic expression. According to punctuationalist theories of adaptive evolution (including the theory of frozen plasticity; Eldredge & Gould, 1972; Flegr, 2013), this leads to long periods of evolutionary stasis, during which the species' ability to respond to directional selection remains limited. This state lasts for most of the species' existence. During evolutionary stasis, organisms can adjust to selective pressures only elastically, by modifying the frequency of existing alleles rather than by fixing or eliminating them. Only consistently and strongly detrimental alleles are removed from the gene pool.

Moreover, once the selection pressure ceases, allelic frequencies and phenotypic traits tend to revert to their previous states.

According to the frozen plasticity theory, a species can undergo irreversible and unidirectional adaptation to new environmental conditions only if a small subset of its original population that carries just part of its genetic diversity becomes isolated. This founder population must persist at low numbers for an extended period, leading to a further reduction of the remaining genetic variability. In such a genetically monomorphic population, newly emerging mutations will always interact with the same allelic background—this ensures that their effect on fitness remains consistent. Consequently, these mutations are either fixed or eliminated, especially as the population size expands and thus allows for a more effective response to natural selection. However, within a relatively short evolutionary timeframe, probably less than 1% of the species' total existence, this process once again leads to an accumulation of numerous distinct alleles and thereby to phenotypic interdependence. At this stage, the species re-enters a state of evolutionary stasis and is once again constrained by the same dynamics that governed its prior stability.

The crucial point is that some alleles exert such a strongly frequency-dependent influence on fitness that their fixation or elimination would require populations to persist at unrealistically low levels for long periods of time. The theory of frozen evolution (Flegr, 2013; Toman & Flegr, 2018a, b) points out that such effectively irreversibly polymorphic alleles accumulate in the gene pool of an evolutionary lineage due to SBS, progressively limiting its capacity to generate new adaptations. Similarly, genes, entire interlinked groups of genes, as well as developmental, functional, or structural modules within organisms that are resistant to evolutionary change tend to accumulate over longer timescales. At the genetic level, this occurs when any, even minuscule, change to a gene renders an organism nonviable or when multiple active backup mechanisms necessitate simultaneous mutations across multiple loci to produce a phenotypic change. Other elements of organismal genetic architecture enhance their stability by integrating additional components or processes, thus becoming structurally or functionally indispensable. As observed already by Riedl (1977, 1978) and Wimsatt (Schank & Wimsatt, 1986; Wimsatt, 2013; Wimsatt & Schank, 2004), the most stable genetic elements are those which are highly integrated within an organism's architecture, play a crucial role in multiple biological processes, are involved in early developmental stages, and are phylogenetically ancient, such as genes regulating cellular physiology, genetic material processing, or embryonic development.

Despite this accumulation of conserved genetic elements, biological evolution has not yet led to a drastic reduction in evolvability, that is, an organism's capacity for adaptive evolution. Instead, the evolution of evolvability has been characterised by an optimisation of responses to selective pressures driven by the development of a modular and robust genetic architecture. This architecture consists of both more and less variable components, reflecting the phenotypic structure of organisms and the selective pressures of their environment (cf. Chap. 16).[3] Nevertheless, these processes had led to an accumulation of the evolutionary constraints that channel the depth and nature of future adaptations. We have described this phenomenon as a reduction of the macroevolutionary potential of evolutionary lineages, that is, a decrease in the likelihood of significant evolutionary innovations over time (Toman & Flegr, 2018a, b). While there are mechanisms by which evolutionary lineages may partially circumvent these constraints over long timescales, such 'macroevolutionary freezing' probably places lineages at a disadvantage. This process favours organisms that can avoid such constraints, for instance by transitioning to a higher level of hierarchical organisation.

SBS, however, operates at all levels of biological organisation. As a result, any escape from these constraints is only temporary and the entire process will eventually repeat itself. This principle helps explain several long-term macroevolutionary trends, including the increase in maximal hierarchical complexity of organisms throughout the history of life on Earth, the reduction in interspecific morphological and functional diversity over time, the decline in intraspecific variability and evolutionary potential in the sexual clades, a progressive diversification and integration of biological elements at lower levels of organisation, machinification of biological structures at the lowest levels, and a gradual acceleration of these trends (Toman & Flegr, 2018a, b). This trend, or 'evolutionary syndrome', is also intriguingly reflected in the increasing semiotic competence of living systems (see Hoffmeyer & Stjernfelt, 2016). Moreover, it aligns with the preservation, advancement, and emergence of functional information understood as a network of signs used by agents to maintain and regulate their functions (Sharov, 2016).

[3] As described in Toman and Flegr (2018b), this is essentially a biosemiotic process. Interestingly, it closely parallels the evolution of biosemiotic capabilities (Hoffmeyer & Stjernfelt, 2016).

15.4 Stability-Based Sorting as an Explanation of Patterns in Cultural Evolution

Above, we saw that SBS manifests itself in various cultural phenomena. But can this process also play a significant role in shaping, or even canalising, the broader trajectory of cultural evolution? Before exploring this question in detail, it is useful to briefly summarise the impact of SBS on biological evolution.

15.4.1 Cultural 'Freezing'

Because SBS operates across all evolving systems, it follows that a gradual accumulation of more stable elements must also take place in cultural evolution. However, it is important to bear in mind that, in this domain, the processes described by the theories of frozen plasticity and frozen evolution (see BOX) most likely intersect. As discussed above, cultural evolution is significantly less constrained and far more dynamic than biological evolution. It is thus unlikely that the accumulation of increasingly stable cultural elements would lead to a complete evolutionary stasis or to an imposition of constraints as rigid as those observed in biological systems. But even if cultural evolution remains highly dynamic, this process could still have significant and illuminating consequences, which might offer new insights into previously unexplained phenomena and yield testable predictions.

One of the most striking similarities between cultural and biological evolution is their shared punctuationalist nature,[4] that is, the alternation between long periods of evolutionary stasis characterised by small, often reversible, and predictable changes, and short periods of abrupt, dramatic, and unpredictable transformation. Such punctuated dynamics is particularly surprising in cultural evolution, because one might expect that, in the absence of additional stabilising forces, its inherently dynamic and largely Lamarckian nature would lead to a relatively continuous and cumulative trajectory of gradual change (Gould, 2002, p. 953). The fact that this is not the case suggests that additional processes must be at play, processes that actively maintain cultural stasis over extended periods before allowing for sudden bursts of change.

Many specific examples of punctuationalist trajectories in cultural evolution—including modern technology, economic cycles, prehistoric tools and art, historical art, social and technological development, the evolution of institutions, ideas, and theories, as well as collective problem-solving and human learning—have been

[4] The alternative mode of biological evolution is a gradualist one, where changes occur slowly but continuously, as originally envisaged by Darwin. In evolutionary history, however, punctuated shifts seem to be the rule rather than the exception.

discussed by Gould (2002, pp. 952–972) and others (Bardeen & Cerpa, 2015; Bax et al., 2004; Gersick, 1988, 1991; Somit & Peterson, 1992). Moreover, it is worth noting that Gould himself acknowledged that the original formulation of punctuated equilibria in palaeontology was significantly influenced by Kuhn's ideas on the development of science, particularly the alternation between long periods of 'normal science' and short periods of crisis and paradigm shifts (Gould, 2002, p. 967).

In this context, let us note that the systems theory describes ultrastability and tipping points, which occur when an ultra-stable system is pushed beyond a critical threshold. For most of its existence, a system can adapt to changes in the external conditions through incremental improvements, but once these changes accumulate to a critical intensity, they reach a breaking point that leads to a systemic collapse. Ultra-stable systems (biological, economic, and technological ones, or from other spheres of cultural evolution) have the ability to preserve at least some of their subunits after such collapse, and that allows them to reorganise into a new metastable state (Bardeen & Cerpa, 2015). Furthermore, the scholars cited above offer valuable comparisons between various aspects of cultural evolution, particularly technological evolution, and biological evolution.

The punctuationalist nature of cultural evolution was also vividly illustrated by the Soviet semiotician and historian of culture Juri Lotman (2009, pp. 7–18, 114–132), who described an alternation between phases of stasis, characterised by slow, continuous, and consistent development devoid of major innovations, and 'explosions', sudden ruptures marked by vast, unpredictable changes capable of restructuring the entire semiotic space of a society. Lotman did not limit the application of these two modes of evolution to the development of society as a whole; he argued that they operate at all levels and across all timescales. One might thus witness an explosion in poetry, fashion, or film while scientific progress in physics at that time remains in stasis. During the explosion stage, cultural systems are still finding their place within the semiotic space of society. In a sense, as soon as these systems become understood, interpreted, criticised, or analysed, they inevitably 'cool' and transition towards stasis. Another notable feature of explosions, including social upheavals, is that they can only be fully explained in hindsight, after they become frozen in time as historical artefacts with observable consequences. Only then can the links be identified and, with their help, retrospective explanations of the explosion found. Once the specific (arbitrarily chosen) causal links are established, the event may seem inevitable, although for those who lived through it, the change felt fluid and impossible to fully comprehend (cf. Chap. 18).

Just as in biological evolution, cultural evolution is likely to accumulate, over time, densely interconnected stable elements which are maintained in a stable state through contextual or frequency-dependent relationships with other elements. These relationships may be incidental, meaning that aside from reducing the variability of certain entities, they do not substantially influence the properties of other parts of the system. More commonly, however, these elements are linked by relationships which, in the short term, enhance the efficiency or functionality of the system or provide other opportunistic advantages. Such advantages can stem for instance from the abovementioned social conservatism, where resistance to change is most pronounced in older individuals, who often hold positions of power and benefit from maintaining the status quo, thereby reinforcing the phenomenon (McAuliffe, 2016, pp. 151–219). The codification of cultural variants in books, textbooks, and curricula also has a preservative function, stabilising certain cultural elements (see Chap. 10). Interestingly, human neophilia, that is, the tendency to favour novel or rare variants, can paradoxically contribute to cultural stability due to negative frequency-dependent selection. A striking example of this are male facial hair preferences: both men and women tend to find eye-catching beards most attractive when clean-shaven faces are predominant—and *vice versa* (Janif et al., 2014).

Such processes contribute to the often discussed 'freezing' of social development, particularly in relation to institutions (Rappaport, 1999, pp. 416, 425–437; Eco, 2000; Markoš, 2014, 2015; Kováč, 2015, p. 26). Immediately following their emergence or a sudden transformation resulting from the collapse of pre-existing relationships, societies may undergo significant shifts in administration, organisation, functioning, or fundamental values. Over time, however, there once again accumulate more stable elements of social organisation and the relationships that structure them, which limit the scope of substantial change. At this stage, transformations tend to take the form of incremental modifications within the existing trajectory of development rather than radical shifts. The early phases of nascent societies illustrate this dynamic. For instance, the pioneers of the American West in the nineteenth century formed societies initially characterised by relatively low social complexity. In such contexts, social structures were fluid, allowing for significant shifts in organisation, profound changes in cultural and religious beliefs, and a highly dynamic adoption of technological innovations.

Over time, both through random processes and due to various forms of selection, societies tend to accumulate contextually stable elements of social organisation, which can undergo changes only with difficulty or not at all. These may include customs, laws, power structures, institutions, and the like, which gradually constrain the scope of further transformation. This increasing rigidity is driven primarily by a growing integration of the social system. In the case of societal evolution, this pertains mainly to the division of labour, specialisation of institutions or geographic regions, establishment of shared norms, development of security and law enforcement systems, the emergence of social welfare structures and similar elements. This process is accompanied by a progressive complexification of the state apparatus necessitated by the need to collect taxes, oversee the population, regulate the economic and social life, and maintain the functionality of institutions. Systems that have undergone this 'freezing' tend to respond to routine challenges more efficiently than more fluid and adaptable systems. This phenomenon is analogous to evolutionary elasticity (see BOX). As observed by Markoš (2014, 2015), such cultural systems are often capable of absorbing even substantial external pressures such as those coming from foreign states, competing cultures, religious movements, or technological innovations. Although they may temporarily deviate from their established trajectory, they tend to revert to their original organisational framework (cf. Chap. 18). Over time, however, the probability of fundamental changes within such systems steadily declines and ultimately approaches zero (Toman & Flegr, 2017b, 2018b; Fig. 15.3).

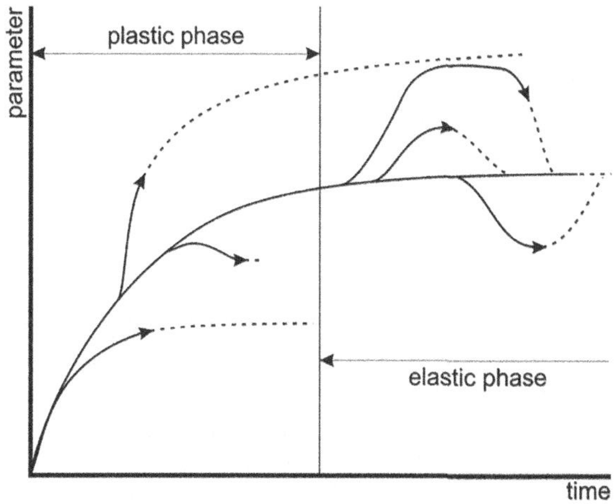

Fig. 15.3 'Freezing' of social development. Societies undergo profound 'plastic' transformations after emergence or major upheavals. This is followed by a gradual accumulation of stability in social organisation and relationships. At this stage, changes become increasingly incremental and reversible. Such 'elastic' changes reflect only variations within prevailing developmental trajectories rather than fundamental shifts. Redrawn from Markoš & Horský (2024)

All institutions exhibit similar patterns of development, with the only significant difference being the timescale at this these processes take place. This is particularly evident in the evolution of businesses, trade associations, and religious organisations. A striking example is the Catholic Church, which had in its early years undergone significant transformations but has since become a highly stable institution. This stands in stark contrast to its reformed branches, which have been historically far more prone to division, adaptation, and dissolution, often lasting only a few centuries or even just decades (see, e.g., Johnson, 1976). Similar patterns of stabilisation can be observed in bureaucratic structures, theoretical frameworks, rituals, symbols, and conceptual systems: they all tend to progress towards formalisation and rigidity over time.

Several scholars have drawn similar conclusions. Ladislav Kováč, a Slovak biologist and politician, examined the evolution of state institutions and argued that morality, culture, laws, and political organisation are all subject to selection for stability (Kováč, 2015, p. 26). Ray Rappaport, an American anthropologist, made a similar claim, suggesting that evolution, and especially social evolution, is fundamentally an attempt to preserve stability. This tendency, he argued, is particularly evident in the emergence, formalisation, and eventual petrification ('freezing') of rituals that define the paradigm within which societies evolve (Rappaport, 1999, pp. 416, 425–437). The American philosopher, mathematician, and semiotician Charles Sanders Peirce described a similar process from the semiotic perspective in his discussion of habit formation and sedimentation (see, e.g., Eco, 2000). He proposed that habits emerge from repeated sign-relationships, solidifying over time into self-reinforcing patterns that shape how individuals perceive and respond to signs in their environment (Hoffmeyer, 1996). Even more significantly, Thomas

Sebeok, a key figure in zoosemiotics, argued that the primary goal of science is to understand how stability is achieved and maintained in living systems. He believed that the key mechanism enabling this stability is semiosis, that is, the continuous interpretation and communication of signs (Sebeok, 1988).

15.4.2 Cultural 'Thawing'

As previously noted, the theory of frozen plasticity explains how sexual species can temporarily regain plasticity and undergo significant transformations, be it in response to selection pressures from the environment or due to other factors. Under certain conditions, cultural systems may exhibit a similar capacity for renewal. While culture-specific mechanisms may also play a role, a clear pathway through which 'frozen' cultural systems can begin to 'thaw' is by reducing the density of relationships between their constituent elements, a process somewhat analogous to the reduction of genetic polymorphism in biological evolution. Such thawing is typically associated with a radical simplification of the system. It may occur following a significant population decline, the secession of a small group (either from the broader society or from a specific cultural domain), or in the aftermath of revolutions and upheavals. Instances of evolutionary thawing may be relatively more frequent in cultural domains not directly tied to biological fitness, such as art, design, and fashion, where the bearers of cultural elements can survive a radical simplification of their system, which allows us to observe them. In contrast, cultural thawing seems less common in systems more directly linked to biological fitness, such as social organisation, religion, or state governance.

The analogy between evolution of cultural phenomena and the theory of frozen plasticity has been explicitly explored by Ostdiek (2011) in his study on theoretical biosemiotics. He argued that the evolution of signs and symbols loosely parallels the behaviour of alleles in sexually reproducing organisms, as predicted by punctuationalist evolutionary theories, particularly the theory of frozen plasticity (Flegr, 2013). While Ostdiek's conclusions are primarily concerned with semiotic processes, they can also be applied more broadly to human cultural evolution, in particular (though not exclusively) to the evolution of language. Still, it is important to realise that meaning attribution is not an exclusively human faculty. According to some biosemioticians, it is in fact a fundamental property of all living entities, inseparable from life itself (see Chaps. 16 and 17), which suggests that some form of sign evolution may also occur in non-human semiotic systems. Semiotic evolutionary models could therefore have applicability going beyond the human culture.

Drawing on and synthesising the ideas of Charles Darwin, Chauncey Wright, Charles Sanders Peirce, Stephen Jay Gould, and Jaroslav Flegr, Ostdiek (2011) describes the solidification of a symbol's meaning as equivalent to a system transitioning into an evolutionary stasis. A new symbol, such as a word, slogan, or graphic element, initially emerges from a state of non-meaning characterised by an absence of established interpretation, a temporary loss of validity, or a breakdown of conventional rules. This is followed by a gradual solidification of meaning driven primarily by the symbol's increasing use within a growing population. As more individuals adopt the symbol, its associations multiply, and it develops a wider network of connotations and relationships with other signs and symbols.

According to Ostdiek, the range of interpretations a symbol can acquire is analogous to genetic polymorphism, while the rich network of mutual references it accumulates corresponds to the network of pleiotropic interactions (see BOX). Both factors can operate across different scales and both resemble the complex interrelations between alleles in the evolution of sexually reproducing organisms. These interconnections contribute to the establishment and solidification of a symbol, while simultaneously limiting its potential for a fundamental semantic change. From the perspective of SBS, a solidified symbol exhibits a greater stability, while in semiotics, it becomes established, habitual, prosaic, and governed by rules and conventions that regulate its use. At this stage, we observe a decline in its plasticity, its ability to acquire a new meaning, function, or use, and its potential for abduction. Paradoxically, however, its elasticity increases, meaning it becomes more widely usable within its established meanings and variations.

Once the meaning of a symbol is established, it remains largely unchanged for the vast majority of its existence. The original plasticity that allowed for the emergence of new meanings can be restored only through a radical loss of most of its acquired meaning, context, or user base. According to Ostdiek, this process is analogous to the separation of a portion of a population in a sexually reproducing species, followed by a reduction in genetic polymorphism and a transition back into a plastic phase. As in biological evolution, most symbols that enter this phase will disappear without ever regaining plasticity, but a minority may succeed in redefining their meaning and function.

Ironically, one of the clearest examples of this phenomenon is the term ‘evolution’ itself. Originally, preformationist theories used it to describe the mere unfolding of biological potential that had been present since the moment of creation; this use was thus fully in line with the religious concept of the world being created by a divine power. By the mid-nineteenth century, however, this early meaning had largely faded from the academic discourse, which allowed Charles Darwin and his contemporaries to recruit the term to refer to a very different concept: the gradual transformation of species through natural selection. The new usage effectively supplanted the older one, with little trace of its theological origin remaining (Rádl, 2015).

According to Ostdiek (2011), all of the aforementioned processes closely mirror biological transitions between the plastic and elastic phases of sexually reproducing species. Moreover, the prevalence of this pattern of cultural evolution—characterised by the absence of gradualist evolution and the predominance of discontinuous development—suggests that the theory of frozen plasticity may serve not only as a model for the evolutionary dynamics of sexually reproducing species but also as a general theory of diachronic semiosis, that is, the emergence and transformation of meaning (see Fomin, 2019).

The possibility that the discontinuous nature of cultural evolution and the punctuationalist trajectory of biological evolution are in fact driven by analogous mechanisms has also been explored by the biosemiotician Anton Markoš (2014). In a later work, Markoš (2015) integrated the ideas of Peirce, Lotman, Rappaport, and Flegr, and argued that they all point to a shared underlying phenomenon. Specifically, he suggested that all semiotic systems (i.e., historical systems with ancestor-descendant relationships, memory, and accumulated experience) exhibit a similar trajectory. Shortly after their emergence, such systems are highly plastic, capable of significant transformation in multiple directions. Over time, their evolutionary trajectory stabilises, which restricts further change to incremental variations within the established course, while fundamental transformations become increasingly improbable. At this stage, the system 'freezes' and enters its elastic phase. Most such systems eventually disappear, while a minority—typically following dramatic disruptions—revert to their original plastic state, once again capable of a profound transformation.

15.4.3 Trends in Cultural Evolution

Another important question is whether cultural evolution undergoes processes analogous to the effectively irreversible evolutionary freezing observed in biological evolution (and described by the theory of frozen evolution; see BOX). In other words, does cultural evolution over long periods of time accumulate elements that become incapable of further transformation even under exceptional circumstances once a cultural system re-enters the plastic phase of its existence? If so, this would suggest a reduced likelihood of significant evolutionary innovations in cultural systems, similar to the emergence of macroevolutionary constraints observed in biological lineages.

What is certain is that cultural evolution is subject to processes comparable to the evolution of evolvability and, more broadly, to genetic architecture as described above. For instance, the cultural attraction theory suggests that cultural elements tend to change in a canalised direction, moving towards an attractor determined by neurological, social, or cultural factors (Sperber, 1996; cf. Chap. 11). A well-known example of neurological canalisation is the *Kinderschema* (see above), which influences aesthetic preferences across cultures. Regardless of the metastability of such attractor, this canalisation of cultural change closely resembles the role of genetic architecture in biological evolution, which evolves towards optimal modularity and robustness. This ensures the development of viable organisms even in complex evolutionary landscapes while introducing constraints that facilitate certain evolutionary changes and inhibit others. The trade-off for this structural robustness is a decreasing likelihood of significant evolutionary novelties over time.

In such cases, evolution begins to resemble variations on a single theme, as exemplified for instance by Vavilov's law of homologous series (Vavilov, 1922), which describes the emergence of corresponding adaptations (such as leaf shapes) in unrelated plant lineages. Similarly, during evolution towards an attractor, cultural elements become more widely accepted, comprehensible, and popular, which increases the likelihood of their transmission and persistence. A clear example of this pattern can be found in popular culture, where new cartoon characters repeatedly develop features conforming to the *Kinderschema*, a phenomenon illustrated by the evolution of Mickey Mouse over time (cf. Gould, 1979). Naturally, this does not apply to characters designed with alternative attractors in mind, where instant likability is not the primary goal. Nevertheless, as in biological evolution, the likelihood of profound transformations within these character designs declines over time—and that is analogous to the decreasing probability of major evolutionary novelties in biological lineages.

On the other hand, the looser and not strictly Darwinian nature of cultural evolution allows for a deliberate planning of changes, including the circumvention of obstacles that would be nearly insurmountable in biological evolution. This manifest itself as the dismantling of large cultural units, such as when horses and the associated service infrastructure were replaced by automobiles. As a result, macroevolutionary freezing in cultural evolution tends to be less pronounced than in biological systems, although exceptions may be found in cultural elements crucial for survival, such as clothing in cold climates or the cultivation of staple crops. In such areas, a fundamental change could increase the risk of a systemic collapse, which would in turn lead to a unidirectional accumulation of stabilising adaptations. When this occurs, the system undergoes freezing, becomes increasingly modular, and its subunits specialise.

As in biological evolution, adaptability can be retained by transition to a higher level of hierarchical complexity. At the societal level, this occurs through the division of labour, increasing social complexity, and a specialisation of businesses and geographic regions (Marengo et al., 2005). At the state level, such transition may take the form of incorporating new population groups for specific functions, as seen in the Roman Empire's integration of Germanic *foederati* in the fourth and fifth century for defensive purposes (cf. Sivan, 1987). These processes can be regarded as analogous to 'transitions to a higher level' in biological systems (Maynard Smith & Szathmáry, 2010). Some involve a multiplication and diversification of units at a lower level, while others entail a merging of independent entities in a manner akin to symbiosis. In both cases, the result is a more modular structure at a new, higher level of organisation. Given that such transitions also reside in (or include) the establishment of a new mode of inheritance, such as the storage, transfer, or (re) interpretation of information, cultural evolution has in this respect a distinct advantage over biological evolution because these phenomena take place far more readily in cultural evolution than in biological systems (Toman, 2020).

Typical examples of cultural modularity include the repetition of motifs in orally transmitted stories, redundancy of language at both the level of meanings and entire speeches (see Chap. 12), the modular construction of artefacts, or for instance the structuring of school curricula. The modular nature of cultural phenomena was analysed in detail by Herbert A. Simon (1962). While modularity in cultural evolution generally arises for different reasons and through different mechanisms than in biological evolution—for instance, not due to an increase in body size but for neurological reasons—its fate is analogous to that of biological modules. Over time, cultural modules tend to differentiate, specialise, and integrate, thus increasing the overall complexity of the cultural element. As a result, there emerge internally integrated, sophisticated complexes capable of self-propagation. The existence of such highly robust complexes is studied especially in memetics, where they are referred to as memetic complexes (see Chap. 10); their typical examples include teachings, religions, or narrative structures. At lower levels of organisation, these cultural elements undergo a process of 'machinification', much like the basic components of complex biological systems (McShea, 2016). As they become more integrated within a larger system, they lose their autonomy and become merely small and almost unchangeable parts of a complicated system (see Chap. 17). Under favourable conditions, elements from different cultural complexes can combine in a process that can be viewed as analogous to horizontal gene transfer or symbiosis in biological evolution. Such phenomena occur at both the material level (as in combining a typewriter, computer, and a display to create the PC) and the conceptual level, where elements from distinct theories or disciplines merge to form new frameworks of understanding.

A key question is whether SBS can in cultural evolution—like in biological evolution—drive the emergence of macroevolutionary trends (Toman & Flegr, 2017b, 2018a, b). Certain statistical patterns in cultural evolution are almost certainly evident, including the decreasing likelihood of large-scale transformations, increasing integration, modularisation and diversification of cultural elements, and the abovementioned 'machinification' of their most basic components. Still, these trends should not be regarded as absolute laws but rather as statistical tendencies. Some biological lineages undergo simplification despite an overall rise in complexity or semiotic competence and cultural evolution can follow a similar course of development that forms an exception to the abovementioned wider patterns. Nonetheless, in at least some cultural systems and at specific stages of development, there has been a clear increase in hierarchical complexity. Duda's chapter provides an example of how Austronesian political systems became more complex during the expansion of these peoples (see Chap. 12), while similar trends can be observed in the evolution of theories, narratives, state structures, bureaucratic systems, and technological innovations. Even so, human cultural evolution is unlikely to exhibit this trend as consistently as biological evolution does, because it is interwoven with other processes, including punctuationalist shifts.

In biological evolution, SBS not only contributes to increasing hierarchical complexity but also seems to accelerate this trend over time. Observational data suggest that each transition to a higher level of complexity leads to a faster accumulation of stable elements and their interrelations and thereby to a progressive reduction of the likelihood of fundamental evolutionary novelties (Toman & Flegr, 2018a). This phenomenon may be grounded in the complex multilevel genetic architecture of biological organisms. The existence of such a structure—and its growing complexity—then results in an increasingly rapid accumulation of contextually stable elements and their interconnections at each successive hierarchical level. The effectively irreversible macroevolutionary freezing associated with these processes may thus accelerate at each subsequent stage of complexity.

In cultural evolution, a similar acceleration can be observed in several domains, such as the exponential growth of information transfer (Moore's Law, Kurzweil's model), increasing energy production, or economic development (Ekstig, 2012). This raises the question of whether the acceleration of the growth of complexity in biological and cultural evolution may be causally linked. We have already seen that cultural evolution exhibits a statistical increase in complexity, which could in principle produce effects analogous to those observed in the hierarchical organisation of biological organisms. This becomes even more intriguing once we take into consideration that cultural evolution is, in essence, a continuation of biological evolution, a specialised cultural domain that follows distinct rules but adheres to the same fundamental principles (see Chaps. 16 and 17)—most notably SBS. This invites speculation as to whether the acceleration the growth of complexity in both biological and cultural evolution might be a manifestation of a universal principle, one of the fundamental laws of living systems (cf. Wong et al., 2023).

15.5 Conclusions

The principle of stability-based sorting (SBS) tells us that, over time, the more stable or frequently emerging elements naturally accumulate in evolving systems, while the less stable elements disappear. We can conclude that SBS plays a significant role in shaping the course of cultural evolution, arguably an even greater than in biological evolution. Social conservatism, literacy, agriculture, and the rise and fall of cultures and civilisations are just a few among many examples of its influence. There are notable isomorphisms between the effects of SBS in biological and cultural evolution, ranging from the fixation of variants with stabilising safeguards and the punctuationalist nature of change, to the emergence of modularity and constraints on the appearance of fundamental innovations, all the way to the mechanisms that restore variability in biological and cultural systems. Nevertheless, SBS also exhibits some distinct manifestations in cultural evolution, such as the persistence of certain variants due to the cognitive architecture of the human brain, social structures, and cultural contexts, the unidirectional movement towards cultural attractors, infeasibility of certain evolutionary changes that might be advantageous from the perspective of individual selection, and even the fixation of variants that reduce biological fitness. On the largest temporal and spatial scales, SBS may ultimately drive the evolutionary trends characterised by an increasing hierarchical complexity of cultural systems.

These patterns strongly resonate with the biosemiotic perspective, which holds that sign processes underpin both biological and cultural dynamics. Life itself can be understood as a semiotic phenomenon in which organisms continuously interpret and stabilise signs in their environment and memory. Applied to cultural evolution, this suggests that SBS is not just a mechanistic process of unit retention but an ongoing interpretive process shaped by the human brain's propensity to assign meaning. The more stable a sign-complex is—be it in the form of a religious ritual, a literary canon, or a legal framework—the greater is its likelihood of persistence and accumulation of additional layers of interpretation which further embed it in the cultural system. Therefore, while some may regard SBS as trivial, it in fact plays a crucial role in explaining a wide array of culturally significant phenomena, shedding light on both the macro-level trajectories and micro-level dynamics of cultural evolution.

References

Bardeen, M., & Cerpa, N. (2015). Editorial: Technological evolution in society-the evolution of mobile devices. *Journal of Theoretical and Applied Electronic Commerce Research, 10*(1), 1–7.
Bax, M., van Heusden, B., & Wildgen, W. (Eds.). (2004). *Semiotic evolution and the dynamics of culture*. Peter Lang.
Bentley, R., Hahn, M., & Shennan, S. (2004). Random drift and culture change. *Proceedings of the Royal Society B-Biological Sciences, 271*(1547), 1443–1450.
Blackmore, S. (1999). *The meme machine*. Oxford University Press.
Boas, F. (1940). *Race, language, and culture*. Macmillan.
Bouchard, F. (2008). Causal processes, fitness, and the differential persistence of lineages. *Philosophy of Science, 75*(5), 560–570.
Bourrat, P. (2014). From survivors to replicators: Evolution by natural selection revisited. *Biology & Philosophy, 29*(4), 517–538.
Boyd, R., & Richerson, P. J. (1985). *Culture and the evolutionary process*. The University of Chicago Press.
Boyer, P. (1998). Cognitive tracks of cultural inheritance: How evolved intuitive ontology governs cultural transmission. *American Anthropologist, 100*(4), 876–889.
Claverie, J. (2020). Fundamental difficulties prevent the reconstruction of the deep phylogeny of viruses. *Viruses, 12*(10), 1130.
Dawkins, R. (1976). *Selfish gene*. Oxford University Press.
Derex, M., & Mesoudi, A. (2020). Cumulative cultural evolution within evolving population structures. *Trends in Cognitive Sciences, 24*(8), 654–667.
Diamond, J. (2005). *Collapse: How societies choose to fail or succeed*. Viking Press.
Doolittle, W. (2014). Natural selection through survival alone, and the possibility of Gaia. *Biology & Philosophy, 29*(3), 415–423.
Doolittle, W. (2017). Darwinizing Gaia. *Journal of Theoretical Biology, 434*, 11–19.
Eco, U. (2000). *Kant and the platypus: Essays on language and cognition*. Houghton Mifflin Harcourt.
Eibl-Eibesfeldt, I. (1989). *Human ethology*. Aldine De Gruyter.
Ekstig, B. (2012). Superexponentially accelerating evolution. *World Futures, 68*(1), 40–48.
Eldredge, N., & Gould, S. (1972). *Punctuated equilibria: An alternative to phyletic gradualism, models in Paleobiology*. Freeman, Cooper and Co.
Feenberg, A. (1999). *Questioning technology*. Routledge.
Flegr, J. (2013). Microevolutionary, macroevolutionary, ecological and taxonomical implications of punctuational theories of adaptive evolution. *Biology Direct, 8*, 1.
Fomin, I. (2019). Memes, genes, and signs: Semiotics in the conceptual interface of evolutionary biology and memetics. *Semiotica, 230*, 327–340.
Gersick, C. (1988). Time and transition in work teams: Toward a new model of group development. *Academy of Management Journal, 31*(1), 9–41.
Gersick, C. (1991). Revolutionary change theories: A multi-level exploration of the punctuated equilibrium paradigm. *The Academy of Management Review, 16*(1), 10–36.
Godfrey-Smith, B. (2009). *Darwinian populations and natural selection*. Oxford University Press.
Gould, S. (1979). Mickey mouse meets Konrad Lorenz. *Natural History Magazine, 51979*, 30–36.

Gould, S. (2002). *The structure of evolutionary theory*. Harvard University Press.
Grand, S. (2001). *Creation: Life and how to make it*. Harvard University Press.
Herrmann-Pillath, C. (2021). Biosemiotic foundations of a Darwinian approach to cultural evolution. *Cultural Science Journal, 13*(1), 16–33.
Hoffmeyer, J. (1996). *Signs of meaning in the universe*. Indiana University Press.
Hoffmeyer, J., & Stjernfelt, F. (2016). The great chain of semiosis: Investigating the steps in the evolution of semiotic competence. *Biosemiotics, 9*(1), 7–29.
Janif, Z., Brooks, R., & Dixson, B. (2014). Negative frequency-dependent preferences and variation in male facial hair. *Biology Letters, 10*(4), 20130958.
Johnson, P. (1976). *History of Christianity*. Simon and Schuster.
Kováč, L. (2015). *Closing human evolution: Life in the ultimate age*. Springer.
Laland, K., Uller, T., Fellman, M., Sterelny, K., Muller, G., Moczek, A., Jablonka, E., & Odling-Smee, J. (2015). The extended evolutionary synthesis: Its structure, assumptions and predictions. *Proceedings of the Royal Society B–Biological Sciences, 282*(1813), 20151019.
Larsen, C. (2006). The agricultural revolution as environmental catastrophe: Implications for health and lifestyle in the Holocene. *Quaternary International, 150*, 12–20.
Lenton, T. (2004). Clarifying Gaia: Regulation with or without natural selection. In S. Schneider, J. Miller, E. Crist, & P. Boston (Eds.), *Scientists Debate Gaia* (pp. 15–25). MIT Press.
Lenton, T., Daines, S., Dyke, J., Nicholson, A., Wilkinson, D., & Williams, H. (2018). Selection for Gaia across multiple scales. *Trends in Ecology & Evolution, 33*(8), 633–645.
Lorenz, K. (1970). *Studies in animal and human behavior*. Harvard University Press.
Lotka, A. (1922a). Contribution to the energetics of evolution. *Proceedings of the National Academy of Sciences of the United States of America, 8*(6), 147–151.
Lotka, A. (1922b). Natural selection as a physical principle. *Proceedings of the National Academy of Sciences of the United States of America, 8*(6), 151–154.
Lotman, J. (2009). *Culture and explosion*. Walter de Gruyter GmbH & Co.
Marengo, L., Pasquali, C., & Valente, M. (2005). Decomposability and modularity of economic interactions. In W. Callebaut & D. Rasskin-Gudman (Eds.), *Modularity: Understanding the development and evolution of natural complex systems* (pp. 383–408). MIT Press.
Markoš, A. (2014). Biosphere as semiosphere: Variations on Lotman. *Sign System Studies, 42*(4), 487–498.
Markoš, A. (2015). The birth and life of species–cultures. *Biosemiotics, 9*(1), 73–84.
Markoš, A., & Švorcová, J. (2019). *Epigenetic processes and the evolution of life*. CRC Press.
Markoš, A., & Horský, J. (2023). *Pohyby semen a tvarů: Vyprávění v dějepisectví a v evoluční biologii. [Movements of seeds and shapes: Narration in historiography and evolutionary biology]*. Pavel Mervart
Maynard Smith, J., & Szathmáry, E. (2010). *The major transitions in evolution*. Oxford University Press.
McAuliffe, K. (2016). *This is your brain on parasites: How tiny creatures manipulate our behavior and shape society*. Eamon Dolan Book Houghton Mifflin Harcourt.
McShea, D. W. (2016). Three trends in the history of life: An evolutionary syndrome. *Evolutionary Biology, 43*, 531–542.
Michod, R. (2000). *Darwinian dynamics: Evolutionary transitions in fitness and individuality*. Princeton University Press.
Mulrooney, M. A., Ladefoged, T. N., Stevenson, C. M., Puleston, C. O., & Haoa, S. (2024). Rethinking collapse on Rapa Nui (Easter Island): A resilient precontact population. *Science Advances, 10*(6), eado1459.
Nicholson, A., Wilkinson, D., Williams, H., & Lenton, T. (2018). Alternative mechanisms for Gaia. *Journal of Theoretical Biology, 457*, 249–257.
Okasha, S. (2006). *Evolution and the levels of selection*. Oxford University Press.
Ong, W. J. (1982). *Orality and literacy: The technologizing of the word*. Methuen.
Ostdiek, G. (2011). Cast in plastic: Semiotic plasticity and the pragmatic reading of Darwin. *Biosemiotics, 4*(1), 69–82.

Pascal, R., & Pross, A. (2014). The nature and mathematical basis for material stability in the chemical and biological worlds. *Journal of Systems Chemistry, 5*, 3.
Pross, A. (2003). The driving force for life's emergence: Kinetic and thermodynamic considerations. *Journal of Theoretical Biology, 220*(3), 396–406.
Pross, A. (2012). *What is life? How chemistry becomes biology*. Oxford University Press.
Radcliffe-Brown, A. R. (1952). *Structure and function in primitive society*. Cohen & West.
Rádl, E. (2015). *The history of biological theories*. BiblioLife.
Rappaport, R. (1999). *Ritual and religion in the making of humanity*. Cambridge University Press.
Riedl, R. (1977). A systems-analytical approach to macro-evolutionary phenomena. *Quarterly Review of Biology, 52*(4), 351–370.
Riedl, R. (1978). *Order in living organisms: A systems analysis of evolution*. Wiley.
Schank, J., & Wimsatt, W. (1986). Generative entrenchment and evolution. *PSA: Proceedings of the Biennial Meeting of the Philosophy of Science Association, Number Two: Symposia and Invited Papers, 1986*, 33–60.
Schmidt, K. (2012). *Göbekli Tepe: A stone age sanctuary in south-eastern Anatolia*. Ex Oriente.
Sebeok, T. (1988). Communication, language and speech: Evolutionary considerations. In M. Herzfeld & L. Melazzo (Eds.), *Semiotic theory and practice: Proceedings of the third international congress of the IASS Palermo* (pp. 1083–1091). Mouton de Gruyter.
Sharov, A. A. (2016). Evolution of natural agents: Preservation, advance, and emergence of functional information. *Biosemiotics, 9*(1), 103–120.
Shcherbakov, V. (2012). Stasis is an inevitable consequence of every successful evolution. *Biosemiotics, 5*(2), 227–245.
Shcherbakov, V. (2013). Biological species as a form of existence, the higher form. In I. Pavlinov (Ed.), *The species problem—Ongoing issues* (pp. 65–91). InTech.
Simon, H. (1962). The architecture of complexity. *Proceedings of the American Philosophical Society, 106*(6), 467–482.
Sivan, H. (1987). On Foederati, Hospitalitas, and the settlement of the goths in AD 418. *American Journal of Philology, 108*(4), 759–772.
Somit, A., & Peterson, S. (1992). *T the dynamics of evolution: The punctuated equilibrium debate in the natural and social sciences*. Cornell University Press.
Sperber, D. (1996). *Explaining culture*. Blackwell.
Tilly, C. (2007). *Democracy*. Cambridge University Press.
Tolkien, J. R. R. (1984). On fairy-stories. In C. Tolkien (Ed.), *The monsters and the critics and other essays* (pp. 109–161). George Allen & Unwin.
Toman, J. (2020). *Evoluce³: Evoluční trendy, evolvabilita a teorie zamrzlé evoluce [evolution³: Evolutionary trends, evolvability and the theory of frozen evolution]*. Academia.
Toman, J., & Flegr, J. (2017a). General environmental heterogeneity as the explanation of sexuality? Comparative study shows that ancient asexual taxa are associated with both biotically and abiotically homogeneous environments. *Ecology and Evolution, 8*(2), 973–991.
Toman, J., & Flegr, J. (2017b). Stability-based sorting: The forgotten process behind (not only) biological evolution. *Journal of Theoretical Biology, 435*, 29–41.
Toman, J., & Flegr, J. (2018a). A virtue made of necessity: Is the increasing hierarchical complexity of sexual clades an inevitable outcome of their declining (macro) evolutionary potential? *Evolutionary Biology, 45*(4), 374–394.
Toman, J., & Flegr, J. (2018b). Macroevolutionary freezing and the Janusian nature of evolvability: Is the evolution (of profound biological novelty) going to end? *Biosemiotics, 11*(2), 263–285.
Tureček, P., Havlíček, J. (2017). Kultura jako předmět a produkt evoluce. [Culture as an object and product of evolution]. In L. Ovčáčková (Ed.) et al., *O původu kultury. Biologické, antropologické a historické koncepce kulturní evoluce [The origin of culture. Biological, anthropological and historical conceptions of cultural evolution]*. (pp. 183–229). Academia.
Tyler, T. (2011). *Memes and the science of cultural evolution*. Mersenne publishing.
Van Valen, L. (1989). Three paradigms of evolution. *Evolutionary Theory, 9*(1), 1–17.
Vavilov, N. (1922). The law of homologous series in variation. *Journal of Genetics, 12*(1), 47–89.

Watson, R., & Szathmary, E. (2016). How can evolution learn? *Trends in Ecology & Evolution, 31*(2), 147–157.

Webb, S. (2002). *If the universe is teeming with aliens... WHERE IS EVERYBODY?* Copernicus Books & Praxis Publishing.

Wimsatt, W. (1980). The units of selection and the structure of the multi-level genome. In P. Asquithand & R. Giere (Eds.), *PSA: Proceedings of the biennial meeting of the philosophy of science association* (pp. 122–183). Philosophy of Science Association.

Wimsatt, W. (2013). The role of generative entrenchment and robustness in the evolution of complexity. In C. Lineweaver, P. Davies, & M. Ruse (Eds.), *Complexity and the arrow of time* (pp. 308–331). Cambridge University Press.

Wimsatt, W., & Schank, J. (2004). Generative entrenchment, modularity, and evolvability: When genic selection meets the whole organism. In G. Schlosser & G. Wagner (Eds.), *Modularity in development and evolution* (pp. 359–394). The University of Chicago Press.

Wong, M. L., Cleland, C. E., Arend, D., Jr., Bartlett, S., Cleaves, H. J., II, Demarest, H., Prabhu, A., Lunine, J. I., & Hazen, R. M. (2023). On the roles of function and selection in evolving systems. *Proceedings of the National Academy of Sciences of the United States of America, 120*(31), e2310223120.

Chapter 16
The Past Lives on with Us: Epigenetics as a Link Between Cultural and Biological Evolution

Jana Švorcová and Karel Kleisner

Abstract This chapter approaches biological and cultural evolution as two complementary processes of the *evolution of life*, processes which are reciprocally interconnected, constantly influence each other, and neither can be postulated as the primary one. To what extent does it even make sense to understand cultural and biological evolution as separate phenomena, each with its own processual dynamics, sources of variability, and specific types of inheritance? The argumentation is based on the concept of dual inheritance, drawing also on the notion of organic memory as it was discussed in in nineteenth-century science. In parallel with modern findings regarding epigenetic inheritance, the authors speculate that, in some cases, cultural memory can be biologically transmitted to the next generation. It is thus ultimately argued that cultural evolution ought to be viewed as just another aspect of biological evolution.

Keywords Epigenetic inheritance · Organic memory · Biocultural evolution

16.1 Introduction

Cultural and biological evolution are often viewed as two fully distinct and separate domains, each with its own processual dynamics, sources of variability, and a specific type of heredity. But is that really so—or is this view merely a conceptual echo of the genocentric paradigm prevalent in the second half of the twentieth century? Does it make sense at all to view the evolution of culture and biological evolution as two different processes? Or is culture rather part of the biological aspect of our existence—and vice versa (cf. Chaps. 9 and 17)?

J. Švorcová (✉) · K. Kleisner
Department of Philosophy and History of Sciences, Faculty of Science, Charles University, Prague, Czech Republic
e-mail: jana.svorcova@natur.cuni.cz; karel.kleisner@natur.cuni.cz

J. Toman et al. (eds.), *Culture Evolving*, Biosemiotics 10,
https://doi.org/10.1007/978-3-032-16124-6_16

In this chapter, we shall build upon the idea of cultural and biological evolution as two generically linked phenomena that can mutually influence each other (both bottom up and top down) without either process being clearly primary. First, we will argue for the notion of biocultural evolution (Boyd & Richerson, 2005), that is, the idea of a close link between our biological heredity and our symbolically structured and socially anchored existence influenced by technologies (cf. Chaps. 7 and 10). But our analysis of the relation between biology and culture will go further than the work of Boyd and Richerson (2005). In particular, we will not limit ourselves to investigating the influence of culture on just the genetic background: we want to highlight also some relatively recent findings pertaining to the epigenetic heredity of behaviour. After the section on epigenetics, several excursions into the past will help us understand how culture and nature has been previously seen as organically isomorphic. Our argumentation follows the premise that the cultural and biological aspects of evolution should be considered an integral whole rather than viewed in separation.

16.2 Biocultural Evolution and Its Cumulative Character

'Biocultural evolution' (sometimes used interchangeably with 'dual inheritance theory') is a concept with roots in the 1960s, which describes the mutual dynamics of evolutionary relations between human nature and culture (Boyd & Richerson, 2005; cf. Chaps. 7, 10 and 12). It emphasises that biology not only enables the emergence of culture but culture then in turn influences the direction of biological evolution. This concept includes an explanation of biological and cultural factors, describes the mutual relations of the two, and explains how they jointly steer human evolution.

Within this discourse, culture is usually viewed as the sum of information that is stored in our minds and transmitted via social learning (observation, imitation, teaching of children by parents, etc.; cf. Chap. 9). We see a similar kind of learning in primates, meerkats, birds, and even fish (for instance regarding the ways of finding food). Nevertheless, in these kinds of animals we do not expect to find cumulative cultural evolution, that is, evolution that grows in complexity with each new generation (as in the cumulative progress in medicine or technologies). In non-human animals, this cumulative effect is known only in a rudimentary form in chimpanzees and in the emergence of dialects in various bird species (Boyd & Richerson, 2005; for more on this subject, see Chaps. 9 and 13). According to this view, the human ability to store and transmit cultural information has a psychological foundation that developed genetically because, at a certain point in human evolution, social learning leading to human cumulative culture turned out to be advantageous. This theory remains within the boundaries of Darwinism and population-genetic thinking, but it also holds that only few differences between (human) populations are in fact due to genes (as evidenced, for instance, by intercultural adoptions of children).

Differences between populations can be much more simply explained as cultural differences.

Even a superficial examination shows that the argument based on cumulation need not apply only to cultural evolution, because biological evolution likewise builds on phylogenetically accumulated adaptations (a phenomenon which Jacques Monod defined using the term 'gratuity'; see Monod, 1972). McShea and Brandon (2010) moreover noted that increases in biological complexity take place as if on their own, without the need for active selection or other agents (see Chap. 17). What we deal with is in effect a *processual* inertia of a system under change. No whole of any complexity can remain quite homogeneous because, with the passage of time, its parts will in the real world increasingly diversify (McShea & Brandon, 2010; cf. Chap. 15). What Richerson and Boyd had in mind is naturally another kind of cumulation of information, but the principle is the same: the historical records (genetic ones or of any other kind) of previous generations are necessary for testing evolutionary novelties and for the creation of new adaptations. And, as we shall see further in the text, living entities carry information about the experiences of previous generations not only in the form of accidental mutations but also via much more plastic mechanisms.

To complete the general picture of the definition of culture, we also ought to mention that simplifications which view culture only as the sum of information in our heads do have their pragmatic raison d'être (cf. Chaps. 10, 11, 12 and 13). Evolutionary theories which study relations between genes and culture within a Darwinian framework are mostly interested in particular traits of populations and their relation to culture: they do not aim at describing the complexity of evolving societies, their traditions, or values (Laland et al., 2010). Human culture turned out to be a concept that is tremendously difficult to pin down and even within humanities, there is no clear consensus on its definition.

Nevertheless, theorists of cultural evolution do seem to be in agreement: based on the work of Boyd and Richerson (2005), they define culture as information capable of influencing the behaviour of an individual that has been acquired from other representatives of the same species via learning, imitation, or other forms of social transfer. In this definition, information is understood in its broadest sense, that is, as including phenomena such as knowledge, thoughts, values, skills, or attitudes (Mesoudi et al., 2006). It seems, therefore, that what determines a specific culture is of a non-material nature and assumes the existence of *some kind of communication*. This characteristic of culture is in effect based on the classical duality of body and mind, which is also why one can assume that the core of cultural manifestations of life will be found in humans. In our closest relatives, we should then be able to find a little bit of culture and in other members of the biota we are likely to encounter only reduced hints to something on the margins of the thus conceived cultural phenomena (cf. Chap. 9). But in living nature, imitation, communication, and other such preconditions of manifestation of culture are fairly ubiquitous. Is it thus not the case that it boils down to the meaning we assign to the observed phenomena?

Similar evolutionary events often lead to different trajectories of development. Human ancestors left the complex three-dimensional space of treetops, this freed their forelimbs, and humans eventually used this potential for work. Human diligence and creativity subsequently led to the discovery of various technologies, and when people managed to produce more than what was strictly necessary, it led to social stratification and division of functional roles in the society, which in turn enabled the development of philosophy, science, and arts, but also advertising, customer satisfaction surveys, bureaucracy, and a subpopulation of individuals who care for proliferation of these aspects of culture.

One group of non-avian dinosaurs, namely ancestors of the current birds, likewise freed its forelimbs but they used this potential not for work but for flight. They went on to occupy the Earth's atmosphere, reached the tops of mountains, penetrated the most secluded valleys, and settled on the coasts and islands. Many mastered the skill of moving and hunting under water, learned to fly through the complicated three-dimensional spaces around treetops, developed navigation systems that use the position of stars and the Earth's magnetic field, created complex vocalisation systems, achieved a level of variety and ornamentation of body surfaces unparalleled among vertebrates, developed elaborate mating rituals and ways of attracting mates and, last but not least, created strategies of care for the offspring in almost any longitude. So how is it? We, the mammals, have culture—and birds likewise have a culture. But theorists of culture would claim that only humans have a culture that is cumulative.

16.3 Examples of Biocultural Evolution

How does the dynamics of coevolution of genes and culture work in human evolution? Let us take a look at some examples: adaptations of cultural evolution, such as the ability to use tools or kindle fire, may have led to completely novel selection pressures, which resulted in the consumption of cooked meat. Cucinivory in turn led to a smaller colon in humans (it represents only 20% of the total volume of the digestive tract compared to about 50% in apes; Milton, 2003). Another adaptation to softer foods is that human have smaller molars and lower bite strength compared to other primates (Eng et al., 2013; Furness et al., 2015). Herbicides, which are often used in agriculture, can affect human semen quality in terms of lower concentration and motility (Swan et al., 2003) or DNA integrity (Knapke et al., 2022), eventually leading to infertility and the release of estrogenic chemicals into waterways is another example of how we exert a possibly strong influence on our evolution (and the evolution of other species) without understanding the full impact.

One often cited example of biocultural evolution in humans is the digestion of lactose. While all children are capable of digesting lactose contained in the mother's milk until 4 years of age, in adults, lactose digestion is governed by a single dominant gene. Continuation of the ability to digest lactose into adulthood seems to be due to a single mutation of the regulation sequence of this gene, which became widespread in central and western Europe about 7500 years ago (Lukito et al., 2015). This mutation has a higher prevalence in populations that have a long tradition of cattle keeping and consumption of milk and dairy products. The situation is different in Africa and in the Middle East, where it has been speculated that the ability to digest lactose in adulthood is due to several point mutations, but even there the correlation between the domestication of cattle and the spread of these mutations is quite apparent (Gerbault et al., 2011). This seems to imply that selection pressures directed by culture are much stronger than pressures of the non-cultural kind. One could explain it by supposing that thanks to social learning, a given cultural trait can spread much faster than the genetic mutation, whereby the more individuals adopt the trait, the greater the selection pressure on genes which support the trait in question.

According to Boyd and Richerson (2005), the scenario of biocultural evolution runs as follows: biological evolution enabled the emergence of highly developed psychic abilities of humans, which in turn facilitated a cumulative evolution of cultural adaptations, one of which was the abovementioned keeping of cattle for meat and dairy products. This cultural adaptation led to an increased prevalence of the lactose-digesting gene in the population. With the spread of the lactose-digestion gene, the possibility of consumption of dairy products grew, which initiated the evolution of other cultural traits, such as the invention of ice cream or cappuccino... and these traits can potentially go on to influence our genes in other ways.

16.4 Niche Construction and Ecological Inheritance

Closely linked to the evolution of culture and to the dynamics between cultural and biological evolution is the notion of niche construction, which describes the dynamic and reciprocal relations between organisms and their environment (see Chap. 13). The idea of niche construction emphasises that selection is influenced also by organisms themselves via changes in their environment (Odling-Smee et al., 2003). A notable example of this phenomenon is the abovementioned influence of domestication of cattle on our genes. Nevertheless, although they have the greatest impact on our planet, people are not unique in their ability to construct niches. Birds build nests, predators excavate lairs, and earthworms chemically change the composition of soil in which they live (they loosen and aerate it), which has an effect on plants and other organisms that depend on the soil. Beavers build dams, which changes the strength of the water current, provides water to other organisms, and changes the appearance of the forest, soil structure, and the structure of root systems. One species of ant, *Myrmelachista schumanni*, which lives in the jungles of Peru, excretes formic acid, which has an herbicidal effect on most trees—but not on *Duroia hirsuta*, which the ants inhabit. The ants not only protect the tree from uninvited guests, such as caterpillars or bush crickets (which they collectively pester until these invaders decide it might be best to leave),[1] but inject the leaves of other, potentially competing, plants with formic acid, which kills them within a few days. That leads to dramatic changes in the tree diversity of the ants' habitat. Ultimately, the ants thus manage to establish tree monocultures, which locals call 'Devil's gardens': the largest known such place encompasses about 320 *Duroia hirsuta* trees growing in an area of about 1300m^2; it is probably over 800 years old.

The concept of niche construction emphasises that the environment not only selects individuals of a given population, but those individuals are in turn capable of changing the environment, which in turn influences the selection pressures acting on them and on other species (see the example of cultural adaptations which influence human evolution). In this complex relationship between organisms and the environment, *reciprocity* plays a key role. A constructed niche is then moreover passed down the generations and belongs to the realm of ecological inheritance, which in turn depends on genes (and other discrete replicators). But in many cases, the individuals who inherit a particular niche need not even be related. The niche is transferred between generations via the environment, not by the process of reproduction.

16.5 Excursion into the Nineteenth Century

Cultural and biological evolution differ above all how the cultural and biological variants are transferred or inherited. Inheritance refers mainly to genetic inheritance, but the term is nowadays increasingly often used to denote also inheritance of epigenetic modifications. When speaking about cultural evolution, the term 'transfer' is usually preferred and what one has in mind is the inheritance of language, alphabet, rituals, or customs. In this description, we are moving within the neo-Darwinian paradigm, where either selection or accident (as in genetic drift) decide which traits become fixed over time. The evolution of cultural variants has been described in similar terms (Boyd & Richerson, 2005), save for subtle differences which we do not need to go into here (cf. Chaps. 10, 11 and 13). In the past, though, at a time when genes as carriers of genetic information had not been known, biological and cultural inheritance were considered to be fully identical processes.

In the nineteenth and early twentieth century, many thinkers were influenced by the theory of inheritance of acquired traits[2] and Haeckel's law of biogenetics, which

[1] BBC document from David Attenborough's series 'Life in the Undergrowth', part 'Intimate relations'.

[2] Lamarck is now best known for the idea of inheritance of acquired characters, although it was something he considered self-evident, dedicated little intellectual effort to it, and neither sought nor expected lasting recognition for it. As Burkhardt Jr. (2013) noted, the concepts Lamarck probably hoped to be remembered for were his comprehensive vision of the progressive emergence of all living forms from the simplest to the most complex and his belief that changes in the behaviour play a central role in driving organismal transformation (see also Corsi, 1989). The concept of

viewed individual development as a recapitulation of evolutionary history. These theories were closely tied to the concept of organic memory, which suggests that living matter can store and transfer experiences across generations. Memory and inheritance were thus in effect considered synonymous. Memory of the brain was viewed as merely one concrete manifestation of this general property of the living matter but various newly acquired abilities, instincts, or habits were all viewed as linked to the ability of living matter to change thanks to sensory impressions and to record these changes for future generations (Hering, 1897, 1921). Ewald Hering, Ernst Haeckel, Richard Semon, Francis Darwin, Samuel Butler, Charles Sanders Peirce, and many other scholars were among those who contributed to the development of this idea and some of them even proposed detailed mechanisms for the transfer of memory traces between generations (Butler, 1878, 1910; Haeckel, 1876; Otis, 1994; Schacter, 2011; Semon, 1921; Švorcová et al., 2023). As Laura Otis (1994) had noted, the concept of organic memory was deeply rooted in the intellectual discourse of the period of 1870–1918 and was used by thinkers in most intellectual regions of not only Europe but also America. The concept of organic memory was the platform where biology met with psychology,[3] criminal science,[4] and literature,[5] but also with some more alarming notions, such as nationalism and race-based identity (cf. Chap. 3). Through the concept of organic memory, culture inherits and transmits traits across generations, paralleling the way individuals do. Thinkers of the time drew parallels between individual and cultural development, likening memory to heredity, and suggesting that organic memory ensures the perpetuation of life through a teleological continuum. Instincts, habits, and memory were all seen as manifestations of living matter's ability to store memory traces. Organic memory posited that an individual harbours not only personal memories but also ancestral history that may offer insights into past experiences, the development of culture, but also for instance the nature of 'race' or the collective unconscious. This perspective equated social and cultural phenomena with natural

inheritance of acquired characters has been emphasised as being Lamarckian in the discussions between Neo-Lamarckians and classical neo-Darwinians.

[3] Sigmund Freud, along with his student Carl Gustav Jung, embraced the concept of organic memory; in particular, they suggested that individuals unconsciously recall not only their own childhood experiences but also those of their ancestors. Although Lamarck's and Haeckel's ideas were already falling out of favour in scientific circles, Freud incorporated them into his psychological theories and proposed that individuals recapitulate the stages of human development, whereby those who experience perversions or neuroses embody certain earlier developmental stages (Otis, 1994).

[4] For instance, Cesare Lombroso (1835–1909) presents in his book *L'uomo delinquente* (1876; *Criminal Man,* Lombroso, 2006) a sort of embryology of crime, because according to his conclusions, there exist occasional criminals but also, so to say, criminals from birth—and those were the main subject of his study. People with innate tendency to crime were in his view as if developmental atavisms. Lombroso even defined them physiologically… but we do not wish to list here any particular characteristics so as not to offend those of our readers who have wild dark hair or a strong jawline.

[5] Especially, Émile Zola's twenty-volume cycle *Rougon-Macquart* subtitled *Natural and social history of a family under the Second Empire* (Otis, 1994).

phenomena, all passed down to future generations by similar processes. Nevertheless, proponents of organic memory struggled to explain how memory traces are transmitted across generations, and they were challenged by Weismann's theory of the continuity of germinal plasm.[6] These problems eventually led to the theory's abandonment.

16.6 The Hunger Winter and Other 'Hereditary Sins'

The Hunger Winter is a term used for a period of starvation that took place in The Netherlands between November 1944 and May 1945. In 1944/1945, Europe was experiencing an especially cold winter, which added to the misery of a continent already devastated by war. The western part of The Netherlands was at that time still controlled by the German armies, which led to an extreme scarcity of food. During that time, the population was surviving on just about 30% of the then common calorie intake. As a result, people turned to eating even grass and stored tulip bulbs and they burned anything that could warm their houses. Before supplies were renewed in May 1945, about 20,000 persons had died. Thanks to the meticulous medical records kept by Dutch physicians, this tragic event led to an enormously interesting study referenced to this day by most scientists who deal with epigenetics. People who survived the famine formed a well-defined group: they all experienced deprivation at the same time and for a similar length of time. Epidemiologists used this to study the impact of the Hunger Winter on the population, and they arrived at most interesting conclusions.

One of the first groups they studied were infants who experienced the famine *in utero*. If a pregnant woman experienced famine during the last months of her pregnancy, while at the time of the child's conception she had good nutrition, children were mostly born with low birthweight. If, however, the woman experienced famine during the first trimester and then her diet was normal, the offspring tended to have average birthweight. Given that the embryo grows the most towards the end of pregnancy, this is not surprising. But the children included in the study were followed throughout their lives, and that revealed further interesting associations. For instance, those who had a low birth weight retained a small stature into adulthood and did not evince any tendency to obesity although for the remainder of their lives they had unlimited access to food. On the other hand, the individuals who were exposed to deprivation *in utero* during the early state of their mother's pregnancy had a much greater tendency to obesity and suffered much more from associated health complications. Interestingly, these health complications—such as cardiac diseases (Roseboom et al., 2000) or elevated blood pressure (Painter et al., 2006)—were inherited by the subsequent generations and some studies report even

[6] The germinal line remains distinctly separate from the somatic line, meaning that the cells that make up our bodies and their encounters throughout an individual's life do not influence our germ cells.

transgenerational effect (Painter et al., 2008; Veenendaal et al., 2013). One such condition was also an increased prevalence of schizophrenia (Hoek et al., 1996). Similar instances of tendency to anxiety and depressive disorders have been observed in the offspring of Holocaust survivors (Lehrner et al., 2014; Yehuda et al., 1998).

It thus turns out that not only events of this kind *per se* but even their timing in relation to an individual's life can have a decades-long impact on future generations. Increased vulnerability during early development is because at this stage, cells of the embryo divide quickly, and epigenetic cellular programming is being set up. With the help of epigenetic modifications, cells reset the expression of genes for metabolism so as to compensate for the energy deficit and to prepare the organism for life in a resource-poor environment. A problem comes if the environment quickly changes and the organism finds itself in a resource-rich environment. It must be noted, though, that severe malnutrition is not the only factor that has an impact on future generations. Similarly unfavourable is the effect of alcohol consumption, drug use, or smoking, and that not only during pregnancy and not only in the mothers. Even things such as poor care for the offspring or fear of the scent of cherry blossom can be inherited (for review, see Švorcová, 2023)—and, as we shall see below, this is not a matter of genetic heredity.

To attempt an explanation of the phenomena outlined above (or other examples of heredity of traumatic experience), we must first take a little detour, this time into epigenetics (see BOX). Readers who have at least some basic knowledge of the subject can skip the box: the explanation offered below is rather general and, in many places, rather simplifying.

Box: Epigenetic Heredity across Generations: An Excursion into Cell Biology

For a long time, biologists thought that deoxyribonucleic acid acquired from parents is the sole carrier biological heredity. But especially in the last two decades, there emerged a growing amount of evidence that not only do there take place in cells various modifications of chromatin (meaning DNA and associated proteins) and different types of RNA molecules, but also that these modifications can be heritable. Epigenetics is a science which studies these phenomena, whereby the prefix epi- means that it investigates 'above' the level of genes. One can find epigenetic inheritance of various kinds in all organisms but, for the moment, let us focus only on eukaryotes and in terms of examples only on mammals, in particular mice and humans. In mammals, epigenetic heredity is even more interesting because it does not fully conform to the notion of a strict separation between germinal and somatic cell lines in the early stages of development, when the somatic line should have absolutely no influence on the reproductive cells.

The abovementioned modifications include above all the methylation of cytosines (and also adenines) of the cell DNA, when methyltransferase transfers a methyl group to cytosines, thus forming a 5-methylcytosine. Methylation thus adds another 'letter of the genetic alphabet', namely meC, to the four basic nucleotides (A, C, G, T). Inclusion of a methylated cytosine often leads to the relevant section not being transcribed in the cell at all: the gene or sequence is present in the cell but remains silent, sometimes even for the entire duration of the individual's life (e.g., the gene for foetal haemoglobin). Nevertheless, methylated cytosines also correlate with active DNA.

Histone modifications are another kind of epigenetic modifications (Fig. 16.1). Histones are cellular proteins that form octameric complexes, so-called nucleosomes, where DNA is wound around the cell nucleus, with 147 pairs of bases on one nucleosome—we can imagine it as a spool on which the DNA molecule is wrapped like a thread. Nuclear DNA is thus not only condensed but it is assumed that the wrapping on nucleosomes (and other associated proteins) also protects the DNA from damage. The tails of histone proteins can be in specific locations acetylated, methylated, phosphorylated, or modified in other ways, for instance by attaching other proteins, such as biotin, ubiquitin, etc. These modifications are then 'read' by other proteins, which can attach themselves to that modification (or a combination of modifications) and either loosen the chromatin or tighten it, thus silencing it for future generations. Other proteins can write these modifications on DNA and histone proteins, while other groups of proteins can read them, and still other proteins have the ability to erase them.

It is important to note that these modifications are reversible and can be removed even in minutes, which is often of key importance for the regulation of transcription. These modifications play a crucial role in cellular differentiation, when totipotent cells become specialised (as, e.g., the cells of the heart, brain, or liver). The specific form of cells and their function are determined by a set of activated as well as deactivated sequences. Modifications are widely spread across the animal kingdom, and they are even mutually dependent. As a consequence, they came to be viewed as a sort of higher-level cellular 'code' (for further details, cf. Markoš & Švorcová, 2009, 2019; Švorcová, 2023).

One area that attracted the attention of scientists is the role of different RNA molecules in the cell: the *ENCODE* project (a follow-up on the Human Genome Project, which had successfully sequenced human genome) had discovered that 80% of human genome has some biochemical function. This means that it is either transcribed into RNA transcripts (60% of those 80% of active genome are thus transcribed) or these are sequences from which transcription is directed, in which case we speak about promotors, enhancers, or silencers. The number of genes (i.e., sequences translated into a protein) is, however, much lower than originally assumed: they form only just under 2%

of the genome (Consortium, 2012). Regulatory RNA molecules must therefore play a much larger role than we used to believe. They can manipulate within the cell both the DNA as such (small interfering RNA or Piwi-interacting RNA), which they silence, or they act in concert with other RNA transcripts (microRNA), which they degrade. Various RNAs can also recruit enzymes, such as methyltransferases, to modify DNA molecule and histone proteins. All three groups of the epigenetic processes outlined above often cooperate and in all three groups, researchers have described cases where these processes were heritable via reproductive cells (e.g., Meaney & Szyf, 2005; Weaver et al., 2005; Rassoulzadegan, 2011; Dias & Ressler, 2014; Siklenka et al., 2015; Gapp et al., 2020; for a review see Švorcová, 2023).

Fig. 16.1 In eukaryotes, DNA is wrapped on histone proteins forming a nucleosome (shown as lizards, DNA shown in white). Both the DNA molecule itself and the histones are subject to epigenetic modifications (flags, stars, and other symbols), and various non-coding RNA molecules can also interact with the whole complex (*). There are about 30 million similar variably modified nucleosome complexes in various cell types of a human body. Redrawn from Markoš and Švorcová (2009)

16.7 Examples of Heredity of Epimutations

What happened to the offspring of people who had survived the wartime famine can be relatively well explained by the *thrifty phenotype hypothesis*, which is supported not only by evidence coming from humans but was also repeatedly tested in mice. The offspring of mice who experienced a shortage of good quality food (with insufficient amounts of, e.g., folate or vitamins B6 and B12) had a tendency to store fat and acquire resistance to insulin once they, unlike their mothers, found themselves in a nutrition-rich environment. On the cellular level, research had shown they had a 20% lower methylation of the PPARα gene, responsible for the metabolism of lipids in the liver, than individuals in the control group (Lillycrop et al., 2007). This took place in conjunction with a different methylation of the promotor gene for the glucocorticoid receptor that plays an important role in the regulation of blood pressure (Burdge et al., 2007).

The gene for glucocorticoid receptors (GR) is involved also in another kind of intergenerational transfer, because its activity levels—jointly with the number of glucocorticoid receptors in the brain—are linked to the intensity of mother's care for its offspring immediately after their birth. The gene for GR is not methylated at birth: it acquires methylation during the first day after birth. Newborn animals who are fortunate enough to receive intensive care from their mothers during the first week of life lose this modification on the enhancer of GR, while in those which do not receive such care the gene remains in their brain cells methylated—and thus silenced. If this gene is active in the brain cells, then (in conjunction with active chromatin) transcription of the gene for GR is also active. As a result, mice that receive maternal care can cope with stress better than mice that did not receive sufficient maternal care (Meaney & Szyf, 2005; Weaver et al., 2004). This example also aptly demonstrates that transcription or silencing of DNA methylation is closely and flexibly dependent on the environment.

Research into the heredity of olfactory preferences in mice had also attracted much attention (Dias & Ressler, 2014). For instance, scientists had regularly exposed males of the F0 generation (prior to the conception of offspring) to electric shocks in the presence of acetophenone, which has an odour reminiscent of cherries or orange blossom. Their male offspring in F1 and F2 generations then evinced fearful reactions when exposed to this scent (and not in the presence of other odours) although they had not been conditioned to fear the odour of acetophenone themselves. In these animals, changes in comparison to the control groups took place not only on an epigenetic and behavioural but even on a neuroanatomical level. Scientists found in the sperm of males of the F0 and F2 generation demethylations of genes for specific receptors of the relevant olfactory pathway, which led to an increased amount of the relevant receptors in the brain, and even to an enlargement of the olfactory bulb for acetophenone, that is, the area in the brain which processes just that specific odour (Dias & Ressler, 2014). Positive about these experiments is that it has also shown that such process can be reversed: traumatised mice exposed to these stimuli with no further electric shocks gradually lost fear of the odour and traumatised phenotypes were not passed on to further generations.

In studies of this kind, it is necessary to eliminate the social transfer of trauma, where a negative behavioural pattern is transferred by upbringing rather than biologically. Such research has been undertaken by Isabelle Mansuy and her team from Switzerland: they separated, at irregular intervals, male mice from their mothers or in various ways traumatised the mouse mothers themselves (Gapp et al., 2020). The mouse mothers then returned to their pups stressed and tended to ignore them. In adulthood, such pups behaved differently than the controls, and these changes were passed on to further (up to five) generations. Traumatised behaviour in mice to some extent resembles depressive behaviour in humans: mice are more anxious, asocial, and do not perform as well in standardised tests (labyrinth, forced swim test or elevated plus maze) as controls do. In order to fully eliminate social transmission, scientists injected RNA from the sperm of traumatised males into zygotes from non-traumatised parents—and the individuals who developed and were born from those zygotes later displayed the same traumatised behaviour patterns (Gapp et al., 2020). Social transmission can also be excluded when traumatised males mate with naïve females and are subsequently excluded from the experiment. A similar purpose is achieved by crossbreeding experiments, when the offspring of a traumatised parent are fostered by naïve females. In all of the abovementioned cases, traumatised behavioural patterns were observed in the subsequent generations, which supports the hypothesis that what we see are instances of transgenerational heredity and not just social transmission via maternal care. Similar results have been reported also in studies by Jawaid et al. (2020), who observed transmission up to the fifth generation, and Van Steenwyk et al. (2018), who reported consistency of transmission up to the fourth generation.

There are dozens of studies which describe neuroanatomical and molecular changes in traumatised phenotypes in generations F1 and F2. Molecular changes often involve the previously mentioned DNA methylation but also modifications to histone proteins, which help regulate DNA transcription. Changes affect mostly those genes in the hippocampus and hypothalamus which are related to stress response, but they also affect for instance brain plasticity, the functioning of long-term memory, or glucose metabolism (Bohacek et al., 2014; Dietz et al., 2011; Franklin et al., 2010; Weiss et al., 2011).

It seems that transmission of information about a possibly traumatic environment is chiefly the task of small non-coding RNA (sncRNA). It turns out that sperm receive their load of sncRNA during their maturation in the epididymis, when somatic tissues produce vesicles (exosomes) that fuse with the sperm cells (Cossetti et al., 2014; Sharma et al., 2018). This is thus an instance of direct communication between the somatic tissue and the germline. The abovementioned RNA then, often during early embryogenesis, regulates also other epigenetic processes, such as DNA methylation or the modification of histone proteins, and can thus function as the main mediator of epigenetic inheritance. It is also known that not only mothers but also fathers contribute their load of RNA to the zygote and the RNAs from both parents then interact with each other during development (Rodgers et al., 2015). If we consider all the abovementioned evidence, it seems that after a century of near-silence, the concept of organic memory is making a comeback.

16.8 Epigenetic Learning: A Missing Link Between Nature and Nurture

The examples of transgenerational epigenetic inheritance show that changes in gene regulation, which can be heritable across several generations, plastically react to the environment. They are the result of selection pressures which we create through our own activity (such as warfare). These modifications then set the norm of reaction—they influence how the phenotype will react in a given environment during specific stages of its development. We can view them as acquired adaptations, but it would be erroneous to speak about acquired traits: these are rather acquired states of gene regulation which determine the norm of the reaction. It would also be wrong to imagine that one can epigenetically inherit any behaviour or experience: so far, studies tended to focus on situations such as exposure to environmental toxins, preparation for energy deficit, or other possible stressful situations. Parodies of Lamarckian inheritance, where highly trained parents have a muscular child (cf. Chap. 10) or a mouse with its tail cut off has pups with the same defect (such experiments were actually conducted by August Weismann), are naturally still beyond the realm of reality.

Readers should thus bear in mind that what we talk about here are no 'intentional' (or voluntaristic) changes in genetic information, which is something that Lamarck has been accused of and how Lamarckism is interpreted nowadays (although quite wrongly, because he did not know about genes, and even the intentionality of any such changes was in his case due to Cuvier's self-promoting eulogy for Lamarck and possibly due to an unfortunate way Lamarck expressed his ideas; see Corsi, 1989). In some sense, though, epigenetic change can lead to a mutation. Methylated cytosines can influence the sequence of DNA thanks to a spontaneous hydrolytic deamination of cytosine to uracil, which can, via several further steps and with the help of repair enzymes, result in a base pair A = T instead of previous G = C pair. Hypermethylated regions of the genome often correlate with a higher number of mutations in those locations (Hernando-Herraez et al., 2015), because the very fact that some manipulation with genes took place and they are open to transcription makes mutation more likely than when chromatin is silenced.

Epigenetic processes should be therefore included into the theory of evolution: they lead to the emergence of adaptive traits through a process that can be much faster and more plastic than response to changes in the environment effected by waiting for an accidental mutation. Epigenetic processes, which plastically react to changes in the environment (such as a change in the nutrition status, stress, or a traumatic event), are thus the missing link in the ongoing nature versus nurture debate.

Another aspect to consider is the intrinsic link between learning and plasticity. What was originally initiated by culture can be, in some cases and to some extent, also biologically internalised (cf. Chap. 18). In evolution, learning is thus not limited to social learning within biocultural evolution: it also extends to non-neuronal organisms. As noted by Švorcová (2024), epigenetic modifications can be seen as representations of the organism's past experiences with its environment. Even mental representations ultimately rely on similar informational processing mechanisms, such as shared developmental patterning modules (Newman & Bhat, 2009). Adaptation, shaped by a dynamic interplay between epigenetics and environmental factors, is not purely random: it is goal-directed, functioning through memory. Such epigenetic changes can be seen as a way in which organisms can anticipate their future environment by drawing on past experiences. Organisms have been evolutionarily fine-tuned to regulate their genome in response to environmental cues, whereby similar signals often are encountered repeatedly.

This is essentially a semiotic learning because it involves a choice between at least two options (Kull, 2015) and, moreover, such learning can become a habit (Švorcová et al., 2023). From the biosemiotic perspective, learning and living are co-extensive with meaning-making about the world through the scaffoldings of previous experiences (Campbell et al., 2019). Scaffolding is a process of directing and constraining the responsiveness of an organism based on its previous choices and decisions. 'Through a recursive process of discovering sign relations in the environment and gradually canalizing these action-habits into scaffolding, organisms achieve new semiotic competences or, more directly, they learn: "[Scaffolding] is the frame for the formation of future habits"' (Kull, 2015, p. 232; see Campbell et al., 2019). Organic memory plays a crucial role in learning about the environment (and in the process of forgetting).

16.9 Canalization of Epigenetic Changes

These processes certainly fit into that part of the evolution script that deals with phenotype plasticity. This approach is sometimes also called the 'phenotype-first' theory; it most general terms, it amounts to a claim that in the course of evolution, changes in the phenotype precede changes in the genotype. This view found its so far most comprehensive and detailed expression in the work of Mary Jane West-Eberhard (2003) and her concept of phenotype and genotype accommodation, whose main tenets are:

1. Organisms are in their nature highly plastic and can react to changes in the environment by changes in the phenotype.
2. These phenotype changes are often adaptive, so that a new phenotypic trait can emerge on the scene (phenotype accommodation).
3. The new phenotype need not be fixed genetically: it can just lead to an epigenetic fixation or changes in the modular organisation of the phenotype. In other words, the evolution of a new phenotype does not require the evolution of new gene complexes but merely changes in the regulation architecture of existing ones.
4. If given environmental conditions are repeated over a number of generations and the phenotypic trait turns out to be advantageous, a genetic change can also take place. That can happen either by waiting for an random mutation and a subse-

quent fixation of the phenotypic trait (inspiration by the Baldwin effect; Baldwin, 1896), by drawing on a so-far cryptic genetic variability (Waddington's assimilation; Waddington, 1952, 1956), or by a quantitative change in gene frequency (genetic accommodation).

This scenario once again brings together biological and cultural evolution and, moreover, it offers an explanation of how biocultural evolution, that is, a cultural adaptation leading to a genetic change, can function. In some cases, cultural evolution can be simply replaced by 'downward causation' (due to changes in behaviour, the Baldwin effect, and the like). This applies for instance to the claim that human genome had in the course of evolution undergone changes in consequence of our cultural behaviour.

A classic example of this is the abovementioned breeding of cattle for milk, which went hand in hand with the emergence of mutations for lactose digestion in adulthood. One could, however, dispute whether one should use the adjective 'cultural' in this context. After all, this is just about a change in dietary habits that eventually led to a physiological adaptation. In a similar way, no one thinks of culture when speaking about the evolution of a feeding strategy that has to do with sucking blood or other fluids in some groups of Hemiptera. Transition to this strategy probably took place 'by mistake', by latching onto animal tissue instead of the usual plant matter (Lehane, 2005; Weible, 2016). It is likely that the same applies to hematophagy in the ancestors of the current vampire bats, who at first just licked bleeding wounds of injured animals and only later evolved further secondary eco-morpho-physiological adaptations. In both cases, we see exaptation caused by changes in behaviour. One might object that, unlike people, vampire bats do not breed cattle as their blood reserve. That is true. But what should we think about ants corralling their aphids underneath a leaf to protect them, carrying aphids to other plants, and taking care of aphid eggs during winter—all in exchange for the sweet honeydew. In this case, the analogy with human pastoral culture is much closer. Still, one could object that unlike the ants who just herd their aphids, people have also created particular breeds of their domesticated animals. But let us not forget that even by just herding animals, one also changes the direction of selection pressures, be it by choosing the environment or by elimination of predation, thus in effect breeding those animals. Is this biocultural evolution or should we rather talk about mutualism and subsequent nutritional or behavioural adaptations? And is there any substantial difference between the two?

16.10 Two Tales of One Evolution

Having gotten this far, we should restate the questions posed at the beginning of this chapter. Does it at all make sense to speak of a cultural evolution and biological evolution in terms of two different and distinct processes (Kleisner & Tureček, 2017)? It is clear that cultural and biological evolution affect each other and are generically linked in the sense of evolution of culture as a daughter of biological evolution. Biological and cultural evolution can, however, influence one another only if they are two distinct processes, which is also how they are usually viewed, as two disparate domains, each with its own processual dynamics and own sources of variability. It has even been claimed that biological and cultural evolution may occasionally head in different or even opposite directions—but conflict is not a definitionally acceptable criterion of such a situation. One cannot simply select all traits at once and expect a full synergy of all tendencies leading to an evolutionary advantage. In a similar way, we find in the classic picture of Darwinian evolution a tension between sexual and natural selection. Sexual selection is often said to move against the natural selection. But natural selection via feedback influences the traits generated by the other process—and vice versa.

If cultural evolution leads to the emergence of traits that would have been maladaptive in some ‘original’ environment, it does not necessarily mean that it is a process which moves counter biological evolution. If these new traits proliferate in a population that lives in an environment determined by a particular culture, that culture is the current environment of those organisms, and all new adaptations are simply adaptations to the new environment regardless of whether we view it as an environment determined by culture. In people, one can nowadays hardly at all speak about any environment that is not also cultural, and this is probably also the case for many other forms of life. Going back to sexual selection, even the Darwinian ‘taste for the beautiful’ on the part of females, or rather one sex’s preference for certain traits in the opposite sex, are rather close to what one would call a cultural habit. The idea that some preference is caused by a point mutation that influences the nervous system or sensory apparatus of the subject is roughly equivalent to claiming that Saturday morning hangover in the ageing European population is due to a dysfunction of alcohol dehydrogenases. Is it not rather so that European culture involves the notion of having fun at the end of the work week and a more fitting explanation should reference a culturally transmitted behavioural custom?

Sometimes, evolution of culture is taken to mean the evolution of artefacts, but evolution of artefacts is not based on natural selection and accidental variability. Rather, it is based on artificial selection and non-accidental variability. One could point to strong parallels, perhaps even processual homologies, between technological innovations and evolutionary novelties. Neither of the two processes takes place based on a fully accidental variability and neither proceeds quite according to a plan. In both cases, phenomena such as exaptation and secondary adaptation, ritualisation, plasticity, neo-functionalisation, and the like, also play a role (cf. Chaps. 11 and 15).

In this context, let us note that one of the main sources of inspiration for Darwin’s theory was the cultural process of domestication, that is, the production of living artefacts from wild forms of organisms. But artificial selection inherently includes the intervention of a conscious subject, the breeder, and that cannot be claimed about natural selection. Is this the difference between cultural and biological evolution? Probably not. Sexual selection also assumes some activity of the living subject who literally determines what is going to appear in the next generation. If sexual selection is a component of biological evolution, how is it different from artificial selection? Domesticated individuals vie for the favour of the breeder and the seeds of cultural plants compete for the favour of the farmer just like individuals of one sex compete among themselves for the favours of the opposite sex. Phenomena such as sexual selection, parental selection, artificial selection, and even some aspects of natural selection (such as predation) involve the activity of another subject. This makes them different from phenomena that involve nothing of that kind, such as for instance environmental selection.

16.11 Conclusions

In this chapter, we have explored cultural and biological evolution as interconnected processes which influence each other bidirectionally, without a clear hierarchy. While agreeing with the concept of biocultural evolution (Boyd & Richerson, 2005), we have extended the discussion beyond genetic influence to include recent findings on the epigenetic heredity of behaviour. Historical perspectives have then helped us illustrate how culture and nature used to be viewed as one organic whole—and such a view can find support also in the model of the plasticity-driven adaptation.

So, what sets apart biological evolution from cultural evolution from the perspective of standard evolutionary biology? It seems that in order to view cultural and biological evolution as two distinct phenomena, we must accept a somewhat extreme interpretation of biological evolution as the result of gene activity or responses to environmental selection and restrict cultural evolution to non-biological human rituals and various human creative undertakings. As long as it remains unclear which trait is cultural and which biological, it also will not be clear where one ought to draw the line between cultural and biological evolution. Or perhaps it is just one process and instead of mutual influence we should rather speak about various symptoms of the evolution of life. Within this view, what we call culture would be present ever since the beginning of life and its evolution.

References

Baldwin, J. M. (1896). A new factor in evolution. *The American Naturalist, 30*, 441–451.

Bohacek, J., Farinelli, M., Mirante, O., Steiner, G., Gapp, K., Coiret, G., Ebeling, M., Durán-Pacheco, G., Iniguez, A. L., Manuella, F., & Moreau, J. L. (2014). Pathological brain plasticity and cognition in the offspring subjected to postnatal traumatic stress. *Molecular Psychiatry, 20*(5), 621–631.

Boyd, R., & Richerson, P. J. (2005). *Not by genes alone: How culture transformed human evolution*. University of Chicago Press.

Burdge, G. C., Slater-Jefferies, J., Torrens, C., Phillips, E. S., Hanson, M. A., & Lillycrop, K. (2007). Dietary protein restriction of pregnant rats in the F0 generation induces altered methylation of hepatic gene promoters in the adult male offspring in the F1 and F2 generations. *The British Journal of Nutrition, 97*(3), 435–439.

Burkhardt, R. W., Jr. (2013). Lamarck, evolution, and the inheritance of acquired characters. *Genetics, 194*(4), 793–805.

Butler, S. (1878). *Life and habit*. Trübner & Co..

Butler, S. (1910). *Unconscious memory*. AC Fifield.

Campbell, C., Olteanu, A., & Kull, K. (2019). Learning and knowing as semiosis: Extending the conceptual apparatus of semiotics. *Sign Systems Studies, 47*(3/4), 352–381.

Consortium H. M. P. (2012). Structure, function and diversity of the healthy human microbiome. *Nature, 486*(7402), 207–214.

Corsi, P. (1989). *The age of Lamarck: Evolutionary theories in France 1790–1830*. Oxford University Press.

Cossetti, C., Lugini, L., Astrologo, L., Saggio, I., Fais, S., & Spadafora, C. (2014). Soma-to-germline transmission of RNA in mice xenografted with human tumour cells: Possible transport by exosomes. *PLoS One, 9*(7), e101629.

Dias, B. G., & Ressler, K. J. (2014). Parental olfactory experience influences behavior and neural structure in subsequent generations. *Nature Neuroscience, 17*(1), 89–96.

Dietz, D. M., Laplant, Q., Watts, E. L., Hodes, G. E., Russo, S. J., Feng, J., Oosting, R. S., Vialou, V., & Nestler, E. J. (2011). Parental transmission of stress-induced pathologies. *Biological Psychiatry, 70*(5), 408–414.

Eng, C. M., Lieberman, D. E., Zink, K. D., & Peters, M. A. (2013). Bite force and occlusal stress production in hominin evolution. *American Journal of Physical Anthropology, 151*(4), 544–557.

Franklin, T. B., Russig, H., Weiss, I. C., Gräff, J., Linder, N., Michalon, A., Vizi, S., & Mansuy, I. M. (2010). Epigenetic transmission of the impact of early stress across generations. *Biological Psychiatry, 68*(5), 408–415.

Furness, J. B., Cottrell, J. J., & Bravo, D. M. (2015). Comparative physiology of digestion. *Journal of Animal Science, 93*(2), 485–491.

Gapp, K., van Steenwyk, G., Germain, P. L., Matsushima, W., Rudolph, K. L. M., Manuella, F., Roszkowski, M., Vernaz, G., Ghosh, T., Pelczar, P., Mansuy, I. M., & Miska, E. A. (2020). Alterations in sperm long RNA contribute to the epigenetic inheritance of the effects of postnatal trauma. *Molecular Psychiatry, 25*(2), 2162–2174.

Gerbault, P., Liebert, A., Itan, Y., Powell, A., Currat, M., Burger, J., Swallow, D. M., & Thomas, M. G. (2011). Evolution of lactase persistence: An example of human niche construction. *Philosophical Transactions of the Royal Society B: Biological Sciences, 366*(1566), 863–877.

Haeckel, E. (1876). *Perigenesis der Plastidule oder die Wellenzeugung der Lebenstheilchen*. Reimer.

Hering, E. (1897). *On memory and the specific energies of the nervous system*. Open Court Publishing Co.

Hering, E. (1870/1921). Über das Gedächtnis als allgemeine Funktion der organisierten Materie. Akademische Verlagsgesellschaft.

Hernando-Herraez, I., Heyn, H., Fernandez-Callejo, M., Vidal, E., Fernandez-Bellon, H., Prado-Martinez, J., Sharp, A. J., Esteller, M., & Marques-Bonet, T. (2015). The interplay between DNA methylation and sequence divergence in recent human evolution. *Nucleic Acids Research, 43*(17), 8204–8214.

Hoek, H. W., Susser, E., Buck, K. A., Lumey, L. H., Lin, S. P., & Gorman, J. M. (1996). Schizoid personality disorder after prenatal exposure to famine. *American Journal of Psychiatry, 153*(12), 1637–1639.

Jawaid, A., Kunzi, M., Mansoor, M., Khan, Z. Y., Abid, A., Taha, M., Rigotti, S., Thumfart, K., Faisal, S., Chughtai, O., & Mansuy, I. M. (2020). Distinct microRNA signature in human serum and germline after childhood trauma. *medRxiv*. Preprint.

Kleisner, K., & Tureček, P. (2017). Cultural and biological evolution: What is the difference? *Biosemiotics, 10*(1), 127–130.

Knapke, E. T., Magalhaes, D. P., Dalvie, M. A., Mandrioli, D., & Perry, M. J. (2022). Environmental and occupational pesticide exposure and human sperm parameters: A navigation guide review. *Toxicology, 465*, 153017.

Kull, K. (2015). Evolution, choice, and scaffolding: Semiosis is changing its own building. *Biosemiotics, 8*(2), 223–234.
Laland, K. N., Odling-Smee, J., & Myles, S. (2010). How culture shaped the human genome: Bringing genetics and the human sciences together. *Nature Reviews Genetics, 11*(2), 137–147.
Lehane, M. J. (2005). *The biology of blood-sucking in insects*. Cambridge University Press.
Lehrner, A., Bierer, L. M., Passarelli, V., Pratchett, L. C., Flory, J. D., Bader, H. N., Harris, I. R., Bedi, A., Daskalakis, N. P., Makotkine, I., & Yehuda, R. (2014). Maternal PTSD associates with greater glucocorticoid sensitivity in offspring of holocaust survivors. *Psychoneuroendocrinology, 40*, 213–220.
Lillycrop, K. A., Slater-Jefferies, J. L., Hanson, M. A., Godfrey, K. M., Jackson, A. A., & Burdge, G. C. (2007). Induction of altered epigenetic regulation of the hepatic glucocorticoid receptor in the offspring of rats fed a protein-restricted diet during pregnancy suggests that reduced DNA methyltransferase-1 expression is involved in impaired DNA methylation and changes in histone modifications. *The British Journal of Nutrition, 97*(6), 1064–1073.
Lombroso, C. (2006 [1876]). *A criminal man*. Duke University Press.
Lukito, W., Slater-Jefferies, J. L., Hanson, M. A., Godfrey, K. M., Jackson, A. A., & Burdge, G. C. (2015). From 'lactose intolerance' to 'lactose nutrition'. *Asia Pacific Journal of Clinical Nutrition, 24*, S1–S8.
Markoš, A., & Švorcová, J. (2009). Recorded versus organic memory: Interaction of two worlds as demonstrated by the chromatin dynamics. *Biosemiotics, 2*(2), 131–149.
Markoš, A., & Švorcová, J. (2019). *Epigenetic processes and the evolution of life*. CRC Press.
McShea, D. W., & Brandon, R. N. (2010). *Biology's first law: The tendency for diversity and complexity to increase in evolutionary systems*. University of Chicago Press.
Meaney, M. J., & Szyf, M. (2005). Environmental programming of stress responses through DNA methylation: Life at the interface between a dynamic environment and a fixed genome. *Dialogues in Clinical Neuroscience, 7*(2), 103–123.
Mesoudi, A., Whiten, A., & Laland, K. N. (2006). Towards a unified science of cultural evolution. *Behavioral and Brain Sciences, 29*(4), 329–347.
Milton, K. (2003). The critical role played by animal source foods in human (*Homo*) evolution. *The Journal of Nutrition, 133*(11), 3886S–3892S.
Monod, J. (1972). *Chance and necessity: An essay on the natural philosophy of modern biology*. Collins.
Newman, S. A., & Bhat, R. (2009). Dynamical patterning modules: A "pattern language" for development and evolution of multicellular form. *The International Journal of Developmental Biology, 53*(5–6), 693–705.
Odling-Smee, J., Laland, K., & Feldman, M. (2003). *Niche construction: The neglected process in evolution*. Princeton University Press.
Otis, L. (1994). *Organic memory: History and the body in the late nineteenth & early twentieth centuries*. University of Nebraska Press.
Painter, R. C., de Rooij, S. R., Bossuyt, P. M., Simmers, T. A., Osmond, C., Barker, D. J., Bleker, O. P., & Roseboom, T. J. (2006). Early onset of coronary artery disease after prenatal exposure to the Dutch famine. *American Journal of Clinical Nutrition, 84*(2), 322–327.
Painter, R., Osmond, C., Gluckman, P., Hanson, M., Phillips, D. I., & Roseboom, T. J. (2008). Transgenerational effects of prenatal exposure to the Dutch famine on neonatal adiposity and health in later life. *BJOG, 115*(10), 1243–1249.
Rassoulzadegan, M. (2011). An evolutionary role for RNA-mediated epigenetic variation? In S. B. Gissis & E. Jablonka (Eds.), *Transformations of Lamarckism: From subtle fluids to molecular biology* (pp. 227–236). MIT Press.
Rodgers, A. B., Morgan, C. P., Leu, N. A., & Bale, T. L. (2015). Transgenerational epigenetic programming via sperm microRNA recapitulates effects of paternal stress. *Proceedings of the National Academy of Sciences USA, 112*(44), 13699–13704.

Roseboom, T. J., van der Meulen, J. H., Osmond, C., Barker, D. J., Ravelli, A. C., Schroeder-Tanka, J. M., van Montfrans, G. A., Michels, R. P., & Bleker, O. P. (2000). Coronary heart disease after prenatal exposure to the Dutch famine, 1944–45. *Heart*(6), *84*, 595–598.

Schacter, D. L. (2011). *Forgotten ideas, neglected pioneers: Richard Semon and the story of memory*. Routledge.

Semon, R. (1921 [1904]). *The mneme*. Allen and Unwin.

Sharma, U., Sun, F., Conine, C. C., Reichholf, B., Kukreja, S., Herzog, V. A., Ameres, S. L., & Rando, O. J. (2018). Small RNAs are trafficked from the epididymis to developing mammalian sperm. *Developmental Cell, 46*(4), 481–494.

Siklenka, K., Erkek, S., Godmann, M., Lambrot, R., McGraw, S., Lafleur, C., Cohen, T., Xia, J., Suderman, M., Hallett, M., Trasler, J., Peters, A. H., & Kimmins, S. (2015). Disruption of histone methylation in developing sperm impairs offspring health transgenerationally. *Science, 350*(6261), aab2006.

Švorcová, J. (2023). Transgenerational epigenetic inheritance of traumatic experience in mammals. *Genes, 14*, 120.

Švorcová, J. (2024). Plastic ontogenesis: Memory, closure, and habitual teleology in development. In J. Švorcová (Ed.), *Organismal agency. Biosemiotics* (Vol. 28). Springer.

Švorcová, J., Ľacková, L., & Fulínová, E. (2023). Evolution by habit: Peirce, Lamarck, and teleology in biology. *Theory in Biosciences, 142*(4), 411–422.

Swan, S. H., Kruse, R. L., Liu, F., Barr, D. B., Drobnis, E. Z., Redmon, J. B., Wang, C., Brazil, C., & Overstreet, J. W. (2003). Semen quality in relation to biomarkers of pesticide exposure. *Environmental Health Perspectives, 111*(12), 1478–1484.

Van Steenwyk, G., Roszkowski, M., Manuella, F., Franklin, T. B., & Mansuy, I. M. (2018). Transgenerational inheritance of behavioral and metabolic effects of paternal exposure to traumatic stress in early postnatal life: Evidence in the 4th generation. *Environmental Epigenetics, 4*(2), dvy023.

Veenendaal, M. V., Painter, R. C., de Rooij, S. R., Bossuyt, P. M., van der Post, J. A., Gluckman, P. D., Hanson, M. A., & Roseboom, T. J. (2013). Transgenerational effects of prenatal exposure to the 1944–45 Dutch famine. *BJOG, 120*(5), 548–554.

Waddington, C. H. (1952). Genetic assimilation of an acquired character. *Evolution, 7*(2), 118–126.

Waddington, C. H. (1956). Genetic assimilation of the bithorax phenotype. *Evolution, 10*(1), 1–13.

Weaver, I. C., Cervoni, N., Champagne, F. A., D'Alessio, A. C., Sharma, S., Seckl, J. R., Dymov, S., Szyf, M., & Meaney, M. J. (2004). Epigenetic programming by maternal behavior. *Nature Neuroscience, 7*(8), 847–854.

Weaver, I. C., Champagne, F. A., Brown, S. E., Dymov, S., Sharma, S., Meaney, M. J., & Szyf, M. (2005). Reversal of maternal programming of stress responses in adult offspring through methyl supplementation: Altering epigenetic marking later in life. *Journal of Neuroscience, 25*(47), 11045–11054.

Weible, D. (2016). *Exaptation: Towards a semiotic account of a biological phenomenon*. University of Tartu Press.

Weiss, I. C., Franklin, T. B., Vizi, S., & Mansuy, I. M. (2011). Inheritable effects of unpredictable maternal separation on behavioral responses in mice. *Frontiers in Behavioral Neuroscience, 5*, 3.

West-Eberhard, M. J. (2003). *Developmental plasticity and evolution*. Oxford University Press.

Yehuda, R., Schmeidler, J., Wainberg, M., Binder-Brynes, K., & Duvdevani, T. (1998). Vulnerability to posttraumatic stress disorder in adult offspring of holocaust survivors. *American Journal of Psychiatry, 155*(9), 1163–1171.

Chapter 17
Abiosphere and Biosphere: An Individual, a Community, a Culture

Anton Markoš

Abstract All living beings can interpret their world—and if so, human language and culture may represent but a species-specific extension of a more general principle of meaning attribution common to all life. Innerness of the internal milieu (a closure towards the outside) may be the basic presupposition required for this ability, while shared communication protocols in the biosphere form another basic precondition. It then follows that the biosphere is a network of symbolically communicating life-forms, attenuating (negotiating) their here-and-now, building their shared world. And given that being a human is the basic state of the author of this text, I draw an image of such biospheric communication based on my intimate experience as a 'languaging' being and try to highlight symbolic communication in other domains of life that do not have language at their disposal. The interpretation abilities of 'differences that make a difference', that is, both transient (olfactory, optical, etc.) and 'inscribed' (e.g. in linear polymers in the form of DNA) messages nevertheless turn out to be the basic powers of all life. This is demonstrated by some examples. I argue that we should finally abandon the divide of nature vs. culture, and treat all life as a cultural phenomenon.

Keywords Biosemiotics · Life: Memory and experience · Biosphere · Genetic script—Interpretation · Casuistics

17.1 Introduction

> But [...] the generality of men [...] could no longer conceive of any manner in which either growth itself or metamorphosis of individual [...] could be determined from within. The appearances were idols. They had no 'within.' Therefore the evolution which had produced

A. Markoš (✉)
Department of Philosophy and History of Sciences, Faculty of Science, Charles University, Prague, Czech Republic
e-mail: anton.markos@natur.cuni.cz

J. Toman et al. (eds.), *Culture Evolving*, Biosemiotics 10,
https://doi.org/10.1007/978-3-032-16124-6_17

> them could only be conceived mechanomorphically as a series of impact of idols on other idols. (Barfield, 1988, p. 63)

Processes in non-living nature, at least in principle and leaving aside quantum processes, can be described by the timeless models of physics and chemistry. With life, however, there comes a closure separating the interior of a living entity from its surroundings, and the composition, dynamics, and 'rules of operation' of this interior radically differ from its surroundings. Aside from straightforward composition, examples of this difference include various 'pre-arranged' codes (Barbieri, 2002) as well as heuristics and a 'machinification' of some processes (McShea, 2016). Most notably, though, life brings with it symbolic abilities that contribute to representation of the external world, such as memory and experience, recognition of signs, interpretation of meanings, etc. (Markoš & Švorcová, 2019; Švorcová et al., 2017). Because ever since the beginning of life, this 'interior' has never ceased to exist, there also exists mutual understanding between organisms from different developmental lines, even organisms belonging to different domains ('levels') of organisation. As Jan Sapp (2016, p. 603) put it, paraphrasing Dobzhansky, 'nothing in biology makes sense except in light of symbiosis.'

We can view the biosphere as an assembly of mutually communicating critters (*sensu* Kauffman, 2000): while in non-living nature we often can formulate 'laws' based on models, in the evolution of the biosphere one can identify only relatively short-term trends of various kinds. Examples of these 'trends' include those described by Haeckel's biogenetic law (cf. Chap. 4), stages and parallelism (cf. Chap. 3), the cumulative 'ratchet principle' (e.g., Kováč, 2015; cf. Chap. 15), and a horizontal transfer of various kinds of information and codes. We can identity such rules against the background of the historical time of evolution, but history itself resists the imposition of law-likeness.

My claim in this chapter is that the closure, representation, and the continuity of communities and individuals form the basis of a new mode of existence, namely life, which ever since its very beginning conforms to the theory of dual inheritance (e.g., Boyd & Richerson, 1985; Richerson & Boyd, 2005; cf. Chaps. 7, 10 and 16). In other words, communities of living creatures form cultures, among which the human one has a special status thanks to language (cf. Chap. 9). Where Jitka Lindová asks whether culture is unique to humans and what she has in mind is an expansion of the concept to encompass some other animals, I believe that culture is unique to all living—as opposed to non-living—entities. Extrapolation of the meme theory (see Chap. 10) to non-language symbolism might be in this context most helpful: it would mean that we would not need to try and estimate the evolution of human culture from the non-culture of *Homo sapiens* predecessors (see, e.g., Bradie & Bouzat, 2016).

The task of cultural anthropology and evolutionary psychology is to discover and describe the evolution of new qualities (language, communication about concepts, abstraction, the concept of time, etc.) in culture conceived of as a phenomenon *shared by all life forms*. Communities of organisms as an analogy of culture is a thread we have been following for a number of years (for some further aspects of these considerations, see, e.g., Markoš, 2016b; Markoš & Švorcová, 2019). Nevertheless, all of the abovementioned topics are, in a nutshell, expressed already in Neubauer's concept of *eidetic biology*,[1] a perspective I endorse, which forms from the start of this text the background of my further exposition. The takeaway message to take for the reader is what my colleagues and I repeat at various places: all life-forms are definitely able to interpret their being-in-the world; in other words, life is a semiotic category.

17.2 Eidetic Biology

> The specific appearance, the shape in the sense of likeness – is a claim of fellowship based on kinship: eidos – species as well as broader genera – stick together via descendence and genealogy. Such a genealogical approach is a fundamental contribution of evolutionism. Genealogy – the "family line" – is corporeal (the embodied story of the genus, its saga). The true incarnation, however, is realized in every living individual. In its own experience, in an original and unrepeatable way, it also represents the evolutionary experience of the whole line. The individual story is rooted in the experience of the lineage, and individual experience grows from these roots and contributes to the whole edifice. History is an expression of evolution in a corporeal present. (Markoš et al., 2009, p. 115)

The line of thought that Zdeněk Neubauer (1942–2016), Czech philosopher and biologist, developed throughout his active life was often misunderstood by his contemporaries. In the eyes of many, Neubauer was a mystic, vitalist, or magus, but not a 'proper' philosopher or scientist. In part, he contributed to this misinterpretation because his intellectual legacy is highly varied, and no one sees clearly into all its corners (Fig. 17.1). When I tried to find the 'first principal component' of this variety, I found it ultimately in his own words:

> …each form of life represents a world, a world of a living species in the sense of its specific way of unifying reality. Analogically, this applies to the ways of thinking, learning, understanding, and seeing. Different species are not compatible, neither logically nor physically. Therefore, they fight and compete among themselves like different forms of the will to power. Sometimes, they supplement and need one another and that is why they ecologically come together, form a community. In other words, they meet, communicate, and in one way or another find mutual understanding. This mutual understanding then evolves – and they evolve with it. (Neubauer, 2005)

I return to the abovementioned understanding later in the text. The 'form of life' can mean an individual, species, nation, a meadow, a religious community, mammals, etc. The Greek term for species is *eidos*, which is where Neubauer's conception takes its name; for the English audience (Markoš et al., 2009), we endorsed a tentative equivalent of 'likeness' or 'appearance'. Appearance is the central concept of Neubauer's teaching: it refers to the self-expression of living entities through the

[1] The writings by Z. Neubauer are mostly available only in Czech. In the following, I therefore leave out references and try to paraphrase the spirit of the message. For a good survey of his work in English, see Markoš et al. (2009), esp. Chaps. 3 and 4.

Fig. 17.1 Zdeněk Neubauer (1942-2016), Czech biologist, philosopher and thinker. Photograph by Hana Rysová. Reproduced with permission

shape of their bodies, behaviour, ornamentation, ontogenesis, or—in people—speech and culture.

A living being engages in all this based on (i) its own individual and evolutionary memory and experience, (ii) its 'indoctrination' by the community it was born into (family, population, biome, schoolmates, etc.), and naturally also (iii) depending on the context and pressures from the physical environment (for more detail, see Markoš & Švorcová, 2019; Horský & Markoš, 2026). It is important to bear in mind that living beings are not mere puppets of those forces: they interpret their situation in the world, their state space of possibilities, and within this space they opt for particular trajectories, modifications of their likeness. Different trajectories lead to different fitness. If a critter produces offspring, it hands down to them also certain aspects of its view of the world and its 'appearance'. This entire activity ultimately forms and changes the 'appearance' of the community by reciprocal formation.

The space of possibilities is best examined under conditions where natural selection is soft and can produce a high variety of appearances, that is, not in an arctic tundra but rather in a tropical jungle or coral reef, eventually an environment where some niches became vacant (such as the rift valley lakes in eastern Africa). This can be contrasted with a similar model proposed for environments with low or non-existent selection by Daniel McShea (McShea, 2016; McShea & Brandon, 2010). His zero-force evolutionary law does not assume any activity on the part of the living entities: they are merely passive objects acted upon by external forces. McShea and Brandon (2010) compare the entire process to an autumn garden with an open garage, where fallen leaves are analogical to life forms. After some time, a heap of leaves will accumulate in the garage, and this 'garage form' of leaves will prevail in the 'autumn evolution' over other forms of accumulation without developing any effort, without actively searching the state space. By now, it is quite clear that the zero-force evolutionary law is the very opposite of Neubauer's proposal.

And one further quotation on evolution, this one from Neubauer, about its specific position in our culture:

> In short, the knowledge involved in and imparted by evolutionary learning is presented in an 'eidetic format' in which reality – unlike concepts and laws – naturally reveals itself. [...] This 'format' corresponds to appearances and stories: we can form notions about them, experience them, [...], learn and integrate them. And that is, I believe, the main reason why so many people have adopted the evolutionary theory and took it for their own. (Neubauer, 2005)

For myself, I would add that living creatures, too, present their view of the world—based on their interpretation of their situation in the world as well as the interpretation of their lineage and community—in such eidetic format.

Below, I consider the circumstances that allow for the emergence of appearances. I will argue that the crucial dividing line is between prebiotic evolution, where the biosphere and symbolic communication did not exist, and biotic evolution, which assumes the existence of a biosphere characterised by interpretation efforts of both its living inhabitants and their no-longer-living ancestors (for more detail, see Markoš & Švorcová, 2019, 2020; Švorcová & Markoš, 2023; Horský & Markoš, 2026). Following this semiotic line of thought, I highlight several necessary assumptions and point out parallels between historiography and the theory of evolution: If the evolution of life is history, then evolutionary biology must be historiography, an interpretation of history.

17.3 The Individual and the Community

In the works referenced above, I pointed out the dialectic of the relationship between the individual and the community, noting that species, populations, interspecies consortia, etc., have characteristics largely analogical to what can be observed in human cultures. Naturally, I do not mean only linguistic communication, but in other respects the isomorphism is readily apparent. On the most general level, it can be found in the relation between the cell and the biosphere but we can be more concrete and look, for instance, at the relationship between an individual and a community or at various ecological relationships. Here is a list of some such isomorphisms:

1. The individual and a community are (jointly) inseparable manifestations of life.
2. Individuals and communities are *born from* self-similar entities, they do not emerge *de novo*. From this perspective, it thus makes no sense to ask about their origins. The lineages of individuals and of the community form a thesaurus of memories and experiences reaching to the most distant past, to the singularity of

the origin of life. Moreover, they can address their current situation using this capacity whose scope vastly exceeds the *current* needs of both individuals and communities (Horský & Markoš, 2026; Markoš & Švorcová, 2019).

3. An individual is born into a community (population, ecosystem, biosphere, nation, etc.), confronts its stores of memories and experiences with the current state of the world, and is 'indoctrinated' by its community in a direction that enables it to find its place in the community. But the individual also has the capacity to resist this indoctrination.
4. This 'ability to resist' makes it possible for the currently living individuals to negotiate the state of their communities and their development towards the *adjacent possible* (Kauffman, 2000), or *horizon of expectation* (Koselleck, 2004). The evolution of life thus involves a reciprocal formation of individuals and their communities, not only a mutation-selection system of propagation of autistic individuals shaped by the external environment into particular forms (as suggested by various 'trees', such as cladograms or phylograms). There is no doubt that this approach could be helpful also in linguistics or in deciphering the history, movements, or genetics of human populations (cf. Chap. 12).
5. The abovementioned negotiation is possible thanks to the fact that denizens of the biosphere have a shared origin and thus always to some extent *understand* each other: they are to some extent capable of interpreting the behaviour of others. A cell (an individual) represents a closure that is metabolically but especially symbolically (via signals and signs) connected with the community (see also Reber et al., 2023). Understanding is enabled by adherence to certain communication protocols which are, on the most general level, shared by the entire biosphere.

This list indicates that many models of cultural evolution, such as memetics, are limited in scope because they deal only with human culture. That is not to imply that they cannot help us clarify our thinking and enable its expansion into other manifestations of life. After all, our own culture is what we know best, which is why anthropomorphic analogies are helpful.

Life, that is, a community (population, biosphere, culture, etc.) of individuals (cellular closures, multicellular organisms, holobionts, ecosystems)[2] cannot be straightforwardly derived from the non-living part of the universe. Only when keeping this in mind can we ask about the beginnings of life. Let me first try to depict the previous era of 'abiosphere', which was characterised by an absence of cellular individuals, absence of closures, and therefore also by absence of the symbolic communication that characterises and constitutes the biosphere.

[2] In Horský and Markoš (2026), we classify living appearances into two categories (closures): One is cellular life, which is 'immortal' in the sense *omnis cellula e cellula*. The second category comprises cell communities and their communities (multicellular organisms, churches, ecosystems, nations, etc.) that may emerge and die. Both types of appearances—naturally in different versions—exist since the very beginnings of life.

The abiosphere featured structures that could emerge *de novo* from their elemental constitutive parts or replicate by growth and subsequent fragmentation (this can be compared with non-human cultures that do not extend beyond elements repeatedly emerging in the population; cf. Chap. 9). Let us call them replicators. I intentionally refrain from claiming that such structures 'spontaneously composed themselves' because, for instance, a molecule of water is made up of oxygen and hydrogen which bind 'spontaneously'—but it is not a replicator. It may seem I contradict myself, but I am convinced that already at this prebiotic stage there must have existed factors, structures in the environment, that canalised and catalysed this spontaneity, such as some rocks that *could* form spontaneously. Regarding replicators, we find only rough and inaccurate analogies in the current biosphere, for instance ribosomes or membranes. This analogy is necessarily inaccurate because the abovementioned structures no longer function as replicators and because the hypothetical replicators of the abiosphere were closely linked to the underlying minerals and telluric forces of the planet (which provided the abovementioned canalisation and catalysis). Their current analogues are largely independent of the planetary 'matrix' and depend instead on biotic processes and structures and on cells, which are *bona fide* replicators. If abiotic structures communicated with other similar structures at all, such communication would have been limited to the sharing of some metabolites. The birth of cellular closures and with it of the biosphere, that is, a community of replicators linked by symbolic communication and jointly negotiating advancement into the adjacent possible (Kauffman, 2000), was then the key state change that amounted to the emergence of life.

17.4 Pieces of the Puzzle of Abiosphere

In Markoš (2016a), I have proposed the notion of abiosphere as the sum of non-living but almost-living or life-like processes linked to the planet. In this place, let me therefore present just a short summary. By abiosphere, I mean neither the import of organic matter from outer space nor the accidental and unorganised organic syntheses (driven by for instance lightning or radiation) that gave rise to various kinds of hypothetical 'primordial soups'. Although such middens of organic substances certainly may have emerged, it is unlikely that this is where life emerged. Organic synthesis must have been relatively early canalised into metabolic pathways that led to only some of the millions of possible compounds and were coupled to particular energy sources, reduction equivalents, and catalysts. Metaphorically speaking, the telluric forces of the planet imposed an order on these rambunctious processes. Such canalised syntheses may have taken place, for instance, on mineral surfaces and at increased temperatures. Günter Wächtershäuser (e.g., 1994) situated the first emerging metabolic pathways on the surface of the freshly emerging pyrite. The teams of Michael Russell and William Martin (e.g., Martin, 2012; Russell et al., 2015) presented what is in my view the best argued theory of metabolism on the interface of oceanic water and (not excessively) hot vents on the ocean floor. And finally, Graham Cairns-Smith (1982) suggested the origin of replicators on clay crystals. From such pathways, evolution may have led to more complicated wholes that resembled current membranes, ribosomes, virions, initial stages of the hypothetical 'RNA world' (a biosphere based on RNA, i.e., without DNA and proteins), and the like.

The planet may have been full of such syntheses on mineral 'chips', but the individual pathways did not mutually communicate. They were not alive. The critical step came with the emancipation of such structures from their mineral foundation and their association into closures, such as cells. They formed a biosphere of cells and, with it, symbolic cellular communication within the community of the biosphere. To repeat: in an abiosphere, structures can emerge *ex nihilo*, by composition or adding of elemental substances based on physical and chemical affinities. There is no need to assume memory in the sense of a thesaurus accessible for context-based evaluation and interpretation based on experience (although memory as a 'database' for algorithms may have existed). Abiosphere required neither an internal representation of the world nor any symbolic communication among different structures.

17.5 Experience and Memory: The Biosphere

As noted in the introduction, one cannot consider the evolution of life based just on the perspective of an individual that is born and dies. One must also take into account the lineage of organisms in which the individual is one of the endpoints and a community of individuals that—like the lineage of organisms—persists over a long time. These characteristics may suggest analogies with communities of people and human cultures, which I explore in more detail below. Such analogies are helpful, although one must constantly 'abstract away' from spoken language unique to humans (which is, ironically, a medium especially well-suited to demonstrating the point).

17.5.1 Exome: A Difference That Makes the Difference

Since the sense of freedom often appeared at its strongest in imagination which were most possessed with the mechanical view of the universe, the paradox is not infrequent – especially in Germany – of philosophers and scientists insisting fiercely on the freedom of thought and using it to deny the possibility of any freedom at all! (Barfield, 1967, p. 199, on the intellectual climate of the nineteenth century)

The DNA in a cell is a linear chain that can be, in some species, up to hundreds of millions bases long. There are four kinds of bases and their sequence is non-random: it carries information. A human being inherits from each parent such DNA strings of about three billion bases, which jointly form a genome. The chain is semantically segmented by a large number of labels, short 'canonical' strings recognised by members of the cell's 'ecosystem', especially thousands of various proteins and RNA. What is recognised are regulatory sequences, including for instance the beginnings and ends of genes. (Activity of those proteins and RNAs adds and changes some 'diacritical marks' but let us leave that aside for the moment.) By 'gene' I mean a segment of DNA that has been recognised as contextually suitable for transcription into RNA, whereby the context is determined by the set, or 'ecosystem', of proteins and structures present both in the cell and outside it. Humans have about 20,000 genes from each parent. They are about 20,000 to 100,000 bases long, which means that they jointly occupy less than one-tenth of the overall length of DNA. The RNA molecules which carry information necessary for protein synthesis in the translation process are usually further modified: they are processed so that long segments of introns are removed and the remaining segments, *exons*, are connected into a much shorter mRNA molecule, which is then translated into the polypeptide chain of the protein.

We can see that identical transcripts can, as a result of alternative processing, lead to numerous mRNAs, often dozens of them, and therefore also to many proteins—once again depending on the need and on the abovementioned community of proteins present at a given place at a particular time. Newly synthesised proteins enter the network of relations within this community, thus changing the character of the entire network and its functioning. Among other things, within the context of other environmental factors and previous development they also influence decisions about which genes will be transcribed into RNA in the next moment, which mRNA will be produced, what amount of which protein will be produced, and how the existing constituents of the networks would be altered (by removal, change of activity, or chemical modification). This entire cycle never starts *de novo*: it has been in place without any interruption since the birth of life about four billion years ago. This brings us to some examples.

Example 1 In August 2016, there appeared an article authored by several dozen authors (Lek et al., 2016; cf. also commentary by Shendure, 2016), which takes a look at a comparative analysis of exomes[3] from 60,706 human individuals from various parts of the world. The analysis revealed a natural variability of transcription of such segments (all in all, the team had identified over 91,000 of them), their geographical distribution, and their links to certain physical traits as well as certain associated hereditary or congenital diseases (although for ethical reasons, these links were not reported). Aside from such switches, the scientists also described various insertions, deletions, and segments with STOP codons that resulted in shorter variants of the resulting proteins.

In this way, the study had reliably described over seven million variants, whereby over 99% of them were found at a frequency of less than 1% and over one-half were found only in a single instance. Some of the variants are well known and have been linked to various metabolic, developmental, mental, or oncological diseases and disorders. For a vast majority of these variants, though, it is unknown whether they are in any way linked to the phenotype (or appearance) and if so, what impact they have. It is likely they are analogical to typographical errors, dialect variants, or

[3] Exome: a sum total of all exons in the genome. Exons: such parts of the gene that may be transcribed into RNA and translated into proteins.

subtle changes of meaning. A large part of these mutations probably does not have any impact on their bearer. Other such mutations may have some impact only under special conditions, for instance via links to other proteins coming from the genetic heritage of the other parent or during some extremes of the environment. Yet other variations may lead to fundamental changes in the structure and function of a protein, but their impact will be compensated by other members of the protein ecosystem. Finally, some mutations do 'shoot up' all the way to the level of the phenotype, usually—but not always—with a deleterious effect. This study could thus help us identify links to some very rare congenital disorders, unusual sport or art abilities, or predispositions to some disorders in a particular context (such as some forms of obesity).

Example 2 Erika Hayden's (2016) short essay introduces us to research on mutations of the PRNP prion protein, whose alternative form can in some instances underlie the Alzheimer disease. The mutations need not take place in the chain of amino acids, because prions have a tendency to change their shape even without them, but in this instance let us focus on the mutations because they can be described at the level of genes and highlight an increased risk. A disease due to a gene mutation affects about two people per million each year. In the analysed sample of 61,000 individuals, one could thus expect about two such mutations. Surprisingly, Hayden actually found 52, which shows that most mutations are in effect 'typos' which rarely or never cause a disease. In a sample of this small size, she found no mutation believed to always lead to a disease. Hayden's study also has a deep ethical dimension: at this point, though, let me just point out that we talk here about diseases caused by a malformation of a single protein. Most congenital diseases are, however, polygenic, meaning that often dozens of various proteins participate in the formation of the observed trait and mutation of each of these proteins, eventually in concert with other mutations, can affect the aetiology of the disorder or disease. It is also known that cells have at their disposal effective means for buffering the effects of mutations, including processes that 'enforce' a correct conformation upon the mutated protein (see, e.g., Rutherford & Lindquist, 1998).

This example opens new vistas into the field of dual heredity where 'genes go along with the memes'. On the one hand, awareness of what might happen to an individual and his or her offspring within a timeframe of even decades will change the individual's own behaviour and the behaviour of the community. On the other hand, with the help of culture, a population can accumulate mutations and epimutations that had no chance to spread in former times, such as hereditary infertility (which can now be tackled by artificial insemination), inability to give birth naturally (which can spread via frequent use of Caesarean section, leading to a toleration of the narrow pelvis phenotype in women), or even the utterly unforeseeable effects of transplantation of the nucleus of a zygote into a foreign egg devoid of its own nucleus.

Example 3 This is actually not a case interpretation but rather a short overview. Pennisi (2016) points out that interspecies hybrids of animals are much more frequent than previously thought (although in plants, this phenomenon has been known for a long time). Hybridisation takes place either repeatedly or rarely, but it does happen and what takes place at the level of genes and proteins is a 'confusion of languages' whose effects are similar to people communicating in related dialects or even in pidgin. In either case, only successful combinations survive, that is, those which either quite naturally or by a sheer chance enabled the survival and reproduction of the hybrid (cf. the case of clawed frogs below).

17.5.2 The Proteome

Based on the previous examples (genome, exome), the reader has certainly guessed that proteome refers to the set of all proteins in the cell. But in what cell and when? While genomes and exomes are identical in almost all cells of the human body (because they are in effect records), the proteome is the outcome of selection from such recorded genetic thesaurus. It thus constantly changes depending on the type of cell and even in individual cells. A cell, or rather its structures that include the proteome, constantly decides about its changes based on its nature, its 'appearance'. Almost every type of cell activates a different set of genes from the shared thesaurus, mutual ratios of various proteins (or their variants caused by alternative splicing of RNA) can vary, and proteins can be part of many different structures. On top of that, there come the 'diacritical marks' acquired post-translationally, epigenetically (see below).

Before proceeding further with our language analogy, a word of warning is in order: a written word on a page or screen is in effect virtual, it does not have a physical shape. Therefore, as long as we make sure that the individual letters can be differentiated, the text can be written in various fonts, in the Morse code, or even by pre-agreed emoticons. What has an 'appearance' is the message. A text that is read also assumes a likeness in the mind or the behaviour of the reader: the written text as such is just an instrument, a tool. We are used to thinking about biological macromolecules as if they were strings of letters but, unlike letters, they are not virtual. They have their own shapes, physical form, and they usually cannot be realised by any altered 'font'.

In a written text, diacritical marks help the speaker or reader decipher the intended form of the speech but in biological macromolecules, the chemical modifications ('diacritics') themselves contribute to a change in conformations (shapes). They are themselves a form of 'speech'. The analogy with written text helps us to imagine these matters but we must bear in mind that it is just an analogy. From the perspective of eidetic biology, to record sequences in the form of letters is enormously reductive—as is, after all, written text in relation to speech (see, e.g., Harris, 2009).

Example 4 The case of yeast vs. humans brings our attention to the necessity of interpreting genetic script in its context, without paying attention to its exact 'spelling'. Kachroo et al. (2015) have replaced over 400 *vitally necessary* genes of yeast by their orthologues (homologues) from humans. In half of the cases, they managed to 'save' the yeast: the surrogate human gene functioned despite the vast length of time over which the two lineages developed separately. The sequences of amino acids in orthologous proteins were different… except for some modules ('word roots') thanks to which even rather foreign proteins could, at least in laboratory conditions, maintain the context of correct cell functioning. One can find similar examples also for homeosis (see below) or for proteins that trigger the formation of the eyes, the limbs, etc.

17.5.3 Homeosis

Homeosis refers to the formation of a body structure in the wrong (ectopic) place of the body. Typical examples include the 'full' blossoms of roses or peonies, where petals form instead of stamens, and flies with legs instead of antennae or with a second pair of wings on an incorrect body segment.

Owen Barfield in his *Poetic diction* (1973) used the example of Latin word *ruo*, whose original meaning was akin to both 'run quickly' and 'fall', to demonstrate how the spirit of this hard-to-translate verb over centuries gave rise to further senses (e.g., 'ruin') while retaining something from its original core meaning and related meanings, such as 'throw to the ground', 'hurry', or 'rush'. What is crucial about it is the generalisation pertaining to the 'soul' of language:

> It will, I think, appear that this 'soul', latent in words, and waiting only to be discovered, is for the most part a kind of buried survival of the old 'given' meaning under later accretions; or, if not of the 'given' meaning itself, then of an old 'created' meaning which has been buried in the same way. For created meanings, once published, are as much subject, of the binding, astringent action of the rational principle as the original given meanings. (Barfield, 1973, p. 115)

Due to the force of custom, words over time lose their animation and petrify, but when 'touched' by the context of other expressions, they can spring to life and engage in new experiments. This 'touch of life' can be imparted by the words and contexts of the constantly evolving language, thanks to which old words and phrases come to be understood differently (for more details, cf. also Barfield, 1967). This can be also mediated by poets and their linguistic experimentation, which can reveal a dormant sense and recruit it, in a given context, into the linguistic corpus.

In connection with biological context, let me quote one more very important sentence: 'The new meaning must be *strange*, not incomprehensible; otherwise the poetry of the whole passage is killed, and the fresh meaning itself will be still-born' (Barfield, 1973, p. 105). Metaphorical, novel, and yet it must carry at least a small remnant of the original meaning because 'no matter how many times it has been carelessly handled for the purposes of false and facile romanticism, the old magic will always be ready to flash out to a touch of true imagination' (Barfield, 1973, p. 108). A word is thus 'waiting, like all the other words in the dictionary, to be kindled into life by a living one; and nothing more is necessary than to surround it with other words (the right ones) from the same museum' (Barfield, 1973, p. 125)—and perhaps this applies also to being surrounded by some completely new words, which give rise to new contexts, ideas, or forms. Let us compare: the poet's relationship to language is one of a creator, while a logician tries to reduce language, as Barfield says in his remark on words:

> The *use* of them is left to the Logician, who, in his endeavour to keep them steady and thus fit them to his laws, is continually seeking to *reduce* their meaning. I say seeking to do so because logic is essentially a compromise. He could only evolve a language, whose propositions would *really* obey the laws of thought by eliminating meaning altogether. But he compromises before this zero-point is reached. (Barfield, 1973, p. 131)

But such reduction is dangerous:

> Of all devices for dragooning the human spirit, the least clumsy is to procure its abortion in the womb of language; and we should recognize, I think, that those and their number is increasing who are driven by an impulse to reduce the specifically human to a mechanical or animal regularity, will continue to be increasingly irritated by the nature of the mother tongue and make it their point of attack. [...] If therefore they succeed in expunging from language all the substance of its past, in which it is naturally so rich, and finally converting it into the species of algebra that is best adapted to the uses of indoctrination and empirical science, a long and important step forward will have been taken in the selfless cause of the liquidation of the human spirit. (Barfield, 1973, p. 23).

Henceforth, let us imagine the genetic record and how it becomes 'embodied' in different forms.

Still, what kind of 'words' are we talking about here? Written or spoken ones? I am afraid that we usually mean the reduction of speech to written language, to which we can ascribe more or less clear orthographic and grammatical rules. In other words, create norms, which are a stepping stone to attempts at a logical description of language and the creation of an ideal language (Eco, 1997) or machine languages. Harris (2009) believes that Aristotle, the founder of formal logic, had hardly any predecessors simply because only at his time people started to consider the sequence of letters: spoken language has a different logic. For example, one cannot ask an illiterate person about the initial letter of their name. Benoît de Maillet, a well-known eighteenth-century thinker, used as his pen name 'Telliamed', his name in reverse, but it could only be formed based on the spelling of his name because his name was pronounced [dema.jɛ]. Christian Nicolas (2019) shows how over the ages and in different communities, the apparently immutable 'Ciceronian' Latin transformed and mutated: it functioned as a *lingua franca* in the written form while oral communication between people from different parts of Europe became almost impossible. (We are witnessing something similar with English, except that it is a living language.) In newly forming nations, diacritical marks then added a new dimension even to the written form of languages.

As linguists from the so-called Distributed Language Group (the abovementioned Roy Harris, but also Stephen Cowley or Nigel Love) point out, expressing oneself is a matter of the entire body. From a written and thus reduced form one must—unless it is logic or a machine language—reconstruct a large part of the message and, what is more, also construct, newly interpret its meaning (for more on this, see Markoš, 2002). Only after that is done, speech is situated within the 'touch' of contexts, words set in their meaning spaces, becoming part of the 'likeness' of the whole. And that applies not only to language but to all activities rooted in history: for instance, in historiography a scarcity of sources can, when confronted with the variety of experience, lead to the discovery of unforeseen connections, to new and faithful explanations of events recent or long past, as Paul Veyne (1984) points out.

This interplay can be extremely interesting but historiographers and we, biologists, deal with these contexts mainly because we seek answers to how we found ourselves in the present situation. Poets, historiographers, or visionaries (prophets, politicians) reveal various versions of stories and investigate to what extent, in the given context, they correspond to reality. Sometimes, they correspond to it only fleetingly, at other times they open new spaces of possibilities, triggering the next chapter of history (e.g., Christianity). Let us try to accept the idea that not only evolutionary biologists but all living entities are 'historiographers', 'poets', 'seekers', critters that search for answers to challenges thrown at them by the environment into which they were born and confront these answers with various forms of their own evolutionary memory and experience (for further details, see Horský & Markoš, 2026). To return to Neubauer: the result is an appearance that represents, compared to a script, a more general form of expression. Speech is just one of its forms.

If we are to take the message of biosemiotics seriously, we must consider the non-speech signs and their meanings, the ancient symbols which are in the course of evolution taken up and re-created, accidentally or purposefully, over and over again. With thus transformed signs (which, however, mediate completely new meanings), evolution enters a new 'state space'. 'And yet it remembers something!', we can declare with Barfield—and *pace* Galilei. Critters use ancient genetic instruments to 'touch' and probe whatever they can, to figure out what the space of possibilities allows to those who can use their particular instruments.

In animals, such instruments include, for instance, a set of regulatory proteins known as Hox or homeotic proteins, which assist in building the anteroposterior axis of all animals, including echinoderms. They function as switches, commands that trigger the morphogenesis of various body parts. For instance, in insect embryos a hypothetical product of the HoxA gene triggers, in the first thoracic segment, functions that lead (in insects) among other things to the formation of the first pair of legs. At the same time, it informs the neighbouring segment behind it that it is the second thoracic segment. Thanks to this information, for instance genes HoxA and HoxB become activated in this neighbouring segment, and the ensuing proteins direct morphogenesis towards the formation of a second pair of legs and the first pair of wings, while also sending information to the next segment about its place in the body. Thanks to this message, HoxC gene is triggered in that segment. About a dozen genes from this set, often in combination, drive the rough outlines of segmentation of the insect body (except for the head). Triggering of a gene ('the use of a word') outside its proper context, that is, in an ectopic position, leads to disruptions, to body parts in wrong places. If some gene from the set is missing, an entire body part can be missing—and in such cases, the development usually does not progress past early embryogenesis. Homeotic proteins are the 'product' of the set of homeotic genes and some of their exons are so conservative that one can, using genetic techniques, substitute them between various animal groups without causing any disruption in morphogenesis (i.e., to 'save' the mutant, as shown also in Example 4 above). Let us for the moment stick with the analogy with ancient word roots such as we saw in Barfield. I do not, however, claim that this is the same!

In further stages of morphogenesis, in other contexts, the same set of tools can trigger completely different functions than the definition of anteroposterior axis, functions that evolved in the course of evolution. In quadrupeds, for instance, these genes are responsible for determining the order of fingers and limb parts (again with a possibility of errors, i.e., homeotic mutations). A good introduction to the role of such triggers both in ontogenesis and in evolution (because homeotic proteins are just one of many such sets of instruments) is found in the work of Sean Carrol (2006).

I would also like to draw attention to the dialectic of strict conservatism and playful creativity: the same sets of genes (and not only homeotic ones) which for hundreds of millions of years strictly adhered to the rules of early morphogenesis can also trigger a wide variety of incredible, almost baroque, manifestations in later morphogenesis; this is especially apparent in the ornamentation of butterfly wings. Like Barfield's poet, we too can observe in different lineages various cases of creative adoption of ancient (but comprehensible) motifs and their transformation into new manifestations. (One can also find the opposite trend, towards atavisms, where ancient and long unused traits are resurrected—for instance wings in stick insects, a tail in humans, three-fingered limb in a horse, or teeth in a chicken.)

Example 5 Let us move away from the homeotic context—textbooks are already full of it—and take a look at a case presented by Session et al. (2016), who uncovered the genome sequence (and its history) of African clawed frog (*Xenopus laevis*). About 34 million years ago, two species of genus *Xenopus* split and for the next 17 million years evolved separately: their genome was changing along with their genes, genetic map, etc. Both species were diploid, meaning they carried two sets of chromosomes, each from one parent. About 17 million years ago, they became 'bastardised', that is, individuals from these two related species have mixed via interspecies breeding. We would not know about it had it not been for the fact the resulting alloploid, which should otherwise have been for many reasons infertile, had doubled its genomic equipment, becoming an allotetraploid capable of reproduction—and the result of further evolution is the current African clawed frog. It is no longer a tetraploid: over time, some genes (paralogues) from one or the other parent species became switched off and transformed into nonfunctional pseudogenes. In other instances, the paralogues shared their function, each taking on just part of its initial role, or they became expressed in different tissues. Finally, in some cases one of the paralogues assumed a completely new function. This opens a variety of questions and issues so vast it could be compared even to the study of Biblical texts.

Example 6 The recruitment of one of the Hox genes for the morphogenesis of fins in ancient fish had introduced into the process new rules that eventually led to five-fingered limbs (Khjerdemil et al., 2016). To put it poetically:

> Each life form has its own unique past, belongs to a story whose plot is merely more or less shared with other forms. This sharing is reflected in the similarity of appearances, in similarities that are essential, part of the nature of beings. This is the core of the descent theory, a perspective from which the origin is the reason why. (Neubauer, 2005)

17.6 Conclusions

In the first part of this chapter, I tried to shed some light on the prebiotic processes which led to the establishment of genetic, metabolic, and structural pathways that may have taken place on rock substrates. Such processes later became emancipated from inorganic surfaces and launched a new dimension of nature—life. Essentially, life exists in two easily recognisable forms, cells and their communities, which interpret their world and take care of themselves, of their appearance.

In the twentieth century, biology increasingly tended to view life as a programmed entity and due to this emphasis, the appearances of life came to be interpreted as mere puppets dependent on such programmes. This development led to an increasingly sharp contrasting of the nature vs. culture paradox (illustrated in many contributions to this volume). In Neubauer's words, we have fallen for the 'linear biology': when it turned out that biological macromolecules can be divided in discrete units, we started replacing them—for simplicity—by letters and came to think of them in terms of letter-like digital entities. Upon closer consideration, we do realise that such digitality does not reflect accurately the natural world, because the shapes of molecules are so changeable that even just telling the discrete units apart requires considerable effort. That is why we, as organisms, undertake this task only when necessary, only when the accuracy of 'transliteration' matters, for instance during replication of nucleic acids and during transcription. In some situations, the level of accuracy can even be predefined. Everything else is a matter of shapes and their changes in various contexts (Markoš et al., 2009). The DNA itself is a massive molecule whose external shape is co-determined by the sequence of monomers but depends also on the bending, torsion, and location of epigenetic modifications ('diacritics')—and, especially, on the composition of the surrounding environment (Markoš & Švorcová, 2019). On the level of proteins, cells, bodies, and communities, the importance of bodily attributes and their spontaneous variations come ever more to the fore.

The main message of this chapter is that culture is inherently shared by all life forms, all living communities; they all play the game of life according to similar rules but in various 'wordings' which depend on momentary contexts, species-dependent habits, and historical contingencies. That is why I like to use analogies with natural languages: just as in languages, appearances of the living definitely do not represent an 'execution' of programs written in a machine script. In this game, which has been ongoing since the beginning of life, there appeared symbolic communication, thanks to which life recognises itself and can be recognised. There also emerged memory and experience thanks to which history can be retold in a way that can be understood here and now. Life is a semiotic category, and all the ways of appearance represent forms of expression of an individual or a community. Some examples are given throughout the text. The takeaway message from this way of viewing the world is a realisation that we ought to discard the antiquated, endlessly repeated, division between nature and culture. Let us be unitarian. Let us equate appearance with culture.

References

Barbieri, M. (2002). *The organic codes. Introduction to semantic biology*. Cambridge University Press.

Barfield, O. (1967). *History in English words*. Lindisfarne Books. (Original work published 1953).

Barfield, O. (1973). *Poetic diction. A study of meaning*. Wesleyan University Press. (Original work published 1928).

Barfield, O. (1988). *Saving the appearances. A study in idolatry*. Wesleyan University Press. (Original work published 1957).

Boyd, R., & Richerson, P. J. (1985). *Culture and the evolutionary process*. University of Chicago Press.

Bradie, M., & Bouzat, J. L. (2016). Patterns and processes in cultural evolution. *Evolutionary Biology, 43*(4), 516–530.

Cairns-Smith, A. G. (1982). *Genetic takeover and the mineral origins of life*. Cambridge University Press.

Carroll, S. B. (2006). *Endless forms most beautiful. The new science of evo-devo*. W. W. Norton & Co.

Eco, U. (1997). *The search for the perfect language*. Fontana Press.

Harris, R. (2009). *Rationality and the literate mind*. Routledge.

Hayden, E. C. (2016). Seeing deadly mutations in a new light. *Nature, 538*(7624), 154–157.

Markoš, A., & Horský, J. (2026). *The seeds of forms and the forms of seeds*. Springer.

Kachroo, A. H., Laurent, J. M., Yellman, C. M., Meyer, A. G., Wilke, C. O., & Marcotte, E. M. (2015). Systematic humanization of yeast genes reveals conserved functions and genetic modularity. *Science, 348*(6237), 921–925.

Kauffman, S. (2000). *Investigations*. Oxford University Press.

Kherdjemil, Y., Lalonde, R. L., Sheth, R., Dumouchel, A., de Martino, G., Pineault, K. M., ... & Kmita, M. (2016). Evolution of Hoxa11 regulation in vertebrates is linked to the pentadactyle state. *Nature, 239*(7627), 89–92.

Koselleck, R. (2004). *Futures past: On the semantics of historical time*. Columbia University Press. (Original work published 1979).

Kováč, L. (2015). *Closing human evolution. Life in the ultimate age*. Springer.

Lek, M., Karczewski, K. J., Minikel, E. V., Samocha, K. E., Banks, E., Fennell, T.,... & Exome Aggregation Consortium. (2016). Analysis of protein-coding genetic variation in 60,706 humans. *Nature, 536*(7616), 285–291.

Markoš, A. (2002). *Readers of the book of life*. Oxford University Press.

Markoš, A. (2016a). *Evoluční tápání [Evolutionary groping]*. Pavel Mervart.

Markoš, A. (2016b). The birth and life of species-cultures. *Biosemiotics, 9*(1), 73–84.

Markoš, A., & Švorcová, J. (2019). *Epigenetic processes and the evolution of life*. CRC Press.

Markoš, A., & Švorcová, J. (2020). Cutting the Porphyrian tree: Objective reality as created by the innerness of living beings. In Ľ. Lacková et al. (Eds.), *Gatherings in biosemiotics XX* (pp. 24–36). University of Tartu Press.

Markoš, A., Grygar, F., Hajnal, L., Kleisner, K., Kratochvíl, Z., & Neubauer, Z. (2009). *Life as its own designer. Darwin's origin and Western thought*. Springer.

Martin, W. F. (2012). Hydrogen, metals, bifurcating electrons, and proton gradients: The early evolution of biological energy conservation. *FEBS Letters, 586*(5), 485–493.

McShea, D. W. (2016). Three trends in the history of life: An evolutionary syndrome. *Evolutionary Biology, 43*(4), 531–542.

McShea, D. W., & Brandon, R. N. (2010). *Biology's first law*. University of Chicago Press.

Neubauer, Z. (2005). Spor o Darwina jako symbol. [The strait for Darwin as a symbol]. *Prostor, 65*, 115–132.

Nicolas, C. (2019). Que restitue-t-on dans la prononciation dite restituée du latin? In M. Bastin-Hammou, F. Fonio, & P. Paré-Rey (Eds.). *Fabula agitur. Pratiques théâtrales, oralisation et didactique des langues et cultures de l'Antiquité* (pp. 173–191). UGA Editions.

Pennisi, E. (2016). Shaking up the tree of life. *Science, 354*(6314), 817–821.

Reber, A. S., Baluška, F., & Miller, W. (2023). *The sentient cell: The cellular foundations of consciousness*. Oxford University Press.

Richerson, P. J., & Boyd, R. (2005). *Not by genes alone: How culture transformed human evolution*. University of Chicago Press.

Russell, M. J., Nitschke, W., & Branscomb, E. (2015). The inevitable journey to being. *Philosophical Transactions of the Royal Society B, 368*(1622), 20120254.

Rutherford, S., & Lindquist, S. (1998). Hsp90 as a capacitor for morphological evolution. *Nature, 396*(6709), 336–342.

Sapp, J. (2016). The symbiotic self. *Evolutionary Biology, 43*(4), 596–603.

Session, A. M., Uno, Y., Kwon, T., Chapman, J. A., Toyoda, A., Takahashi, S.,... & Rokhsar, D. S. (2016). Genome evolution in the allotetraploid frog Xenopus laevis. *Nature, 538*(7625), 336–343.

Shendure, J. (2016). A deep dive into genetic variation. *Nature, 536*(7616), 277–278.

Švorcová, J., & Markoš, A. (2023). Closures as a precondition of life, agency, and semiosis. *Biosemiotics, 16*(1), 45–59.

Švorcová, J., Markoš, A., & Das, P. (2017). Origins of the cellular biosphere. In V. P. Sahi & F. Baluška (Eds.), *Concepts in cell biology—History and evolution* (pp. 271–290). Springer.

Veyne, P. (1984). *Writing history: Essay on epistemology*. Wesleyan University Press.

Wächtershäuser, G. (1994). Life in a ligand sphere. *Proceedings of the National Academy of Sciences USA, 91*(10), 4283–4287.

Chapter 18
Reflections on Analogies Between Cultural and Biological Evolution: Contribution to the Discussion from the Perspective of Historical Sciences

Jan Horský

Abstract This chapter represents a reflective closure of the whole book. It discusses possible analogies between evolutionary and historiographic thinking and methodology, focusing on analogies between the evolution of biological species and human cultures, eventually cultural complexes. Such analogies can be found both on the noetic and the ontological level, bringing directions of interpretation into the chaos of the source materials. The contribution also draws inspiration from possible concordances between biosemiotics and semiotic theories of culture, where the former claims that the meaning of signs is revealed on the phenotypical level of physical manifestations and the latter understands the meaning field as an objectified transcendental subject of knowledge (a spirit, to use Cassirer's terminology). In conclusion, the author asks whether it is possible to study the potential manifestations of somatisation of cultural development, considering it either at the genetic or epigenetic level, or at the level of cultural imprinting during postnatal ontogenesis.

Keywords Cultural historiography · Cultural evolution · Semiosis · Somatisation of cultural development · Nominalism

18.1 Introduction

Previous chapters of this book dealt, among other things, with three basic questions. First of all, to what extent can the principles of biological evolution be applied to explaining the development of culture, be it in the sense of its *evolution* or in the sense of *cultural diffusion*? (This question was reflected in Chaps. 3, 5, 7, 9, 10, 11,

---

J. Horský (✉)
Faculty of Humanities, Charles University, Prague, Czech Republic
e-mail: jan.horsky@fhs.cuni.cz

J. Toman et al. (eds.), *Culture Evolving*, Biosemiotics 10,
https://doi.org/10.1007/978-3-032-16124-6_18

12, and 13). Secondly, when we find analogies between the biological and the cultural evolution, are they analogies on an ontic level or are they perhaps rather just cases of an 'optical illusion' due to the similarity of methods used by evolutionary biologists and historians of culture? (This is addressed in Chap. 14.). And finally, to what extent can culture influence biological heredity (in terms of co-evolution of genes and culture, eventually epigenetic heredity)? (This was the subject of Chaps. 7, 10, 11, and 16) In the following, I address some aspects of these questions from the perspective of (the theory of) historical sciences.

Anton Markoš, in his contribution to this book (Chap. 17) but also elsewhere (Markoš, 2016a, p. 261), says that both historiography and evolutionary biology are explanatory efforts aimed at illuminating the past. Therefore, they approach their sources and try to explain events in similar ways. It thus makes sense to compare some research methods of these fields, especially those relevant to what they say about cultural evolution and its analogy with biological evolution. Thinking about this analogy has received a boost from biosemiotics, which—aside from certain commonly accepted defining features assigned to it by biology—ascribes to life also the ability of (interpretative) *semiosis* (Deely, 2008, pp. 95–192; in the sense of straightforward biosemantics, see Barbieri, 2002), eventually even narrativity (Markoš, 2012, 2015a, b, 2016a).

For our present purposes, let us assume that historians and (semiotic) biologists agree that developmental processes cannot be described purely theoretically and an explanation of development or evolution is also a 'history' in the sense of containing a significant narrative element. Some historians and theorists of historiography moreover agree with biologists who hold these beliefs on the following: not only the *theory* of developmental processes but also the *narration* (narrative historical interpretation) can be *verified*. In other words, one can not only test theories but also distinguish between 'true narratives' and fiction. In exploring these subjects, I will focus on the concept of culture that is common in historiography and discuss categories such as (source) traces, (evolutionary or developmental) lineages and lines, narrativity, the issue of nominalism and essentialism in historiographical and biosemiotic thought, 'disenchantment of the world', and some concrete examples.

In this chapter, I argue that there is no difference in kind, only in degree, between the 'lineages' identified by historians of culture and those delineated by evolutionary biologists in terms of their ontological status. I focus on evaluating the appropriateness of various approaches to drawing analogies between the evolution of biological species and the development of cultures, specifically certain sets of 'cultural contents'. Then I assess the possibility of distinguishing between the plastic and elastic phases in the evolution and development of biological species and of 'human' cultures. This discussion opens new avenues for addressing the extent to which biological and cultural evolution may overlap.

18.2 Culture: Its Conceptualisations and Aims of Studying Its Development

In current historiography, scholars often work with a broad concept of culture (Daniel, 1993, p. 77; Dressel, 1996, pp. 167–188; Habermas & Minkmar, 1992, pp. 7–19; Lorenz, 1997, pp. 127–188) that brings historiography closer to sociology or sociocultural anthropology (cf. Chaps. 5 and 7). According to this conception, culture is understood as a semiotic system of symbols, values, ideas, concepts, and behaviours. According to most conceptions coming from cultural anthropology, what defines culture is that it is transferred not by genes but by social learning (Geertz, 1973; Richerson & Boyd, 2005; cf. Chap. 9).[1] This approach is in principle compatible also with for instance Dan Sperber's conception, according to which culture is a complex network of causal chains of public and mental representations (Sperber, 1996; cf. Chap. 11).

Definitions of culture coming from historical sciences are much broader than those needed when investigating culture in non-human species. For instance, Laland and Hoppitt (cf. Chap. 9) define culture as group-typical behaviour patterns shared by members of a community. These behaviour patterns are based on socially learned and socially transmitted information. Nevertheless, as Jitka Lindová shows (Chap. 9), in concrete cases it is often difficult to decide whether a behavioural pattern really spreads via social learning or is caused by for instance some specific environmental conditions. In relation to defining culture along these lines, some conceptions used by the historical sciences can lead to thinking about 'layers of culture' (including some basic ones, shared between *Homo sapiens* and other species) or about a varying intensity of culture (different abilities of reflexivity, varying amount of polysemicity; cf. below).[2]

The directions in historiography of culture I wish to consider here do not aim primarily at formulating a theory of culture that would enable the investigation, description, and explanation of particular cultural manifestations in the same way Darwin's theory of natural selection does. Such approaches inherently exclude the intentionality of actors aimed at explaining various biological phenomena (cf. for instance Chap. 11). The historiography of culture present in the background of my contribution is rather interested in understanding internal cultural contents (meanings) and their evolutionary change (Lorenz, 1997, pp. 98–125). In doing so, it links (comparative) explanations with interpretative hermeneutics. Using the distinction between the selectionist and the attractionist approach, one can thus conclude that

---

[1] Still, from a biological perspective, culture also has its stores of memory and experience, which can be easily copied and spread.

[2] The emphasis on social learning as the means of spread of culture sometimes (in biological conceptions and in evolutionary anthropology) leads to an identification of 'cultural' and 'social' phenomena. Historiographical conception of culture—but also many sociological theories—nevertheless tends to distinguish between the 'cultural' and the 'social', at least in part. Conceptual differentiation between culture and society, and perhaps also between population and civilisation, can offer finer analytical tools to describe cultural evolution and its factors.

explanations of cultural development based on historical sciences are decidedly closer to the attractionist perspective.

18.3 Traces

In the Ango-American tradition, historiography is often viewed as merely a kind of literature. In the French or German tradition, on the other hand, it is automatically assumed that within the broad field of historiography, there also exist historical sciences. Texts from the German environment commonly distinguish between *Geschichtswissenschaft* (historical science) and *Geschichtsschreibung* (historiography). Historical sciences partly overlap with historiography, and especially where historical sciences deal with long-term development (cultural but also social or civilisational), their methods are similar to those used by evolutionary biology. They approach historical sources not as primary witnesses but rather as indicators or traces (Vašíček, 1994) that point to *immediately inaccessible realities*. If those immediately inaccessible realities belong to the past, then—as Carlo Ginzburg says with reference to Thomas Huxley—historiography, archaeology, geology, astrophysics, or paleoethology apply the same method as Charles Darwin did, namely the method of retrospective prediction (Ginzburg, 2002, p. 38). 'Every source – more exactly, every remnant (*Überrest*) that we transform into a source through our questions – refers us to a history which is either more, less, or in any case something other than the remnant itself' (Koselleck, 2004, p. 150). One should thus ask: Does cultural evolution on its own (some would say ontically, objectively, or beyond doubt) imprint itself onto certain entities as potential traces of such evolution, or do those entities *become* traces because we are trying to sort and categorise the multitude of phenomena using our narrative or evolution-theoretical interpretations (cf. Chaps. 5 and 6)? In other words, to paraphrase Ricoeur's statement: Do the traces form spoors that direct the hunt or is it the hunter who directs the hunt and therefore identifies the traces?[3]

It is also unclear to what extent our ability to identify something as a trace is determined by our participation in the same 'culture' or 'cultural tradition' as those who left the trace. If a culture is indeed a semiotic system of symbols, then the entities (texts, artefacts, generally speaking indications) that form the traces function as traces not only at a discursive or empirical level but also in the semiotic dimension. In other words, they refer to particular meanings.

Let us say that we identify in a layer, say a neolithic one, of an archaeological site, where some remnants are configured in a particular fashion, forming for instance a row of stones marking an astronomically important direction. To be able to treat them as relevant traces within our narrative or evolution-theoretical

[3] Biologists probably would say, following Koselleck and *pace* Ricoeur, that the hunter *creates* the traces. From this perspective, one could thus say that 'objectively speaking' there are no 'traces' before the beginning of the 'hunt' (i.e., research).

interpretation—for instances as traces of prehistoric astronomical thinking or as traces of some solar cult—we consciously or unconsciously import into our interpretation the assumption of at least some *basic cultural similarity*. In that case, we must ask to what extent we have within our archaeoastronomical, mythical, or religious interpretation *discovered* the original cultural meaning of the line of stones and to what extent we have *culturally appropriated* it by our interpretation.

Given the above, it is hard to imagine that some 'cultural content' could be, for instance in the form of an *e*-meme (see Chap. 10), 'stored' as inactive in texts or artefacts. And even if it were possible, one would have to ask whether, if it were after a long time used as part of some 'trace', it could evoke in the investigator the same *i*-meme that was active in the mind of the actor who created the trace. The functioning of traces could represent a special case of (vertical) cultural transmission, eventually a specific case of cultural diffusion. If it were a case of cultural transmission, it would be one that is not connected with (immediate) social learning: it would probably be a case of transmission with transformation (cf. Chap. 11).

18.4 Developmental Lines, Life, and (Cultural) History

Different sciences choose different ways of systematically explaining the variety of phenomena. One such way is based on 'ordering' the multitude of traces by configuring them into developmental lines. The extent to which such lines are interpretatively *imposed on* the variety of traces or, on the contrary, *inferred* from the nature of the set of surviving 'traces' has been the subject of various discussions (cf. Chaps. 5 and 6). It has also been debated whether such lines are, in their nature, *theories* (of development) or rather *narrative configurations* of mutual connections. The lines that appear in descriptions of the evolution of life can rest on the genetic foundations of ontic processual relations (Markoš, 2016a, p. 262), but when it comes to the development—both evolution and diffusion—of culture, the ontological status of developmental lines depends also on whether we treat the investigated reality as a historical reality (thus following Hegel's conception of spirit; cf. Chartier, 1987, p. 121) or merely as a heterogeneous continuum (Kocka, 1986, pp. 23–39).

For the moment being, let us assume the latter, that is, the conception of historical sociocultural realities forming a heterogeneous continuum. In that case, the lines of cultural development constructed by historical sciences can be mere narrative interpretations or form theories of evolution of (partial) processes of cultural change. Theories of this latter kind include for instance the theory of disenchantment of the world (see below). That being said, we should note that historical sciences do not have at their disposal any *general* theory of the entire (cultural) history (Rüsen, 1986, pp. 47–65). Nevertheless, according to Ricoeur (1983, pp. 311–322), historiographers aim at presenting the entire history of particular wholes, for instance the history of prehistory, European history, the history of modern culture, the history of Cold War, in other words—as Ankersmit (1994, 2001) would say—the history of *narrative substances*. According to Ricoeur, as well as Ankersmit, this intention can be fulfilled only by narration. Only narrative operations enable us to configure otherwise incompatible 'traces'. Philosophy of history or grand social theories, such as those presented by Comte, Hegel, Marx, and in a sense also Cassirer, then construct from them what narrativist criticism calls 'master narratives' (cf. Welskopp, 2008, pp. 143–146).

In historical sciences, theories do not meet the criteria of Hampel's fully explicative *covering law model*. In other words, they do not need to be fully explicative like scientific theories. For instance, the school of Historische Sozialwissenschaft emphasises the development of the theoretical component of historiography: its aim is to describe particular features of 'historical theories'. Chris Lorenz (1997, p. 356) quotes Hans Ulrich Wehler's definition, which states that in historical sciences theories are 'such explicit and consistent conceptual systems …., which—without being derived from sources—serve the purpose of identification, discovery, and explanation of historical problems'[4] (Wehler, 1979, pp. 17–18). This is very close to Max Weber's thoughts (Weber, 1904) on the nature and role of theories in culture studies (*Kulturwissenschaften*). But we must also bear in mind that the logical coherence of a theory of developmental process does not have a counterpart in the real connections of cultural phenomena treated by such theories. The logical order of concepts (*logische Ordnung der Begriffe*) must not be mistaken for a proof of empirical organisation of the conceptualised phenomena (*empirische Anordnung des Begriffenes*; Weber, 1904, pp. 77–78; see also BOX).

Nevertheless, Darwin's theory of evolution is also in fact a history (Markoš, 2015a, p. 26) and the claim that biology has at its disposal a general theory of (the development of) life can also be questioned. In particular, one could claim that Darwin's theory of evolution is due to its generality rather a metaphysical research programme within which one can formulate partial theories that can be tested (Wittkau-Horgby, 2002, pp. 89–90). From this perspective, one could perhaps conclude that partial biological theories can be situated within the general evolutionary perspective on the development of life with more precision and less ambiguity than one can situate various partial theories of processes of cultural history into a narrative explanation of the development of culture. In this respect, it seems that the differences between evolutionary biology and historiography of cultural development are a matter of degree but not of principle.

The above holds even if we abandon (Markoš, 2015b, p. 25) some more traditional definitions of life, along the lines of for instance those which state that 'life is a self-sustaining chemical system capable of Darwinian evolution'. We can replace

[4] '…jene expliciten und konsistenten Begriffssysteme […], die—ohne aus der Quellen abgeleitet zu sein—der Identifizierung, Erschließung und Erklärung von historischen Problemen dienen.'

them by a biosemiotic alternative. Anton Markoš (2015b, p. 38; 2016b) defines life as a 'physical, born, semiotic system with history'. Such definition of life includes both biological species and (human) cultures. One could follow up on it with a working definition of culture that brings together Geertz's (1973, pp. 3–54) conception of culture with Cassirer's (1925) theory of symbolic formation and the notion of immanent development of the spirit: Let us thus say that culture is a system of symbols with a history that spreads by social learning. This definition makes no mention of the physical aspect of life. Cassirer transformed the Kantian transcendental subject of knowledge into a historically developing 'spirit' (Scholtz, 1997, p. 26). But this spirit is not understood substantively, as a metaphysical foundation of the continuity of history: rather, it is viewed as a supra-personal space of possible understanding in which the meanings of symbols can become apparent.

18.5 Traces and Lines in Biology and in the History of Culture

Let us assume, as a working hypothesis, that the traces which (evolutionary) biologists work with and those which historians (of culture) investigate are, from the noetic perspective, in principle identical. Paul Ricoeur (1985, pp. 212–228) states that worthy of historians' interests should be only those traces which are somehow imbued with the historicity of being-there (*die Geschichtlichkeit des Daseins*), that is, suffused with the experience of meaningfulness or intentionality of action. Biosemiotics offers the opportunity to see analogies. Morphological traits or phenotypical manifestations which (evolutionary) biologists see as 'traces' have arisen, among other things, based on the semiotic processes of life itself. Once we take aboard that all life is capable of semiosis in Peirce's sense (cf. Chap. 17, also Markoš, 2015b, pp. 38–51; 2016b), we can assume that processes of meaning attribution take place under the superficial 'traces' of both the development of culture and the evolution of life (Švorcová & Kleisner, 2018).

Evolutionary biology defines the evolutionary lineages of particular species. Their branching usually does not take the form of straightforward textbook bifurcations. In individual temporal stages, the branching includes clusters of various forms and based solely on palaeontological records, one cannot foresee which of these clusters would eventually lead to new forms (Markoš, 2012, p. 106). We often have no material evidence or empirical access to those moments of branching, especially if they happened deep in the past. Biology therefore often (re)constructs such branching as 'evolutionary events' using methods analogical to those which historians use to deal with 'events' (Koselleck, 2000, pp. 144–157; Koselleck, 2004, pp. 105–114). Even so, it seems that the developmental lines of 'life' and their branching (as treated by evolutionary biology) are in principle rather evident, at least inasmuch as they have an ontic counterpart in the continuity of spread of genetic information, although even that is from the noetic perspective based on a (narrative) interpretation of statistical models. Historians of culture, on the other hand, are faced with a vast multitude of discrete and diffuse hints and traces (and cultural contents, which are assumed to underlie them), which appear in space and time. Historians then link these traces and construct developmental lines, which can not only branch (in the evolution of culture) but also merge (in cultural diffusion). Various, even mutually competing, developmental lines can lead through one and the same 'trace' (cultural variant). (The same applies also to the traces of palaeontological records, perhaps just less acutely.) The variability of lines of cultural development is moreover potentiated by the fact that they are shaped by the initial cultural values (Weber, 1904; Scholtz, 1997, pp. 24–26) or traditions of interpretation (Gadamer, 1960), which outline how researchers bring some order into the sociocultural field under study. On top of that, the values or traditions that form the researcher's starting point need not be organically based in the subject of research (the conception of a heterogeneous continuum).

'A minimum of "before" and "after" constitutes the significant unity that makes an event out of incidents' (Koselleck, 2004, p. 106). What Koselleck calls a 'collection of incidents' could also be treated as a 'configuration of traces', but with respect to one collection of incidents, we can probably formulate several historical sequences (parts of lines) that would each set a different 'minimum of "before" and "after"'. One and the same collection of incidents can thus be treated as several distinct (evolutionary or historical) events. The sequences determining the 'minimum of "before" and "after"', which are integrated into some initial theory of possible history (see BOX), are thus codetermined by the pre-existing cultural values and traditions of interpretation that are taken as a starting point. This conditionality of values (Weber, 1904) then detracts from the ontological statues of the development lines defined by historical science.

The developmental lines of biologists and historians of culture differ in the strength of their ontological status. The markedly weaker ontological foundation of the developmental lines of historians of culture is reflected in their often noted inclination to nominalism (see BOX below) and, in some cases, even the resignation of some narrativist theorists of historiography (so-called non-representationalists; cf. Zeleňák, 2015) on addressing the relation between a work of historiography and some past reality (Kuukkanen, 2015).

Box: Sociological Nominalism and Nominalism in Historiography

Narrativist critics of historiography, such as Paul Ricoeur (1983, 1985) or Frank Ankersmit (1994), follow in Max Weber's footsteps (1904) and endorse nominalism. But to speak about nominalism (or essentialism, eventually realism) in general would be excessively vague. One must specify to which entities should the nominalist perspective apply. Weber, as a 'methodological individualist' (Lorenz, 1997, pp. 298–292), is a sociological nominalist. In his view, 'society' is no more than a name, *nomen*, applied to a sum of interacting individuals. (In a similar way, Charles Darwin was a nominalist regarding biological species.) In relation to culture, however, Weber is not such a decided nominalist. On the ontic level, he assumes a material culture (Kocka, 1986, p. 37), by which he means a semiotic and axiological system of behaviour of a group of people that enables members of that group to assign meaning to each other's behaviour. On the other hand, Weber draws a strict distinction between material culture and culture in the methodological sense (that is, the culture to which each researcher belongs and on which their research is necessarily based).

Moreover, one must differentiate between sociological nominalism and nominalism in historiography, eventually historical sciences. That is a position which not only Weber but above all the narrativist critics of historiography would endorse. This historiographic nominalism is based on a conviction that past reality as such is not as if an untold story (Mink, 1970). Reality in general does not have the form of a story: it is us, *we* tell stories that help us explain the boundless variety of phenomena (Ankersmit, 1994). Both representationalists and non-representationalists can endorse historiographic nominalism (Kuukkanen, 2015; Zeleňák, 2015), and one can even espouse nominalism with respect to historiography but not sociology (see, e.g., Ricoeur, 1983, pp. 339–348; see also below).

Koselleck's claims regarding the 'theory of possible history' and the 'veto of sources' also have the defining features of historiographic nominalism:

> Thus we need a theory: a theory of possible history. Such a theory is implicit in all the works of historiography; it is only a matter of making it explicit. There is a wide variety of statements on history in its entirety or individual histories which cannot be directly related to the sources, at least in the second phase of study. (Koselleck, 2004, p. 150)

And he goes on to say:

> In principle, a source can never tell us what we ought to say. It does prevent us from making statements that we should not make. The sources have the power of veto. (Koselleck, 2004, p. 151)

18.6 Narrativity

Anton Markoš in his chapter in this book (cf. Chap. 17) talks about the construction of stories through which biological 'societies' interpret the 'texts' of their genetic record in a manner appropriate to the period in question. The stories of evolution are not just narrative constructs spun by biologists. They are constructions told by communities (societies, species) themselves, by life about itself (see Markoš's narrativity of the living, but let us leave aside the degree to which the notion is metaphoric). One thus finds here a kind of duality (Markoš, 2012, 2015a): First of all, stories are constructed by communities of living beings and, in doing so, these communities acquire a particular form within their evolution. Secondly, biologists tell stories about the development of life and, in doing so, they connect theoretical models with narrative forms (Richerson & Boyd, 2005). These two sorts of stories (of the living and about the living) can differ. The aim of biological investigation is to make sure that its story about the living approximates that which communities of the living (species) construct about themselves and by which they are formed. On both levels, it is the story that co-founds the cohesion of the living, both the lineage of life in the ontic sense and the line of the biologist's narrative.

In contrast to this, most current narrative critics of historiography emphasise that historians by their narratives (narrative historical interpretations) bring into history a continuity (of interpretation), but in the case of history of particular units (culture, civilisation, humanity) this continuity and cohesion has no ontic counterpart (Ankersmit, 1994; Šuch, 2017; Zeleňák, 2015; cf. Chaps. 5 and 6). The narrative interpretations of historians—at least according to Ankersmit—project a certain structure onto reality but have no existential implications. They function merely on the level of language. The narrative logic of historical interpretations is thus strictly nominalist (Ankersmit, 1994, p. 39; cf. also BOX), in other words: 'Narrativism [...] recognized that a historical interpretation projects a structure onto the past and does not discover it as if this structure existed in the past itself' (Ankersmit, 1994, p. 36).

Physical embodiment of the narratives of life takes place within a certain range of possibilities that enables some freedom but is also infused with 'norms' (Markoš & Švorcová, 2019). These norms have their counterpart also in the sociocultural sphere, where self-identification actor narratives are formulated. Aside from explicit norms (some of which can be enforced), there also exist various implicit norms, such as cultural traditions, cultural institutions, discursive regularities, current forms of language, etc. Some of these can have an impact on the psychic structure of the actors (see below Elias's notion of habitus). Still, I believe that one can hypothetically assume that these sociocultural norms are *softer* in terms of their action than the norms considered by (biosemiotic or epigenetic; cf. Chap. 16) evolutionary biology. In historical science, we can differentiate between true and fictitious narratives of actors or their groups, but when it comes to their impact on the constitution of identity of a group, the impact of fictitious narratives can be just as great, if not greater, than the effect of 'true' narratives.

A narrative is 'objectified' when it becomes a symbolic instrument of historical and social experience in a group and if—in that group but possibly also outside, in a wider sociocultural field—it becomes comprehensible and acceptable or at least tolerated. One can thus consider, eventually even investigate, how likely it is (Weber, 1972, pp. 1–20) that a narrative will achieve objectification within a particular sociocultural field. The likelihood depends on various explicit as well as implicit norms within a given field but, if we view the sociocultural field as a heterogeneous continuum, it is conceivable that even mutually contradictory narratives may become objectified.

For instance, the 'community' of Lion can hardly present itself, in its particular Dasein (here-and-now), at the same time also as a 'community' of Domestic Cat. But one human community, one particular sociocultural form, can in its Dasein quite readily interchange its (prominent) representations, for instance confessional, ethnic, class-based, xenophobic/nationalist, humanist, etc.

For biologists, just like for historians or people at large, the general semiotic field within which both literal and metaphorical meanings can emerge is the transcendental consciousness (see above Cassirer's historically developing spirit). For living communities viewed from the biosemiotic perspective, this semiotic field is *semiotic–phenotypical*, that is, it is the field of physical manifestations, or *semiotic–morphological*, that is, it is the field of forms (Markoš, 2015a, pp. 19–26), as one could say in connection with Portmann-inspired considerations (Komárek, 2008, pp. 202–232). This sharp distinction between the two types of fields within which the meanings of signs can emerge is nevertheless merely provisional and based on a simplification.

Narratively established facts and forms created by narration and semiosis are, within human cultural history, undoubtedly *less hard* than their counterparts in biology and the field of the living. The world of sociocultural phenomena is markedly more open to a variety of interpretations and, moreover, the signs or metaphors it contains are much more polysemic than phenomena in the world of the (biologically relevant) living. What one ought to ask here is whether this is a matter of degree or a difference of principle. If it is a difference of principle, one should be able to explain it using the tools of biosemiotics. Given the life's general ability of semiosis, human culture would differ from the area of (mere) life by its ability to use, within human spirituality (human cognition), semiotic processes and narratives in a reflected way (Deely, 2008, pp. 95–102; Markoš, 2015b, pp. 76–89). On the other hand, studies that argue for the possibility of culture also in non-human species (cf. Chap. 9 and 13) should make us wonder whether this is a matter of difference between the general semiosis of life on the one hand and human culture on the other, or rather a difference between culture in general, as it appears in different areas of life, and specific cultural layers or specific intensity of culture (which is so far known only in humans). Subsequently, we could ask about the role which the above could play in the search for analogies between the development of species (communities) and the development of (human) cultures.

18.7 Analogies Between (the Development of) Species and Cultures

Analogies between the development of biological species and (human) culture can be used not only to semiotise the explanation of life but also to help apply various explanatory principles from evolutionary biology to explanations of (the development of) culture. One can treat these analogies as existing either merely at the level of explanation or also at an ontic level. Many historiographic nominalists believe that these analogies are largely just explanatory. Biosemiotically oriented biologists, on the other hand, tend to believe that analogies between (the development of) species and cultures do have an ontic foundation (Barbieri, 2002, pp. 217–242; Markoš, 2015b, 2016b). From the perspective of biosemiotics, culture as a special case of a semiotic system simply organically grows from the general semioticity of life.

Search for analogies can be complicated by the fact that biology and historiography do not necessarily mean the same thing by 'development' or 'evolution'. An important characteristic of life is an increase in the complexity of a system or organism, both in terms of epigenetic development (convergent increase of complexity) and evolution (divergent increase of complexity). But when applying this definition of development, one needs to specify on what level of description the increase should take place. Historical sciences—such as Luhmannian sociology—do not need to require an increase in complexity (although for instance Cassirer's conception of the development of spirit would meet this criterion). From the perspective of historiography, evolution (or development) would be defined as an irreversible and gradual change over time, where both the aspect of continuity and the aspect of change are important (Wittkau-Horgby, 2002, p. 88). In historiography, one moreover needs to consider both the development that leads to individuation, that is, to a distinctive character of historical units (nations, institutions, complexes of ideas, etc.), and the development that is unifying (for instance, in connection with the theory of cultural diffusion) and leads to a similarity of character between initially distinctive cultures (e.g., the process of modernisation or demographic transformation).

One example of analogies found between the evolutionary behaviour of species and cultures is the theory of frozen plasticity (Flegr, 2006, 2015), which assumes that a vast majority of species with sexual reproduction shift between the plastic and elastic phase over the course of their evolution. Species that meet the criteria of this theory behave in their evolution 'like a piece of elastic rubber'. Initially, they respond to the (natural and sexual) selection pressures coming from their environment very quickly, 'because they have, right from the beginning, many alleles of each gene' (i.e., high level of genetic polymorphism; Flegr, 2015, p. 22). Changing frequencies of already present alleles facilitates quick reactions, but their mutual (pleiotropic, epistastic, frequency-dependent) interactions imply that the reaction to particular pressures will over time decline, 'until, at a certain point, the species will stop responding to a given selection pressure altogether' (Flegr, 2015, p. 23). After selective pressures cease, the allelic frequencies—and consequently also the phenotype—return to their original state. This elasticity can be disrupted only under particular circumstances, for example after a small subpopulation splits off and it leads to the emergence of a new species. If such subpopulation 'wanders in the desert' in small population numbers for an extended period of time, it loses various alleles of particular genes along with the web of their interactions. This ensures that every new mutation will now occur in the vicinity of the same monomorphic alleles and have the same effect on fitness; in other words, the emerging species becomes capable of plastically reacting to selective pressures. That, however, causes the new mutations (alleles) to accumulate again. They aggregate, plasticity in various traits decreases and 'after a short period of evolutionary plasticity, the new species passes into a state of evolutionary elasticity. In this state, it can respond to changes in the environment faster, but only temporarily and to a small extent' (Flegr, 2015, p. 24). The behaviour of a species in the elastic phase of its development can then be compared to the evolutionary behaviour of (human) cultures (Markoš, 2012, pp. 106–107; Flegr, 2015, pp. 340–349).

Still, it is yet to be seen whether one ought to look for analogies between (a) biological species and human cultures or rather (b) between the phenotypical traits of biological species (or traits present in some higher taxon; since the theory of frozen evolution, which builds on the theory of frozen plasticity, pertains also to phenotypical traits and higher taxonomic units) on the one hand and cultural contents (such as the kind of religiosity, legal discourse, religious ethics in correlation with economic activity, etc.) on the other, or, finally (c) between (entire) biological species on the one hand and (partial) cultural contents or their specific configurations on the other hand (cf. Chap. 15). Here, it seems interpretatively more suitable and analytically more accurate to consider the plasticity (or elasticity) of partial cultural contents, which can be investigated always within a particular group of such contents, and view the whole of culture as constitutive of the environment (approximately in the sense of 'culture space' as presented in Chap. 11) that can exert selection pressures.[5] The 'cultural contents' could then respond to such pressures either plastically or elastically.

[5]The term 'selection pressures' is used here with plenty of reservations because the concept is often viewed in an essentialist way, while we want to use it in a purely nominalist sense. One cannot directly observe a 'selection pressure'– just like one cannot empirically observe the 'force of gravity'. All one can see is a change, which is interpreted as caused by some eventual selection pressure. Still, the explicative force of this explanation is probably smaller than when an observation of centripetal acceleration of some physical body is taken to be the proof of existence of a gravitational field. Ginzburg (2002) seems to be right when stating that evolutionary thinking moves within a 'paradigm of clues' rather than on the ground of the 'Galilean paradigm'.

18.8 Disenchantment of the World

We can take as an example of a cultural analogy to the plastic and elastic phase of evolution of a biological species—and draw attention the associated interpretative difficulties (if not ambiguities)—the process which Max Weber calls 'disenchantment of the world' (*die Entzauberung der Welt*) and the process of its 're-enchantment.' In the sphere of religiosity, such disenchantment takes the form of for instance a gradual rejection of magical means of salvation. Weber (1996, pp. 62–70, 88–105, 114–117, 160–191) identified as key turning points in this process the teaching of Old Testament prophets, Hellenistic learning, and the critical attitude of sixteenth-century European reformation to the Catholic concept of salvation, especially the symbolic conception of sacraments in Protestantism. The disenchantment of the world manifests itself in politics, law, modern science… and the list could go on. Nevertheless, historiography of the (vernacular) culture of Early Modern Era and modern history had started to note that this process did not unequivocally and irrevocably change the world of the West. Scholars started to point to examples of re-enchantment (*Wiederverzauberung*) of the world. One could thus say that we are dealing with a partial change in cultural history that can be used to demonstrate both *evolutionary elasticity* (a tension between Reform efforts aimed at spiritualisation and a removal of magic on the one hand and magic tendencies of part of the vernacular culture, which were boosted by some elements of the Catholic Baroque) and (prevalent) *plasticity*, exemplified in the rationalist/secularising turn of the Enlightenment and apparent for instance in the abolishment of religious offences in criminal law (Tinková, 2004, pp. 157–232). Still, one should also bear in mind that while enlightened legal reforms seem plastic from our current perspective, it is possible that in two or three centuries even those changes will be viewed as manifestations of an elastic evolutionary oscillation.

If we are to assess analogies between the evolution of species and the development of cultures, we need to find within culture an entity that would be in terms of homogeneity comparable to a biological species. If we fail to do so, we will not be able to decide with sufficient accuracy whether the variety of the cultural phenomena under investigation is a manifestation of internal variability of a particular culture or simply the consequence of several cultures intermixing within our area of research. Perhaps the (demographic) category of population could help us: we could, for instance, compare the variability of occurrence of certain phenotypical traits in some species with the variability of occurrence of some cultural contents (or elements, variants, etc.) within a particular human 'population' (cf. Chaps. 5 and 6). But biological species and human populations are not categories of the same order: in both plant and animal species, one can biologically study their particular 'populations' or explain their intraspecies variability by assigning 'culture' to them (cf. Chap. 9). Moreover, many human populations, especially in the Early Modern Era and in modern history, are not 'organically' but rather administratively formed units—unlike the mostly organically formed 'species'. Demography, on the other hand, usually investigates both historical and current 'populations' in relation to some administratively defined area (parish, region, state). The idea of one human population being the carrier of one culture is at this point too simplistic—and it would be excessively simplifying also with respect to a 'community' in the sense in which Anton Markoš considers it (cf. Chap. 5). What we deal with here is thus rather a *degree* of constructivist assumption of cohesion of 'species' and cohesion of a 'culture'.

A culture, especially human culture, is nevertheless multilayered, meaning one can distinguish different degrees or intensity of culturality. This must be taken into account when investigating possible analogies between the development of species and cultures. To wit, one can—at least analytically—distinguish between human behaviour and actions, which are certainly at least in part codetermined by culture, and the occurrence and configuration of some cultural contents within a culture. This is because cultural contents are not just external manifestations (phenotypical traits: behaviours, artefacts) but also meanings, which emerge internally, in the minds of the actors (or within a 'spirit'). That is why preference for some cultural variants and the frequency of their occurrence within a population is not the same as culture. It is rather just an application of culture within the studied society. Preference for certain cultural variants is a behaviourally phenotypical manifestation of a particular society. In a sense, though, it is not an outright manifestation of its culture. From this perspective, it is better to view culture as an in a sense (semi)autonomous semiotic system that can be analytically differentiated both from the sphere of genes and from the sphere of actions and behaviours.

When investigating the analogy between the evolution of species and the development of culture in terms of transition from the plastic stage of development into the elastic one, we therefore should not compare, for instance, the presence of genes, alleles, or phenotypical traits with the occurrence of certain types of behaviour and actions of people who belong to that culture but rather with the frequency and variability of configurations of particular forms of the cultural content we focus on. Further, we can also note that the selectionist interpretation of cultural evolution puts less emphasis on the mental discovery of meanings of 'cultural contents.' It is more interested in the competition between variants, whereby a variant is viewed as successful if it multiplies and forms as many copies as possible (cf. Chap. 10). The attractionist approach to the evolution of culture is more sensitive to the internal, mental dimension of cultural contents, because attractionism emphasises the link between the transmission of cultural variants and their transformation by constant actor (re)interpretations. As Pavlína Hillerová and Petr Tureček show in their chapter (cf. Chap. 11), the selectionist approach is well suited to explaining the evolution of those levels of culture that do not easily lend themselves to an attractionist interpretation. The criterion of stability-based sorting (see Chap. 15) is likely to apply more readily to the layers of culture which are better suited to the selectionist approach: to phenomena which, for instance, Fernand Braudel (1985) calls long-term structures (*longue durée*) or material life (*vie matérielle*) and Max Weber calls customs (*Sitte*). According to Weber, customs are an example of a typically regular behaviour maintained solely by habit and non-reflected imitation. For instance, a large part of economic life is based on such customs (Weber, 1972, p. 187). Customs are sometimes identified with cultural patterns or behaviour patterns (*Verhaltensmuster*; Reinhard, 2017, p. 7).

The principles of stability-based sorting cannot, however, be applied to all layers of (human) culture. Aside from the *vie matérielle*, *Sitten*, and cultural patterns, the history of culture also describes sudden (cultural) changes. For instance, Peter Burke (1990) discusses two key decades (1520–1530 and 1650–1660), when one can observe in the culture of western Europe a profound change in the treatment of symbols. These changes had an impact on the Church and religious life, on the political establishment, but also on the nascent Modern science. Such sudden upheavals take place against the background of more stable layers of culture but have the potential to transform, sometimes quite rapidly, the whole of this culture by reconfiguring the relations between particular cultural contents. For instance, aside from describing the customs (*Sitten*), Max Weber (1972) also suggests analytical categories that can be used to investigate the appearance of individuals who have a charismatic power (*charismatische Herrschaft*). Such 'charismatic man' (*charismatischer 'Herr'*) has the ability to rapidly restructure the culture and society in which he appears. From the perspective of the attractionist conception of culture, one could say that such charismatic individuals reconfigure the basic attractors of the cultural system or establish a new 'meta-attractor'.

18.9 The Nominalism of Historians of Culture Versus the Realism of Biosemioticians?

It should be borne in mind that biology, too, is interested in phenomena one could view as analogical to the internal dimension of cultural contents and their agential experiencing. An older example of this is found in the conception of innerness (*Innerlichkeit*) proposed by Adolf Portmann (Portmann, 1960; Komárek, 2008, pp. 202–232). In this place, though, we are interested in biosemiotics, which must necessarily assume an *inner dimension* of life (living organisms), because it is within this dimension that meaning attribution (Švorcová & Kleisner, 2018), that is, semiotic processes, take place. Nevertheless, despite the abovementioned analogies, we should note the following differences between biosemiotics and historical sciences that deal with culture:

First of all, neither Max Weber's (1904) concept of ideal type (*Idealtypus*) nor the historical narrative interpretation as defined by Ankersmit (1994) have their (exact) ontic counterpart.[6] They are merely applied to a chaotic set of empirical phenomena or sources in order to organise them and thus arrive at an explanation. But even if they turn out to adequately correspond to reality, it does not mean that

[6] The ideal type is a concept defined *a priori* by the researcher in relation to the empirical material, which is the instrument, not the result of knowledge. Historical narrative interpretation is a developmental or event configuration of facts that exists only at the level of language and is an instrument of interpretation. Neither has an ontic counterpart.

the semiotic, logical, or narrative processes assumed in those ideal types or plots actually take place in reality. Sociocultural reality is not unambiguously structured, and its existence in time does not have the nature of an untold story. This is, among other things, due to the considerable heterogeneity of semiosis that takes place in the individual actors within a given sociocultural field. Markoš, on the other hand, argues that narration by biologists—and in his view, Darwin's evolutionary theory is in effect a history of life—does have a (more apparent) apparent ontic counterpart in narrativity of the living (cf. Chap. 17, see also Markoš, 2003, 2012). It is therefore in principle conceivable that, figuratively speaking, the interpretative narration of biologists could overlap with the 'narration' of particular species (societies), which based on their experience and memory interpret the 'text' of their genetic record. The historical 'narration' of evolutionary biology naturally chooses different means (language, concepts, theories) than the 'narration' of the communities investigated by biology. Biologists' identification of these communities does, however, have a more accurate ontic counterpart in the, among other things semiotic, self-formation of the particular 'societies' or communities.

Secondly, even if we take it as our starting point that the 'evolution of cultures and evolution of biological communities are isomorphic', as Anton Markoš (2015a, p. 23) states, it seems that there is still a difference between our ability to state the continuity of developmental line in cultural history and in the history of life (Markoš, 2016a, pp. 261–263). The ontic counterpart of biologist's efforts to grasp the continuity of a developmental line is the real, one could even say physical, continuity of life in that lineage. A species belongs to an evolutionary lineage based on objective kinship relations between organisms (Flegr, 2015, p. 34). The constructions of historical sciences, on the other hand, do not have any such clear ontic counterpart in cultural history, and that holds regardless of whether they are based on ideal types (Weber, 1904) or treated as narrative interpretations (Ankersmit, 1994).

Thirdly, in philosophical or historical analysis of culture, the space in which meanings become apparent is (for instance Cassirer's) 'spirit', a spiritual field. In biosemiotics, on the other hand, the space in which meanings appear is a sort of (semiotically phenotypical) morphological field of forms (cf. eidetic biology as discussed with reference to Zdeněk Neubauer by Anton Markoš in Chap. 17). From a certain perspective, one could thus say that (language, human) cultures can likewise be viewed as merely morphological fields of forms. That would hold if we were considering only artefacts, eventually empirically observable behaviours. But it would be excessively reductive. Let us leave aside the fact that biological evolution, too, works with artefacts, both with creations such as beaver dams and, metaphorically, with those that make up our bodies, such as teeth (Markoš, 2015a, pp. 26–65; Markoš, 2015b, pp. 24–59). What is crucial is that culture cannot be reduced to artefacts because the perception of artefacts is a crucial part of culture (cf. also the attractionistic emphasis on cultural transmission with transformation in Chap. 11). This difference between human culture and other species may be linked to something that Deely (2008) had emphasised, namely the human ability to reflect on semiosis and to perceive abstract relations.

Finally, we are considering categories such as genes, (phenotypical, morphological) traits, species, genera, populations, or taxa on the one side and cultural variants, cultural complexes, cultures, or civilisations on the other. What is important to bear in mind when looking for analogies is whether the units we are trying to compare (for the sake of simplicity let us say species versus cultures) are cohesive to the same degree. The answer to this would be on the part of philosophers and historians of culture probably markedly more nominalist than the answer of evolutionary biologists and biosemioticians. Among the classics of narrativist criticism of historiography, Paul Ricoeur would probably form something of an exception thanks to his conception of quasi-personage (a status ascribed to nations, cultures, civilisations, etc.), which has a basis in both narratology (entities to which behaviours can be ascribed) and social ontology (with help from Mandelbaum's theory of societal facts). Ricoeur is therefore a nominalist with respect to historical sciences but not with respect to sociology (Ricoeur, 1983, pp. 311–396; Lorenz, 1997, pp. 294, 301–302).

The fact that some biosemioticians incline to realism (or essentialism) in their treatment of semiosis, while some philosophers or historians of culture incline to nominalism may have to do with John Deely's influence on the development of biosemiotics. Following up on medieval realism, he emphasised the reality of relations in semiosis. According to Deely (2008, pp. 95–102), semiosis goes beyond perception, and one could thus ask whether semiosis is just a noetic or also, and perhaps chiefly, an ontic process. Biosemioticians would (and not only under Deely's influence) respond that biosemiotics is also an ontic process, while historians of culture probably would not be able to give an unequivocal answer regarding historical processuality.

18.10 Can Cultural Evolution Somatise?

By ascribing the ability to engage in semiosis to all life, biosemiotics relaxes the borders between 'nature' and 'culture', which subsequently makes it easier to ask whether culture, or its development, might also be somehow inscribed into the course of biological evolution. (Anton Markoš in Chap. 17 ascribes 'double heredity', i.e. genetic and semiotic one, to life in general.) One can imagine numerous physiological, behavioural, and even cognitive properties which are phenotypical traits of (not only) humans but could also be relatively closely linked to particular cultural contents. These properties must have been somehow established by evolution. What is not clear is whether, and if so to what extent, this process took place within biological evolution or as part of cultural development (evolution and diffusion of culture) and whether, and if so to what extent, the biological and cultural development influenced or permeated each other. In this connection, we can distinguish at least five ways in which such development may have taken place:

(a) Only in the sense of *biological evolution.*
(b) In the sense of *coevolution of genes and culture*. This would include the processes of dual heredity or heredity of genes, culture, and the environment (cf. Chaps. 7, 11, 16; see also Richerson & Boyd, 2005). Also included would be the development described by the theory of human self-domestication, that is, a development in which 'culture' is the source of selection for cooperativity and nonaggression. This theory claims that non-cooperative and aggressive individuals are pushed out of the process of reproduction, which gradually changes the representation of genes and their alleles in the population.
(c) One can also consider possible *epigenetic fixation* of particular *cultural variants* (cultural contents). Jana Švorcová and Karel Kleisner (cf. Chap. 16), with reference to Mary Jane West-Eberhard point in their chapter to the *phenotype first* principle, which is endorsed by some evolutionary biologists. This principle assumes that in evolution, changes in the phenotype—whereby it is yet to be seen whether this includes culturally conditioned changes in human behaviour and cognition—precede changes in the genotype. It is also taken in consideration that new phenotypical traits (phenotypical accommodations) do not need to be genetically fixed: in some cases, they can be fixed only temporarily, by epigenetic mechanisms. What is meant here by epigenetics is the study of short-term or permanent structures and relations which in some way participate in the regulation of gene expression and building of bodies (Markoš & Švorcová, 2019).
(d) The properties we are considering here have formed by *cultural development* that took place on a deeper level and within a wider range (on the level of 'texture' of behaviour, to use Ricoeur's terminology) than the development of language ('texts' about behaviour; cf. above). This could apply for instance to Geertz's conception of culture as an external program that enables the human brain to function like a human brain (Geertz, 1973, pp. 55–86). From this perspective, a cultural variant can be inherited—either solely or also—culturally.
(e) Finally, we could also consider the possibility that the abovementioned qualities have emerged though evolution (only) on the level of (cultural) development of language and are inherited only via its social learning.

Different levels of mixing of biological and cultural evolution can be demonstrated using various examples. Let us take a look at three of them: (1) the grandmother hypothesis, (2) the theory of the civilising process, and (3) linguistic (cultural) conditioning of colour perception.

Example 1 The so-called grandmother hypothesis among other things assumes that female menopause associated with relatively long post-reproductive phase of life became established in humans because children's fitness, that is, their chance of future reproductive success, improves when not only mothers but also (maternal) grandmothers care for young children during their early childhood. Specifically, it assumes that for a significant part of a woman's life the benefits of helping to rear her grandchildren outweigh the eventual benefits of having her own children (the theory of inclusive fitness; Hawkes, 2003). Purely hypothetically, it is possible that something that used to be transmitted only by cultural transfer (the cultural institution of grandmothers) transited onto the level of genetic heredity (the evolutionary establishment of menopause). We are thus assuming a possible ancient evolutionary event whose symptoms we can try to find for instance by a demographic analysis of some Early Modern (seventeenth to nineteenth century) local populations. In some, one can indeed find a 'grandmother effect.' In particular, we find that children who have received the care of not only their mother but also (maternal) grandmother had a lower infant mortality and their mothers shorter intervals between births compared to the rest of the population (Fox et al., 2017; Havlíček et al., 2021). In exceptional cases, one can in particular local populations observe a positive effect of maternal grandmothers simultaneously with a negative effect of paternal grandmothers, where their presence increased infant mortality (Voland & Beise, 2005). Cultural phenomena (demographic behaviour) can hypothetically indicate earlier biological evolutionary changes in the sense of point (a) above, but that does not necessarily mean that cultural development really participated in that change, much less so in the sense of the abovementioned assumption. If it were so, it would be probably only in the sense of the relation described in (b) and it would be most likely excessively daring to assume an influence of culture on the emergence of menopause in the biological development in the sense of abovementioned point (c).

Example 2 Norbert Elias in his theory of civilising process (*Prozeß der Zivilisation*), which influenced the research methods of historical anthropology and gave rise to what is known as the new cultural history (Chartier, 1997, pp. 107–144), assumes a gradual historical stabilisation of habitus (cf. Chap. 15). By habitus, Elias means the subconsciously functioning apparatus of individual human self-control. His starting assumption is that anything that initially presents itself as external (explicit) prohibition or pressure over time becomes internalised by becoming from the actor's perspective unconsciously and unintentionally part of the functional psychic structures, that is, of the habitus. Elias also assumes that with increasing density of the network of social fabric, as the social figurations become ever more complex, there also increases the stability of the habitus of people who live in these figurations. And finally, the process of a historical increase of stability of habitus goes hand in hand with increase of the 'threshold of sensitivity.'

According to Elias, a notable 'habitus-firming' and 'civilising' transformation took place in Europe in the sixteenth to eighteenth century. At that time, one can observe an increase in intensity of the experience of shame and repugnance. This took place as part of an evolutionary process characterised by increased differentiation of psychic and social entities, that is, a fragmentation of individual mental makeup. One can thus say that 'Both [...] are different aspects of the growing split in the individual personality that occurs with the increasing division of functions, they are different aspects of the growing differentiation between drives and drive-controls, between "id" and "ego" or "superego" functions...' (Elias, 1982, p. 146). The initial sociogenic pressure (from the authorities, morally prescriptive religious and secular texts, ingrained cultural behaviour patterns) thus according to Elias transformed into a psychogenic pressure. But advanced differentiation between the id and the (super)ego leads to a tension between the two, and this tension can—and since the Early Modern Era does—transform into psychosomatic manifestations of shame and physical experience of repugnance.

Elias's theory of the 'civilising' process describes similar (assumed) phenomena as the theory of self-domestication does, for instance, the historically increasing human ability to control their own aggressivity. But the civilising process is described as taking place within much shorter spans of time, mostly across centuries, than the length of time over which the self-domestication of humans is supposed to have taken place, which was in the order of thousands or dozens of thousands of years. It therefore seems most unlikely that the evolutionarily emerging characteristic, in this case the increased ability to differentiate between the id and the (super)ego, could have had any impact on the culturally influenced preference for certain genes and alleles (see above point (b)). What is not clear is whether this characteristic would have been inherited only socially and culturally, as in point (d), or—purely hypothetically—also in some organic way, as in point (c). Elias's conception of the civilising process (and similar conceptions) thus create space for the formulation of further hypotheses and testable theories.

Example 3 Numerous studies have investigated the correlation between languages and colour perception (among European languages, for instance a comparison between German and Slovak; Šuchová, 2015). Lenneberg, in connection with Boas's cultural and linguistic relativism, for example showed that monolingual Zuni speakers do not differentiate, and in their cognitive processes seem unable to perceive the difference, between yellow and orange, while bilingual Zuni speakers (who also speak English) usually have this ability (Lenneberg & Roberts, 1956; Šuchová, 2015, p. 47). Even linguistic universalists (Berlin & Kay, 1969; Kay & McDaniel, 1978; Kay & Maffi, 2000; Šuchová, 2015, pp. 48–77) admit in their theories of basic colour terms that members of some cultures find it difficult to differentiate between green and blue and their languages likewise do not have separate terms for these colours. One could thus purely hypothetically ask whether a native English speaker who would learn some Inuit language could also start 'seeing' the different colours of snow which are differentiated in Inuit languages. In that case, the ability to differentiate would be solely due to language in the sense of the abovementioned point (e). Alternatively, such ability could be acquired only if the English speaker learned Inuit in childhood, at a time when the brain is still highly malleable. In that case, the ability to differentiate between various colours would be based on the impact of a language (or wider culture) on psychic structures, which could correspond to point (d). One could also hypothetically consider the possibility that this impact of language on cognitive structures in inherited not only culturally but also organically. In that case, we could consider processes described in the abovementioned point (c). Purely hypothetically, one could thus entertain the possibility that the Inuit ability to differentiate between various colours of the snow has some genetic foundation. If it were so, such phenomenon might be explained by some variant of the theory of self-domestication. It would state that the influence of the environment and food acquisition strategy led to a selection of individuals for their ability to maximally finely differentiate between snow of different qualities (and therefore also colours). This would then fall under the abovementioned point (b).

In all three of these examples, and in similar cases, the main challenge is to identify to what extent one can differentiate between what is inherited somatically and what is inherited culturally. As biosemiotics increasingly opens the possibility of considering an evolutionary somatisation of the results of development of (human) culture, the answer is becoming ever more complicated. 'A Sign is a Representamen of which some Interpretant is a cognition of a mind. Signs are the only representamens that have been much studied' (Peirce, 1903, p. 291). Biosemiotics seems to be replacing cognition of the mind by the ability of all living organisms to attribute meanings in semiotic processes. Let us say that people in their cultural development create symbolic mental images (Cassirer, 1923, p. 12) or signs (Peirce, 1903) and with their help, they can take control of what emerges as a feeling or experience with some urgency. It thus becomes possible to grasp and manipulate those feelings or experiences. An indifferent irritation is transformed into the id, which the ego can differentiate from itself as its content. It is transformed into elements of human innerness that can be potentially reflected or into experiences one can become aware of. This takes place on the interface between the neurophysiological plane and culture (Cassirer, 1923; Krois, 2004). These cognitive characteristics may have been fixed also for instance epigenetically or even pass onto the level of genetic heredity or, alternatively, they may have been inherited only culturally, by social learning. Moreover, both epigenetic and cultural transfer to the next generation may have been the result of semiotic processes and, on top of that, it is conceivable that epigenetic and cultural heredity could in some cases potentiate one another (cf. Chap. 16).

Do we have at our disposal methods that could help us decide in this matter? Does it not ultimately mean that if we search, within a particular range of 'cultural contents' for plasticity, we will find it, and if we look for elasticity, we will find it too (cf. Chap. 14)? To conclude with a question that could inspire further thoughts on the subject: Is the analogy between species and cultures merely explicatory or is it also ontic? Or, to be precise, with respect to what layer of reality and with respect to which layer of phenomena is it ontic and with respect to which is it merely an explicatory principle—whereby even if we admit that it is 'merely an explicatory principle', it most certainly does not mean that the analogy is unproductive. One thing is clear. Evolutionary biology and historical sciences about culture can not only inspire each other in their search for answers to these questions: with respect to some interesting problems, they can even join forces and work in concert.

18.11 Conclusions

In this chapter, we saw that the distinctions between lineages made by historians and evolutionary biologists are not fundamentally different in kind; rather, they vary in degree concerning their ontological status. Both this chapter and the whole concluding section of this book show that there are various approaches to drawing analogies between the evolution of biological species and the development of cultures. Such discussions open new perspectives on the extent to which biological and cultural evolution may intertwine. And more importantly, not only the conception of culture but also life as such as semiotic systems lay the foundation for formulating more general hypotheses. Such hypotheses can in turn form the basis for developing testable theories that would deal with possible correlations and mutual influences between cultural and biological relevant entities.

References

Ankersmit, F. R. (1994). *History and tropology. The rise and fall of metaphor*. University of California Press.

Ankersmit, F. R. (2001). *Historical representation*. Stanford University Press.

Barbieri, M. (2002). *The organic codes: An introduction to semantic biology*. Cambridge University Press.

Berlin, B., & Kay, P. (1969). *Basic color terms. Their universality and evolution*. University of California Press.

Braudel, F. (1985). *La dynamique du capitalisme*. Flammarion.

Burke, P. (1990). Historians, anthropologist, and symbols. In E. Ohnuki-Tierney (Ed.), *Through times: Anthropological approaches* (pp. 268–288). Stanford University Press.

Cassirer, E. (1923). *Philosophie der symbolischen Formen. Teil 1: Die Sprache*. Bruno Cassirer Verlag.

Cassirer, E. (1925). *Philosophie der symbolischen Formen. Teil 2: Das mythische Denken*. Bruno Cassirer Verlag.

Chartier, R. (1987). L'histoire ou le récit véridique. In R. Chartier (Ed.), *Philosophie et histoire* (pp. 115–135). Centre Georges Pompidou.

Chartier, R. (1997). *On the edge of the cliff: History, language and practices*. Johns Hopkins University Press.

Daniel, U. (1993). "Kultur" und "Gesellschaft". Überlegungen zum Gegenstandsbereich der Sozialgeschichte. *Geschichte und Gesellschaft, 19*, 69–99.

Deely, J. (2008). *Descartes & Poinsot. The crossroad of signs and ideas. (Vol. 2 of postmodernity in philosophy Poinsot trilogy)*. University Scranton Press.

Dressel, G. (1996). *Historische Anthropologie. Eine Einführung*. Böhlau Verlag.

Elias, N. (1982). *The civilizing process II*. Blackwell.

Flegr, J. (2006). *Zamrzlá evoluce aneb Je to jinak, pane Darwin. [Frozen evolution, or: It's different, Mr Darwin]*. Academia.

Flegr, J. (2015). *Evoluční tání aneb O původu rodů. [Evolutionary thawing, or: On the origin of genera]*. Academia.

Fox, J., Willführ, K., Gagnon, A., Dillon, L., & Voland, E. (2017). The consequences of sibling formation on survival and reproductive success across different ecological contexts: A comparison of the historical Krummhörn and Quebec populations. *The History of Family, 22*(2-3), 364–423.

Gadamer, H.-G. (1960). *Wahrheit und Methode: Grundzüge einer philosophischen Hermeneutik*. J. C. B. Mohr.

Geertz, C. (1973). *The interpretation of cultures*. Basic Books.

Ginzburg, C. (2002). *Spurensicherung. Die Wissenschaft auf der Suche nach sich selbst*. Verlag Klaus Wagenbach.

Habermas, R., & Minkmar, N. (Eds.). (1992). *Das Schwein des Häuptlings. Sechs Aufsätze zur Historische Anthropologie*. Wagenbach Verlag.
Havlíček, J., Tureček, P., & Velková, A. (2021). One but not two grandmothers increased child survival in poorer families in west bohemian population, 1708–1834. *Behavioral Ecology, 32*(6), 1138–1150.
Hawkes, K. (2003). Grandmothers and the evolution of human longevity. *American Journal of Human Biology, 15*(3), 380–400.
Kay, P., & Maffi, L. (2000). Color appearance and the emergence and evolution of basic color lexicon. *American Anthropologist, 101*(4), 743–760.
Kay, P., & McDaniel, C. (1978). The linguistic significance of the meaning of basic color terms. *Language, 54*(3), 610–646.
Kocka, J. (1986). *Sozialgeschichte. Begriff—Entwicklung—Probleme, 2nd expanded edition*. Vandenhoeck & Ruprecht.
Komárek, S. (2008). *Příroda a kultura. Svět jevů a svět interpretací. [Nature and culture: The world of phenomena and the world of interpretations]*. Academia.
Koselleck, R. (2000). *Vergangene Zukunft. Zur Semantik geschichtlicher Zeiten*. Suhrkamp.
Koselleck, R. (2004). *Futures past. On the semantic of historical time*. Columbia University Press.
Krois, J. M. (2004). Kultur als Zeichensystem. In F. Jaeger & B. Liebsch (Eds.), *Handbuch der Kulturwissenschaft, Band 1: Grundlagen und Schlüsselbegriffe* (pp. 106–118). Verlag J. B. Metzler.
Kuukkanen, J.-M. (2015). Why we need to move from truth-functionality to performativity in historiography. *History and Theory, 54*(2), 226–243.
Lenneberg, E., & Roberts, J. M. (1956). *The language of experience: A study in methodology*. Waverly Press.
Lorenz, C. (1997). *Konstruktion der Vergangenheit. Eine Einführung in der Geschichtstheorie*. Böhlau Verlag.
Markoš, A. (2003). *Tajemství hladiny. Hermeneutika živého. [The secret of the surface: Hermeneutics of the living]*. Dokořán.
Markoš, A. (2012). Narativita v biologii a narativita živého. [Narrativity in biology and the narrativity of the living]. In J. Horský & J. Šuch (Eds.), *Narace a (živá) realita* (pp. 101–112). Togga.
Markoš, A. (2015a). *Co je nového v biologii. Vynalézaví obratlovci [What's new in biology. Ingenious vertebrates]*. Nová beseda.
Markoš, A. (2015b). *Znaky a významy v evoluci [Signs and meanings in evolution]*. Nová beseda.
Markoš, A. (2016a). *Evoluční tápání [Evolutionary groping]*. Pavel Mervart.
Markoš, A. (2016b). The birth and life of species-cultures. *Biosemiotics, 9*, 73–84.
Markoš, A., & Švorcová, J. (2019). *Epigenetic processes and evolution of life*. CRC Press.
Mink, L. O. (1970). History and fiction as models of comprehension. *New Literary History, 1*(3), 541–558.
Peirce, C. S. (1903). Nomenclature and divisions of triadic relations as far as they are determined. In Peirce Edition Project (Ed.), *Essential Peirce: Selected philosophical writings 2 (1893–1913)* (p. 1998). Indiana University Press.
Portmann, A. (1960). *Neue Wege der Biologie*. Piper Verlag.
Reinhard, W. (2017). In P. Burschel & M. von Lüneburg (Eds.), *Geschichte als Anthropologie*. Böhlau Verlag.
Richerson, P. J., & Boyd, R. (2005). *Not by genes alone: How culture transformed human evolution*. University of Chicago Press.
Ricoeur, P. (1983). *Temps et récit. I. L'intrigue et le récit historique*. Éditions du Seuil.
Ricoeur, P. (1985). *Temps et récit III. Le temps raconté*. Édition du Seuil.
Rüsen, J. (1986). *Rekonstruktion der Vergangenheit. Grundzüge einer Historik II: Die Prinzipien der historischen Forschung*. Vandenhoeck & Ruprecht.
Scholtz, G. (1997). Zum Strukturwandel in den Grundlagen kulturwissenschaftlichen Denkens (1880–1945). In W. Küttler, J. Rüsen, & E. Schulin (Eds.), *Geschichtsdiskurs. Vol. 4:*

Krisenbewußtsein, Katastrophenerfahrungen und Innovationen 1880–1945 (pp. 19–50). Fischer Verlag.

Sperber, D. (1996). *Explaining culture*. Blackwell.

Šuch, J. (2017). The constructivist understanding of history and its ethical dimension. *Historický časopis, 65*(5), 799–808.

Šuchová, J. (2015). *Farbwörter im Deutschen und Slowakischen. Eine kognitiv-schemantische Untersuchung*. Verlag Dr. Kovač.

Švorcová, J., & Kleisner, K. (2018). Evolution by meaning attribution: Notes on biosemiotic interpretations of extended evolutionary synthesis. *Biosemiotics, 11*(2), 231–244.

Tinková, D. (2004). *Hřích, zločin, šílenství v čase odkouzlování světa. [Sin, crime, madness in the age of the disenchantment of the world]*. Argo.

Vašíček, Z. (1994). *L'archéologie, l'histoire, le passé: chapitres sur la présentation, l'épistémologie et l'ontologie du temps perdu*. Kronos.

Voland, E., & Beise, J. (2005). Bilanzen des Alters—oder: Was lehren uns ostfriesischen Kirchenbücher über die Evolution von Großmüttern? *Historical Social Research, 30*(3), 205–218.

Weber, M. (1904). Die "Objektivität" sozialwissenschaftlicher und sozialpolitischer Erkenntnis. *Archiv für Sozialwissenschaft und Sozialpolitik, 19*(1), 22–87.

Weber, M. (1972). *Wirtschaft und Gesellschaft. Grundriss der verstehenden Soziologie. (5th* rev. ed.). Mohr.

Weber, M. (1996). *Die protestantische Ethik und der "Geist" des Kapitalismus. Edited and with introduction by K. Lichtblau and J. Weiß* (2nd ed.). Beltz Athenäum Verlag.

Wehler, H.-U. (1979). Anwendung von Theorien in der Geschichtswissenschaft. In J. Kocka & T. Nipperdey (Eds.), *Theorie und Erzählung in der Geschichte* (pp. 17–40). Deutscher Taschenbuch Verlag.

Welskopp, T. (2008). Theorien in der Geschichtswissenschaft. In G. Budde, F. Freist, & H. Günther-Arndt (Eds.), *Geschichte. Studien—Wissenschaft—Beruf* (pp. 138–157). Akademie Verlag.

Wittkau-Horgby, A. (2002). Evolution. In *Lexikon Geschichtswissenschaft. Hundert Grundbegriffe, herausgegeben von Stefan Jordan* (pp. 88–90). Philipp Reclam.

Zeleňák, E. (2015). Two versions of a constructivist view of historical work. *History and Theory, 54*(2), 209–225.

Author Index

J. Toman et al. (eds.), *Culture Evolving*, Biosemiotics 10,
https://doi.org/10.1007/978-3-032-16124-6

Subject Index

J. Toman et al. (eds.), *Culture Evolving*, Biosemiotics 10,
https://doi.org/10.1007/978-3-032-16124-6

M

Q

R

The manufacturer's authorised representative in the EU is Springer Nature Customer Service Centre GmbH, Europaplatz 3, 69115 Heidelberg, Germany. If you have any concerns regarding our products, please contact ProductSafety@springernature.com

Printed and bound by CPI Group (UK) Ltd, Croydon, CR0 4YY
16/09/2026
02217791-0013